Fine Needle Aspiration Cytology of the Thorax and Abdomen

With a contribution by

MICHAEL B. KODROFF, M.D.
Clinical Associate Professor
Department of Radiology
East Carolina University School of Medicine
Greenville, North Carolina

Fine Needle Aspiration Cytology of the Thorax and Abdomen

JAN F. SILVERMAN, M.D.
Professor and Vice Chairman
Director of Cytology
Department of Pathology and Laboratory Medicine
East Carolina University School of Medicine
Greenville, North Carolina

KIM R. GEISINGER, M.D.
Professor and Director of Surgical Pathology and Cytology
Department of Pathology
Bowman Gray School of Medicine of Wake Forest University
Winston-Salem, North Carolina

CHURCHILL LIVINGSTONE

New York, Edinburgh, London, Madrid, Melbourne, San Francisco, Tokyo

Library of Congress Cataloging-in-Publication Data

Distributed in the United Kingdom by Churchill Livingstone, Robert Stevenson House, 1–3 Baxter's Place, Leith Walk, Edinburgh EH1 3AF, and by associated companies, branches, and representatives throughout the world.

Acquisitions Editor: *Kerry Willis*
Assistant Editor: *Marc Strauss*
Production Supervisor: *Sharon Tuder*
Desktop Coordinator: *Robb Quattro*
Cover Design: *Jeannette Jacobs*

Printed in Hong Kong

Bound in China

First published in 1996 7 6 5 4 3 2 1

We want to dedicate this book to our parents,
Ann Silverman and Karl and Florence Geisinger;
to our spouses, Mary and Lori, and
to our children, Mischell, Jeff, Laura, Kristen, and Brian, and
grandchild, Jessica; and to their friends, Squeaky, Speedy, and Buster.

Preface

The morphologic spectrum of benign and malignant mass lesions encountered in cytologic aspirates of the deep organs is both intellectually stimulating and clinically challenging. We strongly believe that to optimize diagnostic yield, pattern recognition at low magnification and detailed high-power examination of cytomorphologic features must always be assessed. Accordingly, the fundamental goals of *Fine Needle Aspiration Cytology of the Thorax and Abdomen* include emphasizing the surgical pathologist's approach to fine needle aspiration biopsy of these sites. Therefore, a working knowledge of the underlying histopathologic bases for the cytologic findings in Papanicolaou- and Diff-Quik–stained aspiration smears is essential.

We have attempted to portray our own style of practice throughout the text, charts, and photomicrographs. As detailed in this book, specific diagnoses can often be rendered after integrating clinical, radiologic, cytomorphologic, and, at times, ancillary diagnostic studies. When finalizing our diagnoses, we generally use terminology identical to that used in conventional surgical pathology reports. Occasional aspirates, however, permit generation of only a differential diagnosis that may be embellished as a comment in the cytology report. This approach, hopefully, will permit the pathologist to avoid false-negative and especially false-positive diagnoses, again based on specific cytologic criteria coupled with clinical and radiologic findings. Thus, a pivotal aspect in the evaluation of these aspirates is a working relationship with our colleagues in radiology, pulmonology, and surgery; this includes use of an immediate interpretation of the aspirate (analogous to the frozen section in surgical pathology). This "quick read" permits not only the determination of specimen adequacy but also allows a preliminary, rapid diagnosis to be rendered in many cases. In the more difficult examples, additional material may be requested to be used for extra direct smears, electron microscopy, immunocytochemistry, molecular biologic assays, and cytogenetic testing.

We truly hope that readers find this integrated approach to be readily applicable in all types of pathology practice settings.

Jan F. Silverman, M.D.
Kim R. Geisinger, M.D.

Acknowledgments

We would like to thank the following for their contributions to this book: Dr. Phil Feldman, Charlottesville, VA; Dr. Bernard Naylor, Ann Arbor, MI; Dr. Celeste Powers, Syracuse, NY; Drs. Jo Ann Shaw and Thomas Zirker, Greensboro, NC; Dr. George Thomas, Richmond, VA; Dr. Paul Wakely, Jr., Richmond, VA; Dr. Sherry Woodhouse, Lenoir, NC, and Dr. Fred Worsham, Charleston, SC.

We would also like to thank Kathy Carnes and Kelly Matthews for typing the manuscript. Our appreciation is also expressed to Dr. Kerry Willis and others from Churchill Livingstone for their efforts and patience in making this book possible.

Contents

1 Lung

The major indication for transthoracic FNA biopsy is the cytomorphologic evaluation of a localized pulmonary nodule that cannot be diagnosed readily by exfoliative respiratory cytologic procedures.[1–12] FNA biopsy is especially useful in the specific clinical situations listed on page 2.[13] Potential contraindications to pulmonary FNA are also detailed on page 2.[13]

Complications of pulmonary FNA biopsy may include pneumothorax, pulmonary hemorrhage with or without hemoptysis, air embolism, needle tract seeding of tumor cells, and death.[14] Except for pneumothorax and hemoptysis, the other complications are quite rare. By far, pneumothorax is the most frequent complication of percutaneous transthoracic FNA biopsy, occurring in 5 to more than 50 percent of patients, with an average incidence of approximately 25 percent.[15] Fortunately, most instances of pneumothorax require no therapy. Yet in 4 to 10 percent of lung FNAs, chest tube insertion is required to treat a symptomatic or large pneumothorax.[15] The incidence of pneumothorax decreases with the experience of the operator and with the need for fewer needle passes, accomplished by immediate assessment of each pass for adequacy as discussed below. Unfortunately, a few reports claim a higher pneumothorax rate in current series due to aspiration of smaller and thus more challenging lesions that may require more needle passes for successful sampling.[15]

The main value of pulmonary FNA biopsy is the diagnosis of cancer. The procedure has reported accuracy rates of 80 to 95 percent, a false-negative rate of 7 to 10 percent, and a false-positive rate of 1 to 2 percent.[8,16–18] The accuracy of tumor classification based on correlating the cytologic diagnosis with the subsequent surgical pathology examination remains in debate. However, using a single binary classification scheme of small cell carcinoma and non-small cell carcinoma, the histologic-cytologic concordance rates for small cell carcinoma range from 71 to 100 percent and for non-small cell carcinoma from 94 to 100 percent, based on our recent review.[19] We believe that such a binary classification scheme currently provides very useful clinical data, as small cell carcinomas will be treated with multimodality therapy, while non-small cell carcinomas are primarily treated with surgery with or without adjuvant therapy.[19] The diagnosis of infectious and other inflammatory pulmonary diseases by FNA biopsy has recently emerged coinciding with the acquired immunodeficiency syndrome (AIDS) epidemic and with an increased number of immunosuppressed patients having opportunistic infections secondary to intensive chemotherapy or irradiation, or both. Infectious and inflammatory lesions are also amenable to FNA biopsy.

PROCEDURE AND TECHNIQUE

At our institutions, for superficial lesions, a 22- to 25-gauge (0.6-mm) 1-inch or 1.5-inch needle is attached to a disposable 20-ml syringe fitted into a commercially available syringe holder allowing one-handed aspiration.[20,21] FNA of lung, mediastinum, and deep abdominal masses is generally aspirated percutaneously with a 3.5-inch, 22-gauge Chiba biopsy needle under fluoroscopic, ultrasound, or computed tomography (CT) guidance. Two or more aspirations ("passes") are performed

Clinical Situations in Which Pulmonary FNA Biopsy Is Recommended

- Preoperative diagnosis of a small peripheral lung nodule
- Diagnosis of inoperable or metastatic malignancies
- Diagnosis of small cell carcinoma, best treated with chemotherapy and/or irradiation
- Diagnosis of infectious processes, especially in immunocompromised patients
- Diagnosis of complex diffuse pulmonary diseases of unknown etiology
- Intraoperative diagnosis of pulmonary mass lesions

Potential Contraindications in the Use of Pulmonary FNA Biopsy

- When patient is uncooperative
- Conditions in which an undue risk of significant bleeding exists (e.g., severe recent hemoptysis, bleeding diathesis, severe pulmonary hypertension, vascular lesions—aneurysm, pulmonary arterial venous malformation, pulmonary varix)
- When patient is severely debilitated (especially those having chronic obstructive pulmonary disease (COPD) with PO_2 under 60 mmHg, less than 50% predicted outflow)
- When patient has large bullae, pulmonary hydatid cyst, or only one functioning lung

in most cases. The aspirated material is rapidly expressed onto slides and compressed with another slide similar to the preparation of a bone marrow aspirate. Approximately one-half of smears are immediately wet-fixed with either spray or immersion in 95 percent ethanol for subsequent Papanicolaou staining, while the air-dried smears are stained by a modified rapid Wright-Giemsa stain (Diff-Quik stain). In almost all cases, two air-dried smears from each separate aspirate are chosen by the pathologist for rapid Diff-Quik staining (Fig. 1-1). This stain can be performed in the laboratory or on site within 30 seconds and thus lends itself to an immediate interpretation of the material. Specimens obtained by clinicians or radiologists are immediately brought to the laboratory, stained, and

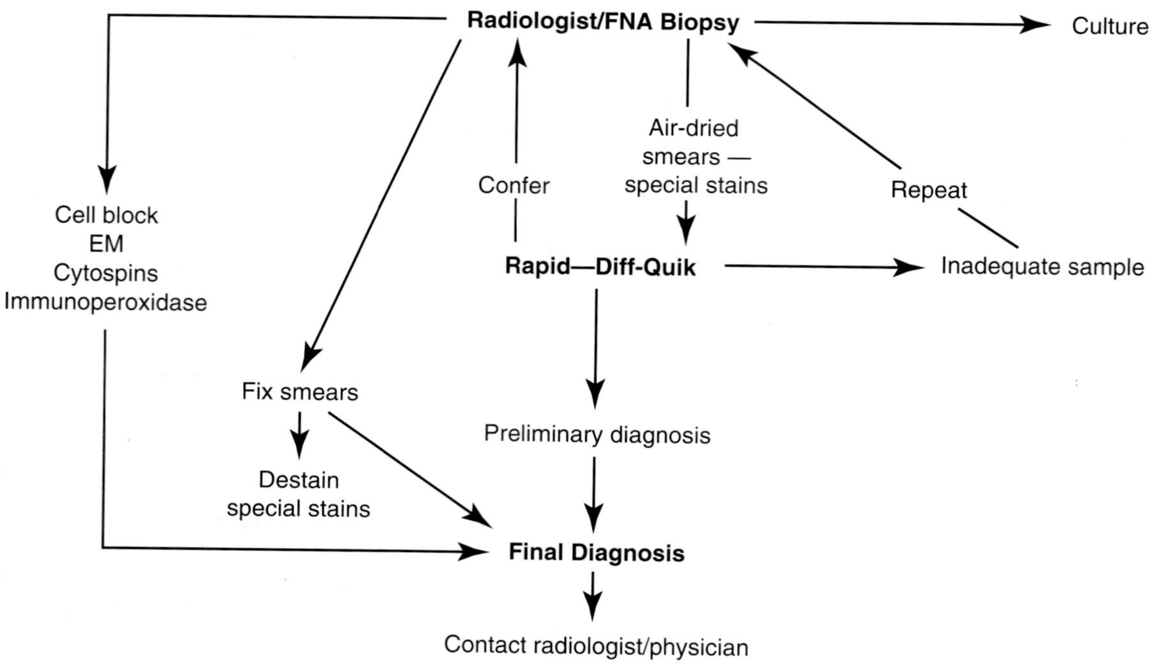

Fig. 1-1. Approach to FNA of the lung.

examined microscopically with an immediate report given by telephone or by a quick trip to the radiology department (where films and clinical details can be reviewed). When the FNA biopsy of superficial lesions is performed by the pathologist, the immediate examination is facilitated by a mobile cart (taken to the site) containing stains and a microscope. Transbronchial FNA at bronchoscopy can be used as a diagnostic and staging procedure for peribronchial, paratracheal, mediastinal, and pulmonary submucosal lesions.[22,23]

An early critical step in the workup of any FNA biopsy is the immediate examination of the Diff-Quik or other rapid stains. The immediate interpretation may answer a number of questions including

1. Is the specimen adequate or representative?
2. Is the lesion inflammatory or neoplastic?
3. Is the lesion benign or malignant?
4. Is the malignant lesion primary to the site or metastatic?
5. Are additional passes needed to obtain material for a definitive diagnosis?
6. Are additional passes needed for ancillary studies such as electron microscopy, immunocytochemistry, flow cytometry, cytogenetics, microbiology, or molecular biology studies? (Fig. 1-1).

These principles apply not only to pulmonary FNA, but to aspiration biopsies from other sites as well, as discussed in this book.

NORMAL CYTOLOGY

Expected normal (or residential) cellular elements include benign bronchial epithelial cells of columnar, goblet, and reserve types, alveolar macrophages, and occasional benign hematopoietic cells[24] (Fig. 1-2). Benign respiratory cells have a columnar shape with apical terminal bar and cilia. The basal portion of the cell has a tapered configuration. Alveolar macrophages have eccentric oval to reniform shape nuclei and a moderate to abundant pale cytoplasm. Occasionally, intracytoplasmic anthracotic pigment will be present (Fig. 1-2). Benign mesothelial cells are often also seen in FNA biopsies. Mesothelial cells are usually arranged in sheets and consist of cells having uniform-appearing nuclei with small nucleoli and occasional nuclear grooves and surrounding pale, well-defined cytoplasm (Fig. 1-3). Rarely, lung aspirates may contain reactive and atypical mesothelial cells that could be confused with carcinoma if the architectural arrangement of the cells is overlooked. Although nuclear enlargement and hyperchromasia may be noted, the mesothelial cells will usually be arranged in flat or folded sheets with well-defined cell spaces "windows." Occasional nuclear irregularities and grooves may be seen (Fig. 1-3). In rare instances, mesenchymal cells such as smooth muscle cells, fibroblasts, adipocytes, and chondrocytes may be present in the smears.

A

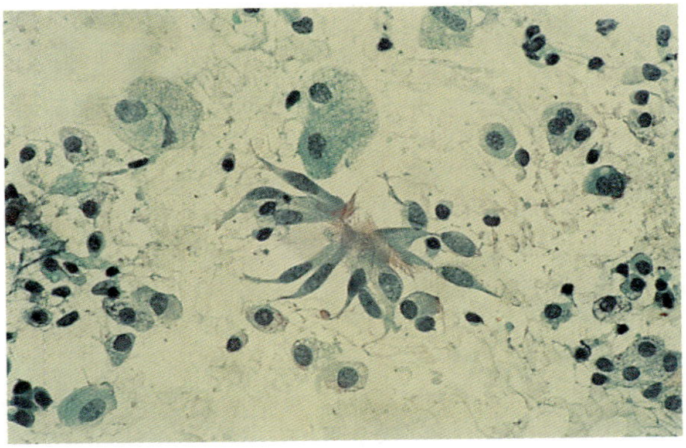

B

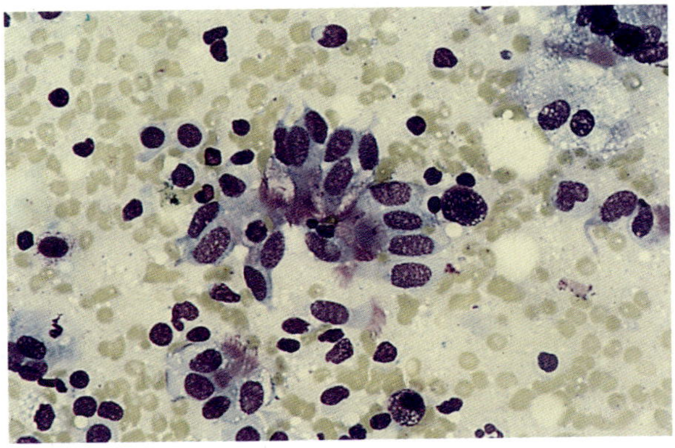

Fig. 1-2. (**A**) Benign respiratory cells in center of field have columnar shapes with tapered basal cytoplasm and cilia arising from the apical terminal bar. A few scattered alveolar macrophages having small oval to reniform shape nuclei and abundant finely vacuolated, pale cytoplasm are present along with a scattering of some neutrophils in the background (Papanicolaou). (**B**) Bronchial lining cells having a diagnostic terminal bar with cilia. A few scattered inflammatory cells and alveolar macrophages are also present (Diff-Quik).

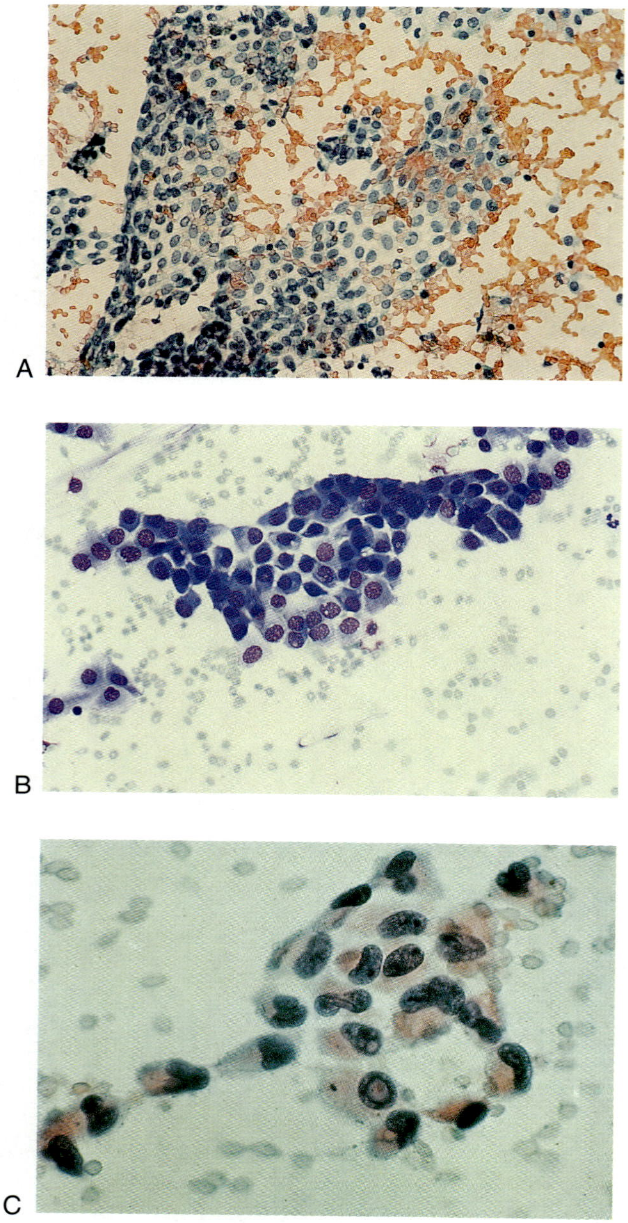

Fig. 1-3. (**A**) Sheets of benign mesothelial cells having uniform appearing nuclei with well defined cytoplasm and no loss of polarity (Papanicolaou). (**B**) Group of mesothelial cells having a characteristic folded, sheet-like arrangement (Diff-Quik). (**C**) Reactive and slightly atypical mesothelial cells having a characteristic folded, twisted pattern. Note the nuclear irregularity and an intranuclear inclusion. However, the cells are spaced with well-defined cell borders and no loss of polarity (Papanicolaou).

NON-NEOPLASTIC CONDITIONS

Although FNA of the lung was first used by Leyden in 1883 to diagnose pneumonia,[25] only recently have a sizable number of reports describing the FNA cytologic features of a variety of inflammatory lung lesions been published[1,2,25–35] (Tables 1-1 and 1-2). Microscopically, lung abscesses consist of tannish yellow or green, semisolid, purulent material. The major cytologic feature consists of sheets of neutrophils, many undergoing liquefactive digestion. Additionally, necrotic cells, cell debris, fibrin, and histiocytes are intermixed.[36] In some cases, bacteria will be identified in the Diff-Quik–stained smears, both within the cytoplasm of phagocytes and free in the smear background (Fig. 1-4). A major differential diagnosis is cavitary squamous cell carcinoma. The extensive necrosis and acute inflammation seen in some squamous cell carcinomas can simulate an abscess both radiologically and cytologically.[37–39] Smears of "carcinomatous abscess" often demonstrate a well-developed acute inflammatory cell component with an occasional foreign body reaction to the keratinous type material (Fig. 1-5). A false-negative diagnosis for squamous cell carcinoma is possible, especially when only rare intact malignant cells are present. A clue for squamous cell carcinoma includes the presence of keratinous fragments or ghost cells, or both. The diagnosis is much more challenging in the Diff-Quik–stained smears, since the cytoplasmic orangeophilia characteristic of Papanicolaou-stained smears is not evident. However, the diagnosis should be suspected when irregularly shaped cells having pyknotic nuclei, dense cytoplasm, and sharp borders are detected (Fig. 1-6).

Granulomatous processes are the most common types of inflammatory lesions to be sampled by FNA biopsy.[25] Although Robicheaux believed that culture of the aspirated material was most useful for making a specific diagnosis of tuberculosis,[40] Dahlgren and Ekstrom[41] emphasized the diagnostic FNA cytomorphologic features of pulmonary tuberculosis. Dahlgren and Lind[42] also reported the enhanced sensitivity of FNA biopsy of the lung in diagnosing tuberculosis vis-à-vis routine sputum cytology. We concur with Bailey et al.,[43] who reported that the examination of both FNA smears and bacteriologic culture is the most effective way of diagnosing tuberculosis. Rajwanshi et al.[44] noted that acid-fast bacilli (AFB) could be identified in nearly 41 percent

Table 1-1. Identification of Bacteria

Microorganism	*Morphology*	*Stains*	*Organs Affected*	*DifferentialDiagnosis*[a]
Mycobacterium tuberculosis	Slender, sometimes slightly curved rods averaging 4 μm in length	Acid-fast stain	Disseminated	Other mycobacteria
Mycobacterium avium intracellulare	Morphologically indistinguishable fromTB	Acid-fast stain, negative images with Romanovsky stain	Primarily lungs and lymph nodes	Other mycobacteria
Mycobacterium kansasii	Longer and broader than TB; usually occurs as intracellular bacilli grouped in bundles; also seen free in smear	Acid-fast stain	Lungs	Other mycobacteria
Mycobacterium leprae	Morphologically similar to TB; usually occurs as intracellular bacilli grouped in bundles; also seen free in smear	Acid-fast stain	Liver, kidneys, subcutaneous tissues, spleen, skin, lymph nodes	Other mycobacteria
Legionella sp.	Short rods within the cytoplasm of neutrophils; extracellular rods also seen	Fite's acid-fast stain, direct fluorescent antibody stain	Lungs	Mycobacteria
Actinomyces israelli	Delicate, nonseptate filaments 1–1.5 μm in diameter; variable in length and degree of branching; sulfur granules present	Papanicolaou, H&E, Gram stains, methenamine silver	Disseminated	*Nocardia*, other bacteria
Nocardia asteroides	Delicate, nonseptate filaments 0.5–1.0 μm in diameter and 10–20 μm in length; right-angle branching; sulfur granules usually absent	Gram stains, Fite's acid-fast stain, Gram–Weigert, Glemsa, methenamine silver	Central nervous system, lungs	*Actinomyces*, other bacteria
Other bacteria	Gram-positive and -negative bacilli and cocci	Gram stains	Disseminated	By culture

[a]Differential diagnosis based on appearance in cytologic preparations.
(From Silverman,[36] with permission.)

of FNA biopsies of tuberculosis. AFB positivity of the smears was highest in specimens with necrotic, caseous-type material in the smears compared to those aspirates demonstrating cytologic evidence of granulomatous inflammation.[43,44] The absence of AFB in the smears with otherwise typical cytologic features of tuberculosis should not deter the pathologist from suggesting this possibility.[36,44] This is based on the fact that 10,000 to 100,000 AFB/ml are needed to be present before the *Mycobacterium* can be microscopically recognized.[45] FNA cytomorphology of granulomatous lesions demonstrates variably cohesive clusters of epithelioid histiocytes. These cells have elongated nuclei with straight, curved, or bent contours and amphophilic cytoplasm (Fig. 1-7). Cell borders are indistinct. Diagnostic cytologic considerations include sarcoid, tuberculous, and fungal diseases. Occasionally, granulomatous reaction to a neoplasm can be seen in the FNA smears (Fig. 1-8). Tuberculosis and some fungal infections such as histoplasmosis may cytologically have a background of caseous necrosis consisting of fluffy, amorphous, to finely grained necrotic material with a few scattered epithelioid cells, multinucleated Langhans-type giant cells and occasional small calcified fragments (Fig. 1-9). Special stains for organisms including Gomori methenamine silver (GMS) for fungi and AFB stain for *Mycobacterium* are crucial along with culture of the aspirated material.

The cytomorphologic features of many fungal infections are well described[36,46,47] and outlined in Table 1-2. The diagnosis should be suspected whenever there is cytologic evidence of granulomatous inflammation in smears from a pulmonary aspirate.[1,25,36,48] Cryptococcosis is a fungal infection of worldwide distribution, with the respiratory tract serving as the primary portal of entry, which may disseminate to the central

Table 1-2. Identification of Fungi

Microorganism	*Morphology*	*Stains*	*Organs Affected*	*Differential Diagnosis*[a]
Aspergillus fumigatus	Septate hyphae 3–6 μm in diameter with uniform, acute-angle, dichotomous branching	Papanicolaou, H&E, PAS, methenamine silver, immunoperoxidase	Disseminated	*Candida* sp.
Blastomyces dermatitidis	Spherical organisms (8–20 μm) with sharply defined, thick, refractile cell walls; broad-based budding	Papanicolaou, H&E, PAS, methenamine silver	Lungs, bone, skin, other organs	*Cryptococcus, Histoplasma, Coccidioides*
Candida sp.	Pseudohyphae with budding yeast forms (3–4 μm)	Papanicolaou, H&E, Wright, Diff-Quik, methenamine silver, immunoperoxidase	Disseminated	*Aspergillus*
Coccidioides immitis	Nonbudding spherules 5–100 μm in diameter with thick refractile walls; contain numerous small endospores (2–5 μm); septate hyphae can also be seen	Papanicolaou, H&E, PAS, methenamine silver	Lungs, bone, skin, subcutaneous tissues,viscera	Pollen grains, parasites, vegetable material, *Blastomyces, Cryptococcus*
Cryptococcus neoformans	Round or oval organisms (5–10 μm) surrounded by a clear thick mucopolysaccharide capsule; single, thin-walled, narrow-based budding	PAS, mucicarmine, Alcian blue, methenamine silver, immunoperoxidase	Lungs, central nervous system, skin, viscera	*Blastomyces, Histoplasma, Coccidioides*
Histoplasma capsulatus	Budding oval yeast (2–5 μm) surrounded by rigid cell wall; usually seen within macrophages	PAS, methenamine silver	Disseminated	*Blastomyces, Cryptococcus, Coccidloides*
Phaeohyphomycosis	Pseudohyphae composed of chains of round, oval, or rectangular bodies resembling strings of beads; true septate hyphae, branching, single budding forms also found	Papanicolaou, H&E, methenamine silver	Subcutaneous cysts	Other hyphal fungi
Zygomycetes	Very broad, rarely septate, ribbon-like hyphae that vary greatly in thickness and shape; branching common but angle and shape varied	Papanicolaou, H&E, poor sensitivity to fungal stains	Orbit, nasal sinuses, lungs, central nervous system, GI tract, blood-vessels	Other hyphal fungi, contaminants

[a]Differential diagnosis based on appearance in cytologic preparations.
(From Silverman,[36] with permission.)

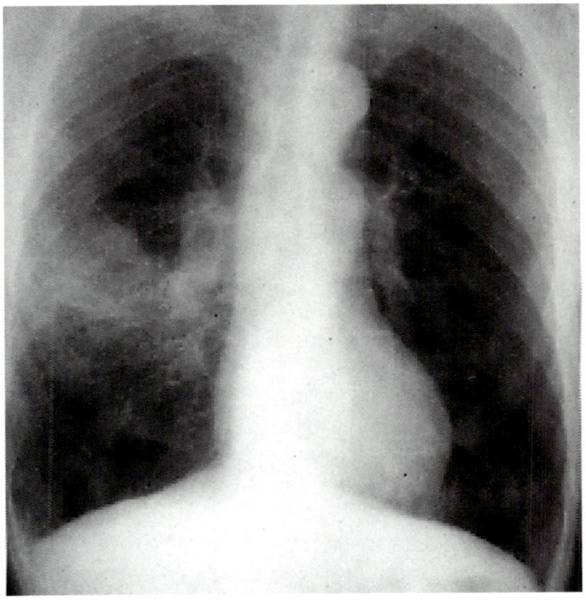

A

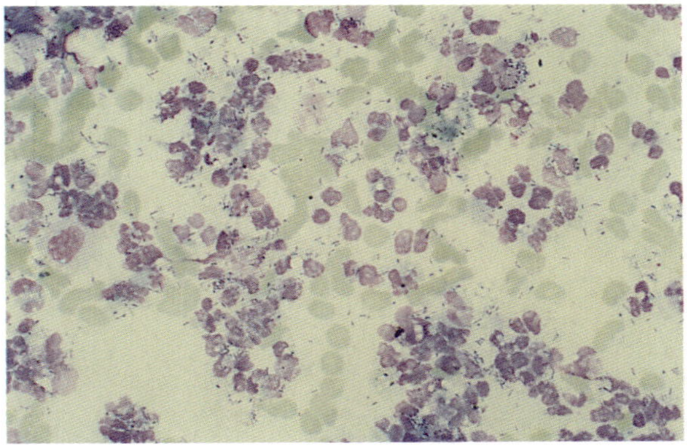

B

Fig. 1-4. (**A**) Radiograph of lung abscess showing central cavity. (**B**) FNA of lung abscess showing sheets of neutrophils and numerous rod-shaped bacteria in the background (Diff-Quik).

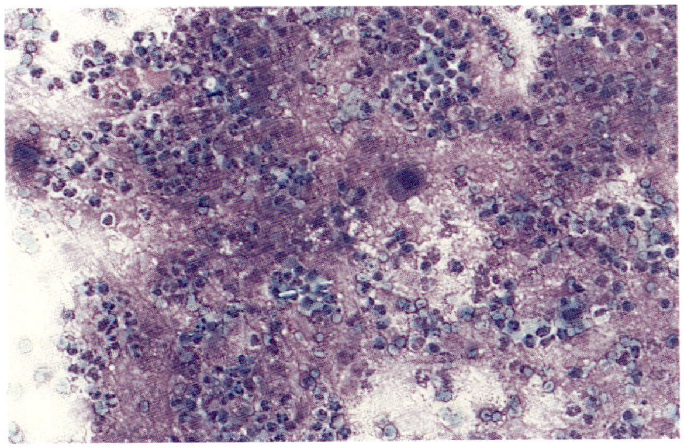

A

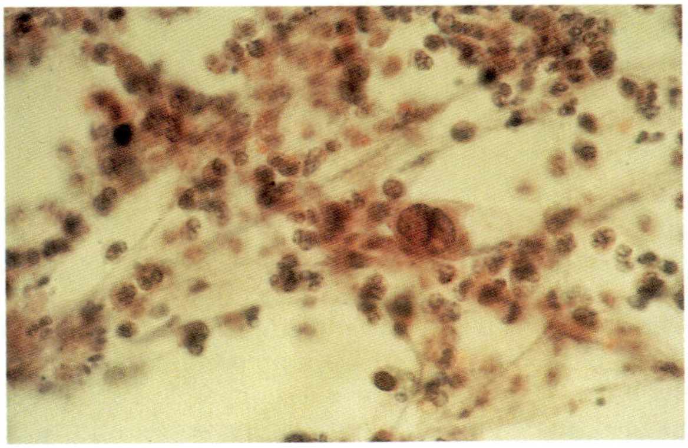

C

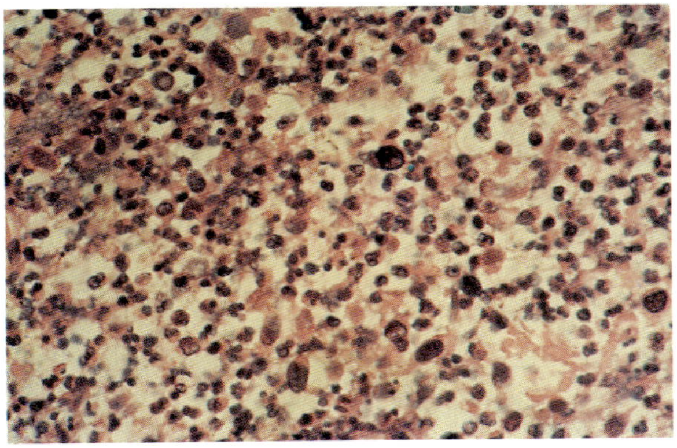

B

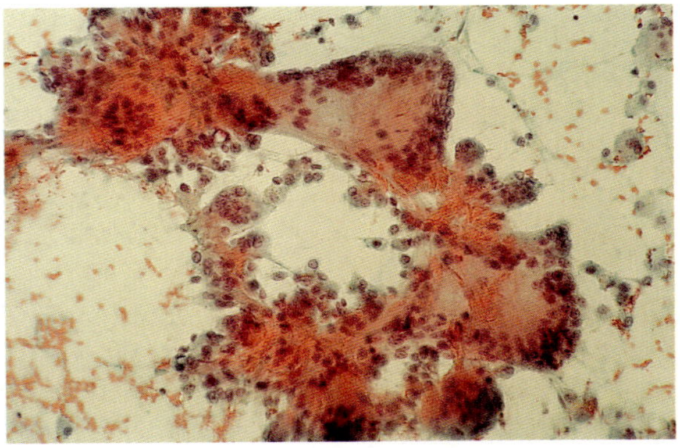

D

Fig. 1-5. (**A**) FNA of "carcinomatous abscess" consisting of sheets of neutrophils with rare keratinized squamous cells (Diff-Quik). (**B**) Another case of keratinizing squamous cell carcinoma with extensive acute inflammation cytologically simulating an abscess. Note the few scattered atypical squamous cells (Papanicolaou). (**C**) Carcinomatous abscess consisting of sheets of neutrophils along with diagnostic malignant squamous cells showing multinucleation and high nuclear to cytoplasmic ratios (Papanicolaou). (**D**) FNA of keratinizing squamous cell carcinoma of the lung demonstrating a foreign body reaction to keratinous material (Papanicolaou).

A

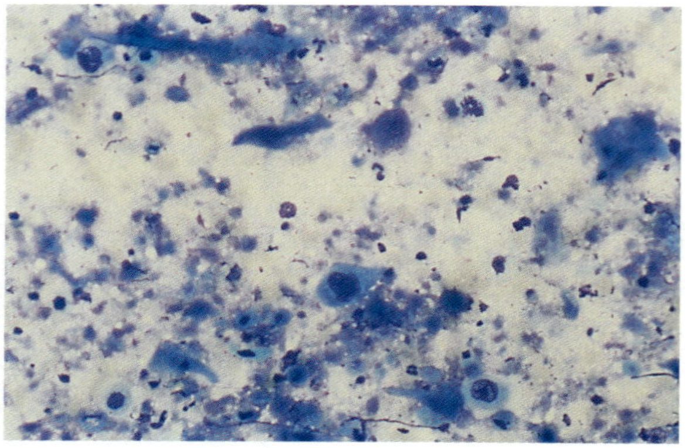

B

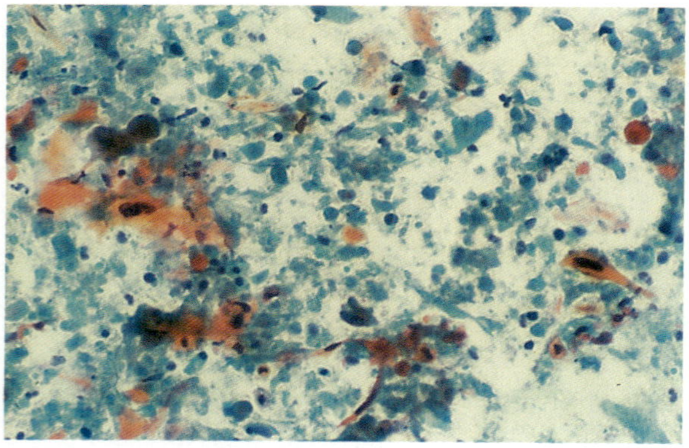

Fig. 1-6. (**A**) FNA of keratinizing squamous cell carcinoma of the lung in Diff-Quik preparation. The diagnosis is more difficult than in comparable Papanicolaou stained material, although the presence of irregular dense keratinized fragments along with rare atypical cells should suggest the diagnosis (Diff-Quik). (**B**) The diagnosis of keratinizing squamous cell carcinoma with extensive necrosis and acute inflammation is made more readily in the Papanicolaou-stained smears due to the striking orangeophilia of the malignant squamous cells (Papanicolaou).

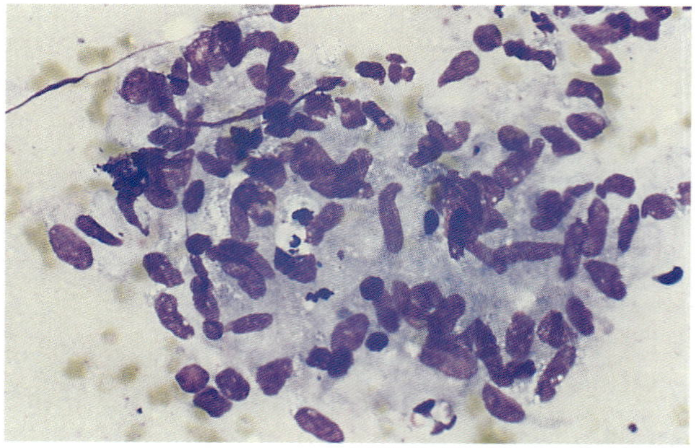

Fig. 1-7. FNA of granulomatous process of the lung demonstrating a cluster of epithelioid histiocytes. The cells have elongated nuclei with straight or bent contours and amphophilic cytoplasm arranged in a syncytial fashion. A search for organisms including use of special stains should be performed when granulomatous inflammation is appreciated (Diff-Quik).

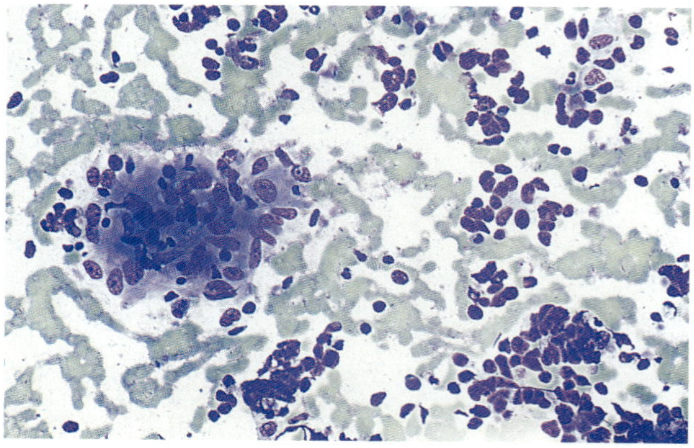

Fig. 1-8. FNA of small cell carcinoma of the lung associated with a single cohesive granulomatous fragment (Diff-Quik).

nervous system (CNS) in susceptible individuals.[36,49,50] FNA cytology of cryptococcosis will vary with the stage of the disease. FNA cytology of early lesions consists of grossly gelatinous masses that contain yeast cells with little inflammation, fibrosis, or granuloma formation.[49,53] The yeast measures 5 to 20 μm in diameter and has a characteristic single narrow teardrop-shaped bud, with a surrounding capsule of variable thickness[10,50–57] (Fig. 1-10). Unusual morphologic forms of *Cryptococcus* may include chains of budding yeasts and pseudohyphae.[58] Special stains highlighting the mucincarminophilic capsule permit a specific diagnosis.[58,59] In older lesions, granulomatous inflammation may be recognized in the smears without visible organisms.[1,48]

Histoplasmosis is found throughout North America and is endemic to parts of the United States, especially the Ohio-Mississippi Valley.[1,60] The yeast phase of *H. capsulatum* measures 2 to 4 μm in diam-

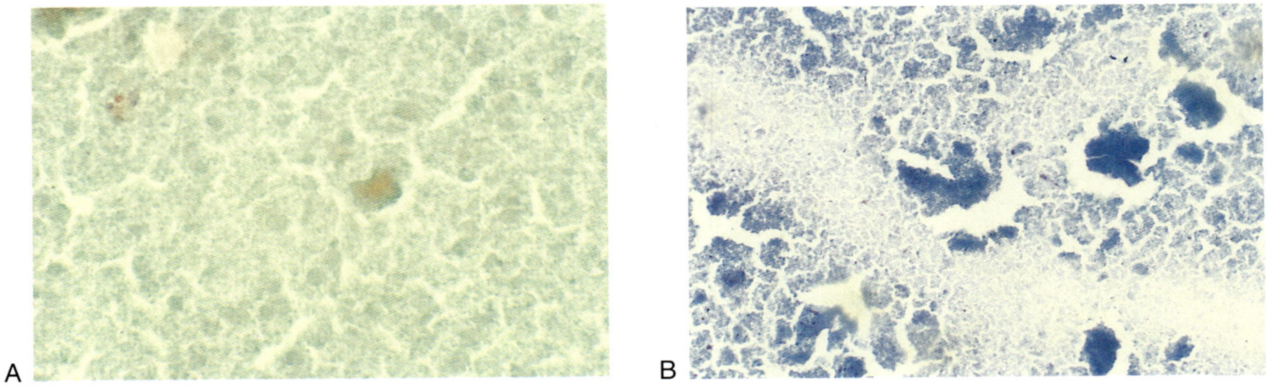

Fig. 1-9. (**A**) FNA of caseous necrosis demonstrating amorphous, fluffy necrotic material (Papanicolaou). (**B**) Caseation necrosis in Diff-Quik–stained material (Diff-Quik).

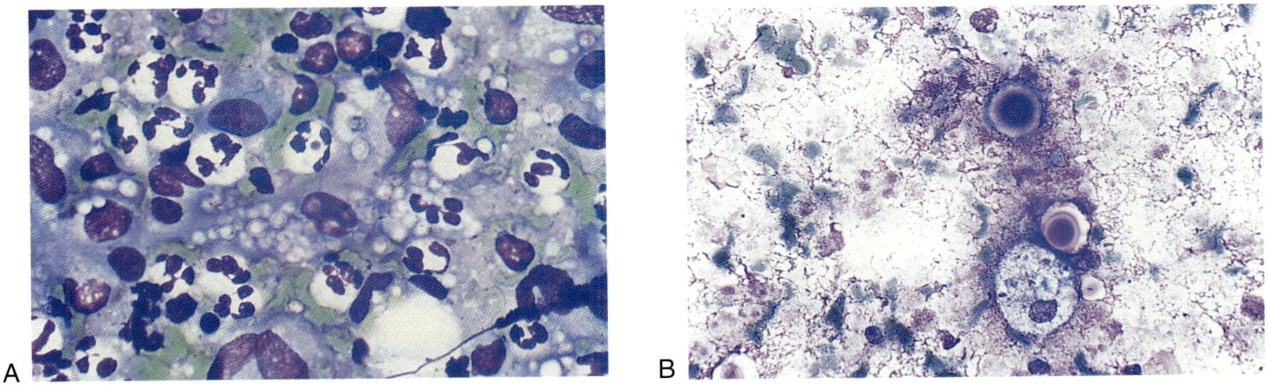

Fig. 1-10. (**A**) FNA of cryptococcosis of lung demonstrating numerous fungal yeasts forms within histiocytes, as well as individually scattered fungi in the background (Diff-Quik). (**B**) FNA of cryptococcosis revealing individually scattered organisms having metachromatically staining capsules in the Diff-Quik preparation (Diff-Quik).

eter and has a rigid cell wall. In contrast to other fungi, the yeast of *H. capsulatum* can often be difficult to appreciate in routine smears. However, the yeast is readily recognized with special stains[1,36] (Fig. 1-11). GMS stain is preferred, as it clearly distinguishes the organisms from calcified background debris.[1,36]

Coccidioidomycosis is endemic to the southwestern United States and is therefore a common cause of solitary pulmonary nodules in individuals living or traveling through that region.[36,61–63] In immunocompromised patients from endemic regions, coccidioidomycosis is the most common fungal infection.[61,62] The cytomorphologic feature of coccid-

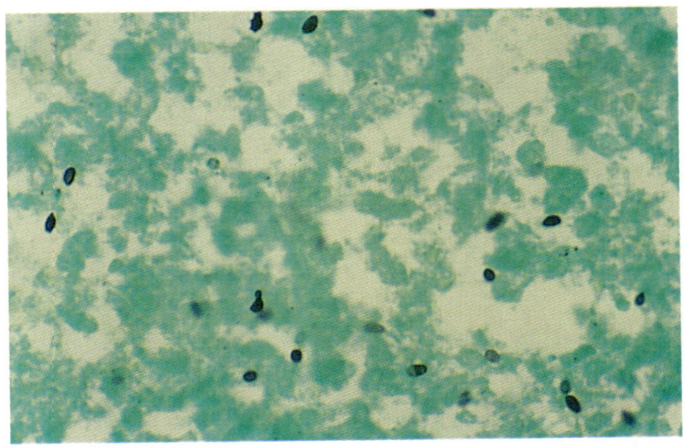

Fig. 1-11. FNA of histoplasmosis showing numerous small fungal yeasts scattered throughout the smear. Note the background of caseous necrosis (GMS).

ioidomycosis includes round spherules with double walls measuring 5 to 100 μm in diameter, which are orange to reddish purple with the Papanicolaou stain. Mature spherules contain endospores that measure 2 to 5 μm each.[63] Unfortunately, with the Diff-Quik stain, the spherules are colorless and refractile.[36] Older nonviable spherules may be empty with a folded, fractured, or collapsed structure that is occasionally reminiscent of a cracked "ping-pong ball"[63] (Fig. 1-12). The latter are most often seen in chronic, cavitary coccidioidomycosis. Mycelia elements are characterized by branched septate hyphae measuring 2 μm in thickness and having spaced barrel-shaped arthroconidia ranging 2 to 4 μm in width and 3 to 6 μm in length.[63] Other fungi and related organisms reported in the FNA literature include *Blastomyces*[64] and *Actinomyces*[65] (Fig. 1-13).

Pulmonary *Aspergillus* can be invasive or noninvasive and occur in both immunocompromised and immunocompetent patients.[30,33,66,67] (Fig. 1-14). Cytologic features include characteristic septated, acute angle branching hyphae (Fig. 1-15). The hyphae of *Aspergillus* can be confused with other fungi such as *Pseudallescheria, Fusarium, Trichosporon,* dematiaceous *Hyphomycetes, Candida,* and *Zygomycetes*[36] (Fig. 1-15). In occasional cases, a possible aid in diagnosing an *Aspergillus* infection is the appreciation of calcium oxalate crystals in the FNA cytologic smears.[68–70]

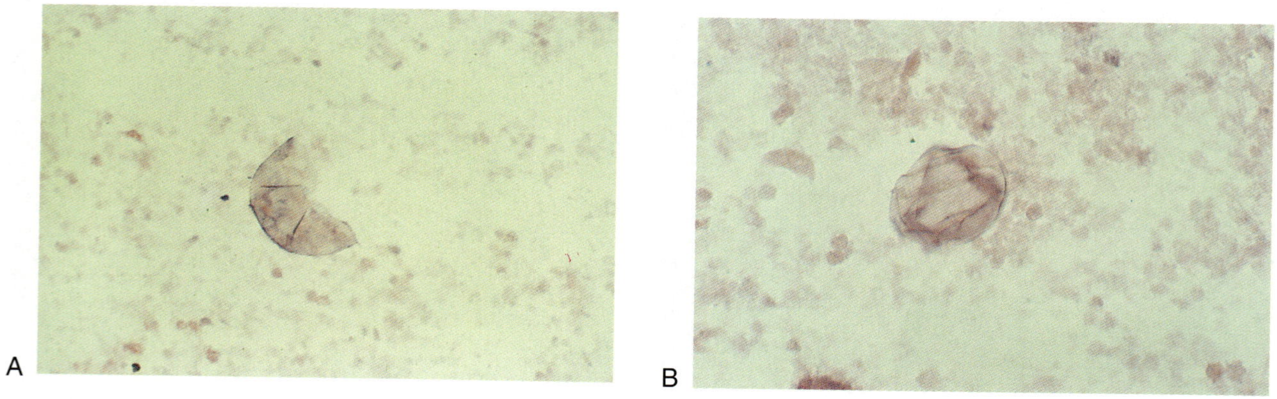

Fig. 1-12. (**A**) Spherule of *Coccidioides immitis* in FNA biopsy having a cracked "ping-pong ball" appearance (Papanicolaou). (**B**) Collapsed spherule of *Coccidioides immitis* (Papanicolaou).

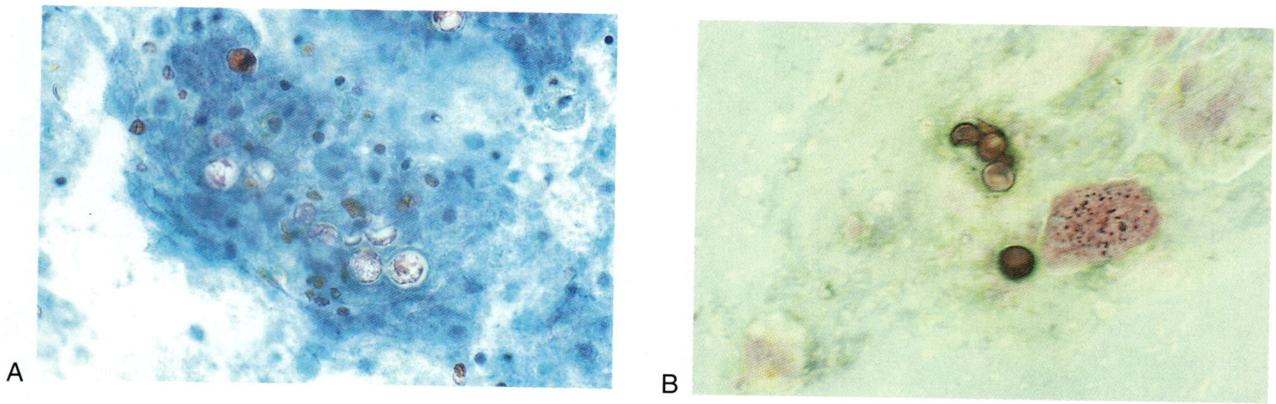

Fig. 1-13. (**A**) FNA of pneumonic process of lung revealing scattered yeasts forms including some showing broad-based budding and well-defined cell walls with internal structure consistent with blastomycosis (Papanicolaou). (**B**) GMS stain showing chain of fungal yeasts with broad-based budding consistent with blastomycosis (GMS).

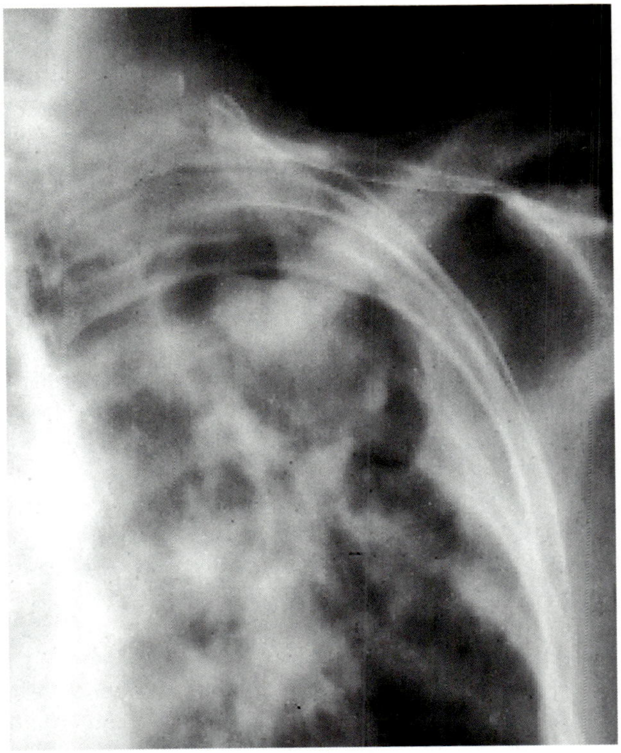

Fig. 1-14. Radiograph of characteristic fungal ball involving left upper lobe of lung in immunocompetent patient. FNA biopsy confirmed diagnosis of *Aspergillus* infection.

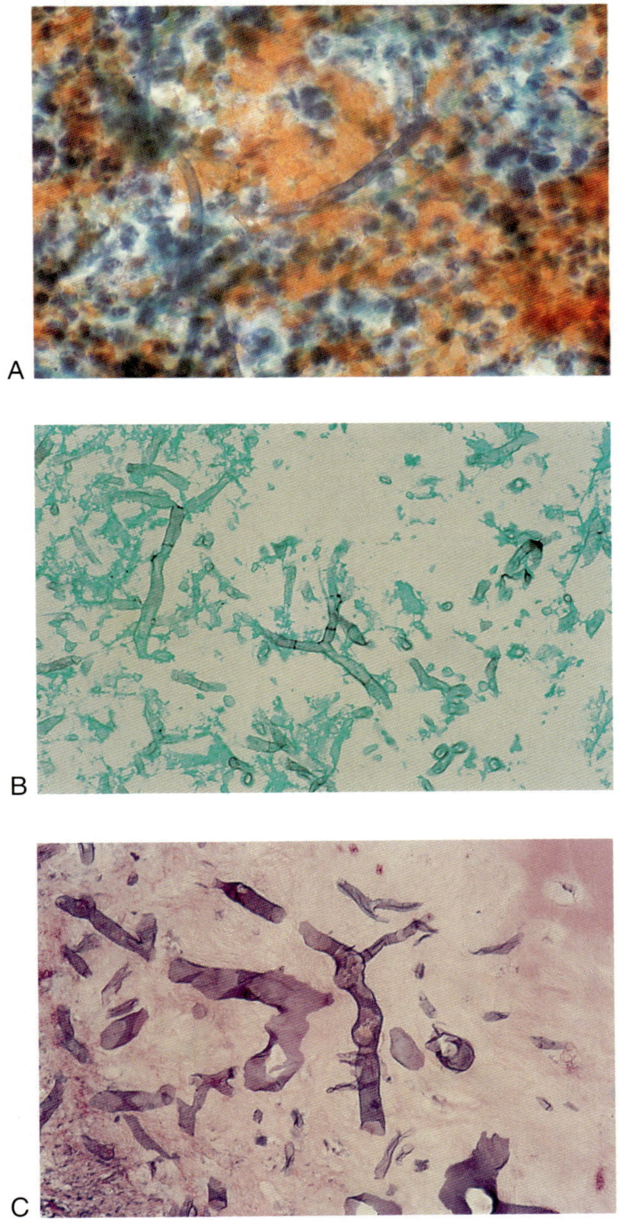

Fig. 1-15. (**A**) FNA of aspergilloma revealing characteristic septated, acute angle branching hyphae along with extensive acute inflammation in the background (Papanicolaou). (**B**) Cell block from FNA of aspergilloma showing acute angle branching and septated hyphael forms-consistent with *Aspergillus* (GMS). (**C**) Note thick, non-septated hyphae in FNA of *Zygomycetes* infection (cell block, H&E).

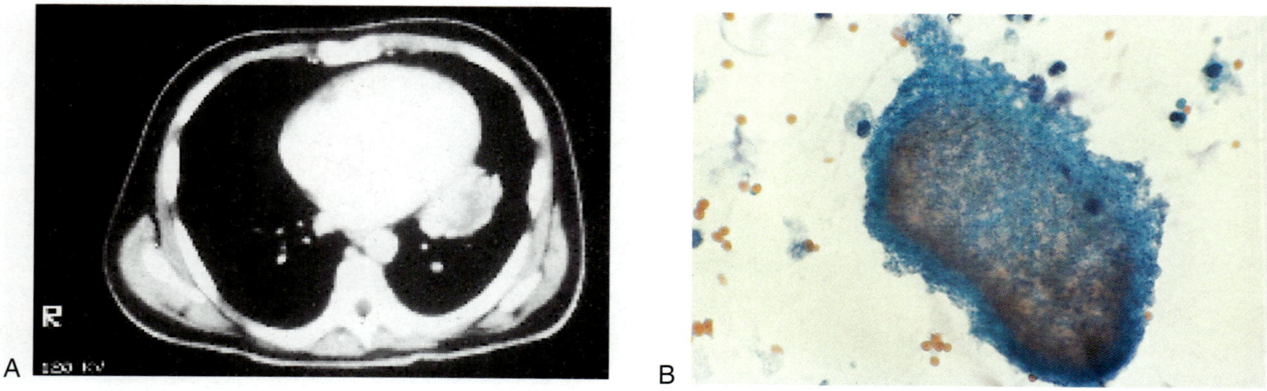

Fig. 1-16. (**A**) CT scan of mass lesion of *Pneumocystis carinii* pneumonia (PCP) in AIDS patient. (**B**) Aspirate of *P. carinii* pneumonia revealed numerous honeycomb casts of organisms (Papanicolaou).

Pneumocystis carinii pneumonitis is the most common life-threatening opportunistic infection in AIDS patients, occurring in up to 85 percent of this population; a 50 percent mortality rate may be seen.[36,71] Although bronchoalveolar lavage is the procedure of choice for the diagnosis of *P. carinii*,[72] occasional infections will have an atypical presentation as a solitary cavitary or nodular mass.[31,38,73–77] (Fig. 1-16). The cysts of *P. carinii* can be identified by stains including GMS, Gram-Weigert, Grocott, periodic acid-Schiff (PAS), and toluidine blue, which outline the cyst wall (Table 1-3). Trophozoites are best visualized with cresyl violet, Diff-Quik, Giemsa, May-Grünwald-Giemsa, PAS, and Wright-Giemsa stains.[38,77–79] *P. carinii* can be rapidly identified with Diff-Quik staining of the aspirated smears,[36] in which up to eight intracystic trophozoites can be seen within the nonstaining cysts. The Papanicolaou-stained FNA smears may contain the characteristic honeycomb casts[4] (Fig. 1-16). In many cytology laboratories, the GMS is the special stain of choice to confirm *P. carinii*.[80] Recently, immunofluorescence and immunoperoxidase stains and/or hybridization probes have been used to diagnose *P. carinii*.[72,81–84]

Especially in immunocompromised patients, cytomegalovirus (CMV) and herpes simplex virus (HSV) can cause pulmonary lesions amenable to FNA biopsy.[85,86] However, there have been only a few FNA reports of viral infections[36,85] (Table 1-5). The differential diagnosis of viral infections is broad and includes reactive alveolar cells, drug reactions, repair, and/or malignancy due to the misinterpretation of the intranuclear viral inclusions for large nucleoli of reactive or neoplastic cells[10] (Fig. 1-17). Other organisms that have been recognized in FNA biopsy are parasites such as *Echinococcus, Dirofilaria* (Fig. 1–17), lung flukes, and amoebae[87–89] (Table 1–5).

In any infectious or inflammatory process, inflammatory atypia of epithelial cells, fibroblasts and histiocytes can create a potential source for a false-positive diagnosis of cancer[9,10,90–92] (Tables 1-4 and 1-5; see also the box on p. 17). Cytologic features favoring an inflammatory or infectious process include a relative predominance of inflammatory cells and/or specific organisms with only a limited number of atypical cells showing suggestive "malignant" cytologic features. In general, the atypical cells may also be poorly preserved or degenerating. Metaplastic cells in cavitary lesions (e.g., tuberculosis, aspergilloma) may also be mistaken for malignant cells due to nuclear atypia, including irregular, clumped chromatin and occasionally large nucleoli.[87] An important clue is that the metaplastic squamous cells will maintain low nuclear to cytoplasmic ratios (Fig. 1-18). Regeneration and repair in aspirated material will possess features similar to those seen in exfoliative respiratory specimens (Fig. 1-19, Table 1-5). Degenerating histiocytes may also have atypical features, including hyperchromasia and larger than usual nucleoli. An important differential

Table 1-3. Identification of Parasites and Viruses

Microorganism	*Morphology*	*Stains*	*Organs Affected*	*Differential Diagnosis*[a]
Parasites				
Pneumocystis carinii (may be a fungus)	Single round, cup-shaped or comma-shaped organisms; cysts (5–8 μm) with refractile walls containing 6–8 trophozoites (1–2 μm); granular eosinophilic or basophilic background material	Toluidine blue, Wright's, methenamine silver, Gram-Weigert, Immunoperoxidase; Papanicolaou, H&E for background material	Lungs, extrapulmonary sites (rare)	Hemosiderin, cell debris (background material)
Filariae	Slender, long, threadlike nematodes with filiform end (microfilariae)	Papanicolaou, H&E, Wright, Diff-Quik	Disseminated	Other nematodes, contaminants
Trichinella spiralis	Coiled or spiral larval forms 0.8–1.0 mm long	Papanicolaou, H&E, Wright-Giemsa	Muscle	Other nematodes, contaminants
Echinococcus granulosus	Invaginated or everted scolices with hooklets; detached hooklets	Papanicolaou, H&E, Wright-Giemsa	Liver and other organs	Contaminants
Entamoeba histolytica	Round or oval basophilic organisms (18–20 μm) with single, usually eccentric nucleus and cytoplasm containing ingested red cells	Papanicolaou, H&E, PAS, immunoperoxidase	GI tract, liver, disseminated	Contaminants
Viruses				
Cytomegalovirus	Enlarged cells with single, large intranuclear inclusion surrounded by halo; small cytoplasmic inclusions	Papanicolaou, H&E, immunoperoxidase	Lungs, kidney, liver, central nervous system, salivary glands, pancreas	Other viruses, degenerative changes in cells
Herpes simplex	Single and multinucleated cells having enlarged nuclei with bland, ground-glass appearance or with large, single intranuclear inclusion surrounded by clear halo; no cytoplasmic inclusions	Papanicolaou, H&E, immunoperoxidase	Lungs, skin, central nervous system, disseminated	Other viruses, degenerative changes in cells

[a]Differential diagnosis based on appearance in cytologic preparations.
(From Silverman,[36] with permission.)

diagnosis is adenocarcinoma due to the pale foamy nature of histiocytic cytoplasm. Morphologic features favoring histiocytes include eccentric, reniform nuclei, delicate nuclear membranes, relatively small nucleoli, and lower nuclear to cytoplasmic ratios than are typical of adenocarcinoma. Bronchioloalveolar cell carcinoma (BAC) is the most common malignancy confused with atypical degenerating histiocytes. Features in favor of BAC include greater cellularity with three-dimensional and/or cohesive groups of cells demonstrating a significant degree of nuclear atypically and higher nuclear to cytoplasmic ratios than histiocytes.

Patients who have previously received prior irradiation or chemotherapy, or both, may have a lung aspirate to evaluate a new lesion. A potential for a false-positive diagnosis of malignancy is due to the presence of atypical cells of pulmonary epithelial or mesothelial origins.[36] Clues to the identification of chemotherapeutic or irradiation-induced alterations are repair-like arrangements of the atypical cells with large hyperchromatic irregular nuclei, but with a degenerative or smudged quality to the chromatin (Fig. 1-20). By contrast, malignant cells have distinctly granulated chromatin. Irradiated cells also demonstrate cytoplasmic and nuclear vac-

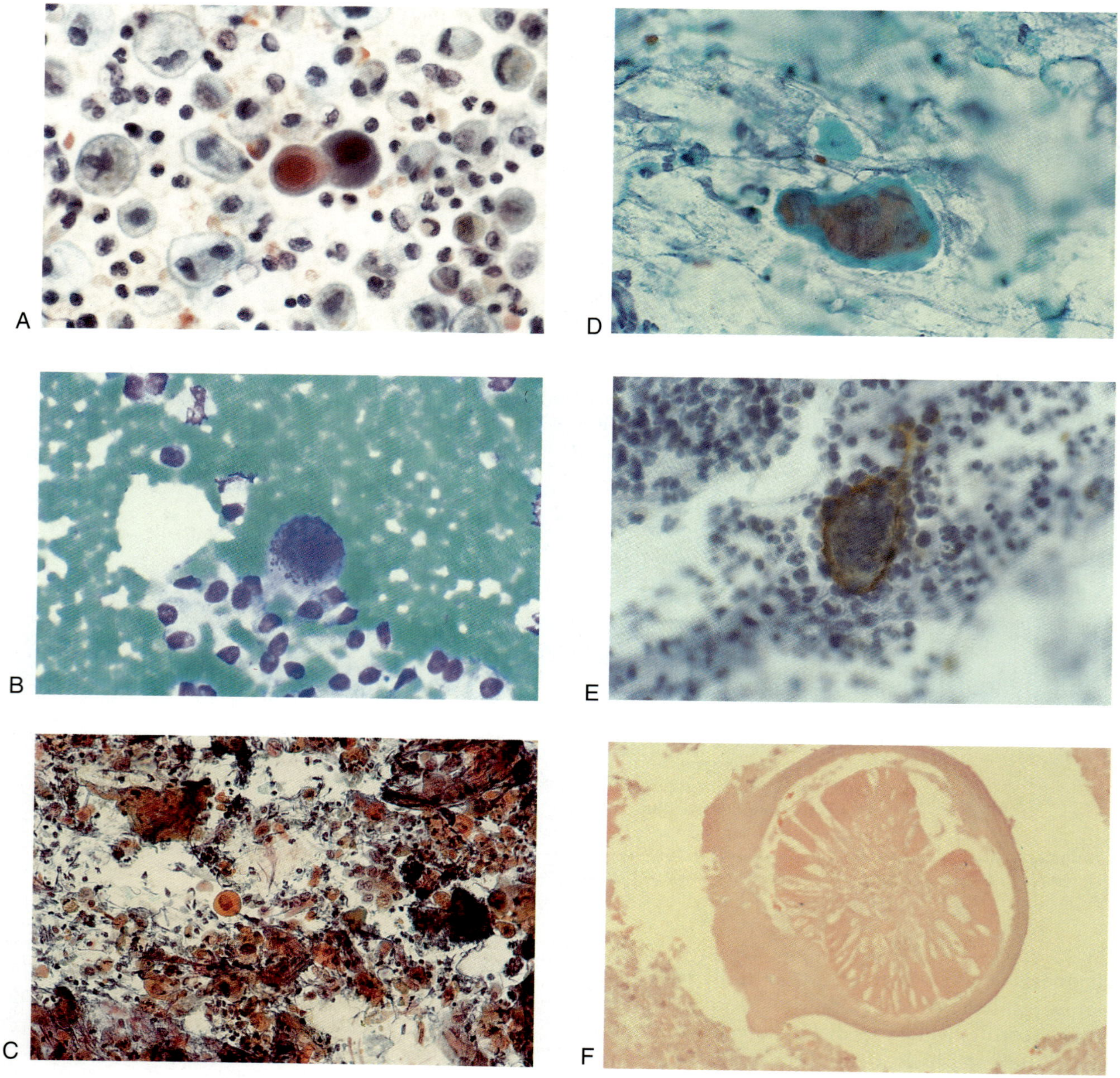

Fig. 1-17. (**A**) FNA of pneumonic process of lung in patient with AIDS showing scattered epithelial cells with characteristic large intranuclear viral inclusion consistent with cytomegaloviral infection (Papanicolaou). (**B**) CMV in Diff-Quik–stained smear. Note prominent cytoplasmic inclusions (Diff-Quik). (Case courtesy of Dr. Celeste Powers, Syracuse, NY.) (**C**) FNA of pneumonic process of lung in patient with prior history of squamous cell carcinoma. Atypical herpes-infected keratinized cells in acute inflammatory background could potentially simulate a keratinizing squamous cell carcinoma (Papanicolaou). (**D**) The correct diagnosis of herpetic infection was suggested by scattered multinucleated cells having characteristic nuclear changes including homogenization of the chromatin with margination of the chromatinic rim (Papanicolaou). (**E**) Diagnosis of herpetic infection was confirmed with immunoperoxidase stains for herpes (immunoperoxidase). (**F**) FNA of dirofilaria of the lung (cell block, H&E).

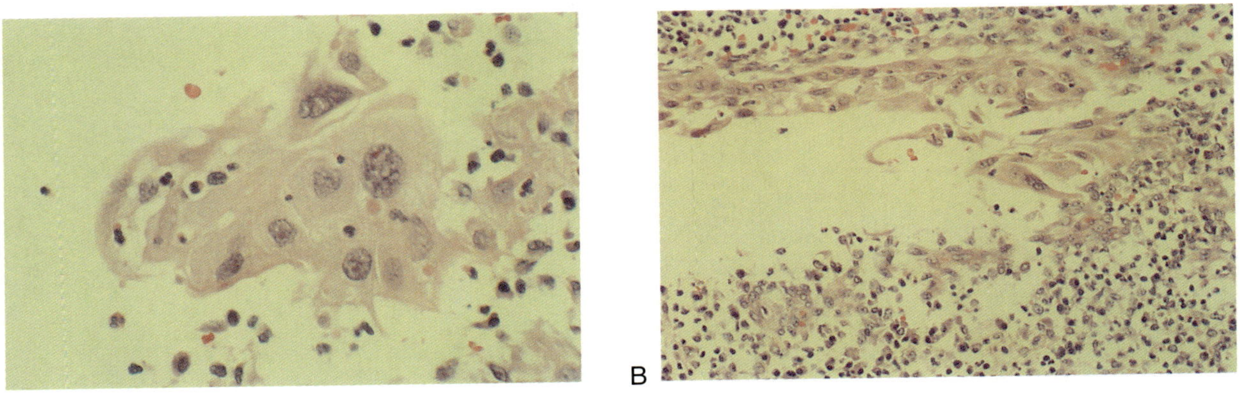

Fig. 1-18. (**A**) FNA of emphysematous cavity containing atypical squamous metaplastic cells that could potentially be misdiagnosed as malignant. Note the low nuclear to cytoplasmic ratios with cells having well-defined cell borders and no loss of polarity (Papanicolaou). (**B**) Resected emphysematous cavity lined by metaplastic squamous cells with extensive acute inflammation (H&E). (Case courtesy of Dr. Fred Worsham, Roper Hospital, Charleston, SC.)

Table 1-4. Pitfalls in Respiratory Cytology

Benign	*Malignancy (False-Positive Diagnosis)*
Repair	Aceno or squamous cell carcinoma
Squamous metaplasia	Squamous cell carcinoma
Irritated (hyperplastic) bronchial cells	Adenocarcinoma
"Creola bodies"	Small cell carcinoma
Reserve cells	Bronchioloalveolar cell carcinoma
Pneumocytes	

(From Silverman,[36] with permission.)

Table 1-5. Cytomorphologic Differential Diagnostic Features of Epithelial Repair Versus Malignancy

Epithelial Repair	*Pulmonary Malignancy*
Groups of cells	Groups of cells
Rare single cells	Many single cells
Flat metaplastic sheets, streaming	Syncytial arrangement
Distinct cell borders	Variable to indistinct cytoplasmic borders
Cellular polarity maintained	Loss of polarity
Enlarged nuclei	Enlarged nuclei
Round or oval nuclei with smooth nuclear membrane	Irregular nuclear borders
Uniform nuclei	Variable nuclei
Vesicular, hypochromatic nuclei with even chromatin distribution	Opaque to hyperchromatic nuclei with irregular chromatin distribution and parachromatin clearing
Nucleoli	Nucleoli (often irregular in shape)
Normal mitotic figures	Abnormal mitotic figures
No diathesis	Clear background or diathesis

(From Silverman,[36] with permission.)

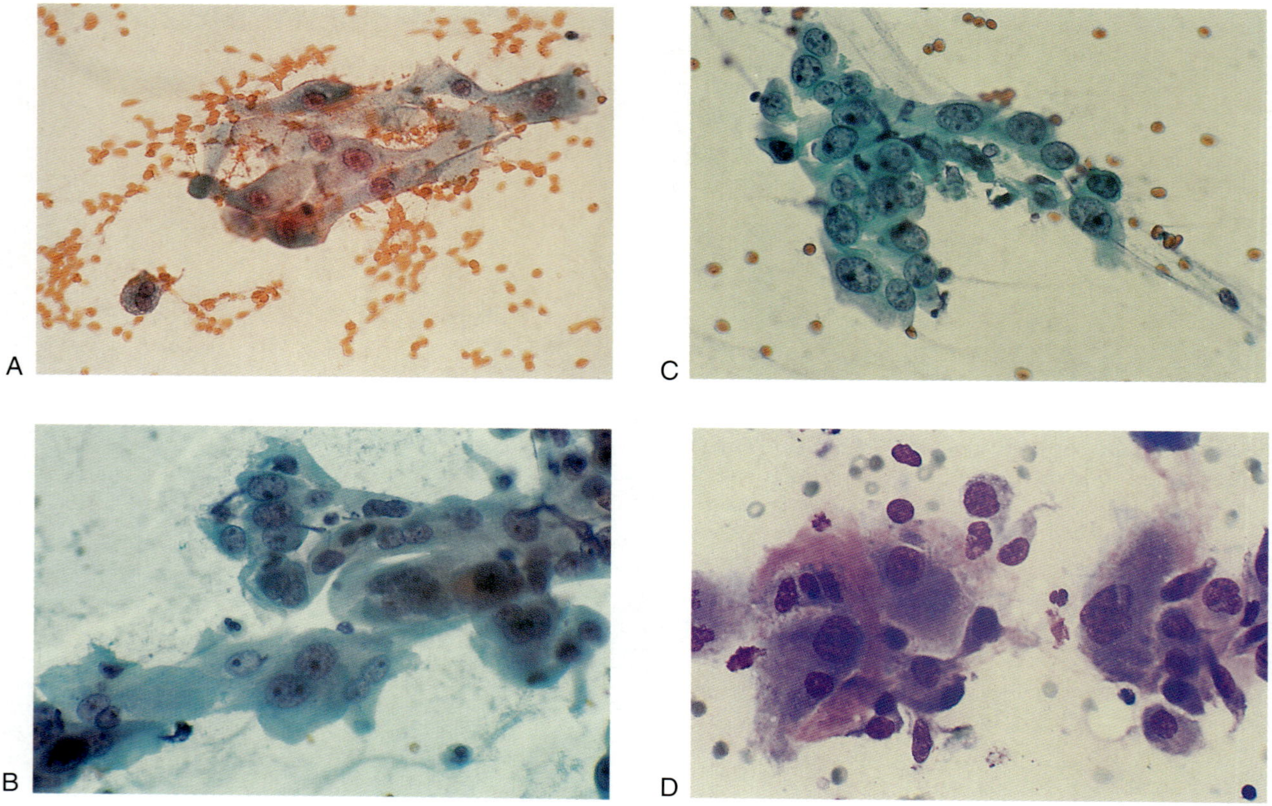

Fig. 1-19. (**A**) Clump of atypical cells consistent with repair reaction in aspirate of lung. Note the monolayer of cells having slightly enlarged nuclei with nucleoli, but low nuclear to cytoplasmic ratios. Well-defined cell borders and no loss of polarity is noted (Papanicolaou). (**B**) Repair reaction showing slightly atypical features including nuclei close together with higher nuclear to cytoplasmic ratios. However, overall pattern of repair is maintained and no individually scattered atypical cells were noted (Papanicolaou). (**C**) Atypical repair reaction demonstrating a greater degree of nuclear enlargement and more prominent nucleoli with cells close together. However, a repair pattern is maintained (Papanicolaou). (**D**) Mesenchymal repair reaction in FNA lung. Note loose cluster of atypical cells having mesenchymal features including stellate to spindle shapes with frayed cytoplasmic borders. Although some of the cells posses nucleoli, the nuclear to cytoplasmic ratios are low with no significant hyperchromasia (Diff-Quik).

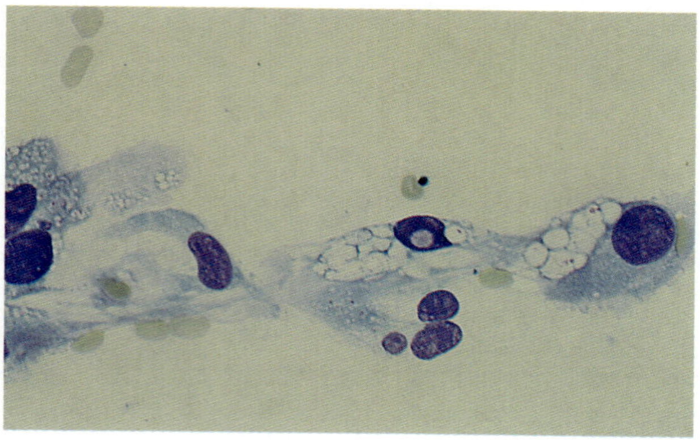

Fig. 1-20. FNA of patient irradiated for squamous cell carcinoma of the lung. An ill-defined recurrent lesion was aspirated, which reveals cells consistent with radiation effect. Note the repair pattern reaction with cytologic features of radiation including nuclear and cytoplasmic vacuolization (Diff-Quik). (From Silverman,[36] with permission.)

Pitfalls for False-Positive Diagnosis of Malignancy in Respiratory Cytology

- Reserve cell hyperplasia
- Irritation form of bronchial cells ("Creola bodies")
- Atypical squamous metaplasia
- Atypical bronchoalveolar cell hyperplasia
- Regeneration and repair
- Radiation/chemotherapy effect
- Viral cytopathic changes
- Megakaryocytes
- Vegetable cells
- Degenerating macrophages

(From Silverman,[36] with permission.)

Cytologic Changes Due to Radiation and Chemotherapy

- Nuclear and cytoplasmic enlargement
- Nuclear and/or cytoplastic vacuolization
- Binucleation and multinucleation
- Anisonucleosis
- Karyorrhexis, karyolysis
- Macronucleoli, multiple
- Amphophilia, pale staining
- Cytophagocytosis
- Loss of sharp cytoplasmic border

(From Silverman,[36] with permission.)

uolization. Features of irradiation and chemotherapy are presented above.

Clinically and radiologically, a solitary pulmonary infarct can masquerade as a peripheral lung neoplasm.[93] Cytomorphologically, a pulmonary infarct may result in a false-positive diagnosis of cancer secondary to benign atypical bronchioloalveolar cells arranged in repair-like clusters and sheets[10,93–97] (Fig. 1-21). Other nonspecific cytomorphologic findings of pulmonary infarct include squamous metaplasia and histiocytes with refractile, intracytoplasmic coarse pigment consistent with hemosiderin.[36]

Noncellular material reported in FNA biopsies include ferruginous bodies, corpora amylacea, calcospherites, hemosiderin, silicate crystals (talc), and amyloid.[36,98–100] Covell and Feldman[101] reported a case of aspiration pneumonia (phytopneumonitis) diagnosed by FNA biopsy. The cytologic features of aspiration pneumonia include acute and/or chronic inflammation associated with abundant vegetable matter; this material was variably sized and had

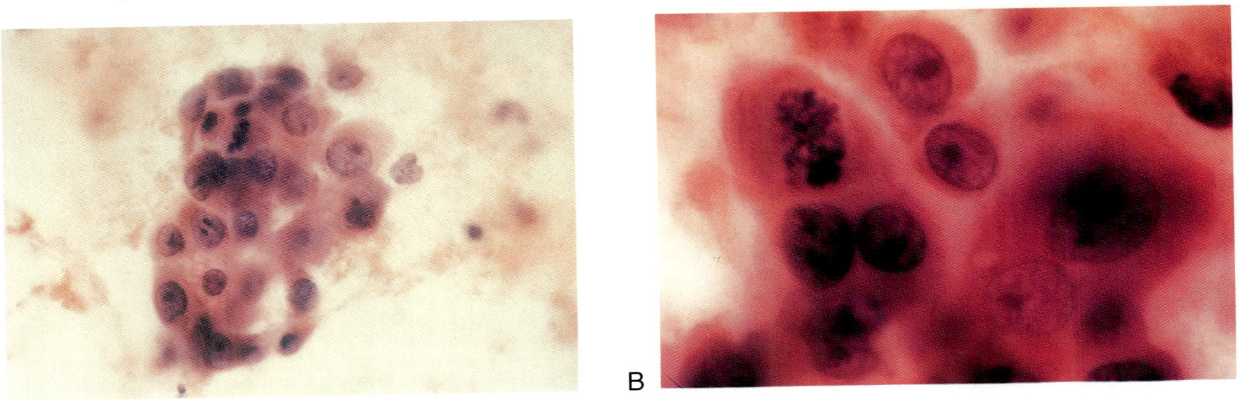

A B

Fig. 1-21. (**A**) FNA of pulmonary infarct that potentially could be confused with adenocarcinoma if the features of a repair reaction are overlooked. Worrisome features include mitotic figures and enlarged nuclei with prominent nucleoli. However, note the low-power pattern of repair as well as cells having low nuclear to cytoplasmic ratios. The smear had low cellularity with only a few scattered individual cells (Papanicolaou). (**B**) High-power view of repair reaction secondary to pulmonary infarct. Note the enlarged nuclei, prominent nucleoli and mitotic figure (Papanicolaou).

Non-neoplastic Pulmonary Lesions That Might Be Sampled by FNA Biopsy

- Abscess
- Granulomatous processes (e.g., sarcoid, fungi)
- *Pneumocystis carinii* (pneumocytoma)
- Viral infections
- Other parasites
- Irradiation and/or chemotherapy-related abnormalities
- Pulmonary infarcts
- Pseudolymphoma and other inflammatory pseudotumors
- Amyloid
- Pneumoconiosis
- Aspiration pneumonia
- Wegener's granulomatosis

(From Silverman,[36] with permission.)

angulated contours and thick refractile cell walls and internal round basophilic bodies[36] (Fig. 1-22). These angulated structures could potentially be mistaken for fungi or malignant cells.[36]

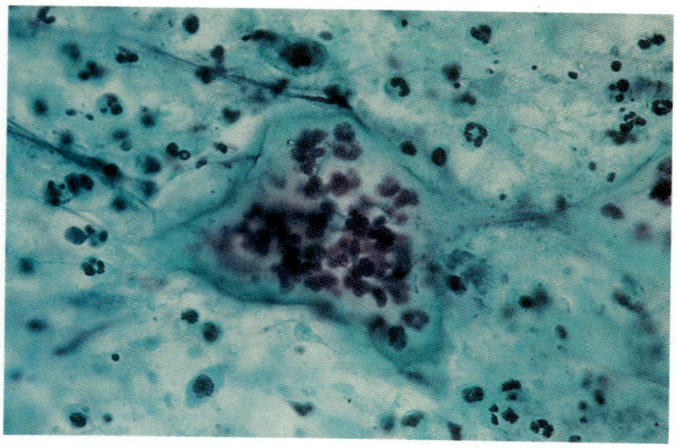

Fig. 1-22. FNA of aspiration pneumonia in which scattered vegetable matter was present along with extensive acute inflammation. Note the angulated contours and thick refractile walls with internal basophilic round structures of the vegetable material (Papanicolaou). (Courtesy of Dr. Phil Feldman, University of Virginia, Charlottesville, VA.)

Inflammatory pseudotumors can clinically and radiologically mimic a neoplastic lesion.[102] Although traditionally diagnosed by open lung biopsy, pulmonary inflammatory pseudotumors can be sampled by FNA biopsy.[36,103–106] Recently, Machicao et al.[102] reported four patients diagnosed by FNA cytology. The cytomorphologic features include the mixture of uniform spindle-shaped fibroblasts in small clusters and uniform sheets of alveolar pneumocytes. Isolated atypical large histiocytes and fibroblasts may have abundant vacuolated cytoplasm, large nuclei, and macronucleoli, which could be misinterpreted as malignant cells (Fig. 1-23). Aspirates of lesions involving the right

A

B

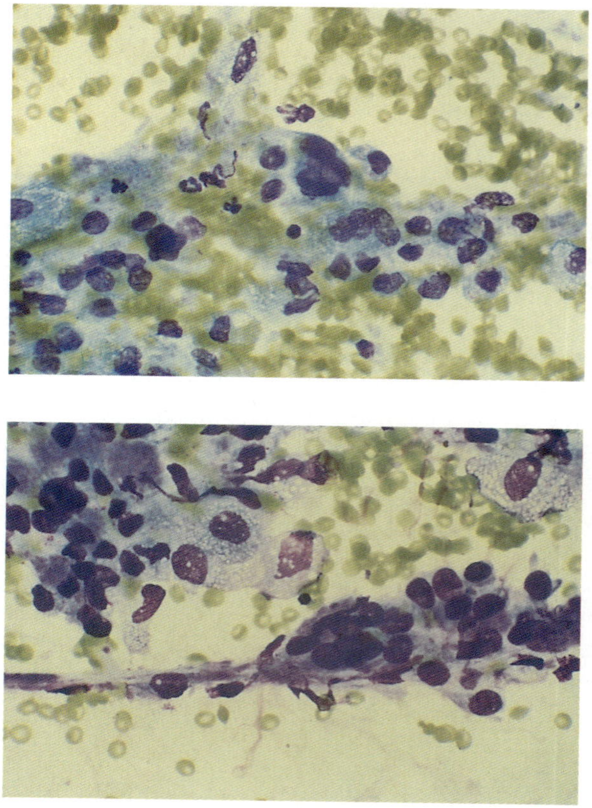

Fig. 1-23. (**A**) Aspirate of xanthogranuloma of lung consisting of scattered enlarged, atypical histiocytes and fibroblasts that potentially could by confused with malignancy (Diff-Quik). (**B**) Aspirate of xanthogranuloma of lung consisting of scattered histiocytes having enlarged nuclei along with mildly atypical spindle-shaped fibroblastic cells (Diff-Quik).

lower lobe of the lung overlying the diaphragm can inadvertently sample the liver. In these "right-sided aspirates," liver cells can be potentially confused with malignant cells (Fig. 1-24).

Amyloidosis involving the thoracic region may be diagnosed by FNA cytology[107,108] (Fig. 1-25). One must recognize the clumps of amorphous material that stain metachromatically with the Romanowsky stains and are Congophillic with characteristic apple green birefringence following polarization[36] (Fig. 1-25). Occasionally a foreign body reaction to the amyloid material can be appreciated.[109]

In FNA biopsy, Wegener's granulomatosis features necrosis, mixed inflammatory cell infiltrates, and histiocytes, including multinucleated giant cells.[36, 110–112] These findings, of course, are not specific. Other unusual non-neoplastic pulmonary processes sampled by FNA biopsy include extramedullary hematopoiesis,[113] lipid pneumonia,[114] and lymphomatoid granulomatosis.[115] The types of non-neoplastic

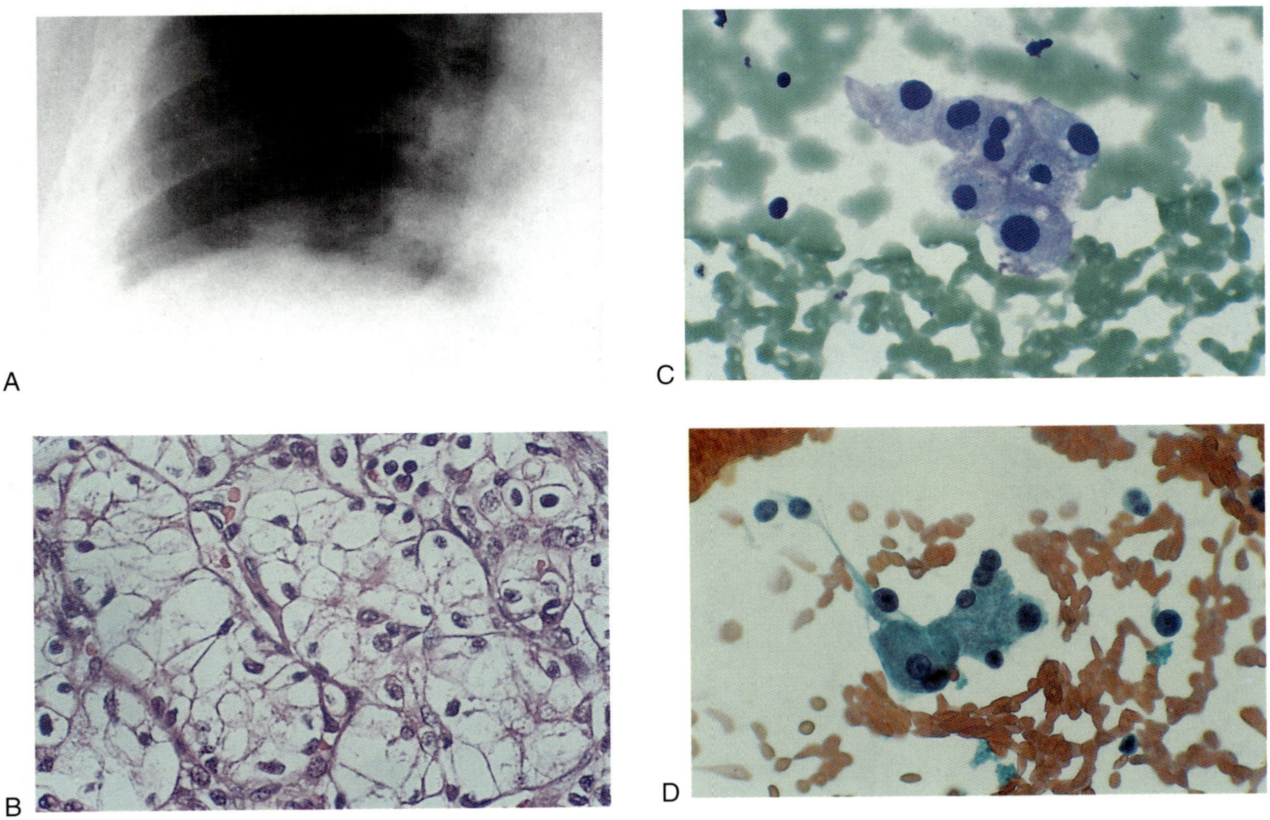

Fig. 1-24. (**A**) Radiograph of nodule overlying diaphragm in right lower lobe of lung in patient with history of renal cell carcinoma. Other scattered nodules were present in both lung fields. (**B**) Histologic examination of patient's primary renal cell carcinoma, clear cell type, for comparison with FNA findings (H&E). (**C**) FNA of nodule in right lower lobe of lung consisting of uniform pale cells which potentially could be confused with metastatic renal cell carcinoma. Cytologic features of liver cells in this "right-sided aspirate" include cuboidal shape of cells with central nuclei and granular to slightly vacuolated cytoplasm (Diff-Quik). (**D**) The cytologic features of benign liver cells are enhanced in Papanicolaou-stained smear showing the granular cytoplasm of the cells (Papanicolaou).

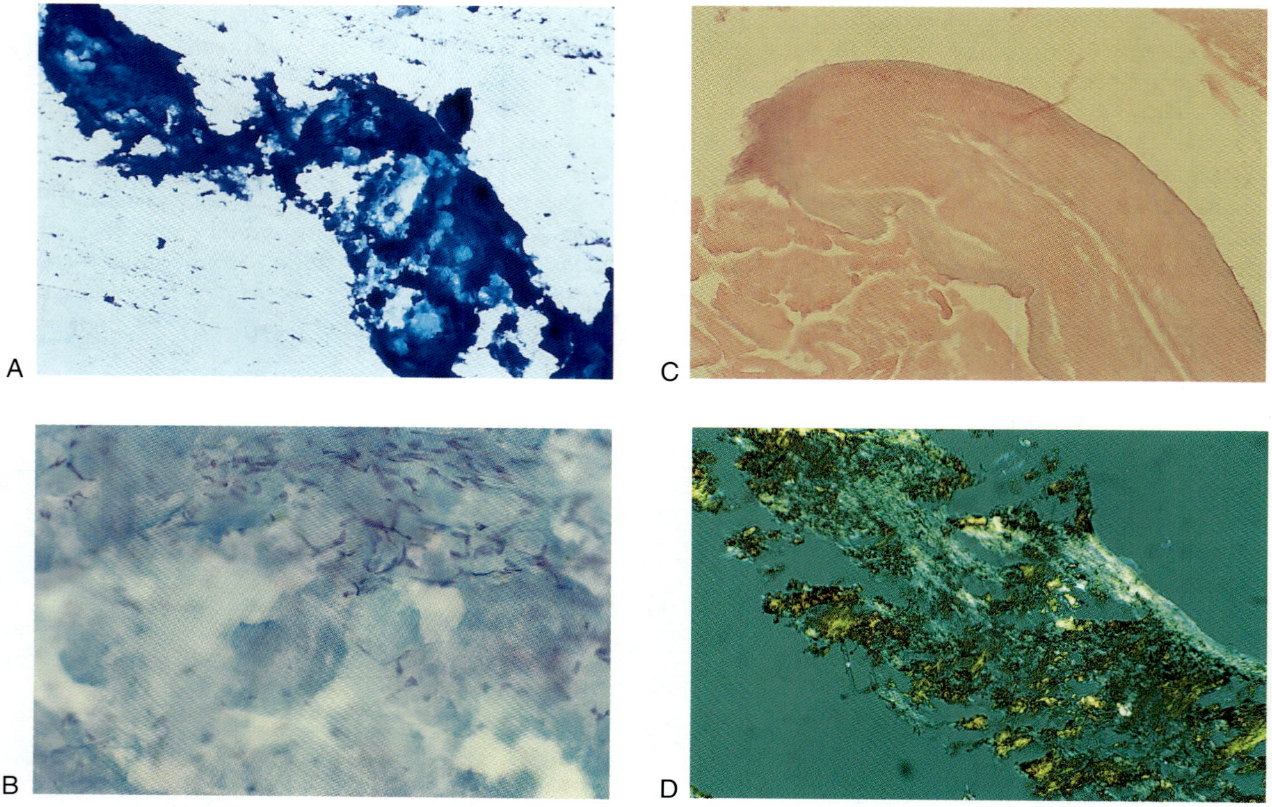

Fig. 1-25. (**A**) Low-power view of aspirate of amyloid tumor of lung revealing clumps of amorphous, metachromatically staining amyloid (Diff-Quik). (**B**) FNA of amyloid tumor of lung characterized by amorphous, homogeneous material with a few interspersed scattered stromal cells (Papanicolaou). (**C**) Cell block revealing eosinophilic homogenous amorphous material consistent with amyloidoma of lung (H&E). (**D**) Amyloid material demonstrating characteristic apple-green birefringence following polarization (Congo red).

lesions that can be sampled and diagnosed by FNA biopsy are presented on page 18.

LUNG TUMORS

In the World Health Organization (WHO) histologic classification scheme of lung tumors, most lung carcinomas are divided into four categories: adenocarcinoma (50 percent of all primary lung carcinomas), squamous cell carcinoma (30 percent), large cell undifferentiated carcinoma (less than 5 percent), and small cell undifferentiated carcinoma (15 percent).[116,117] Although this scheme implies differences in tumor histogenesis, an evolving concept is that most pulmonary carcinomas may exhibit multidirectional differentiation based on light microscopic, electron microscopic (EM), and immunohistochemistry findings.[118–128] Thus, several investigators have claimed that the WHO categorization is not valid.[19,120,129] Therefore, either new categories must be included reflecting differences in tumor histogenesis (e.g., neuroendocrine carcinomas), or exiting categories must be combined to reflect similarities in histogenesis (e.g., large cell carcinoma, adenocarcinoma, and/or squamous cell carcinoma).[19] This book describes the cytologic features of the four main types of lung carcinoma, as well as less common neoplasms. Dividing lung carcinomas into only two types, small cell and non-small cell, has significant prognostic and therapeutic implications, as does clinical stage.[19,130] Although the surgical excision for low-stage, non-small cell carcinoma is the treatment of choice (regardless of subtype), differences in survival based on histologic classification may exist.[131,132] For example, stage II carcinomas may have different recurrence patterns with squamous cell carcinomas more apt to recur locally, whereas with adenocarcinomas more distant sites are likely.[130,133] For all lung carcinomas, the overall 5-year survival rate varies from 10 percent to 13 percent; median survival of untreated patients is less than 6 months.[134,135] This poor survival is probably attributable to a high proportion of cases being detected initially at a high stage. Classification of lung carcinomas by FNA biopsy into the WHO types may have some value in situations in which a primary malignancy needs to be distinguished from a metastatic extrathoracic carcinoma. Pertinent facts include 4 percent of all solitary pulmonary nodules are metastases from an extrapulmonary carcinoma and 25 percent of patients with an extrapulmonary carcinoma and pulmonary metastasis will have only a solitary pulmonary nodule at presentation.[136,137]

ADENOCARCINOMA

The incidence of bronchogenic adenocarcinoma is increasing in the United States, especially in women.[138] Relative to squamous cell carcinoma, adenocarcinoma does not appear to have as strong association with smoking.[137] Most adenocarcinomas of the lung are peripherally situated and present as discrete round to irregular nodules measuring 2 to 5 cm.[139] Central lesions tend to be symptomatic as a result of obstructive effects, whereas peripheral lesions are more likely to be asymptomatic when discovered. Peripheral adenocarcinomas frequently spread to lymph nodes, and more than one-half invade the pulmonary arterial system. This predisposes extrathoracic dissemination, especially to the brain.[138] In the lung cancer study group analysis of 1,121 patients undergoing apparent complete resection of non-small cell carcinoma, adenocarcinoma was shown to have a poorer outcome when compared to squamous cell carcinoma in every TMN subset.[140] BAC presenting as a peripheral solitary nodule, in contrast, may have a slightly improved survival when compared with other types of adenocarcinomas.

The cytomorphologic features of adenocarcinoma include malignant cells generally arranged in three-dimensional ball-like clusters, acinar groupings, or side-by-side alignment[141] (Fig. 1-26), as listed below. Cellular contours vary from columnar to cuboidal to round. Occasionally, the cells are aligned to form a luminal border (Fig. 1-27). In more poorly differentiated adenocarcinomas, the malignant cells tend to have a syncytial arrangement (Fig. 1-28). Relative to normal bronchial elements, the cells have increased nuclear to cytoplasmic ratios with oval to round nuclei with irregularly distributed, finely granular chromatin and often prominent nucleoli (Fig. 1-28). Basophilic cytoplasm with fine or coarse vacuolization may be seen. Either a necrotic or clean background may be present. BAC, a subtype of adenocarcinoma, may include papillary groupings composed of uniform malignant cells having prominent depth of focus with a lack of significant nuclear molding related to more abundant cytoplasm[132,142,143] (Fig. 1-29), as listed on page 26. Other features of BAC include cells resembling atypical alveolar macrophages and benign bronchial lining cells[144,145] (Fig. 1-30). Occasionally, aspirates have malignant cells situated along alveolar septae with hobnail-shaped nuclei (Fig. 1-31). These septal

Features of Adenocarcinoma

Cellular pattern

- Three-dimensional groups, cell balls, papillary, acinar arrangement
- Side-by-side arrangement, palisading
- Syncytial groups in poorly differentiated carcinomas
- Single cells

Smear background

- Clear or diathesis

Cytoplasm

- Basophilic, lacy, finely to coarsely vacuolated columnar, cuboidal or round cells
- Eccentrically placed nuclei

Nucleus

- Finely granular, irregular chromatin distribution
- Prominent nucleoli (single or multiple)
- Nuclear molding

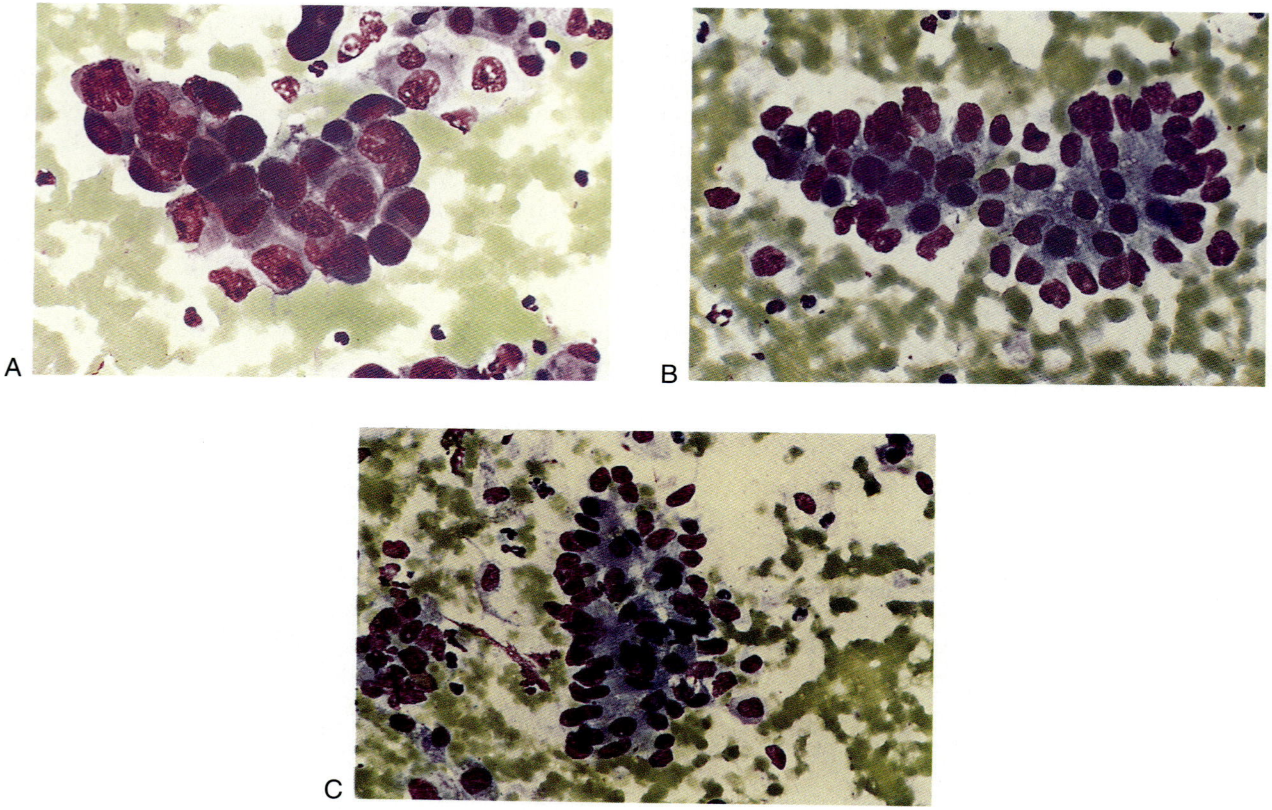

Fig. 1-26. (**A**) FNA of adenocarcinoma consisting of three-dimensional cluster of cells (Diff-Quik). (**B**) Adenocarcinoma of lung demonstrating acinar arrangement of cells (Diff-Quik). (**C**) FNA of adenocarcinoma revealing side-by-side alignment of tumor cells (Diff-Quik).

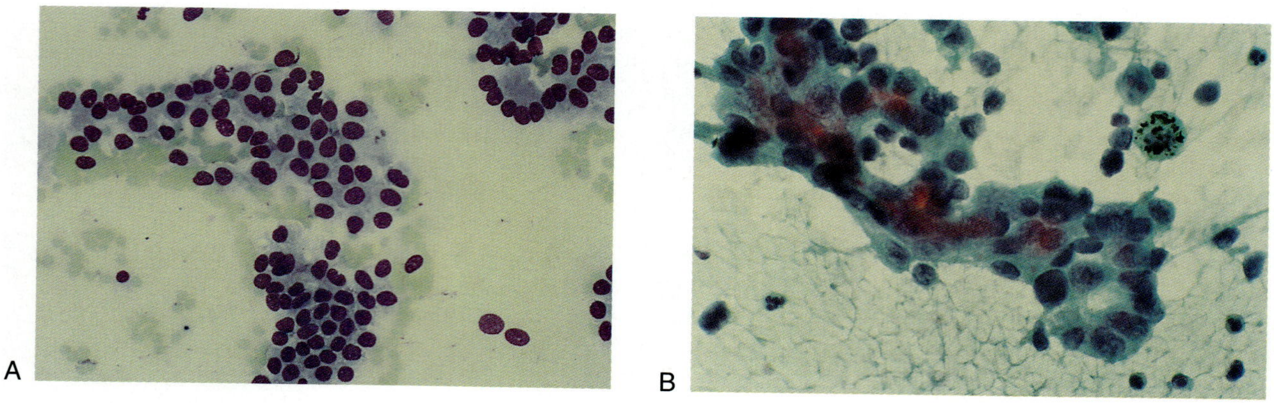

Fig. 1-27. (**A**) FNA of adenocarcinoma of the lung revealing both sheet-like grouping of tumor cells as well as others showing partial acinar formation (Diff-Quik). (**B**) Acinar arrangement of tumor cells in adenocarcinoma of the lung (Papanicolaou).

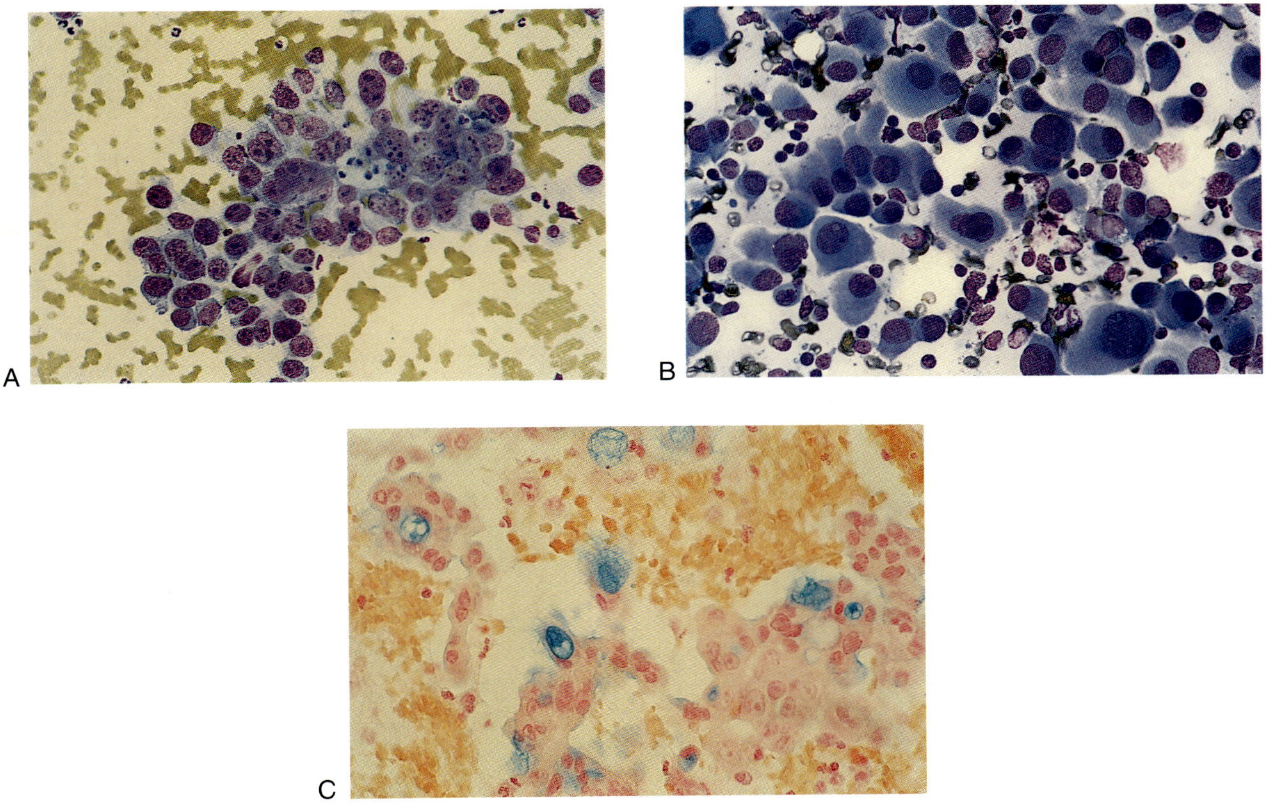

Fig. 1-28. (**A**) FNA of poorly differentiated adenocarcinoma of the lung consisting of loose cluster of malignant cells with high nuclear to cytoplasmic ratios (Papanicolaou). (**B**) FNA of poorly differentiated adenocarcinoma of the lung having a dissociated pattern but revealing features of adenocarcinoma including columnar shapes of many of the tumor cells along with others having eccentrically placed nuclei (Diff-Quik). (**C**) Mucin stain performed on poorly differentiated adenocarcinoma of the lung demonstrating intracytoplasmic mucin (Alcian blue).

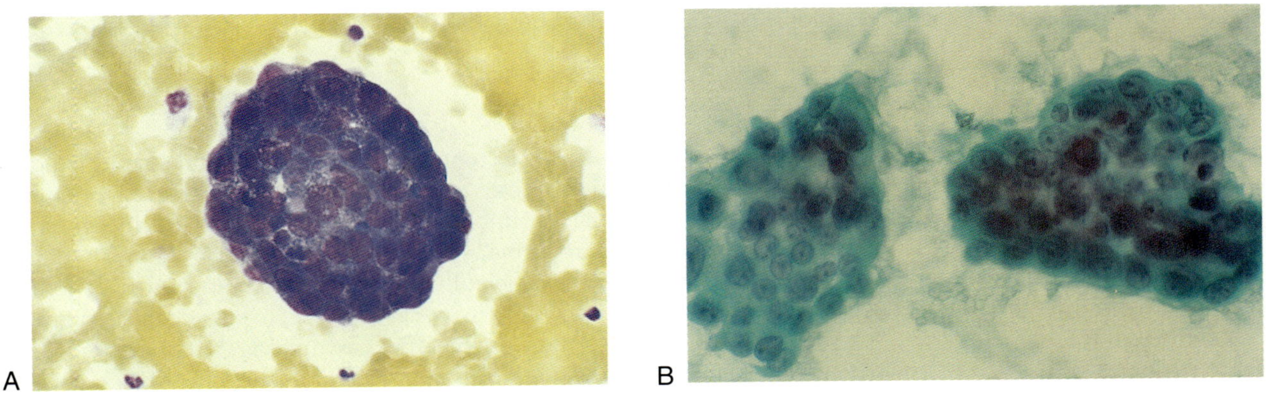

Fig. 1-29. (**A**) FNA of bronchioloalveolar cell carcinoma (BAC) revealing papillary grouping of uniform cells having prominent depth of focus without significant nuclear molding (Diff-Quik). (**B**) Aspirate of BAC revealing group of uniform cells with depth of focus (Papanicolaou).

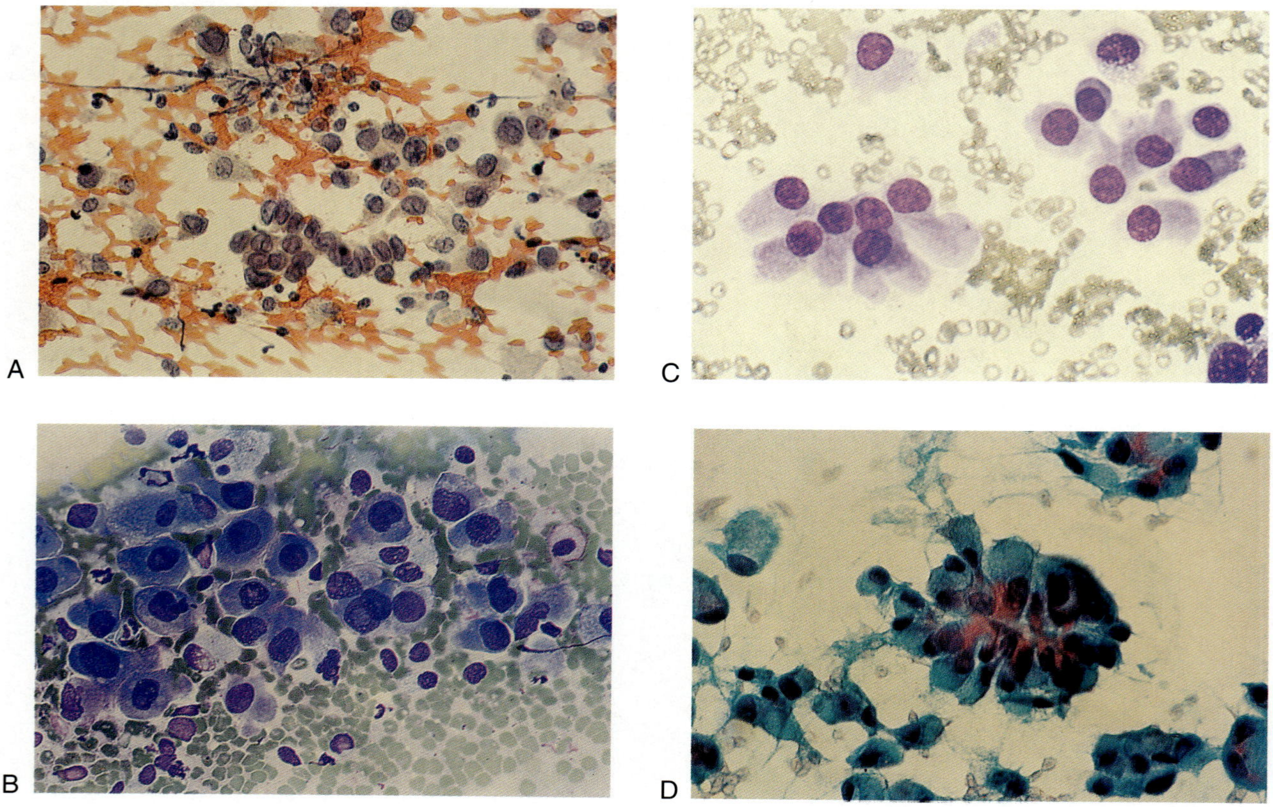

Fig. 1-30. (**A**) Aspirate of bronchioloalveolar cell carcinoma (BAC) consisting of atypical cells resembling alveolar macrophages. Note, however, the increased nuclear to cytoplasmic ratios with nuclear irregularity and greater cellularity than would be appreciated in a benign aspirate (Papanicolaou). (**B**) Scattered atypical cells in aspirate of BAC resembling alveolar macrophages. Note the nuclear irregularity, intranuclear inclusions, and higher nuclear to cytoplasmic ratios than what would be appreciated in an aspirate consisting of benign alveolar macrophages (Diff-Quik). (**C**) Aspirate of BAC having cells resembling benign bronchial lining cells based on the columnar configuration of the cells. However, the overall cellularity in this aspirate was much greater than what would be seen in a benign aspirate (Diff-Quik). (**D**) Aspirate of BAC consisting of atypical cells resembling benign bronchial lining cells (Papanicolaou).

arrangements are seen only in aspirates and not in bronchoscopic cytologic specimens. Cell groups having a radiating "cartwheel" or "flower-petal" pattern have also been described.[118,143] Although occasionally present in conventional adenocarcinoma, psammoma bodies and/or optically clear nuclei are more characteristic of the BAC morphologic variant[146] (Fig. 1-32). Ultrastructurally, features of Clara cells or rarely type II pneumocytes may be seen. The differential diagnosis of bronchogenic adenocarcinoma includes chemotherapy/irradiation effect, viral pneumonitis, organizing pneumonia, pulmonary infarct, large cell undifferentiated carcinoma, and metastatic adenocarcinoma[36] (Fig. 1-33). Nucleoli in adenocarcinoma are generally smoother in contour than in large cell carcinoma. Certain metastatic carcinomas have cytomorphologic features suggestive of a specific origin, such as elongated nuclei and dirty necrosis in colonic carcinoma and metastatic colonic carcinoma are

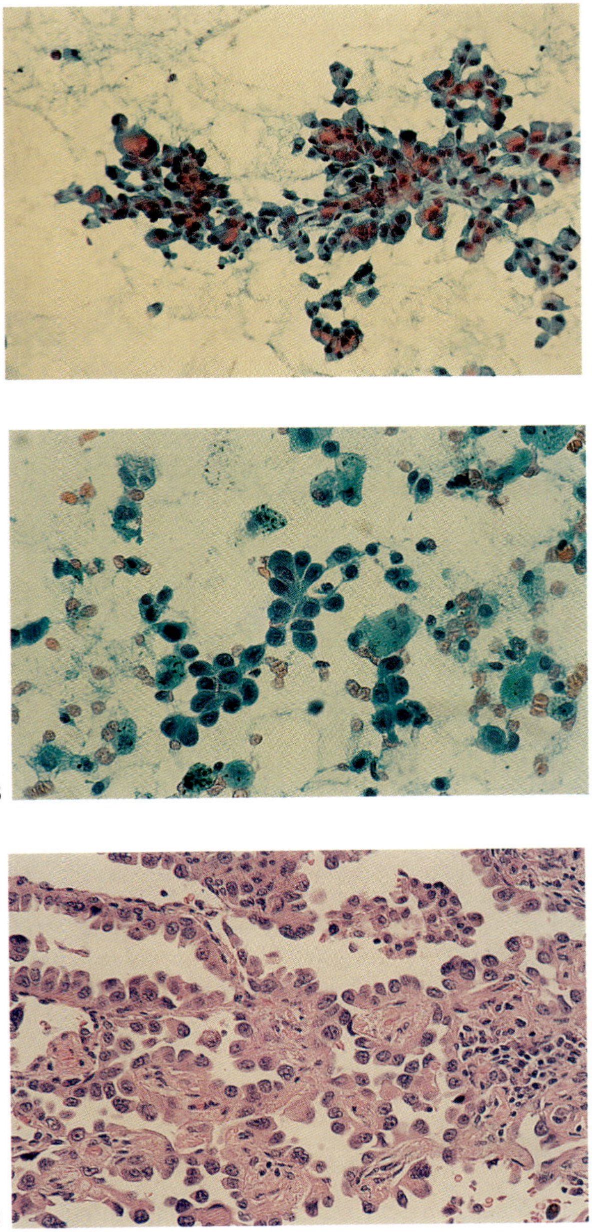

Fig. 1-31. (**A**) Aspirate of bronchioloalveolar cell carcinoma (BAC) revealing neoplastic cells aligned along alveolar septae (Papanicolaou). (**B**) Aspirate of BAC revealing malignant cells aligned along alveolar septae. Note the hobnail-shaped nuclei (Papanicolaou). (**C**) Histologic correlate revealing malignant bronchioloalveolar cells aligned along alveolar septae in resected BAC (H&E).

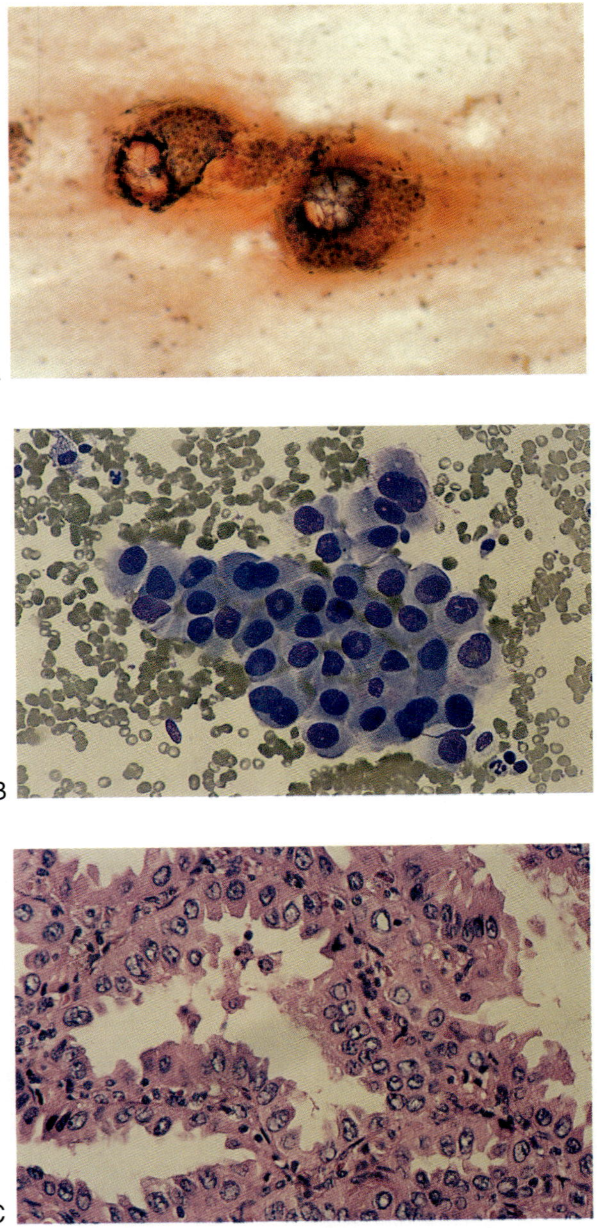

Fig. 1-32. (**A**) FNA biopsy of bronchioloalveolar cell carcinoma (BAC) consisting of three-dimensional clumps of cells in which psammoma bodies were present (Papanicolaou). (**B**) Cluster of malignant bronchioloalveolar cells, including some possessing intranuclear inclusions (Diff-Quik). (**C**) Histologic correlate demonstrating numerous cells having intranuclear inclusions in resected BAC (H&E).

Features of Bronchioloalveolar Cell Carcinoma

Cellular pattern
- Three-dimensional cell balls
- Side-by-side arrangement
- Cells aligned along septae
- Single cells

Smear background
- Clear to diathesis

Cytoplasm
- Basophilic, finely vacuolated
- Occasional psammoma bodies
- Oval to columnar-shaped cells

Nucleus
- Round to slightly irregular contours
- No nuclear molding
- Uniform size and shape
- Eccentric nuclei
- Nuclear pseudoinclusions

immunocytochemically different from primary lung carcinomas[147] (Fig. 1-34). However, the cytologic distinction remains difficult in many instances.[147]

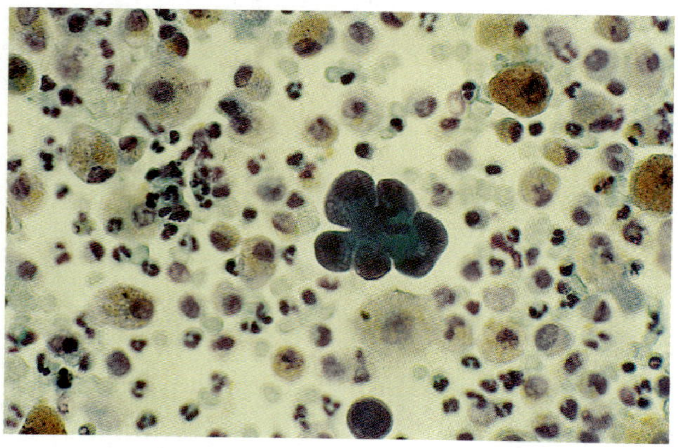

Fig. 1-33. FNA of pneumonic process of lung containing a cluster of a few atypical bronchioloalveolar cells having a somewhat scalloped pattern that could potentially be misinterpreted as malignant. However, note the extensive background inflammation and low nuclear to cytoplasmic ratios of the reactive cells. The patient clinically was thought to have an organizing pneumonia (Papanicolaou).

A

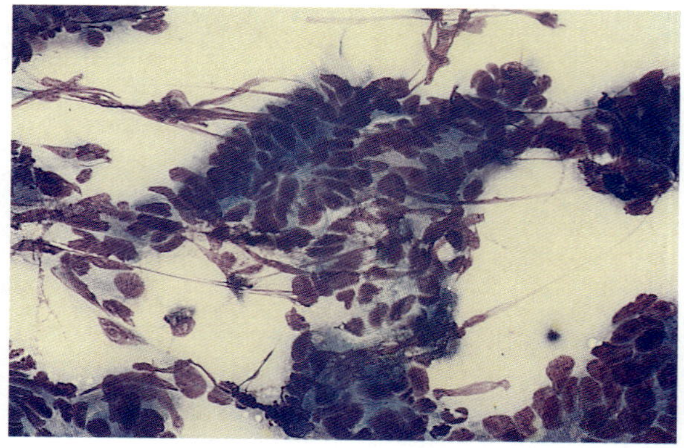

B

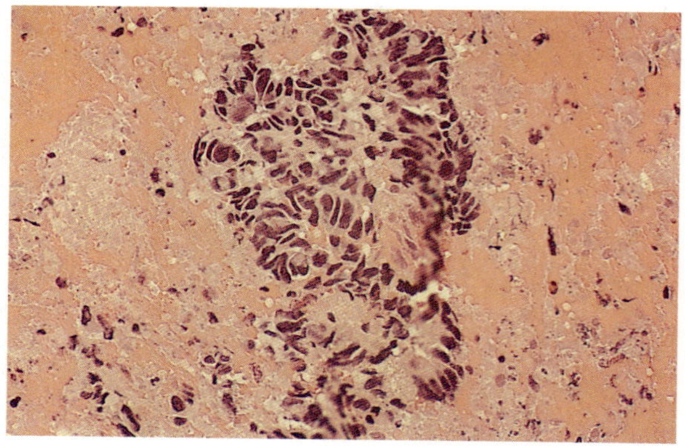

Fig. 1-34. (**A**) FNA of metastatic colon cancer revealing characteristic but not pathognomonic features of malignant acini with elongated palisading nuclei. Other fields show characteristic dirty background (Diff-Quik). (**B**) Cell block from FNA of metastatic colon carcinoma to the lung revealing star-shaped configuration of the tumor cells arranged in a pseudopalisading fashion. Extensive necrosis and hemorrhage are present in the background (H&E).

SQUAMOUS CELL CARCINOMA

Although recent reports from the United States and Japan indicate that adenocarcinoma is more common, squamous cell carcinoma has been considered the most common form of lung cancer in many countries.[137] Most squamous cell carcinomas arise centrally and often result in nonproductive cough, increased mucous secretion, loss of appetite and weight, chest pain, and/or obstructive pneumonitis.[137] Pancoast's syndrome and Horner's syndrome are most often due to squamous cell carcinomas followed by small cell carcinoma. Paraneoplastic states including hypercalcemia and clubbing are also associated with squamous cell carcinoma.[148] The clinical outcome for patients with squamous cell carcinoma may be better than for adenocarcinoma[139] in stage I and II malignancies, but not stage III disease.

In aspirates, squamous cell carcinoma typically consists of malignant cells in sheets and singly[10] (Fig. 1-35), as listed below. In moderately or well-differentiated squamous cell carcinoma, the malignant cells will have pleomorphic shapes with cytoplasmic keratinization (orangeophilia) and/or Herxheimer's spirals[149] (Fig. 1-35). Both nuclei and cytoplasm vary considerably in size and shape, including spindle and tadpole cells. Concentric arrangements (pearl formation), intercellular bridges, and keratin rings may be seen. In more poorly differentiated carcinomas, the cells have cyanophilic cytoplasm (Fig. 1-36). Syncytial groups of malignant cells with flattened, spindle-shaped malignant cells at the periphery is a feature of squamous cell carcinoma (Fig. 1-37). In general, squamous cell carcinoma will have coarsely granular, irregularly distributed chromatin. Prominent nucleoli are not a feature of small cell carcinoma but will be present in poorly differentiated squamous cell carcinoma. Occasional cells will have large nuclei possessing an opaque, India ink appearance. Nuclei with "squared-off" corners or sharp angulated contours suggest squamous cell carcinoma. A prominent tumor diathesis is usually present, and some cases of squamous cell carcinoma will have associated acute inflammation or foreign body reaction, or both. The major differential diagnoses of squamous cell carcinoma include the intermediate type of small cell carcinoma and large cell carcinoma. Regeneration and repair may also enter into the differential diagnosis. A repair reaction is characterized by a flat well-polarized sheet-like arrangement of cells that have enlarged nuclei with vesicular chromatin pattern and prominent nucleoli.

Features of Squamous Cell Carcinoma

Cellular pattern
- Sheets, syncytial groups
- Single cells
- Little nuclear overlapping or molding

Smear background
- Tumor diathesis (necrosis)
- Acute inflammation
- Occasional foreign body reaction to keratinous debris

Cytoplasm
- Bizarre shapes (e.g., spindle, tadpole)
- Dense cytoplasm—orangeophilic to cyanophilic; glossy and refractile
- Sharp cell borders
- Keratin rings

Nucleus
- Coarse granular to opaque chromatin
- Nucleoli more apparent in poorly differentiated carcinoma
- Irregular to bizarrely shaped

LARGE CELL CARCINOMA

Large cell carcinoma accounts for 10 to 20 percent of all lung carcinomas. By definition, large cell carcinoma is a non-small cell carcinoma that does not show morphologic evidence of differentiation in hematoxylin and eosin (H&E)-stained sections or mucin positivity with special stains.[137] Typically, large cell carcinoma is aggressive in behavior and rapidly fatal.[137] The anaplastic variant of large cell carcinoma (giant cell carcinoma) and the newly recognized large cell neuroendocrine carcinoma are special subtypes.[150] Large cell neuroendocrine carcinoma has large cells, brisk mitotic activity, and necrosis and demonstrates light microscopic features associated with neuroendocrine tumors such as organoid, palisading, and/or trabecular growth patterns, as well as immunohistochemical and EM evidence of neuroendocrine differentiation. By contrast, large cell carcinoma with neuroendocrine differentiation lacks light microscopic evidence of neuroendocrine differentiation but does have immunohistochemical or EM neuroendocrine features. To the best of our knowledge, no FNA cytologic studies have reported the consistent cytologic features of the neuroendocrine types of large cell carcinoma. Clinically, it appears that these neuroendocrine types of large cell carcinoma share the same poor prognosis with the other general types of large cell carcinoma, although specialized treat-

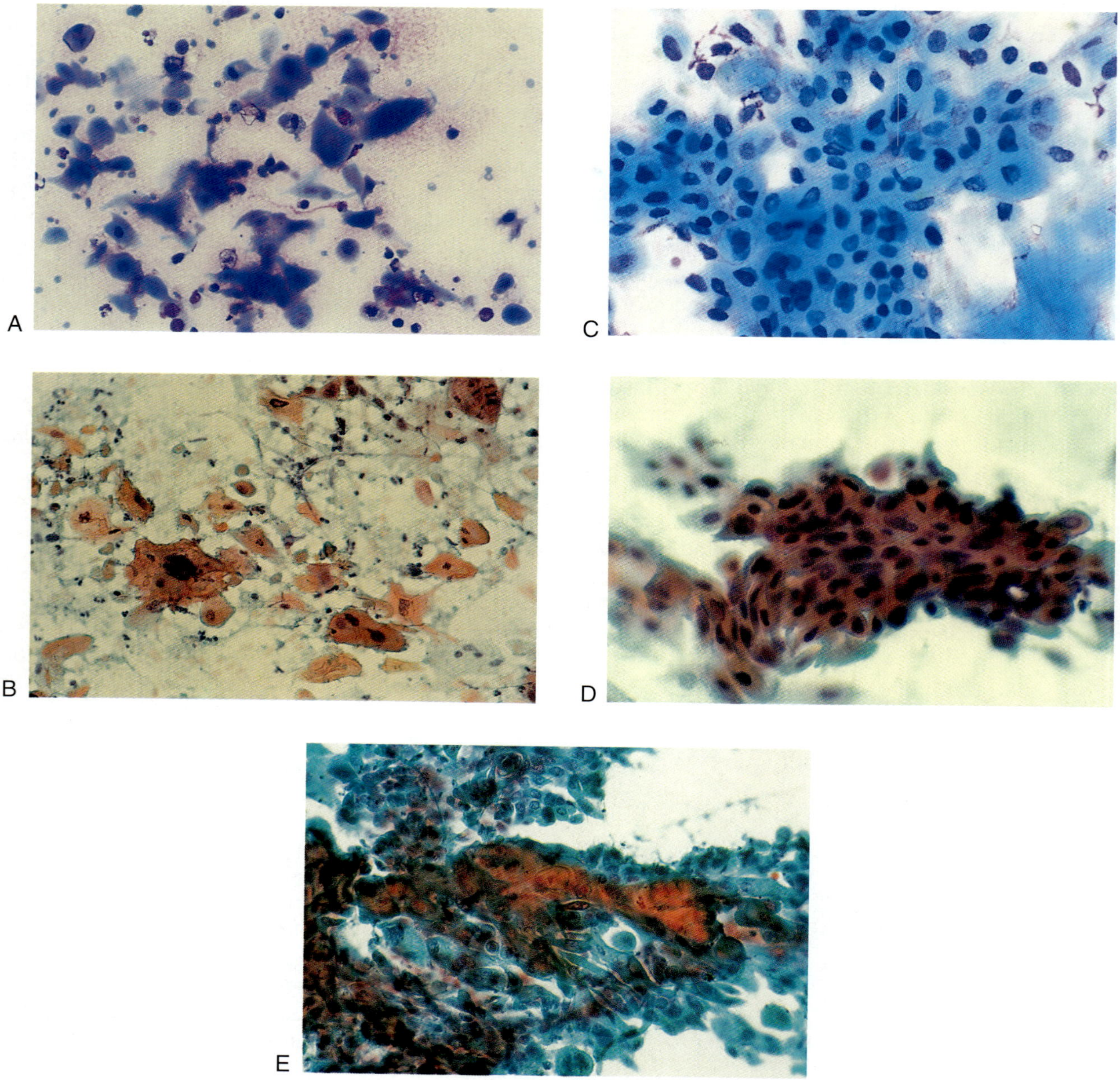

Fig. 1-35. (**A**) Aspirate of squamous cell carcinoma in Diff-Quik preparation revealing numerous scattered irregular fragments of keratinized cells, including many lacking nuclei (Diff-Quik). (**B**) FNA of keratinizing squamous cell carcinoma. Note the bizarre shapes, orangeophilia, dense cytoplasm, and opaque nuclei (Papanicolaou). (**C**) Markedly atypical keratinized squamous cells arranged in a sheet-like fashion. Some of the cells have irregular opaque nuclei (Diff-Quik). (**D**) Clump of markedly atypical keratinized squamous cells in aspirate of squamous cell carcinoma of the lung. Note the irregular shapes, opaque nuclei, and cytoplasmic keratinization (Diff-Quik). (**E**) Aspirate of keratinizing squamous cell carcinoma consisting of clumps of malignant keratinized cells as well as other cells demonstrating tumor necrosis (Papanicolaou).

Fig. 1-36. FNA of poorly differentiated squamous cell carcinoma consisting of syncytial group of malignant cells. Differentiation from poorly differentiated adenocarcinoma would be quite difficult (Diff-Quik).

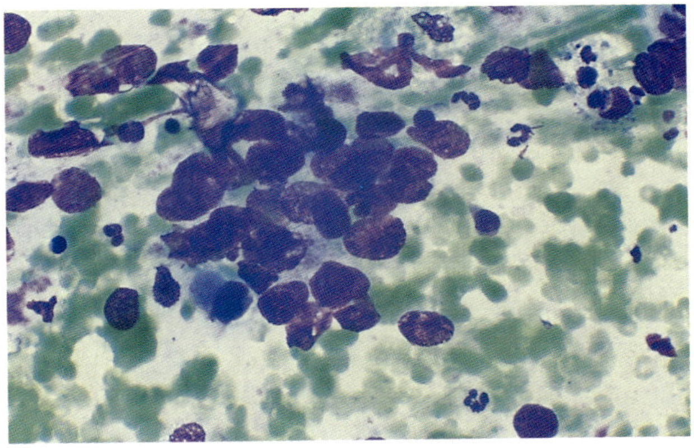

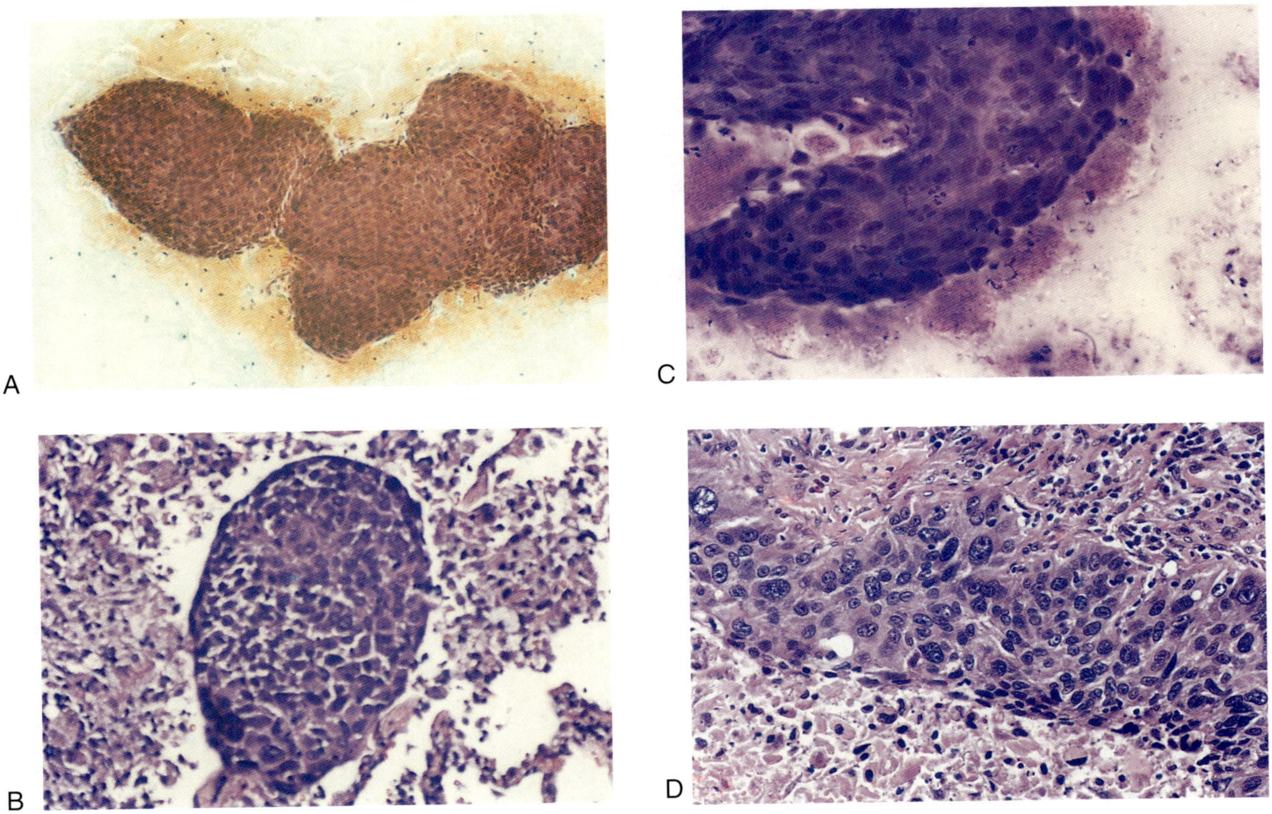

Fig. 1-37. (**A**) FNA of poorly differentiated squamous cell carcinoma consisting of three-dimensional clumps of malignant cells that could be confused with adenocarcinoma. Helpful features for the diagnosis of squamous cell carcinoma are the surface flattening of the cells and the presence of malignant spindle-shaped cells (Papanicolaou). (**B**) Histologic correlate of squamous cell carcinoma depicted in Fig. **A**, demonstrating rounded three-dimensional clumps of malignant cells (H&E). (**C**) Aspirate of poorly differentiated squamous cell carcinoma demonstrating surface flattening and spindling of the malignant cells, which are helpful diagnostic features (Diff-Quik). (**D**) Histologic correlate showing surface flattening and spindling of the malignant squamous cells. Also note the extensive tumor necrosis (H&E).

ment protocols may be considered.[126] Large cell carcinomas generally form large lesions with extensive areas of necrosis and hemorrhage.

Cytomorphologically, large cell carcinoma consists of numerous malignant cells arranged predominantly individually and in a few syncytial groupings (Fig. 1-38), such as those listed on page 31. The tumor cells have markedly increased nuclear to cytoplasmic ratios. Finely to coarsely granular irregularly distributed chromatin, one or more prominent irregular nucleoli, and considerable variation in size and shape of the characteristic nuclei are present. Cytoplasm is basophilic with no evidence of cytoplasmic keratinization, gland formation or histochemical evidence of cytoplasmic mucin. Giant cell carcinoma consists of predominantly individually scattered huge malignant cells that are multinucleated and/or spindled and often show evidence of neutrophilic cytophagocytosis (Fig. 1-39). Large cell carcinoma will usually contain abundant necrotic debris in the background. The differential diagnosis of large cell carcinoma includes primary poorly differentiated adenocarcinoma, small cell carcinoma of intermediate type, and sarcomatoid squamous cell carcinoma.[144] In fact, a large percentage of specimens classified cytologically as large cell carcinoma in cytologic specimens will be reclassified subsequently as another primary cell type when a more complete examina-

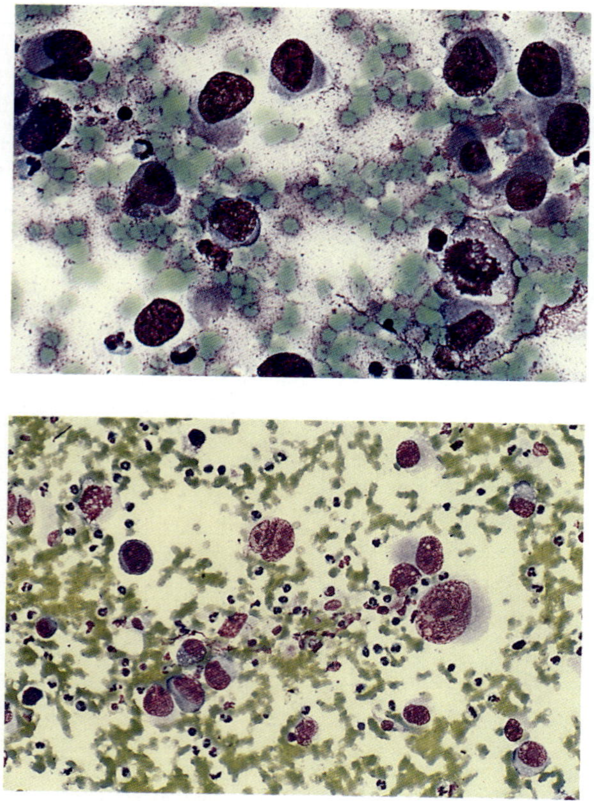

Fig. 1-38. (**A**) Aspirate of large cell carcinoma of the lung consisting of individually scattered bizarre cells with high nuclear to cytoplasmic ratios (Diff-Quik). (**B**) FNA of large cell carcinoma consisting of individually scattered malignant cells including a few tumor giant cells along with acute inflammation are present in the background (Diff-Quik).

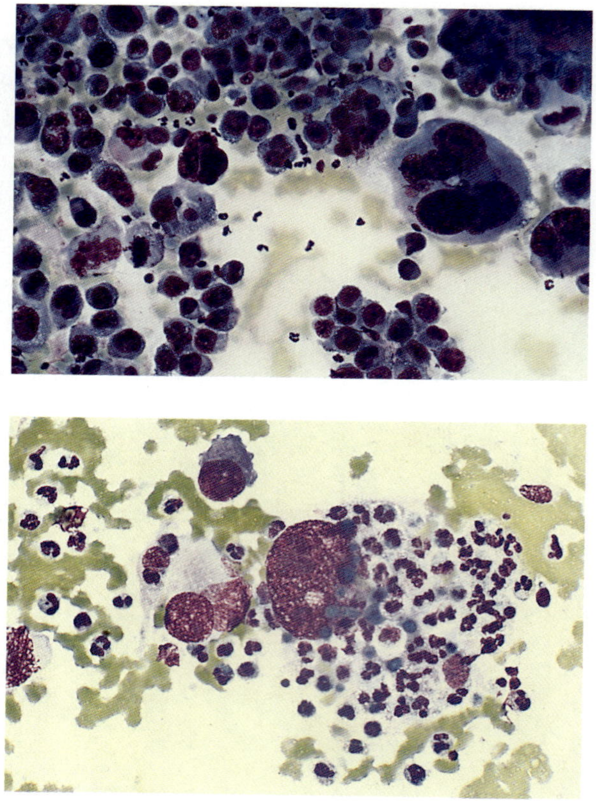

Fig. 1-39. (**A**) Aspirate of giant cell carcinoma of the lung consisting of scattered bizarre tumor giant cells, as well as other smaller poorly differentiated malignant cells (Diff-Quik). (**B**) Aspirate of giant cell carcinoma of the lung in which a few scattered malignant tumor giant cells are present, including some showing striking cytophagocytosis of neutrophils (Diff-Quik).

Features of Large Cell Carcinoma

Cellular pattern
- Predominantly single cells
- Syncytial groups

Smear background
- Tumor diathesis
- Occasional neutrophilic exudate

Cytoplasm
- Cyanophilic, occasional finely granular
- Occasional cytophagocytosis
- Occasional bizarre shapes

Nucleus
- Anisonucleosis, pleomorphic, oval to irregular shapes
- Finely to coarsely granular chromatin
- Multinucleated (tumor giant cells)
- Single to multiple prominent large nucleoli

tion of the resected tumor is performed.[10] Other diagnostic considerations include sarcoma, melanoma, and metastatic carcinoma. Pertinent clinical information and ancillary studies are often needed to arrive at a correct diagnosis (Table 1-6).

SMALL CELL CARCINOMA

Small cell carcinoma represents 10 to 25 percent of all primary lung tumors.[138] Although there is a striking male predominance, more recently small cell carcinoma has been rising in women.[139] Small cell carcinoma most often occurs centrally and has a dismal prognosis with very few long-term survivors. Yet, 15 to 25 percent of the few patients with limited-stage disease survive 2 years and can be considered cured if free of tumor at that time.[19] The median survival for patients with limited-stage disease and extensive-stage disease is 14 to 18 months and 9 to 11 months, respectively.[19,151] Patients with small cell carcinoma are often symptomatic. Occasional patients will have clinical evidence of ectopic hormone production, including adrenocorticotropic hormone (ACTH) secretion syndrome, antidiuretic hormone (ADH) secretion, and/or neurologic disorders such as Lambert-Eaton syndrome.[137] At diagnosis, two-thirds of patients with small cell carcinoma have detectable metastatic disease.[137] Common metastatic sites include bone, liver, brain and spinal cord, and mediastinal lymph nodes. Small cell carcinoma is the most common cause of the superior vena cava syndrome.

Table 1-6. Pleomorphic Large Cell Malignancies of the Lung

	Cytology	*Immunocytochemistry*	*Electron Microscopy*
Giant cell carcinoma Large cell carcinoma	Giant and spindle cells Cytophagocytosis Dissociative pattern	Cytokeratin + Vimentin ± Neuroendocrine markers ±	Epithelial features[a] Rarely—neuroendocrine granules
Sarcomatoid carcinoma	Spindle and giant cells Squamous features ±	Cytokeratin + Vimentin +	Epithelial features[a]
Malignant melanoma	Giant and spindle cells Polygonal cells Intranuclear inclusions Prominent nucleoli Binucleation Cytoplasmic pigment ±	S-100 + HMB-45 +	Melanosomes Premelanosomes
Sarcoma	Spindle and/or giant cells Frequent multinucleation Background matrix Dissociative pattern	Variable, depending on type (e.g., actin, vimentin, S-100)	Variable, depending on type

[a]Epithelial EM feature: (1) extracellular lumens with well-formed intercellular junctions, (2) intracytoplasmic lumens, (3) grouped microvilli, (4) granules/vacuoles, (5) tonofilaments; (6) basement membrane around groups of cells.

(From Silverman,[35] with permission.)

Cytomorphologically, aspirates of small cell undifferentiated carcinoma consist of numerous small cells arranged singly and in syncytial groups with prominent nuclear molding (Fig. 1-40). Nuclei vary from round or oval to angular in contour. Nuclear chromatin is finely granular without prominent clumping and parachromatin clearing. In general, nucleoli are inconspicuous or small (Fig. 1-40). The tumor cells have extremely high nuclear to cytoplasmic ratios with a scant amount of surrounding indistinct cytoplasm.[152,153] The intermediate type of small cell carcinoma will have slightly larger nuclei with a finely granular chromatin and slightly more cytoplasm. Paranuclear blue cytoplasmic inclusions were recently described as a feature in air-dried Romanowsky-stained smears.[154,155] Although strongly associated with small cell carcinoma, they are not specific, as up to 9 percent of non-small cell carcinomas and lymphomas will also demonstrate these inclusions.[154] These structures probably represent phagocytic lysosomes.[155] Occasionally, small cell carcinoma have scattered larger malignant cells with more abundant cytoplasm and more prominent nucleoli. Small foci of squamous or glandular differentiation may also be found. When a large cell component is found in addition to a small cell carcinoma, both should be diagnosed; these tumors will generally be treated as a small cell carcinoma (Fig. 1-41). The background of small cell carcinoma may show extensive necrotic debris and strands of basophilic material representing extravasated smudged nuclear material ("crush artifact") (Fig. 1-42).

The differential diagnosis of small cell carcinoma includes benign and malignant lymphoid processes

Features of Small Cell Carcinoma

Cellular pattern
- Single cells
- Syncytial groups
- Nuclear molding, linear pattern

Smear background
- Diathesis
- DNA streaking
- Necrotic cells

Cytoplasm
- High nuclear to cytoplasmic ratios (scant cytoplasm)
- Slight amount of cyanophilic cytoplasm
- ± Blue bodies

Nucleus
- Oval to angulated nuclei
- Molding
- Variable-sized nuclei
- Dense "salt and pepper" chromatin
- No or ± small nucleoli
- No large nucleoli

A

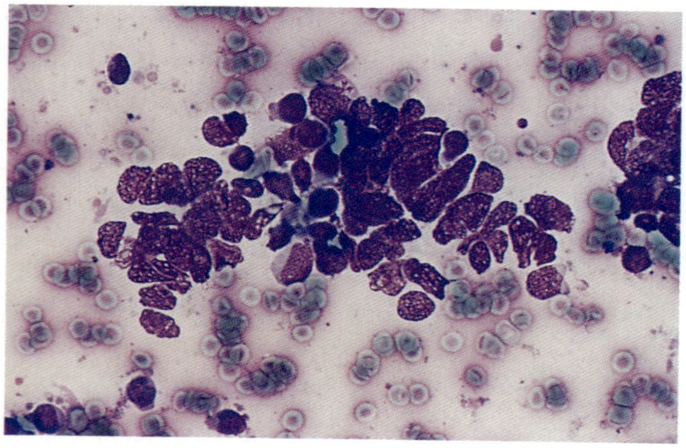

B

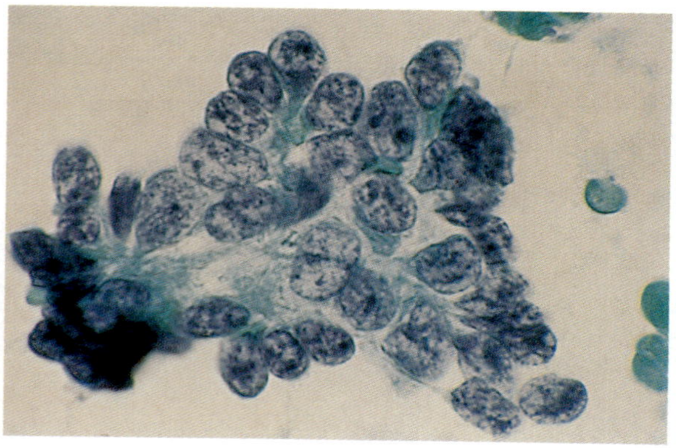

Fig. 1-40. (**A**) FNA of small cell carcinoma showing tumor cells with high nuclear to cytoplasmic ratios, nuclear angularity, and molding (Diff-Quik). (**B**) Aspirate of small cell carcinoma showing prototypical "salt and pepper" chromatin pattern and the lack of prominent nucleoli. Note the nuclear angularity and molding. Some of the tumor cells have slight amount of cytoplasm, but the overall features of this aspirate are that of small cell carcinoma (Papanicolaou).

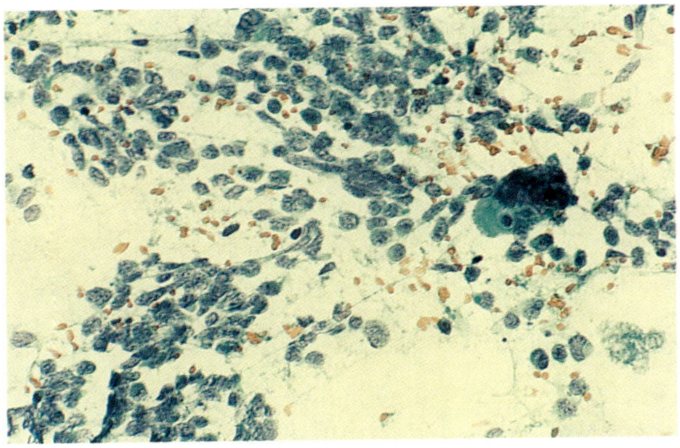

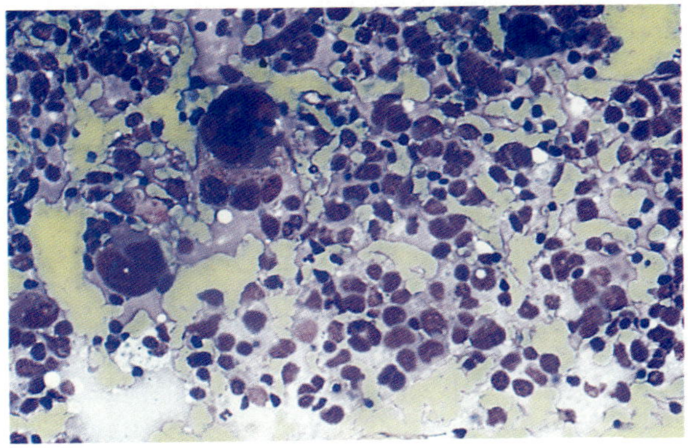

Fig. 1-41. (**A**) FNA of mixed small-large cell carcinoma of the lung consisting of numerous individually scattered small tumor cells having high nuclear to cytoplasmic ratios, along with a few scattered larger malignant cells (Papanicolaou). (**B**) Aspirate of mixed small–large cell carcinoma showing both components in this field (Diff-Quik).

in which true intercellular cohesion or nuclear molding will not be evident. Occasionally, however, pseudomolding can be seen in aspirates of malignant lymphoma (Fig. 1-43). Lymphoid lesions will also contain basophilic cytoplasmic fragments termed lymphoglandular bodies in the background of the Diff-Quik–stained smears. The Diff-Quik stain is superior to Papanicolaou stain in demonstrating the hematopoietic nature of the cells. However, occasional cases will necessitate immunocytochemical and/or EM studies for a correct diagnosis (Fig. 1-44). Small cell carcinoma will show cytokeratin cytoplasmic positivity along with neuroendocrine markers staining for neuron-specific enolase (NSE), while non-Hodgkin's lymphoma will be positive with hematopoietic markers such as leukocyte common antigen (LCA) and/or specific B- or T-cell markers (Fig. 1-45). Leu-7 may be positive in both malignancies. EM will demonstrate neurosecretory granules in many cases of small cell carcinoma. Other neoplasms in the differential diagnosis include poorly differentiated non-small cell carcinoma and carcinoid tumor. The former occasionally may have cells in the size range of small cell carcinoma, but the nuclear features will be different. Squamous cell carcinoma will have a more vesicular chromatin with prominent nucleoli, although some cells may have more opaque and

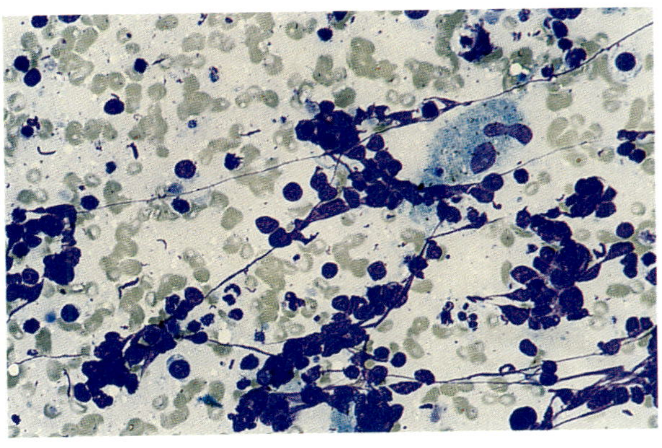

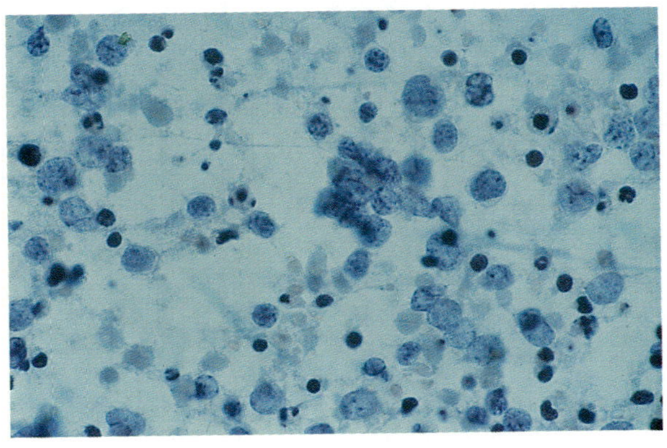

Fig. 1-42. (**A**) Characteristic crush artifact in FNA of small cell carcinoma of the lung (Diff-Quik). (**B**) Individual cell necrosis in aspirate of small cell carcinoma of the lung (Papanicolaou).

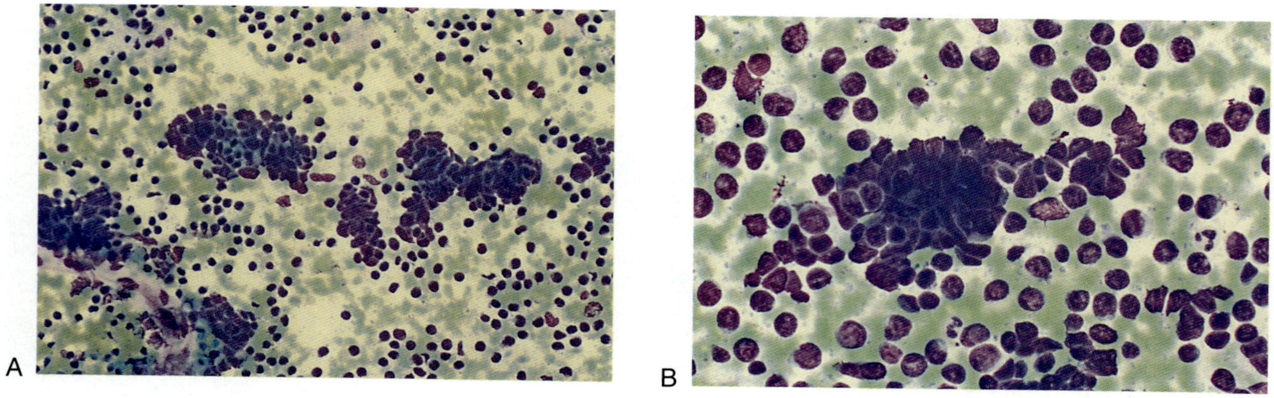

Fig. 1-43. (**A**) Aspirate of poorly differentiated lymphocytic lymphoma of lung showing pseudoclustering ("molding") of the cells (Diff-Quik). (**B**) High-power view of aspirate of poorly differentiated lymphocytic lymphoma of lung showing pseudomolding that could potentially be confused with a small cell carcinoma. However, note the individually scattered cells having lymphoid features, as well as numerous lymphoglandular bodies in the background (Diff-Quik).

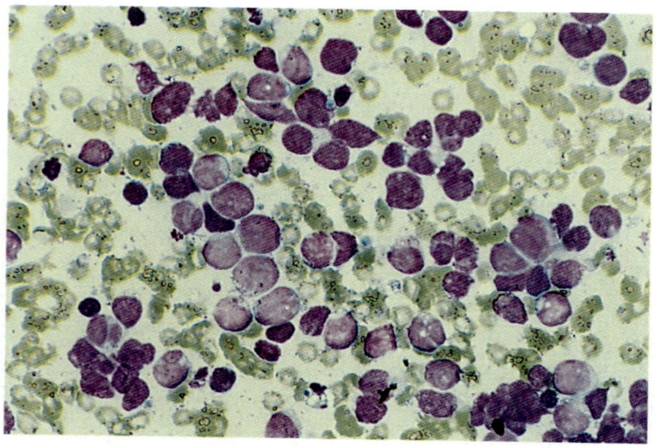

Fig. 1-44. Aspirate of small cell carcinoma of the lung consisting of individually scattered oval cells that could potentially be confused with a malignant lymphoma. Note that a few groups of cells show nuclear molding and there are no lymphoglandular bodies in the background (Diff-Quik).

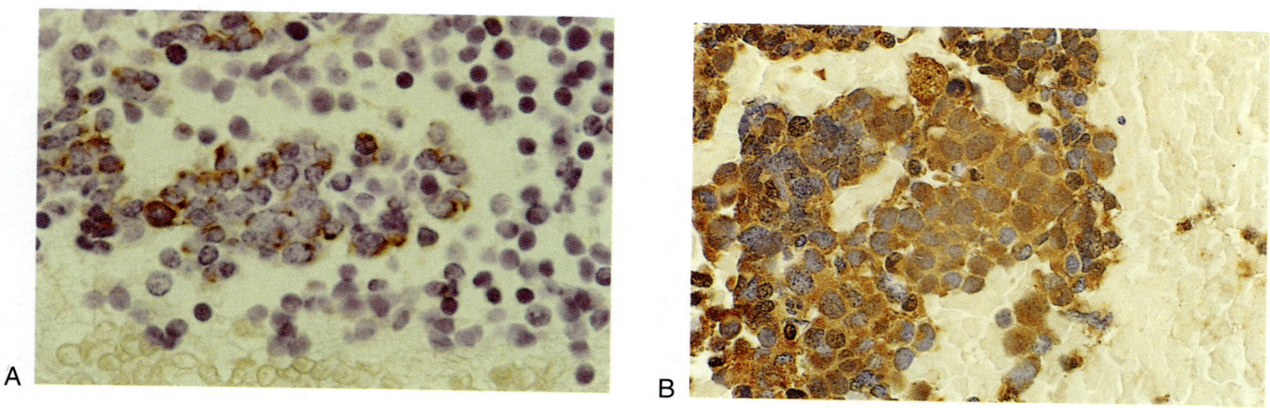

Fig. 1-45. (**A**) Aspirate of small cell carcinoma of the lung in which the differential diagnosis was malignant lymphoma. Note punctate cytoplasmic staining for cytokeratin in many of the neoplastic cells (immunoperoxidase). (**B**) Positive staining of small cell carcinoma of the lung for synatophysin in FNA cell block (immunoperoxidase).

Table 1-7. Features of Small Cell Malignancies of the Lung

	Cytology	*Immunocytochemistry*	*Electron Microscopy*
Small cell carcinoma	Single cells and syncytial groups Nuclear molding High nuclear to cytoplasmic ratios "Salt and pepper" chromatin No prominent nucleoli Oval to angular nuclei Perinuclear blue bodies ±	NSE + Chromogranin ± Leu 7 ± Synaptophysin ±	Few neurosecretory granules
Carcinoid	Single cells, acini, sheet Loosely cohesive groups Plasmacytoid with D/Q Granular cytoplasm "Salt and pepper" chromatin Round to oval nuclei Occasionally stripped nuclei Tumor cells associated with capillaries	NSE + Chromogranin ±	Numerous neurosecretory granules
P.D. Squamous cell carcinoma	Spindle cells Opaque nuclei Dense cytoplasm	High- and low-molecular weight cytokeratin positive	Desmosomes, tonofilaments
Lymphoma	Dispersed pattern Pseudoclustering ± Lymphoglandular bodies	LCA, B- and T-cell markers Occasionally, Leu 7 +	No specific findings

Abbreviations: D/Q, Diff-Quik; LCA, leukocyte common antigen; NSE, neuron-specific enolase.
(From Silverman,[35] with permission.)

coarsely granular chromatin. Generally, some cells possess greater volumes of cytoplasm than are seen in small cell carcinoma. Carcinoid tumors share similar histogenetic and cytologic features with small cell carcinomas but consist of populations of neoplastic cells with a more uniform appearance, having lower nuclear to cytoplasmic ratios and more abundant granular cytoplasm. Carcinoid tumors lack nuclear molding, significant atypia, and background necrosis. In addition, blood vessels may be seen in tissue fragments of carcinoids. The different features of small cell neoplasms are presented in Table 1-7.

CARCINOID TUMOR

Carcinoids account for approximately 2 percent of all primary lung tumors.[138] The average age of diagnosis is 50 years.[137] Bronchial carcinoids appear to be unrelated to smoking, occupation, or other environmental causes. Although occasional patients may present with neuroendocrine symptoms, most patients are asymptomatic.[137] As most carcinoid tumors are quite vascular, bronchoscopic biopsies may not be done because of the risk of extensive bleeding. By contrast, transbronchial FNA biopsy may be successfully employed to make the specific diagnosis without this risk.[156] Patients with typical carcinoid tumor have an excellent prognosis, although 5 to 20 percent of carcinoids have regional lymph node metastasis. By contrast, atypical carcinoid tumors will metastasize to regional lymph nodes and distant sites in 40 to 48 percent of cases. The 5- and 10-year survival of carcinoid and atypical carcinoid is 69 percent and 52 percent, respectively.[157–160]

Aspiration smears of carcinoid tumors include neoplastic cells arranged predominantly singly, although occasional small sheets and groups with possible acinar formation may be seen (Fig. 1-46), as listed on page 37. The tumor cells have uniform round to oval nuclei with evenly distributed, finely granular chromatin and inconspicuous nucleoli. The nuclear to cytoplasmic ratio is relatively high, with slight to moderate amounts of surrounding

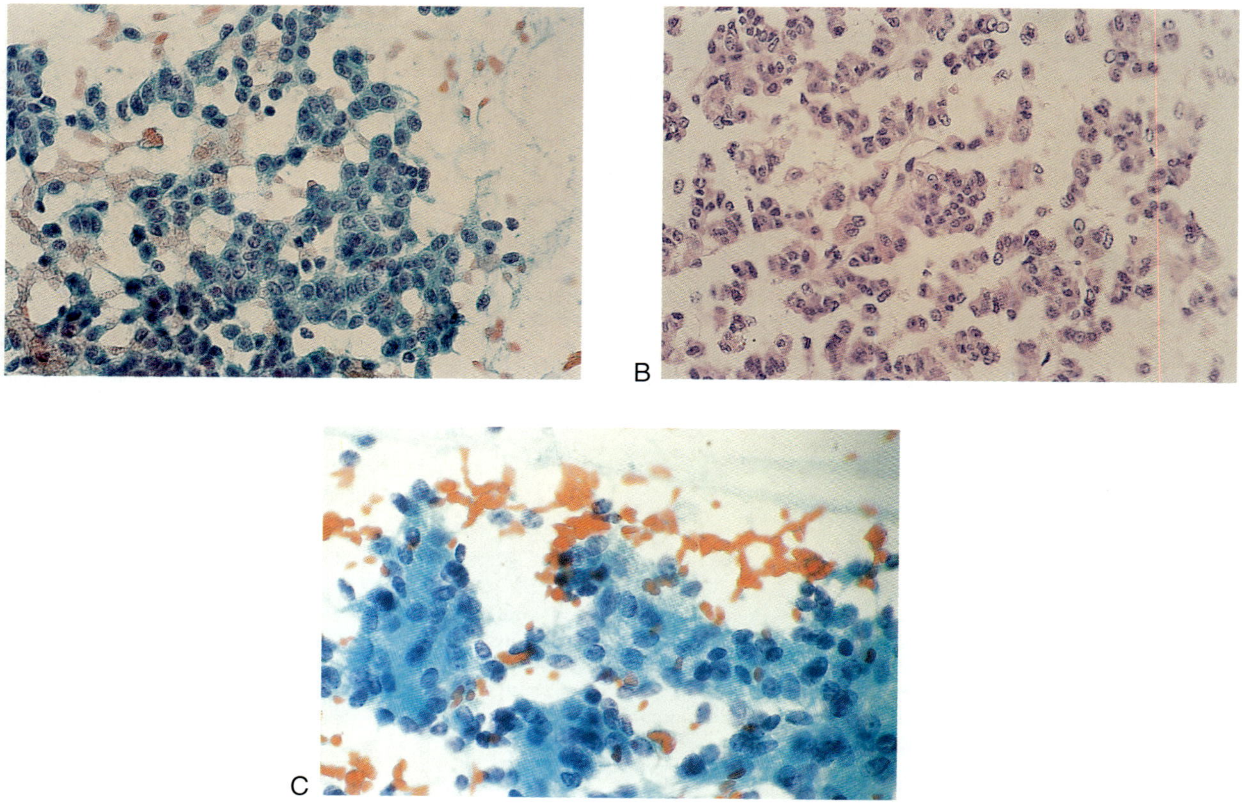

A B C

Fig. 1-46. (**A**) Aspirate of carcinoid tumor of the lung consisting of loose ribbons and cords of uniform neoplastic cells having prototypical neuroendocrine nuclei and a slight amount of granular cytoplasm (Papanicolaou). (**B**) Cell block from same case showing similar features of thin cords and nests of uniform cells from aspirate of carcinoid tumor (H&E). (**C**) Aspirate of carcinoid tumor consisting of aggregates of cells arranged in an acinar fashion that could be confused with an adenocarcinoma. However, attention to nuclear characteristics of a carcinoid tumor as well as cytoplasmic granularity should enable a correct diagnosis to be made (Papanicolaou).

granular amphophilic to basophilic cytoplasm. In the Diff-Quik stain, the cells of carcinoid tumor take on a somewhat plasmacytoid appearance but lack a perinuclear hof[161,162] (Fig. 1-47). Peripheral carcinoids can have a spindle cell pattern in FNA biopsy (Fig. 1-48). Occasinally, tumor cells will be arranged along capillary channels or a stripped (naked) nuclei pattern will be present (Fig. 1-49). The stripped nuclei could cause confusion with a small cell carcinoma if the other cytologic feature of carcinoid are overlooked. Atypical carcinoid may be suspected in FNA biopsy when greater variability in nuclear size and shape, higher nuclear/cytoplasmic ratios, greater nuclear hyperchromasia, more prominent nucleoli, and a necrotic background are present.[163–167] Immunocytochemical and EM demonstration of neuroendocrine differentia-

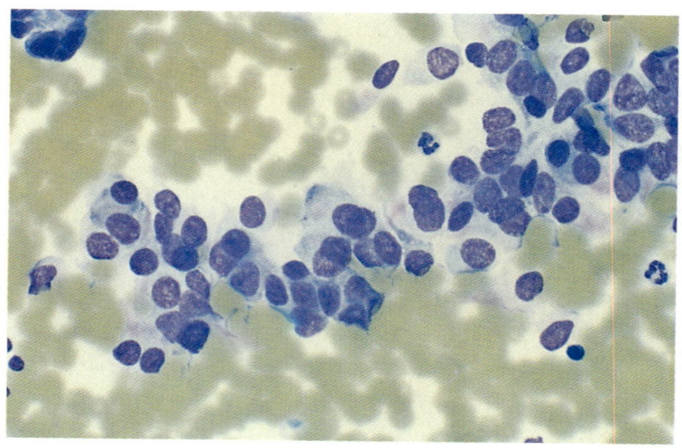

Fig. 1-47. Aspirate of carcinoid tumor showing "plasmacytoid" features of the cells in Diff-Quik preparation.

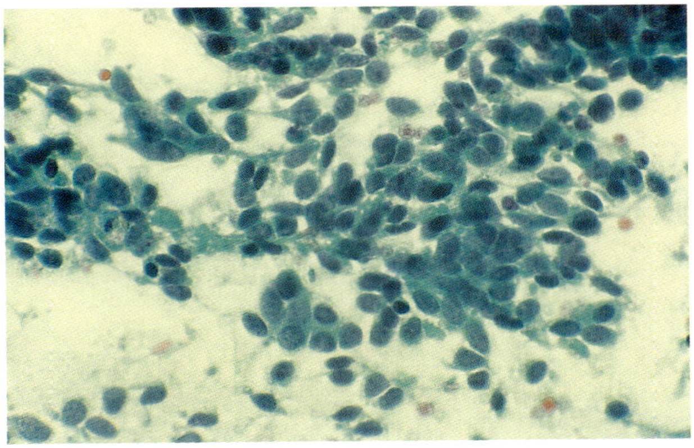

Fig. 1-48. Aspirate of peripheral carcinoid tumor having a more spindled configuration of the cells (Papanicolaou).

Features of Carcinoid Tumors

Cellular pattern
- Single cells
- Loose groups, acinar formation
- ± Stripped nuclei
- Vascularized fragments

Smear background
- Clear to granular debris

Cytoplasm
- Oval-cuboidal shapes
- Granular, cyanophilic to basophilic
- Plasmacytoid with Diff-Quik stain

Nucleus
- Round to oval
- Uniform size
- "Salt and pepper" chromatin
- No or ± small nucleoli

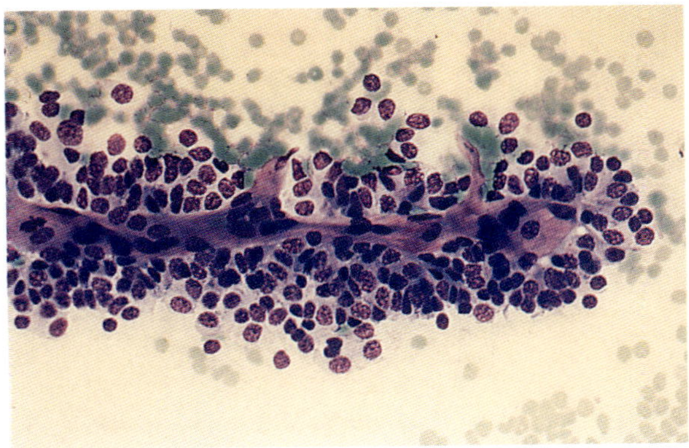

A

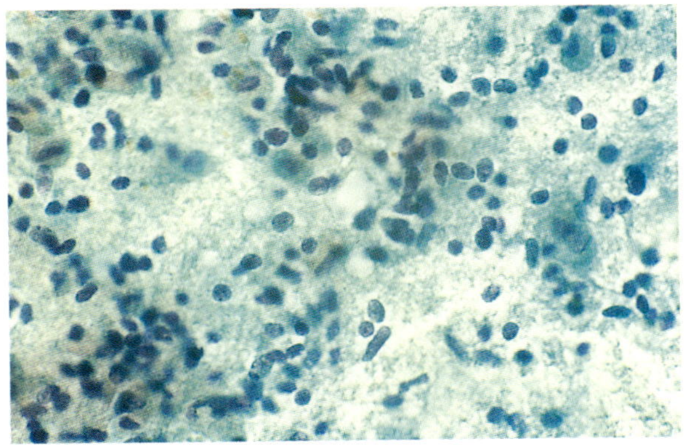

B

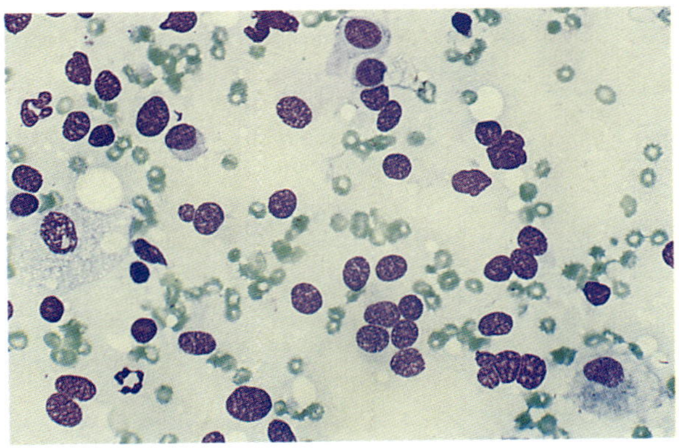

C

Fig. 1-49. (**A**) Aspirate of carcinoid tumor showing presence of tumor cells aligned along delicate capillary-type channels (Diff-Quik). (**B**) Stripped nuclei pattern in aspirate of carcinoid tumor. The potential exists to confuse the stripped nuclei with tumor cells from a small cell carcinoma due to the simulation of high nuclear to cytoplasmic ratios (Papanicolaou). (**C**) High-power view of stripped nuclei pattern seen in aspirate of carcinoid tumor (Diff-Quik).

tion is confirmatory but seldom needed. Other unusual bronchial gland tumors that have been diagnosed by pulmonary FNA biopsy include adenoid cystic carcinoma, mucoepidermoid carcinoma, and pleomorphic adenoma.

PULMONARY HAMARTOMA

Solitary pulmonary hamartoma is the most common benign pulmonary tumor, found in approximately 0.25 percent of patients at autopsy, usually in the periphery of the lung.[138] Most are round, well-circumscribed, solid nodules measuring less than 4 cm in maximum dimension. Typically, hamartomas are asymptomatic and grow slowly.[138] FNA of pulmonary hamartoma may contain fibromyxoid tissue, having bland spindle-shaped fibroblasts embedded in a fibrillary matrix, along with fragments of hyaline cartilage, adipocytes, and epithelial cells[168–171] (Fig. 1-50). The level of cellularity does not approach that seen in FNA biopsy of myxoid sarcomas.

METASTATIC NEOPLASMS

Cancers metastatic to the lung are a frequent clinical scenario that must be differentiated from primary lung carcinoma. Knowledge of a previously diagnosed or suspected extrapulmonary malignancy and the radiologic evidence of multiple nodules are helpful but are not always available or definitive. Occasional metastatic tumors present as a solitary lesion, whereas some primary lung malignancies such as bronchioloalveolar carcinoma can be multifocal. The cytomorphologic features of certain

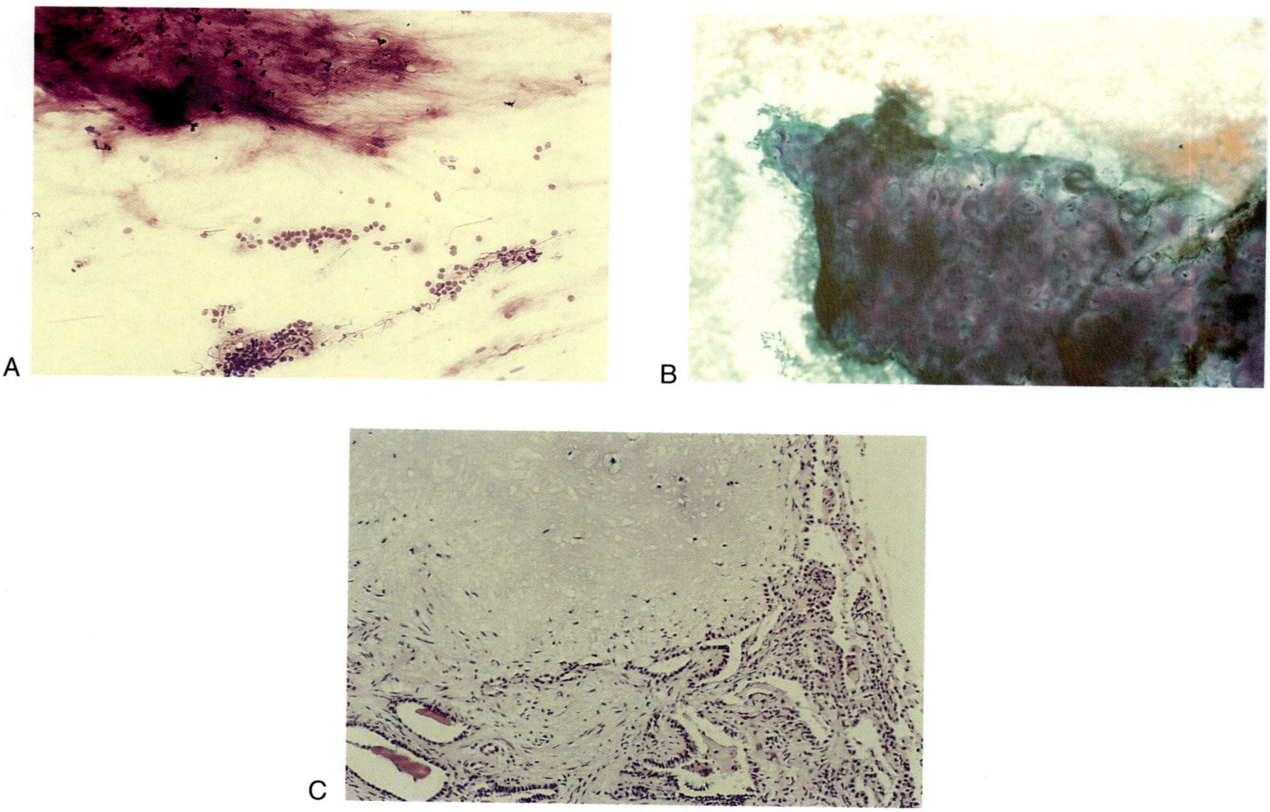

Fig. 1-50. (**A**) Aspirate of pulmonary hamartoma consisting of uniform groups of epithelial cells along with a metachromatically staining myxoid stroma in the background (Diff-Quik). (**B**) A few cartilaginous fragments also present in the aspirate of a pulmonary hamartoma. However, the presence of cartilage could represent aspiration of the bronchial cartilage (Papanicolaou). (**C**) Resected hamartoma demonstrating the presence of fibromyxoid stroma, hyaline cartilage, and benign epithelial cells (H&E).

metastatic cancers may suggest the correct diagnosis.[10] Metastatic melanoma is characterized by numerous bizarre cells, arranged individually or in loose aggregates, often with macronucleoli and/or intranuclear cytoplasmic pseudoinclusions (Fig. 1-51). Large binucleated or multinucleated tumor cells and spindle-shaped cells may be seen. The presence of intracytoplasmic melanin pigment is diagnostic when present, although melanoma can be suspected when the other cytomorphologic features are appreciated in the absence of melanin. Immunocytochemical staining for S-100 and HMB-45 with negative staining for epithelial markers are strongly supportive evidence, as is ultrastructural evidence of premelanosomes and/or melanosomes (Fig. 1-51). Metastatic renal cell carcinoma consists of loose aggregates of tumor cells that often have abundant pale cytoplasm related to abundant intracytoplasmic glycogen and lipid. In the better differentiated tumors, the chromatin is finely granular with small nucleoli, while poorly differentiated metastatic renal cell carcinoma will show a greater degree of nuclear pleomorphism, hyperchromasia, and prominence of nucleoli. Tumor cells are often attached to vascular channels. However, it is important to remember that occasionally a primary non-small cell carcinoma can have a clear cell component. Metastatic well-differentiated prostatic carcinoma and follicular carcinoma of the thyroid will show uniform cells arranged in small, acinar groupings. Immunoperoxidase stains for prostatic acid phosphatase and prostate-specific antigen (PSA) can be a helpful ancillary study for the diagnosis of prostate carcinoma as well as positive thyroglobulin staining for a

A

B

C

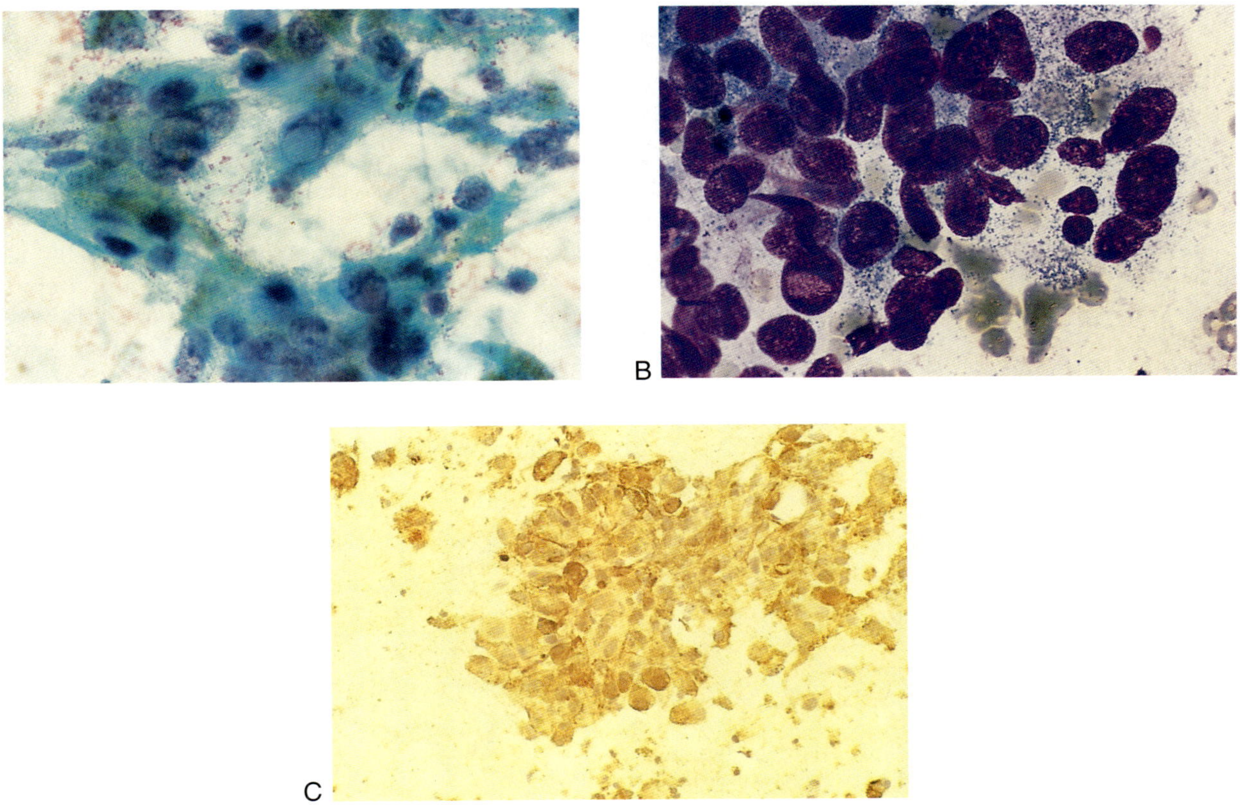

Fig. 1-51. (**A**) Aspirate of metastatic malignant melanoma to lung showing polygonal to spindle-shaped cells, including some with intracytoplasmic pigmentation and others demonstrating intranuclear inclusions (Papanicolaou). (**B**) Diff-Quik stain of metastatic malignant melanoma showing fine to coarsely granular intracytoplasmic melanin pigment, as well as some nuclei possessing intranuclear inclusions. The tumor cells also have prominent nucleoli (Diff-Quik). (**C**) Aspirate of a metastatic malignant melanoma to lung demonstrating positive staining of the tumor cells for HMB-45 in the smears (immunoperoxidase).

thyroid primary. Metastatic ductal carcinomas of the breast consist of sheets and clusters of malignant cells that share cytologic features with adenocarcinoma of the lung (Fig. 1-52). Occasional cases of metastatic ductal and lobular carcinoma of the breast will have subtle bland cytologic features. Although not specific, immunocytochemical studies for estrogen receptor protein support the diagnosis of metastatic breast carcinoma in those cases of primary breast tumors positive for this antigen. Adenocarcinomas from many other sites may share similar cytologic features with pulmonary adenocarcinoma. Comparison with the histology of the primary tumors is essential for a correct diagnosis. Squamous cell and/or transitional cell carcinoma of the cervix, bladder, esophagus, and head and neck may be indistinguishable from primary pulmonary squamous cell carcinoma.

A

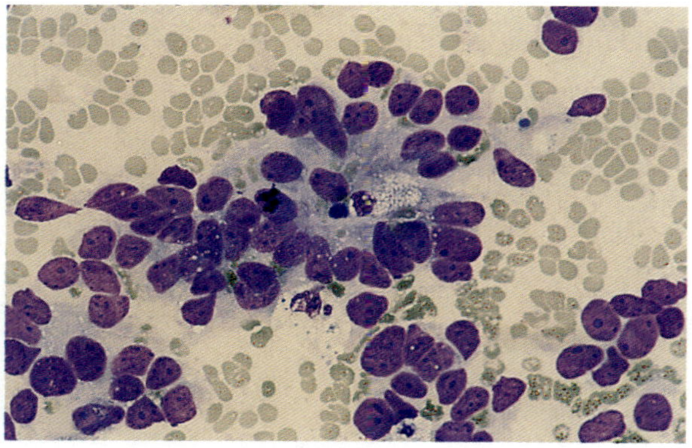

B

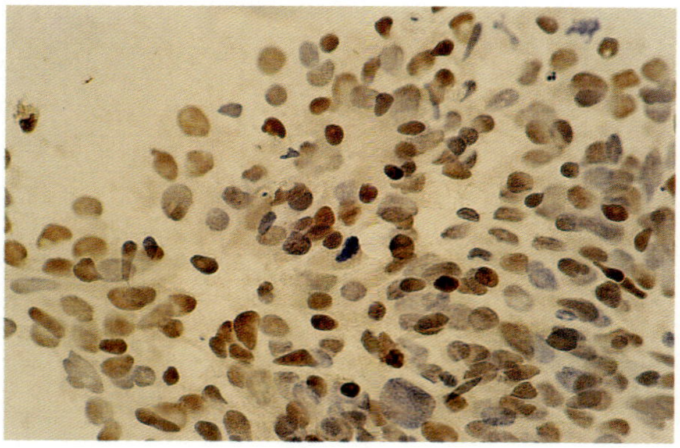

Fig. 1-52. (**A**) Aspirate of metastatic adenocarcinoma of the breast to the lung in patient with a history of breast carcinoma. Differential diagnosis includes primary lung adenocarcinoma (Diff-Quik). (**B**) Immunoperoxidase stain for estrogen receptor protein reveals positive intranuclear staining of tumor cells supporting the diagnosis of a metastasis from a breast primary in this case (immunoperoxidase).

SARCOMAS

Primary pulmonary sarcomas occur generally as large solitary lesions in contrast to the much more common metastatic sarcomas, which often produce multiple small nodules. Sarcomas can either be endobronchial or parenchymal in location.[172] Malignant fibrous histiocytoma appears to be the most common primary sarcoma of the lung.[172] Cytologic features of sarcomas are quite variable, depending on the histogenesis and differentiation of the mesenchymal malignancy.[173–176] Cytologic diagnostic features suggesting a sarcoma include the presence of malignant spindle cells arranged singly or in small groups (Fig. 1-53). The spindle cell sarcomas such as leiomyosarcoma, fibrosarcoma, malignant schwannoma, and synovial sarcoma have cells with hyperchromatic nuclei and surrounding basophilic wispy cytoplasm. Cytologic clues that may suggest the correct classification include the presence of medium-sized oval to spindle-shaped cells with more pointed nuclei in fibrosarcoma, round epithelioid cells arranged in small clusters or cords in synovial sarcoma, long spindle-shaped oval cells with

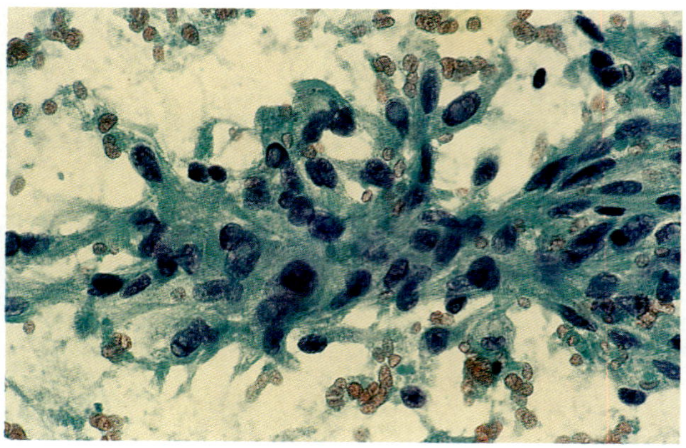

Fig. 1-53. FNA of primary sarcoma of the lung consisting of loose aggregate of malignant spindle-shaped cells. Histologic examination of the resected tumor revealed a malignant fibrous histiocytoma (Papanicolaou).

blunt cigar-shaped nuclei in leiomyosarcoma, and thin, elongated, wavy spindle cells in malignant schwannoma. Malignant fibrous histiocytoma can demonstrate a variety of cell types, including spindle-shaped cells, multinucleated pleomorphic giant cells, and undifferentiated mesenchymal-type cells. Specific cytologic features such as evidence of lipoblasts can suggest a correct diagnosis of liposarcoma. Vascular patterns may be seen in angiosarcoma, while extracellular matrix material such as osteoid in osteosarcoma and chondromyxoid stroma in chondrosarcoma are also useful findings. Immunocytochemical studies or EM, or both, can be quite beneficial in making the correct diagnosis of sarcoma and often permit subclassification when specific diagnostic features are found. A more complete discussion of the FNA cytologic features and ancillary findings of sarcoma is presented in Chapter 7. Finally, separation of sarcomas from primary and metastatic epithelial malignancies with sarcomatoid features is critical.[172] Entities to be considered include spindle cell squamous cell carcinoma of the lung and metastatic epithelial malignancies such as sarcomatoid renal cell carcinoma and malignant melanoma.[172] Attention to subtle cytologic features of epithelial differentiation as well as immunocytochemical and EM studies can be helpful.

Carcinosarcoma is a rare aggressive pulmonary malignancy that occurs predominantly in older men. The malignancy consists of both epithelial clusters and bizarre spindle-shaped cells arranged in loose groups and microtissue fragments along with scattered single cells.[177] The most common epithelial component is squamous cell carcinoma, followed by adenocarcinoma and undifferentiated carcinoma.[172] The malignant cells co-express cytokeratin and vimentin, which unfortunately does not allow distinction from other pulmonary malignancies demonstrating dual staining such as large cell carcinoma.[178]

PULMONARY BLASTOMA

Pulmonary blastoma (embryoma) can occur in children or adults. This tumor is usually peripherally located and consists of well-differentiated malignant glands set in an undifferentiated cellular stroma containing spindle cells, reminiscent of both Wilms' tumor and fetal lung. The stromal component may show differentiation toward skeletal muscle, cartilage, or bone. Pulmonary blastoma is similar to carcinosarcoma, except that the epithelial component of carcinosarcoma is usually squamous cell in type. Tumors similar to pulmonary blastoma but lacking the sarcomatous component have been referred to as pulmonary endodermal tumor or adenocarcinoma of fetal type. There is a very limited literature describing the cytologic findings of either pulmonary blastoma or pulmonary adenocarcinoma of fetal type.[179–181] Cosgrove et al.[182] reported the FNA cytologic features of a case of pulmonary blastoma consisting of malignant epithelial and stromal elements. The cell block showed the biphasic pattern of pulmonary blastoma. Lee and Cho[183] reported the FNA findings of pulmonary adenocarcinoma of fetal type. These smears contained relatively small, uniform malignant cells arranged individually and in cohesive groups including rosettes or acini. Clusters of larger malignant cells with eosinophilic cytoplasm and prominent nucleoli were also present. Immunohistochemical studies performed on the resected tumor demonstrated positive staining of the malignant cells for α-fetoprotein, NSE, and somatostatin; EM examination demonstrated neuroendocrine granules.

LYMPHOMA

Primary pulmonary lymphoma is uncommon except in patients with AIDS. Most primary pulmonary lymphomas are immunophenotypically B cell.[138] Aspirates from non-Hodgkin's lymphoma are generally hypercellular and consist of uniform populations of dispersed atypical lymphoid cells. The presence of lymphoglandular bodies, a feature of both benign and malignant lymphoid processes, is helpful to distinguish lymphoma (and benign lymphoid processes) from small cell carcinoma (Fig. 1-54). In non-Hodgkin's lymphoma, the cytologic features of the tumor cells will vary depending on the type of lymphoma. Well-differentiated lymphocytic lymphoma consists of uniform small round lymphocytes, while poorly differentiated lymphocytic lymphoma will show somewhat larger lymphocytes with irregular nuclear folds or cleavage but without prominent nucleoli. Large cell lymphoma can be either cleaved or no cleaved. Noncleaved lymphomas often possess prominent nucleoli. Immunoblastic lymphomas will have

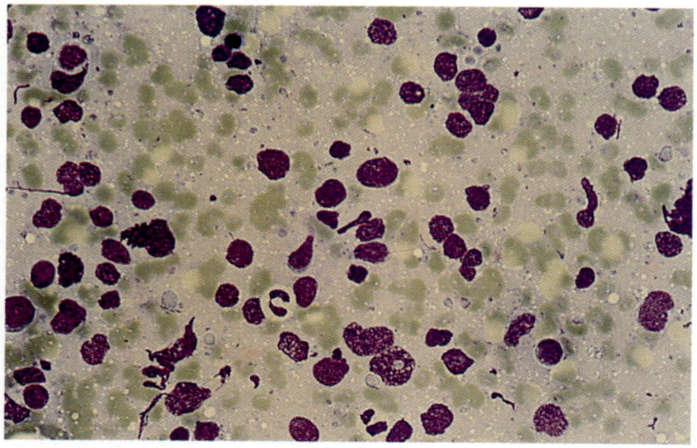

Fig. 1-54. Aspirate of large cell lymphoma of lung consisting of dissociative pattern of malignant lymphoid cells as well as numerous lymphoglandular bodies in the background (Diff-Quik).

more abundant amphophilic cytoplasm, possibly with plasmacytoid features. Lymphoblastic lymphomas consist of a uniform population of intermediate size cells with irregular to oval nuclei, finely granular chromatin pattern, and inconspicuous nucleoli. Aspirates of Hodgkin's disease are characterized by a polymorphic population of cells, including small lymphocytes, eosinophils, neutrophils, histiocytes, and plasma cells. Associated granulomatous fragments may be present. The diagnosis is based on identifying diagnostic Reed-Sternberg cells, although mononuclear variant Hodgkin cells may be present.

MESOTHELIOMA

Mesotheliomas, well known for their strong association with asbestos exposure, occur most often in patients 40 to 70 years of age. Therapy is usually only palliative, and survival is exceptional.[138] The three main types of malignant mesothelioma are (1) epithelial, (2) sarcomatous, and (3) biphasic or mixed. The definitive cytologic diagnosis of malignant mesothelioma can be difficult.[184–186] The differential diagnosis often includes primary lung adenocarcinoma, metastatic malignancy, and reactive mesothelial cell proliferations. Radiologic findings of a pleural-based lesion diffusely involving the pleural surface with an associated effusion is more supportive of mesothelioma than lung carcinoma. Epithelial and mixed epithelial-fibrous mesothelioma share cytologic features with adenocarcinomas metastatic to the pleura and peripheral adenocarcinoma of the lung. A zone of pallor in the periphery of the cytoplasm with "fuzzy" cell boundaries favors mesothelioma over carcinoma. Immunocytochemistry and/or EM performed on the aspirated material can be helpful (Fig. 1-55, Table 1-8). A panel of immunoperoxidase stains is recommended with positive staining for carcinoembryonic antigen (CEA), epithelial membrane antigen (EMA), Leu M-1, Ber EP4, and/or B72.3 favoring adenocarcinoma. Both adenocarcinoma and mesotheliomas are AE1/3 positive, and many cases of mesothelioma also demonstrate EMA positivity (Fig. 1-56). Reactive

Table 1-8. Morphologic Differences Between Adenocarcinoma and Mesothelioma

Adenocarcinoma	*Mesothelioma*
Three-dimensional groups with scalloped or smooth borders	Flat sheets with cells having dense cytoplasm
Acinar formation	Scalloped, knobby peripheral borders of groups
Side-by-side arrangement of cells	Uniform cell population
Syncytial groupings	Central acellular core in papillary groups
Eccentrically placed nuclei	Cytoplasmic blebs (best seen in Diff-Quik)
Nuclear molding	Dispersed spindle cells (uncommon variant)
Lacy, finely to coarsely vacuolated cytoplasm	Hard edge vacuoles containing hyaluronic acid
Mucicarmine positive	Neg for CEA, B72.3, Leu-M1, BER-EP4
Positive for CEA, B72.3, Leu-M1 BER-EP4, EMA, and/or AE1/3	Can be positive for AE1/3, EMA, Vimentin
EM—variable, depending on histogenesis	EM—long microvilli

(From Silverman,[35] with permission.)

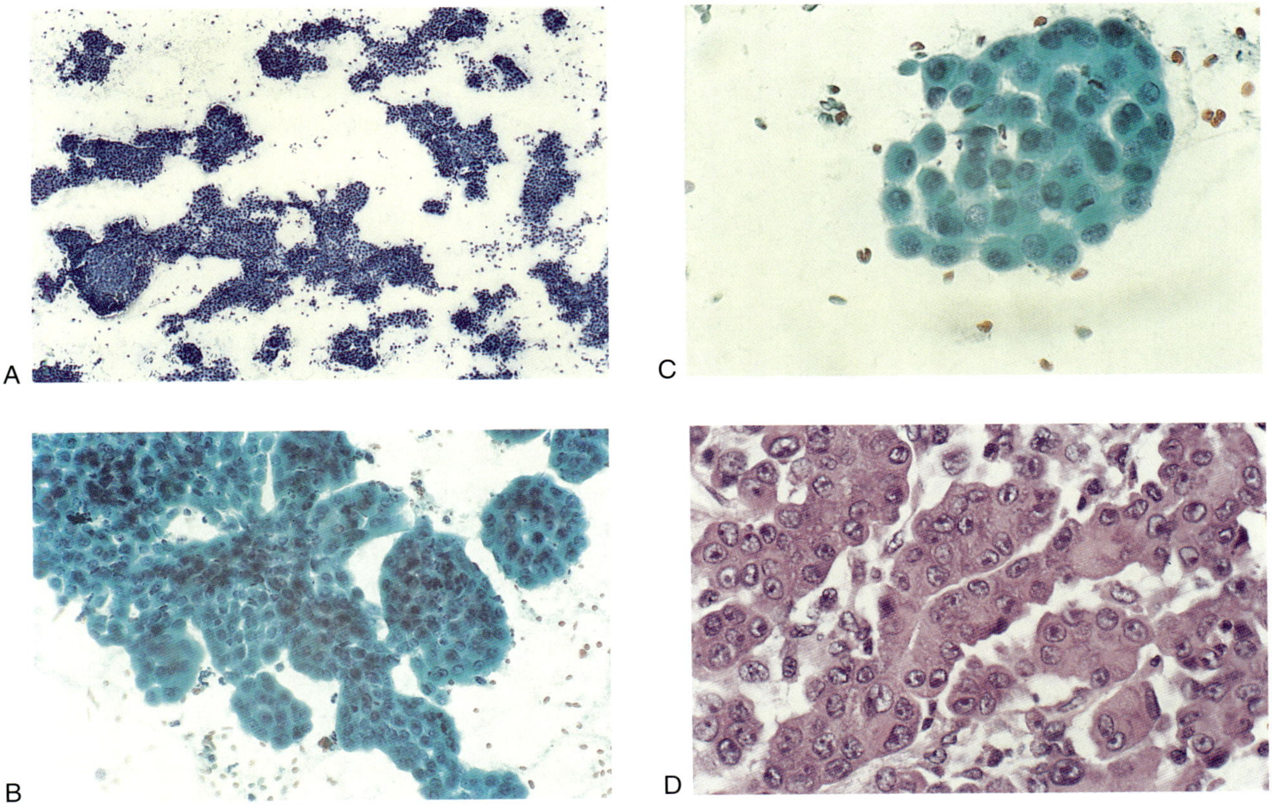

Fig. 1-55. (**A**) Aspirate of mesothelioma revealing hypercellular specimen consisting of numerous clusters of atypical cells (Diff-Quik). (**B**) Papillary and three-dimensional clusters of tumor cells, including some maintaining intercellular windows (Papanicolaou). (**C**) High-power view of group of malignant cells from aspirate of mesothelioma showing prominent intercellular windows (Papanicolaou). (**D**) Histologic examination of mesothelioma showing cords and three-dimensional clusters of atypical mesothelial cells (H&E).

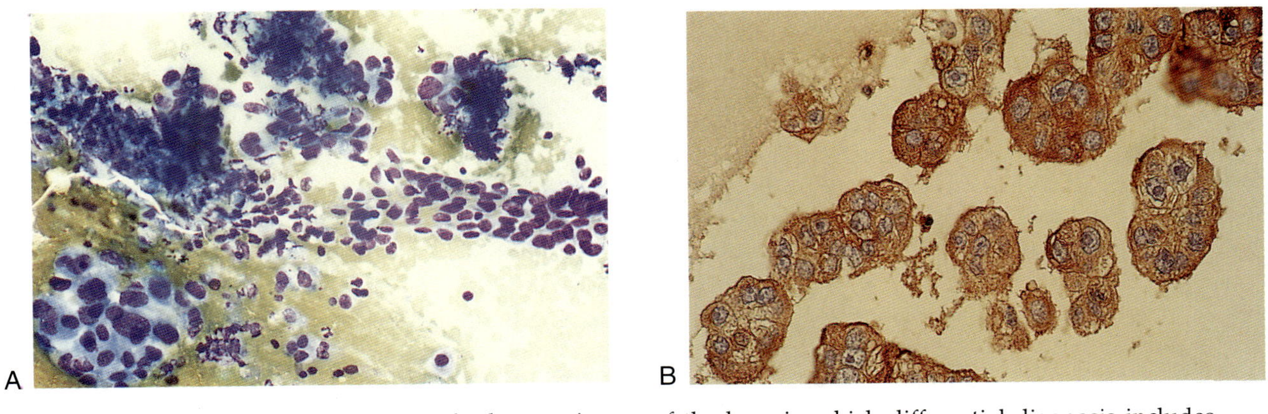

Fig. 1-56. (**A**) Aspirate of peripheral adenocarcinoma of the lung in which differential diagnosis includes mesothelioma. In this field a cluster of malignant cells is seen along with another group of mesothelial cells arranged in a monolayer (Diff-Quik). (**B**) Immunoperoxidase stain for CEA decorates numerous clumps of adenocarcinoma cells (immunoperoxidase).

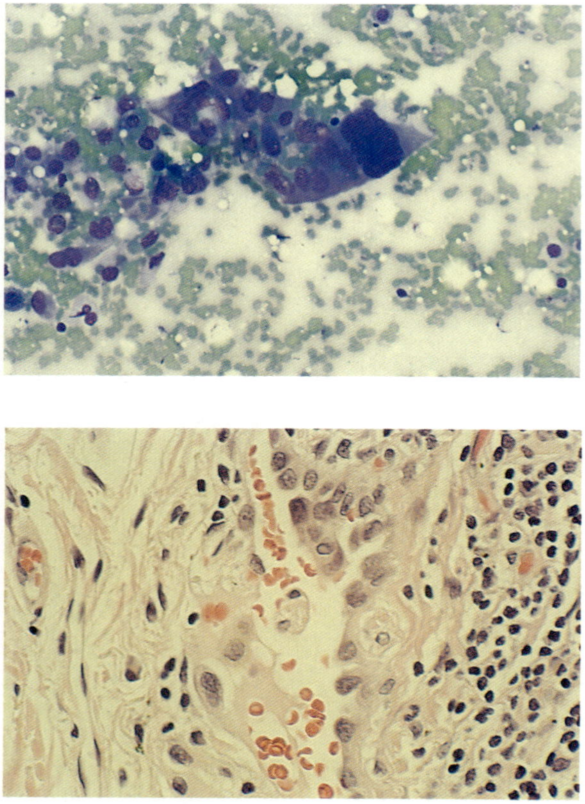

A

B

Fig. 1-57. (**A**) Aspirate of peripheral lung lesion showing atypical cells having low-power pattern of mesothelial cells. Note the large nuclei in a few of the cells (Diff-Quik). (**B**) Resected peripheral lung lesion proved to be a granuloma. Overlying atypical but reactive mesothelial cells are present (H&E).

mesothelial cells are seldom EMA positive in our experience[187] (Fig. 1-57). Ultrastructural demonstration of long slender microvilli favors a mesothelioma. Often, the correct diagnosis is based on the gross findings of diffuse involvement of the pleura, necessitating a more extensive histologic examination, rather than a definitive FNA cytologic study.

THORACIC PNET

A neoplasm related to neuroblastoma is the peripherally located primitive neuroectodermal (neuroepithelial) tumor (PNET), a small round cell malignancy of soft tissue and bone occurring predominantly in older children and adolescents.[188] Central PNETs are those that occur in the CNS. Primary sites of involvement of peripheral PNET include the chest wall, trunk, retroperitoneum, pelvis, and extremities. Askin et al.[189] initially reported a unique clinicopathologic entity in 20 children characterized by a small cell malignancy arising in the soft tissue of the chest or peripheral lung. The average age of the patients was 14.5 years. Local recurrence was common without widespread metastasis, but the median survival was still only 8 months. The tumors consisted of small primitive cells with high nuclear to cytoplasmic ratios. Rosettes were not seen in all cases, although the tumor cells were positive for NSE. Many cases have cells showing cytoplasmic positivity with the PAS stain. EM examination usually demonstrate dense-core neurosecretory granules and cell processes with intermediate filaments supportive of a neuroectodermal origin of the tumor. Most believe that the Askin tumor represents a PNET with a possible origin from an intercostal nerve. PNET and Ewing's sarcoma both may share the same neuroectodermal histogenesis based on the presence of the same reciprocal chromosomal abnormality (11;22 translocation) and positive staining with monoclonal antibody HBA-71 to the p 30/32^{MIC2} antigen, which occurs almost exclusively on the cell membranes of PNET and Ewing's sarcoma, but not other small round cell tumors of childhood.[188]

There have been a few case reports describing the FNA cytomorphologic appearance of peripheral PNET.[190–193] We recently described the cytomorphologic features and ancillary studies of eight FNA biopsies from three pediatric patients.[188] Two of the biopsies suggested the initial diagnosis of PNET of the chest wall, while the remaining six documented metastatic or locally recurrent disease. A primary diagnosis of PNET was made by FNA biopsy in one case enabling the pediatric oncologist to provide specific therapy and achieve remission. FNA cytologic findings included numerous malignant cells having high nuclear to cytoplasmic ratios with scant amounts of cytoplasm (Fig. 1-58). Within some clusters, nuclear molding was identified. Homer Wright rosettes were seen in two aspirates. The tumor cells had nuclei with evenly dispersed, finely granular chromatin. Prominent nucleoli were not appreciated. Pseudorosette formation can also be present,

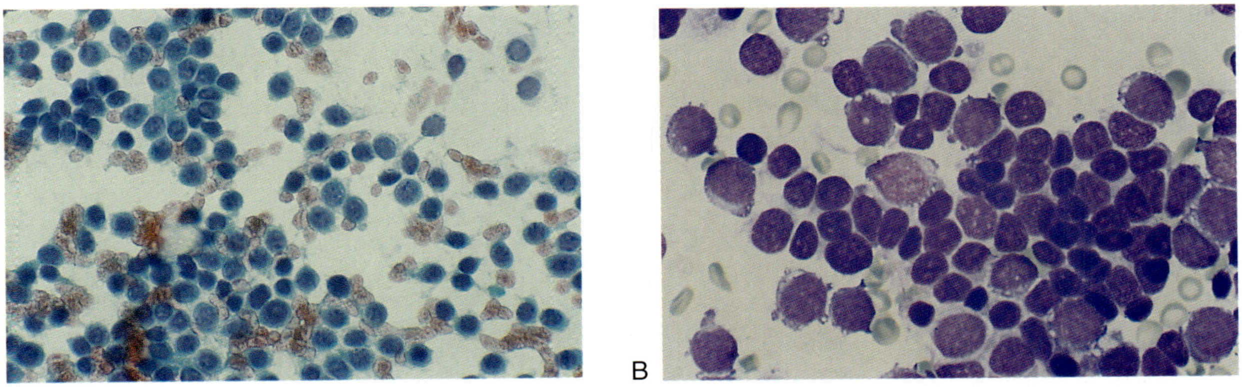

A B

Fig. 1-58. (**A**) Aspirate of PNET involving chest wall and lung demonstrating individually scattered tumor cells with high nuclear to cytoplasmic ratios (Papanicolaou). (**B**) High-power view of PNET consisting of tumor cells having high nuclear to cytoplasmic ratios. Some cytoplasmic vacuolization is present (Diff-Quik).

characterized by malignant cells adhering to branching blood vessels. The Diff-Quik–stained smears showed fine cytoplasmic vacuolization associated with membranous cytoplasmic blebs in a few cases. In the Papanicolaou-stained smears, occasional cells had fine small unipolar cytoplasmic tags. In our experience, the major feature separating PNET from neuroblastoma is the lack of obvious neuropil in PNET, in contrast to its presence in neuroblastoma. PNET also demonstrates prominent cytoplasmic vacuoles, while Homer Wright rosettes are less often seen. Neoplastic ganglion cells are not present in PNET. When a predominantly dispersed pattern of single cells is present, the differential diagnosis of PNET also includes malignant lymphoma.

REFERENCES

1. Pe LM, Johnston WW, Dunnick NR: Percutaneous transthoracic needle aspiration: a review. AJR 152:451, 1989
2. Alonxo P, Sanchex S, Ramirez E et al: Transthoracic needle biopsy in neoplastic and nonneoplastic pathology: experience in a general hospital. Diagn Cytopathol 2:284, 1986
3. Zavala DC, Schoell JE: Ultrathin needle aspiration of the lung in infectious and malignant disease. Am Rev Respir Dis 123:125, 1981
4. Schwinn CP, Sargent EN, Turner AF et al: Cytopathology of percutaneous pulmonary needle aspiration biopsy. p. 325. In Wied GL, Koss LG, Reagan JW (eds): Compendium on Diagnostic Cytology. Tutorials of Cytology, Chicago, 1983
5. Lalli AF, McCormack LJ, Zelch M et al: Aspiration biopsies of chest lesions. Radiology 127:35, 1978
6. Diefenthal HC, Tashjian J: The role of plain films, CT, tomography, ultrasound, and percutaneous needle aspiration in the diagnosis of inflammatory lung disease. Semin Respir Infect 3:83, 1988
7. Gibney RTN, Man GCW, King EG et al: Aspiration biopsy in the diagnosis of pulmonary disease. Chest 80:300, 1981
8. Johnston WW: Percutaneous fine needle aspiration biopsy of the lung: a study of 1,015 patients. Acta Cytol 28:218, 1984
9. Bonfiglio TA: Fine needle aspiration biopsy of the lung. Pathol Annu 16:159, 1981
10. Johnston WW, Frable WJ: Diagnostic respiratory cytology. p. 96. In Johnston WW (ed): Masson Monographs in Cytopathology. First volume. Masson Publishing USA, New York, 1979
11. Khouri NF, Stitik FP, Erozan YS et al: Transthoracic needle aspiration biopsy of benign and malignant lung lesions. AJR 144:281, 1985
12. Cappellari JO, Thompson EN III, Wallenhaupt SL: Utility of intraoperative fine needle aspiration biopsy in the surgical management of patients with pulmonary masses. Acta Cytol 38:707, 1994
13. Silverman JF: Respiratory tract cytology including FNA. p. 137. In Atkinson B (ed): Atlas of Diagnostic Cytopathology. WB Saunders, Philadelphia, 1992
14. Sinner WN: Complications of percutaneous transthoracic needle aspiration biopsy. Acta Radiol Diagn 17:813, 1976

15. Johnsrude IS, Silverman JF, Weaver MD et al: Rapid cytology to decrease pneumothorax incidence after percutaneous biopsy. AJR 144:793, 1985
16. Zarbo RJ, Fenoglio-Preiser CM: Interinstitutional database for comparison of performance in lung fine-needle aspiration cytology. Arch Pathol Lab Med 116:463, 1992
17. Thornbury JR, Burke DP, Naylor B: Transthoracic needle aspiration biopsy: accuracy of cytologic typing of malignant neoplasms. AJR 136:719, 1981
18. Cagle PT, Kovach M, Ramzy I: Causes of false results in transthoracic fine needle lung aspirates. Acta Cytol 37:16, 1993
19. Raab SS, Silverman JF: The clinical utility of cytologic typing of lung tumors. Diagn Cytopathol 10:376, 1994
20. Silverman JF, Finley JL, O'Brien KF et al: Diagnostic accuracy of immediate interpretation of fine needle aspiration biopsy specimens from various sites. Acta Cytol 33:791, 1989
21. Silverman JF, Frable WJ: The use of the Diff-Quik stain in the immediate interpretation of fine-needle aspiration biopsies. Diagn Cytopathol 6:366, 1990
22. Gay PC, Brutinel WM: Transbronchial needle aspiration in the practice of bronchoscopy. Mayo Clin Proc 64:158, 1989
23. Wagner ED, Ramzy I, Greenberg SD, Gonzalez JM: Transbronchial fine-needle aspiration. Reliability and limitations. Am J Clin Pathol 92:36, 1989
24. Bibbo M: Comprehensive Cytopathology. WB Saunders, Philadelphia, 1991
25. Silverman JF, Marrow HG: Fine needle aspiration cytology of granulomatous diseases of the lung, including nontuberculous *Mycobacterium* infection. Acta Cytol 29:535, 1985
26. Morales CF, Patefield AJ, Strollo PJ, Schenk DA: Flexible transbronchial needle aspiration in the diagnosis of sarcoidosis. Chest 106:709, 1994
27. Klein JO: Diagnostic lung puncture in the pneumonias of infants and children. Pediatrics 44:486, 1969
28. Garcia de Olarte D, Trujillo H, Uribe A et al: Lung puncture-aspiration as a bacteriologic diagnostic procedure in acute pneumonias of infants and children. Clin Pediatr 10:346, 1971
29. Arroyo J, Gordan V, Postic B: Transthoracic needle aspiration in the management of pulmonary infections. J SC Med Assoc 77:427, 1981
30. Bhatt ON, Miller R, Le Riche J et al: Aspiration biopsy in pulmonary opportunistic infections. Acta Cytol 21:206, 1977
31. Bandt PD, Blank N, Castellino RA: Needle diagnosis of pneumonitis. Value in high-risk patients. JAMA 220:1578, 1972
32. Vernon SE: Nodular pulmonary sarcoidosis. Diagnosis with fine needle aspiration biopsy. Acta Cytol 29:473, 1985
33. de Vivo F, Pond GD, Rhenman B et al: Transtracheal aspiration and fine needle aspiration biopsy for the diagnosis of pulmonary infection in heart transplant patients. J Thorac Cardiovasc Surg 96:696, 1988
34. Lyon AB: Bacteriologic studies of 165 cases of pneumonia and postpneumonic empyema in infants and children. Am J Dis Child 23:72, 1922
35. Silverman JF: Inflammatory and neoplastic processes of the lung: differential diagnosis and pitfalls in FNA biopsies. Diagn Cytopathol, in press
36. Silverman JF: Fine-needle aspiration cytology of infectious and inflammatory diseases and other nonneoplastic disorders. p. 201. In Kline T (ed): Guides to Clinical Aspiration Biopsy Series. Igaku-Shoin, New York, 1991
37. Koss LG, Woyke S, Olszewski W: Aspiration biopsy. Cytologic Interpretation and Histologic Bases. Igaku-Shoin, New York, 1984
38. Wallace RJ Jr, Cohen A, Awe RJ et al: Carcinomatous lung abscess. Diagnosis by bronchoscopy and cytopathology. JAMA 242:521, 1979
39. Bernhard WF, Wylie RH, Malcolm JA: The carcinomatous abscess. A clinical paradox. N Engl J Med 266:914, 1962
40. Robicheaux G, Moinuddin SM, Lee LH: The role of aspiration biopsy cytology in the diagnosis of pulmonary tuberculosis. Am J Clin Pathol 83:719, 1985
41. Dahlgren SE, Ekstrom P: Aspiration cytology in the diagnosis of pulmonary tuberculosis. Scand J Respir Dis 53:196, 1972
42. Dahlgren SE, Lind B: Comparison between diagnostic results obtained by transthoracic needle biopsy and by sputum cytology. Acta Cytol 16:53, 1972
43. Bailey TM, Akhtar M, Ali MA: Fine needle aspiration biopsy in the diagnosis of tuberculosis. Acta Cytol 29:732, 1985
44. Rajwanshi A, Bhambhani S, Das DK: Fine-needle apsiration cytology diagnosis of tuberculosis. Diagn Cytopathol 3:13, 1987
45. Pagel W, Simmonds FAH, Macdonald J et al: Pulmonary Tuberculosis. 4th Ed. Oxford University Press, London, 1964, p. 245
46. Johnston WW, Schlein B, Amatulli J: Cytopathologic diagnosis of fungus infections. I. A method for the preparation of simulated cytopathologic material for the teaching of fungus morphologic in cytology specimens. II. The presence of fungus in clinical material. Acta Cytol 13:488, 1969
47. Johnston WW: The cytopathology of mycotic infections. Lab Med 2:34, 1971

48. Silverman JF, Johnsrude IS: Fine needle aspiration cytology of granulomatous cryptococcosis of the lung. Acta Cytol 29:157, 1985
49. Binford CH, Dooley JR: Diseases caused by fungi and actinomycetes. p. 551. In Binford CH, Connor DH (eds): Pathology of Tropical and Extraordinary Diseases. Vol. II. Armed Forces Institute of Pathology, Washington, DC, 1976
50. Littman ML, Walter JE: Cryptococcosis: current status. Am J Med 45:922, 1968
51. Gordonson J, Birnbaum W, Jacobson G et al: Pulmonary crytococcosis. Radiology 112:557, 1974
52. Hatcher CR, Jr, Sehdeva J, Waters WD et al: Primary pulmonary cryptococcosis. J Thorac Cardiovasc Surg 61:39, 1971
53. Walts AE: Localized pulmonary cryptococcosis: diagnosis by fine needle aspiration. Acta Cytol 27:457, 1983
54. Dormer BA, Scher P: Tumor of lung due to *Cryptococcus histolyticus*. Clin Proc 6:269, 1947
55. Whitaker D, Sterrett GF: *Cryptococcus neoformans* diagnosed by fine needle aspiration cytology of the lung. Acta Cytol 20:105, 1976
56. Lalli AF, McCormack LJ, Zelch M et al: Aspiration biopsies of chest lesions. Radiology 127:35, 1978
57. Linsk JA, Franzen S: Aspiration cytology of metastatic hypernephroma. Acta Cytol 28:250, 1984
58. Williamson JD, Silverman JF, Mallak CT, Christie JD: Atypical cytomorphologic appearance of *Cryptococcus neoformans*. Acta Cytol, in press
59. Prolla J: The detection of *Cryptococcus neoformans* in sputum cytology. Acta Cytol 14:87, 1970
60. Sinner WN: Needle biopsy of histoplasmosis. Fortschr Geb Rontgenstr Nuklearmed Erganzungsbd 133:590, 1980
61. Feedman SI, Ang EP, Haley RS: Identification of coccidioidomycosis of the lung by fine needle aspiration biopsy. Acta Cytol 30:420, 1986
62. Forseth J, Rohwedder JJ, Levine BE et al: Experience with needle biopsy for coccidioidal lung nodules. Arch Intern Med 146:319, 1986
63. Raab SS, Silverman JF, Zimmerman KG: Fine-needle aspiration biopsy of pulmonary coccidioidomycosis: spectrum of cytology findings in seventy-three patients. Am J Clin Pathol 99:582, 1993
64. Trumbull ML, Chesney TM: The cytological diagnosis of pulmonary blastomycosis. JAMA 245:836, 1981
65. Valicenti JF, Daniell C, Gobien RP: Thin needle aspiration cytology of benign intrathoracic lesions. Acta Cytol 25:659, 1981
66. Colby TV, Weiss RL: Current concepts in the surgical pathology of pulmonary infections. Am J Surg Pathol 11:25, 1987
67. McCalmont TH, Silverman JF, Geisinger KR: Cytologic diagnosis of aspergillosis in cardiac transplantation. Arch Surg 126:394, 1991
68. Reyes CV, Kathuria S, MacGlashan A: Diagnostic value of calcium oxalate crystals in respiratory and pleural fluid cytology. A case report. Acta Cytol 23:65, 1979
69. Farley ML, Mabry L, Munoz LA et al: Crystals occurring in pulmonary cytology specimens. Association with *Aspergillus* infection. Acta Cytol 29:737, 1985
70. Lee SH, Barnes WG, Schaetzel WP: Pulmonary aspergillosis and the importance of oxalate crystal recognition in cytology specimens. Arch Pathol Lab Med 110:1176, 1986
71. Fauci AS, Masur H, Gelmann EP et al: NIH conference. The acquired immunodeficiency syndrome. An update. Ann Intern Med 102:800, 1985
72. Bedrossian CWM, Mason MR, Gupta PK: Rapid cytologic diagnosis of Pneumocystis: a comparison of effective techniques. Semin Diagn Pathol 6:245, 1989
73. Batra P, Wallace JM, Ovenfors C-O: Efficacy and complications of transthoracic needle biopsy of lung in patients with *Pneumocystis carinii* pneumonia and AIDS. J Thorac Imaging 2:79, 1987
74. Jacobs JB, Vogel C, Powell RD et al: Needle biopsy in *Pneumocystis carinii* pneumonia. Radiology 93:525, 1989
75. Castellino RA: Percutaneous pulmonary needle diagnosis *Pneumocystis carinii* pneumonitis. Natl Cancer Inst Monog 43:137, 1976
76. Saldana MJ, Mones JM: Cavitation and other atypical manifestations of *Pneumocystis carinii* pneumonia. Semin Diagn Pathol 6:273, 1989
77. Strigle SM, Gal AA: A review of pulmonary cytopathology in the acquired immunodeficiency syndrome. Diagn Cytopathol 5:44, 1989
78. Moas CM, Evans DA, Stein-Streilein J et al: Cresyl violet: a rapid, simple, easily interpretable stain for detecting *Pneumocystis carinii* in sputum. South Med J 82:957, 1989
79. Chandra P, Delaney MD, Tuazon CU: Role of special stains in the diagnosis of *Pneumocystis carinii* infection from bronchial washing specimens in patients with the acquired immune deficiency syndrome. Acta Cytol 32:105, 1988
80. Shimono LH, Hartman B: A simple and reliable methenamine silver stain for *Pneumocystis carinii* and fungi. Arch Pathol Lab Med 110:855, 1986
81. Baughman RP, Strohofer SS, Clinton BA et al: The use of an indirect fluorescent antibody test for detecting *Pneumocystis carinii*. Arch Pathol Lab Med 113:1062, 1989

82. Blumenfeld W, Kovacs JA: Use of a monoclonal antibody to detect *Pneumocystis carinii* in induced sputum and bronchoalveolar lavage fluid by immunoperoxidase staining. Arch Pathol Lab Med 112:1233, 1988
83. Kovacs JA, Ng VL, Masur H et al: Diagnosis of *Pneumocystis carinii* pneumonia: improved detection in sputum with use of monoclonal antibodies. N Engl J Med 318:589, 1988
84. Gal AA, Klat EC, Koss MK et al: The effectiveness of bronchoscopy in the diagnosis of *Pneumocystis carinii* and cytomegalovirus pulmonary infections in the acquired immunodeficiency syndrome. Arch Pathol Lab Med 111:238, 1987
85. Buchanan AJ, Gupta RK: Cytomegalovirus infection of the lung: cytomorphologic diagnosis by fine-needle aspiration cytology. Diagn Cytopathol 2:341, 1986
86. Falguera M, Nogues A, Ruiz-Gonzalez A et al: Transthoracic needle aspiration in the study of pulmonary infections in patients with HIV. Chest 106:697, 1994
87. Kline TS: Handbook of Fine Needle Aspiration Biopsy Cytology. 2nd Ed. Churchill Livingstone, New York, 1988
88. Hawkins AG, Hsiu J-G, Smith RM et al: Pulmonary dirofiliariasis diagnosed by fine needle aspiration biopsy: a case report. Acta Cytol 29:19, 1985
89. Rangdaeng S, Alpert LC, Khiyami A et al: Pulmonary paragonimiasis. Report of a case with diagnosis by fine needle aspiration cytology. Acta Cytol 36:31, 1992
90. Flower CDR, Verney GI: Percutaneous needle biopsy of thoracic lesions: an evaluation of 300 biopsies. Clin Radiol 30:215, 1979
91. Sinner WN: Pulmonary neoplasms diagnosed with transthoracic needle biopsy. Cancer 43:1533, 1979
92. Johnson RD, Gobien RP, Valicenti JF, Jr: Current status of radiologically directed pulmonary thin needle aspiration biopsy: an analysis of 200 consecutive biopsies and review of the literature. Ann Clin Lab Sci 13:225, 1983
93. Bewtra C, Dewan N, O'Donahue WJC: Exfoliative sputum cytology in pulmonary embolism. Acta Cytol 27:489, 1983
94. Sasahara AA, Sharma GVRK, Barsamian EM et al: Pulmonary thromboembolism. JAMA 249:2945, 1983
95. Johnston WW, Frable WJ: The cytopathology of the respiratory tract. Am J Pathol 84:372, 1976
96. Scoggins WG, Smith RH, Frable WJ et al: False positive cytologic diagnosis of lung carcinoma in patient with pulmonary infarct. Ann Thorac Surg 24:474, 1977
97. Frable WJ: Pulmonary infarcts. Check Sample Program. Cytopathology: No. C-17. ASCP Commission on Continuing Education. American Society of Clinical Pathology, Chicago, 1974
98. Roggli VL, Johnston WW, Kaminsky DB: Asbestos bodies in fine needle aspirates of the lung. Acta Cytol 28:493, 1984
99. Tao L-C, Morgan RC, Donat EE: Cytologic diagnosis of intravenous talc granulomatosis by fine needle aspiration biopsy. A case report. Acta Cytol 28:737, 1984
100. Kung ITM, Johnson FB, So S-Y et al: Blue bodies in cytology specimens in a case of pulmonary talcosis. Am J Clin Pathol 81:675, 1984
101. Covell JL, Feldman PS: Fine needle aspiration diagnosis of aspiration pneumonia (phytopneumonitis). Acta Cytol 38:77, 1984
102. Machicao CN, Sorensen K, Abdul-Karim FW et al: Transthoracic needle aspiration biopsy in inflammatory pseudotumors of the lung. Diagn Cytopathol 5:400, 1989
103. Tchertkoff V, Bok YL, Wagner BM: Plasma cell granuloma of lung. Dis Chest 44:440, 1963
104. Bahadori M, Liebow AA: Plasma cell granulomas of the lung. Cancer 31:191, 1973
105. Carter D, Eggleston J: Fascicle on Tumors of the Lower Respiratory Tract. Armed Forces Institute of Pathology, Washington, DC: 1974, p. 300
106. Berardi RS, Chen HP, Stines GJ: Inflammatory pseudotumors of the lung. Surg Gynecol Obstet 156:89, 1983
107. Hsiu J-G, Stitik FP, D'Amato NA et al: Primary amyloidosis presenting as a unilateral hilar mass: report of a case diagnosed by fine needle aspiration biopsy. Acta Cytol 30:55, 1986
108. Tomashefski JF, Cramer SF, Abramowsky C et al: Needle biopsy diagnosis of solitary amyloid nodule of the lung. Acta Cytol 24:224, 1980
109. Silverman JF, Dabbs DJ, Norris HT et al: Localized primary (AL) amyloid tumor of the breast. Cytologic, histologic, immunocytochemical and ultrastructural observations. Am J Surg Pathol 10:539, 1986
110. Frable WJ: Thin-Needle Aspiration Biopsy. Major Problems in Pathology. Vol. 14. WB Saunders, Philadelphia, 1983
111. Fekete PS, Campbell WJ, Jr, Bernardino ME: Transthoracic needle aspiration biopsy in Wegener's granulomatosis. Morphologic findings in five cases. Acta Cytol 34:155, 1990

112. Fauci AS, Wolff SM: Wegener's granulomatosis: studies in eighteen patients and a review of the literature. Medicine (Baltimore) 52:535, 1973
113. Walker AN, Feldman PS, Walker GK: Fine needle aspiration of thoracic extramedullary hematopoiesis. Acta Cytol 27:170, 1983
114. Nathanson L, Frenkel D, Jacobi M: Diagnosis of lipoid pneumonia by aspiration biopsy. Arch Intern Med 72:627, 1943
115. Williams WL, Clark DA, Saiers JH: Fine needle aspiration diagnosis of lymphomatoid granulomatosis. A case report. Acta Cytol 36:91, 1992
116. The World Health Organization histological typing of lung tumours. Second edition. Am J Clin Pathol 77:123, 1987
117. Martini N, Bains MS, McCormack PM et al: Surgical treatment in non-small cell carcinoma of the lung: the Memorial Sloan-Kettering experience. p. 111. In Hoogstraten B, Addis BJ, Hansen HH et al (eds): Treatment of Lung Tumors. Springer-Verlag, Heidelberg, 1987
118. Johnston WW, Elson CE: Cytology of other body sites: respiratory tract. p. 320. In Bibbo M (ed): Comprehensive Cytopathology. 1st Ed. WB Saunders, Philadelphia, 1991
119. Hammar SP: Common neoplasms. p. 727. In Dail DH, Hammer SP (eds): Pulmonary Pathology. 2nd Ed. Springer-Verlag, New York, 1993
120. McCue PA, Finkel GC: Small cell lung carcinoma: an evolving histopathologic spectrum. Semin Oncol 20:153, 1993
121. Hirsch FR, Osterlind K, Hansen HH: The prognostic significance of histopathologic subtyping of small cell carcinoma of the lung according to the classification of the World Health Organization. A study of 375 consecutive patients. Cancer 52:2144, 1983
122. Hirsch FR, Matthews MJ, Aisner S et al: Histopathologic classification of small cell lung cancer: changing concepts and terminology. Cancer 62:973, 1988
123. Horie A, Ohta M: Ultrastructural features of large cell carcinoma of lung with reference to the prognosis of patients. Hum Pathol 12:423, 1981
124. Hammar S: The use of electron microscopy and immunohistochemistry in the diagnosis and understanding of lung neoplasms. Clin Lab Med 7:1, 1987
125. Dingemans KP, Moori WJ: Ultrastructure of squamous cell carcinoma of the lung. Pathol Annu 19:244, 1984
126. Wick MR, Berg LC, Hertz M: Large cell carcinoma of the lung with neuroendocrine differentiation. A comparison with large cell "undifferentiated" pulmonary tumors. Am J Clin Pathol 97:796, 1992
127. Sundaresan V, Reeve JG, Stenning S et al: Neuroendocrine differentiation and clinical behavior in non small cell lung tumors. Br J Cancer 64:333, 1991
128. O'Reilly PE, Brueckner J, Silverman JF: Value of ancillary studies in fine needle aspiration cytology of the lung. Acta Cytol 38:144, 1994
129. Marchevsky AM: Definitive answers to old questions about the pathology of non-small cell carcinoma of the lung? Am J Clin Pathol 97:163, 1992
130. Martini N: Operable lung cancer. CA Cancer J Clin 43:201, 1993
131. The Lung Cancer Study Group: Effects of postoperative mediastinal radiation on completely resected stage II and stage III epidermoid cancer of the lung. N Engl J Med 315:1801, 1986
132. Smith JH, Frable WJ: Adenocarcinoma of the lung. Cytologic correlation with histologic types. Acta Cytol 18:316, 1974
133. Holmes EC, Gail M: Surgical adjuvant therapy for stage II and III adenocarcinoma and large-cell undifferentiated carcinoma. J Clin Oncol 4:710, 1986
134. Martini N, Burt ME, Bains MS et al: Survival after resection of stage II non-small cell lung cancer. Ann Thorac Surg 54:460, 1992
135. Boring CC, Squires TS, Tong T: Cancer statistics 1993. CA Cancer J Clin 43:7, 1993
136. Karsell PR, McDougall JC: Diagnostic tests for lung cancer. Mayo Clin Proc 68:288, 1993
137. Coppage L, Shaw C, Curtis AM: Metastatic disease to the chest in patients with extrathoracic malignancy. J Thorac Imaging 2:24, 1987
138. Mackay B, Lukeman JM, Ordonez NG: Tumors of the Lung. Major Problems in Pathology. Vol. 24. Philadelphia, WB Saunders, 1991
139. Kamiya M, Uei Y, Shimosato Y: Cytologic features of peripheral squamous cell carcinoma of the lung. Acta Cytol 39:61, 1995
140. Mountain CF, Lukeman JM, Hammar SP et al: Lung cancer classification: the relationship of disease extent and cell type to survival in a clinical trials population. J Surg Oncol 35:147, 1987
141. Bonfiglio TA: Cytopathologic interpretation of transthoracic fine-needle biopsies. p. 79. In WW Johnston (ed): Masson Monographs in Diagnostic Cytopathology. Masson Publishing USA, New York, 1983
142. Silverman JF, Finley JL, Park HK et al: Fine needle aspiration cytology of bronchioloalveolar cell carcinoma of the lung. Acta Cytol 29:887, 1985
143. Elson CE, Moore SP, Johnston WW: Morphologic and immunocytochemical studies of bronchioloalveolar carcinoma at Duke University Medical Center, 1968–1986. Anal Quant Cytol Histol 11:261, 1989

144. Colby TV: Malignancies in the lung and pleura mimicking benign processes. Semin Diagn Pathol 12:30, 1995
145. Lozowski W, Hajdu SI: The cytologic diagnosis of primary bronchioloalveolar carcinoma. Acta Cytol 30:569, 1986
146. Silverman JF, Finley JL, Park HK et al: Psammoma bodies and optically clear nuclei in bronchioloalveolar cell carcinoma: diagnosis by fine needle aspiration with histologic and ultrastructural confirmation. Diagn Cytopathol 1:205, 1985
147. Flint A, Lloyd RV: Pulmonary metastases of colonic carcinoma. Arch Pathol Lab Med 116:39, 1992
148. Patel A, Peters SG: Clinical manifestations of lung cancer. Mayo Clin Proc 68:273, 1993
149. Frost JK: The cell in health and disease. In Wied GL (ed): Monographs in Clinical Cytology, Vol. 2. S Karger, Basel, 1986
150. Travis WD, Linnoila RI, Tsokos MG et al: Neuroendocrine tumors of the lung with proposed criteria for large cell neuroendocrine carcinoma. An ultrastructural, immunohistochemical, and flow cytometric study of 35 cases. Am J Surg Pathol 15:529, 1991
151. Hinson JA, Perry MC: Small cell lung cancer. CA Cancer J Clin 43:216, 1993
152. Lee T-K, Horner RD, Silverman JF et al: Implications of nuclear diameter in small cell carcinoma of the lung. Acta Cytol 34:449, 1990
153. Lee T-K, Silverman JF, Horner RD et al: Overlap of nuclear diameters in lung cancer cells. Anal Quant Cytol Histol 12:275, 1990
154. Mullins RK, Thompson SK, Coogan PS, Shurbaji MS: Paranuclear blue inclusions: an aid in the cytopathologic diagnosis of primary and metastatic pulmonary small-cell carcinoma. Diagn Cytopathol 10:332, 1994
155. Wittchow R, Laszewski M, Walker W, Dick F: Paranuclear blue inclusions in metastatic undifferentiated small cell carcinoma in the bone marrow. Mod Pathol 5:555, 1992
156. Givens CD, Marini JJ: Transbronchial needle aspiration of a bronchial carcinoid tumor. Chest 88:152, 1985
157. Arrigoni MG, Woolner LB, Bernatz PE: Atypical carcinoid tumors of the lung. J Thorac Cardiovasc Surg 64:413, 1972
158. Mills SE, Cooper PH, Walker AN, Kron IL: Atypical carcinoid tumor of the lung. A clinicopathologic study of 17 cases. Am J Surg Pathol 6:643, 1982
159. Warren WH, Memoli VA, Gould VE: Well differentiated and small cell neuroendocrine carcinomas of the lung. Two related but distinct clinicopathologic entities. Virchows Arch B Cell Pathol 55:299, 1988
160. Warren WH, Farber LP, Gould VE: Neuroendocrine neoplasms of the lung. A clinicopathologic update. J Thorac Cardiovasc Surg 98:321, 1989
161. Craig ID, Finley RJ: Spindle-cell carcinoid tumor of lung: cytologic, histopathologic and ultrastructural features. Acta Cytol 26:495, 1982
162. Dube VE: Peripheral bronchial carcinoid with a spindle-cell pattern. Arch Pathol 89:374, 1970
163. Szyfelbein WM, Ross JS: Carcinoids, atypical carcinoids, and small-cell carcinomas of the lung: differential diagnosis of fine-needle aspiration biopsy specimens. Diagn Cytopathol 4:1, 1988
164. Wilkins EW, Grillo HC, Moncure AC et al: Changing times in surgical management of bronchopulmonary carcinoid tumor. Ann Thorac Surg 38:339, 1984
165. Wilson RA: An unusual secon primary tumor: report of a case of atypical carcinoid of the lung. Acta Cytol 22:362, 1978
166. Frierson HF Jr, Covell JL, Mills SE: Fine needle aspiration cytology of atypical carcinoid of the lung. Acta Cytol 31:471, 1987
167. Horan DC, Bonfiglio TA, Patten SF, Jr: Fine needle aspiration cytopathology of bronchial carcinoid tumors. An analytical study of the cells. Anal Quant Cytol 4:105, 1982
168. Dunbar F, Leiman F: The aspiration cytology of pulmonary hamartomas. Diagn Cytopathol 5:174, 1989
169. Ludwig ME, Otis RD, Cole SR et al: Fine needle aspiration cytology of pulmonary hamartomas. Acta Cytol 26:671, 1982
170. Ramzy I: Pulmonary hamartomas: cytologic appearances of fine needle aspiration biopsy. Acta Cytol 20:15, 1976
171. Suen KC, Quenville NF: Fine needle aspiration cytology of uncommon thoracic lesions. Am J Clin Pathol 75:803, 1981
172. Dail DH: Uncommon tumors. p. 1279. In Dail DH, Hammar SP (eds): Pulmonary Pathology. Springer-Verlag, New York, 1988
173. Kim G, Naylor B, Han IH: Fine needle aspiration cytology of sarcomas metastatic to the lung. Acta Cytol 30:688, 1986
174. Perry MD, Furlong JW, Johnston WW: Fine needle aspiration cytology of metastatic dermatofibrosarcoma protuberans: a case report. Acta Cytol 30:507, 1986
175. Silverman JF, Weaver MD, Gardner N et al: Aspiration biopsy cytology of malignant schwannoma metastatic to the lung. Acta Cytol 29:15, 1985
176. Crosby JH, Høeg K, Hager B: Transthoracic fine needle aspiration of primary and metastatic sarcomas. Diagn Cytopathol 1:221, 1985
177. Finley JL, Silverman JF, Dabbs DJ: Fine-needle aspiration cytology of pulmonary carcinosarcoma with

immunocytochemical and ultrastructural observations. Diagn Cytopathol 4:239, 1988

178. Upton MP, Hirohashi S, Tome Y et al: Expression of vimentin in surgically resected adenocarcinomas and large cell carcinomas of lung. Am J Surg Pathol 10:560, 1986
179. Roth TA, Elguezabal A: Pulmonary blastoma evolving into carcinosarcoma. Am J Surg Pathol 2:407, 1978
180. Spahr J, Draffin RM, Johnston WW: Cytopathologic findings in pulmonary blastoma. Acta Cytol 23:454, 1979
181. Non DP, Lang WR, Patchefsky A et al: Pulmonary blastoma: cytopathologic and histopathologic findings. Acta Cytol 20:381, 1976
182. Cosgrove MM, Chandrasoma PT, Martin SE: Diagnosis of pulmonary blastoma by fine-needle aspiration biopsy: cytologic and immunocytochemical findings. Diagn Cytopathol 7:83, 1991
183. Lee KG, Cho NH: Fine-needle aspiration cytology of pulmonary adenocarcinoma of fetal type: report of a case with immunohistochemical and ultrastructural studies. Diagn Cytopathol 7:408, 1991
184. Dusenbery D, Grimes MM, Frable WJ: Fine-needle aspiration cytology of localized fibrous tumor of pleura. Diagn Cytopathol 8:444, 1992
185. Obers VJ, Leiman G, Girdwood RW et al: Primary malignant pleural tumors (mesotheliomas) presenting as localized masses: fine needle aspiration cytologic findings, clinical and radiologic features and review of the literature. Acta Cytol 32:567, 1988
186. Tao L-C: Aspiration biopsy cytology of mesothelioma. Diagn Cytopathol 5:19, 1989
187. Singh HK, Silverman JF, Berns L et al: The significance of epithelial membrane antigen in the workup of problematic serous effusions. Diagn Cytopathol 13:3, 1995
188. Silverman JF, Berns LA, Holbrook CT et al: Fine needle aspiration cytology of primitive neuroectodermal tumors. A report of three cases. Acta Cytol 36:541, 1992
189. Askin FB, Rosai J, Sibley RK, Dehner LP: Malignant small cell tumor of the thoracopulmonary region in childhood. Cancer 43:2438, 1979
190. Auger M, Bedard YC, Keating S: Diagnosis of a peripheral neuroectodermal tumour by fine needle aspiration. Cytopathology 1:243, 1990
191. Neuhold N, Artlieb U, Wimmer M et al: Aspiration cytology, immunocytochemistry and electron microscopy of a malignant peripheral neuroectodermal tumor. A case report. Acta Cytol 33:74, 1989
192. Silverman JF, Dabbs DJ, Ganick DJ et al: Fine needle aspiration cytology of neuroblastoma, including peripheral neuroectodermal tumor, with immunocytochemical and ultrastructural confirmation. Acta Cytol 32:367, 1988
193. Kumar PV: Fine needle aspiration cytologic findings in malignant small cell tumor of the thoracopulmonary region (Askin tumor). Acta Cytol 38:702, 1994

2 Mediastinum

The mediastinum can be defined as those tissues and organs that occupy the thoracic cavity between the lungs and the pleural cavities laterally, between the sternum anteriorly and the vertebral column posteriorly, and from the thoracic inlet down to the diaphragm. The mediastinum can actually be considered a box that contains a variety of normal anatomic structures, all of which may give rise to benign or malignant neoplasms. Foremost among these are the thymus and lymph nodes, as these are the sites of a large proportion of primary and metastatic neoplasms. Classically, the mediastinum is divided into several compartments on the basis of external and internal anatomic landmarks. The divisions are most often determined with a lateral chest radiograph. Although imaginary, these compartmental divisions provide useful data in formulating a differential diagnosis, as specific tumor types show strong tendencies to occur within specific compartments (Table 2-1). Most often, primary mediastinal neoplasms present clinically secondary to the effect of a mass lesion compressing or invading adjacent structures. At other times, they are discovered as an incidental mass lesion on chest radiographs. In either situation, these masses are amenable to fine needle aspiration (FNA) biopsy. In our experience, mediastinal masses are usually approached under radiologic guidance with a transthoracic FNA biopsy procedure.[1–6] Mediastinal structures, in particular lymph nodes, may also be sampled by transbronchial FNA biopsy during a bronchoscopic examination. We believe that FNA biopsy is a valuable first step in the evaluation of mediastinal masses. Often, the cytologist is able to provide very useful diagnostic and prognostic information. In addition to a careful evaluation of the aspiration cytomorphology, the cytologist needs to integrate all relevant clinical, radiographic, ultrastructural, and immunocytochemical data to formulate a meaningful final diagnosis.

THYMOMAS

The most frequent primary neoplasms to arise in the mediastinum are thymomas.[7–17] These tumors are defined as neoplasms of thymic epithelium that are variably associated with reactive or non-neoplastic lymphoid cells. The vast majority occur in the anterosuperior compartments.

Thymomas may affect individuals of all ages, including rare incidences in the pediatric population. However, the vast majority occur in adults with a median age of about 50 years, involving men and women with relatively equal frequency. Clinical signs and symptoms may include chest pain, cough, dyspnea, dysphagia, and the superior vena cava syndrome. A peculiar aspect of thymomas is their strong and frequent association with paraneoplastic syndromes, the most common one being myasthenia gravis. These may also include a number of benign hematologic conditions, most frequently pure red cell aplasia. Thymoma is also strongly associated with the development, either concurrently or subsequently, of extrathymic neoplasms, many of which are lymphoreticular in type.[18] Finally, thymoma will be discovered as an incidental finding on a chest radiograph in a sizable proportion of all patients.

Table 2-1. Common Mediastinal Locations of Tumors by Compartment

Anterosuperior	*Middle*	*Posterior*
Thymoma Lymphoma Germ cell tumors Carcinoid tumor Thyroid lesions Parathyroid lesions Benign cysts	Lymphoma Benign cysts	Nerve sheath tumors Paraganglioma Ganglioneuroblastic tumors Benign cysts

Cytomorphologic Features of Thymoma

- Moderate to high cellularity
- Dual population of intermixed uniform epithelial tumor cells and lymphocytes
- Variable intercellular cohesion among tumor cells
- Contour of tumor cells usually polygonal to round, rarely spindled
- Neoplastic nuclei: bland chromatin, small nucleoli

Almost all thymomas are totally encapsulated and thus clinically benign. There are two types of malignant neoplasms of thymic epithelium.[7–17,19–21] The more common of these two is termed malignant thymoma, defined grossly and microscopically by infiltration of the tumor capsule. On a cell-for-cell basis, these neoplasms are indistinguishable from the benign tumors. By contrast, thymic carcinomas are cytologically malignant. The distinction between benign and malignant forms of thymic epithelial tumors is important, in that their prognosis and therapeutic approaches differ considerably.

A number of different histologic classifications of thymomas have been offered. The most widely used classification is based on the relative proportions of neoplastic epithelium and benign reactive lymphocytes and on the predominant contour of the tumor cells.[7,8] However, a number of studies have failed to demonstrate a significant relationship between the histologic subtypes and the biologic potential for malignancy. Recently, Müller-Hermelink and colleagues have provided an alternative histologic approach, the foundation of which is the resemblance of thymomas to different compartments and components of the normal thymus.[9,19] Within recent years, several published investigations have confirmed the ability of this classification scheme to predict more accurately a malignant clinical course and, to an extent, the occurrence of certain paraneoplastic syndromes.[12,16,17]

Aspiration smears of thymomas are moderately to highly cellular and usually contain a characteristic dual population of cells, consisting of an admixture of neoplastic epithelial cells and benign lymphocytes[6,22–26] (Figs. 2-1 to 2-3). The tumor cells may vary in both size and appearance from thymoma to thymoma, but within a given neoplasm, they are generally morphologically uniform. In most instances, thymoma cells are round to polygonal. Rarely, the tumor cells have a distinctly spindled contour. Intercellular cohesion is variable.[26] In some specimens, large sheets and aggregates of neoplastic epithelium predominate (Fig. 2-1). At the other end of the spectrum, most tumor cells appear as isolated elements (Fig. 2-3). In most cases, some degree of intercellular cohesion can be recognized in the smears (Fig. 2-2). Usually, there are distinct cell membranes. It is important to recognize that the nuclei of both benign and malignant thymomas do not possess features characteristically associated with clinically aggressive neoplasms.[6,25–29] That is, the nuclei usually are round to oval with smooth delicate membranes (Figs. 2-2 and 2-4). Quite frequently, thymoma cells with spindled contours and elongated nuclei dominate an aspirate (Fig. 2-5). Furthermore, chromatin is usually finely granular and evenly distributed throughout the nucleus. Nucleoli are generally small. However, as the chromatin is often pale-staining, these nucleoli may be readily apparent.

One very characteristic and diagnostically helpful feature is the intimate association of the lymphocytes with the neoplastic epithelium, which includes lymphocytes admixed within the epithelial aggregates (Figs. 2-1 to 2-4). One notable feature is that small, mature-appearing lymphocytes will

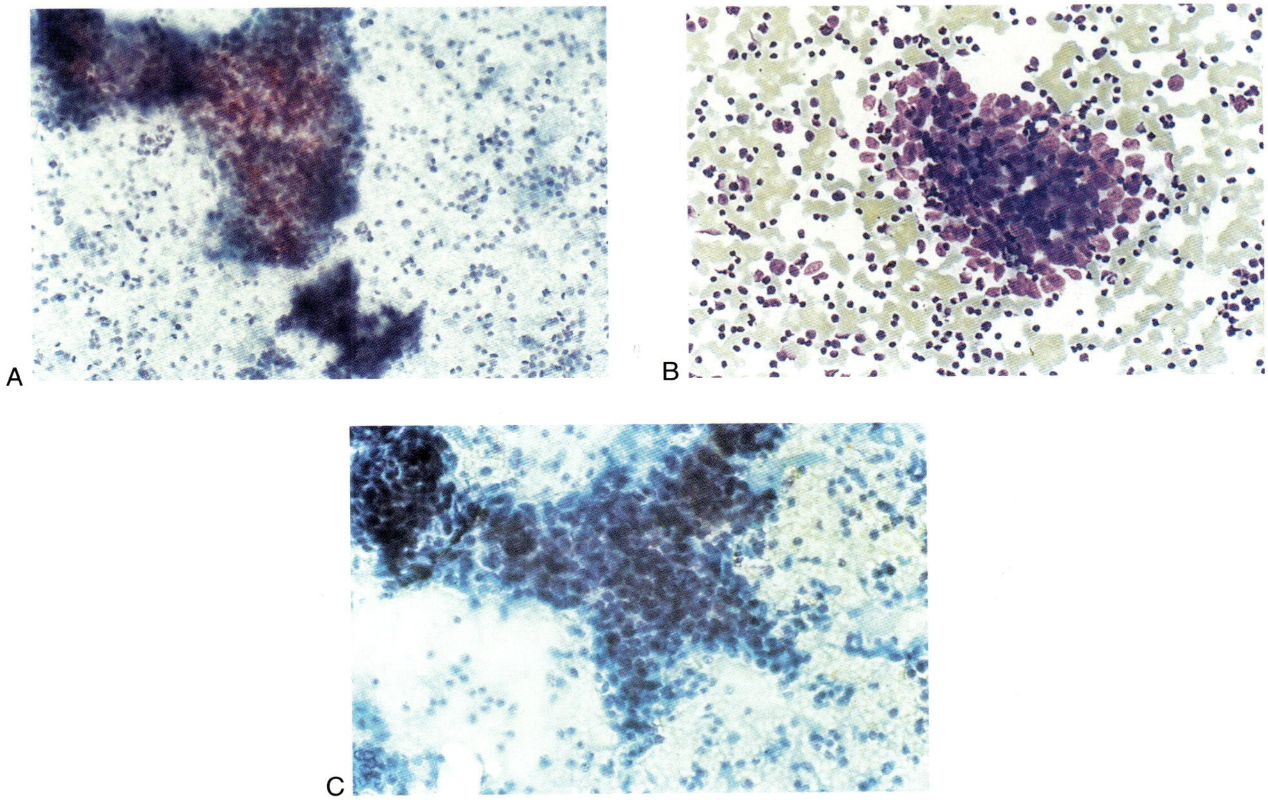

Fig. 2-1. (**A–C**) A high level on intercellular cohesion is evident among the aspirated thymoma cells in these smears. Figures **A** and **B** are from one patient, whereas Fig. **C** is from a different individual. The specimens are characterized by huge, sharply defined cellular aggregates that have a three-dimensional quality. (**A**) A peripheral palisading of the tumor cells is suggested. The background of smears contains numerous individually dispersed lymphocytes. Even at these low magnifications, the difference in sizes between thymoma cells and lymphocytes is apparent. In addition, lymphocytes are intermingled among the epithelial cells within the aggregates; this is best seen in Fig. **B** (**A & C**, Papanicolaou; **B**, Diff-Quik).

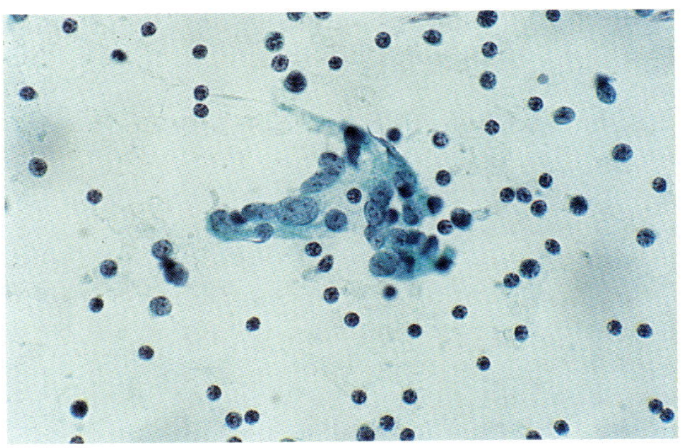

Fig. 2-2. Compared to Figure 2-1, intercellular cohesion is greatly reduced. The neoplastic cells are present in a small flat sheet. Their nuclei are ovoid and mildly pleomorphic in size. The chromatin is finely granular and uniformly distributed throughout the nuclei. Only very small nucleoli are evident, and nuclear membranes are delicate. Cytoplasm is cyanophilic and cell borders are indistinct. Small mature-appearing lymphocytes are sprinkled within the aggregate as well as scattered around it. This is the classic biphasic appearance of thymoma (Papanicolaou).

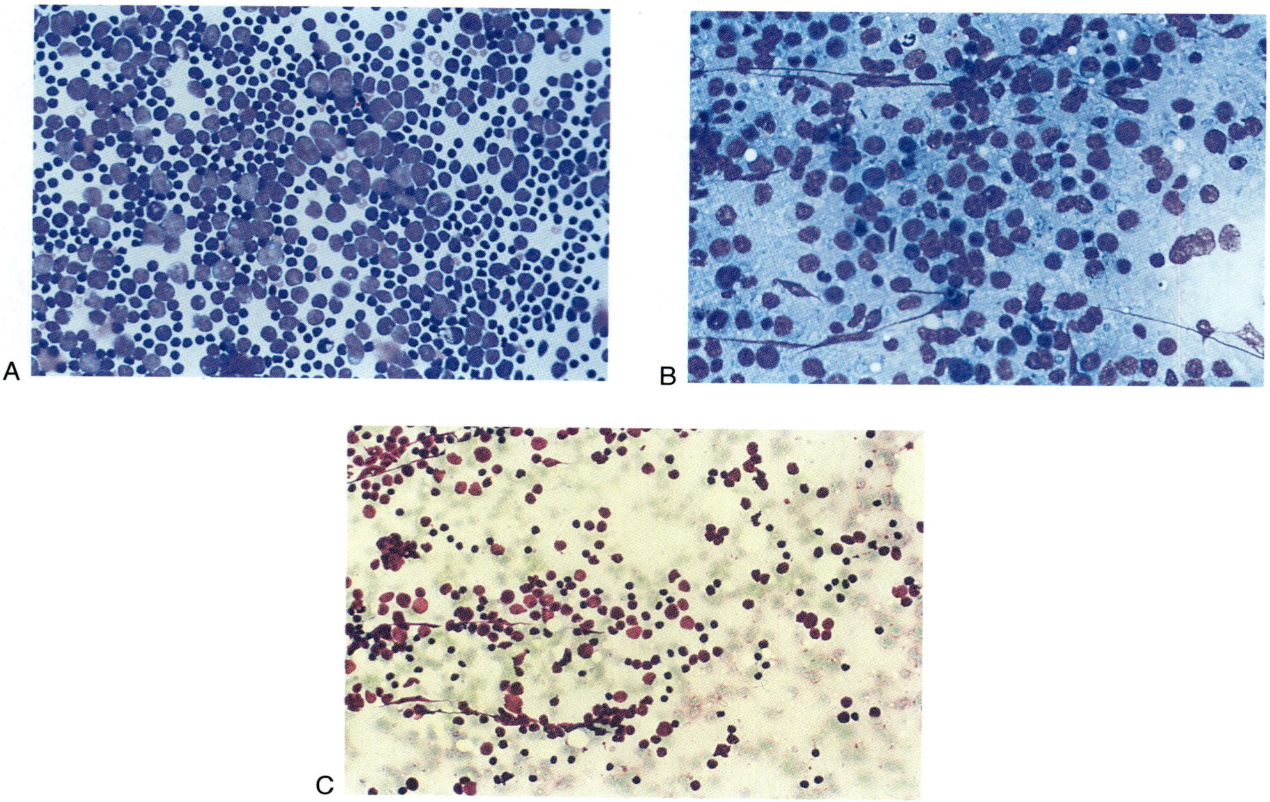

Fig. 2-3. (**A–C**) In these examples, intercellular cohesion is so reduced that most of the neoplastic cells are present in the smears as isolated elements. They are intimately admixed with the smaller lymphoid cells. The dual cellular population of thymoma is obvious. When the cohesion is this reduced, the differential diagnosis needs to include malignant lymphoma and seminoma (Diff-Quik).

predominate numerically. Somewhat larger lymphoid cells with slightly irregular nuclear contours are present in some thymomas (the cortical type of Müller-Hermelink). Other mediastinal lesions may also yield a large number of lymphocytes within aspiration biopsy specimens, generating a lengthy differential diagnosis in some specimens.

We are unaware of any cytomorphologic means to distinguish reliably benign from malignant thymomas.[6,25–29] In both entities, the tumor cells have rather bland chromatin and small nucleoli (Fig. 2-6). Analogous to the situation in thyroid follicular neoplasms, histologic evidence of invasion of the neoplasm capsule needs to be identified in order to recognize the tumor as biologically aggressive. Capsular invasion cannot be evaluated in aspiration biopsy smears. It has recently been demonstrated that DNA ploidy analysis may assist in the distinction between benign and malignant thymomas[30]; we are unaware of a published attempt to apply this specific technology to aspiration biopsies of these tumors. By contrast, thymic carcinomas are composed of cytologically malignant cellular elements[28] (Fig. 2-7). Although many of these may not show any specific form of epithelial differentiation, others appear as squamous cell carcinoma, adenocarcinoma, lymphoepithelioma-like carcinoma, small cell carcinoma, and sarcomatoid carcinoma. In most of these, the malignant nuclei will be larger and much more irregular in contour than those of

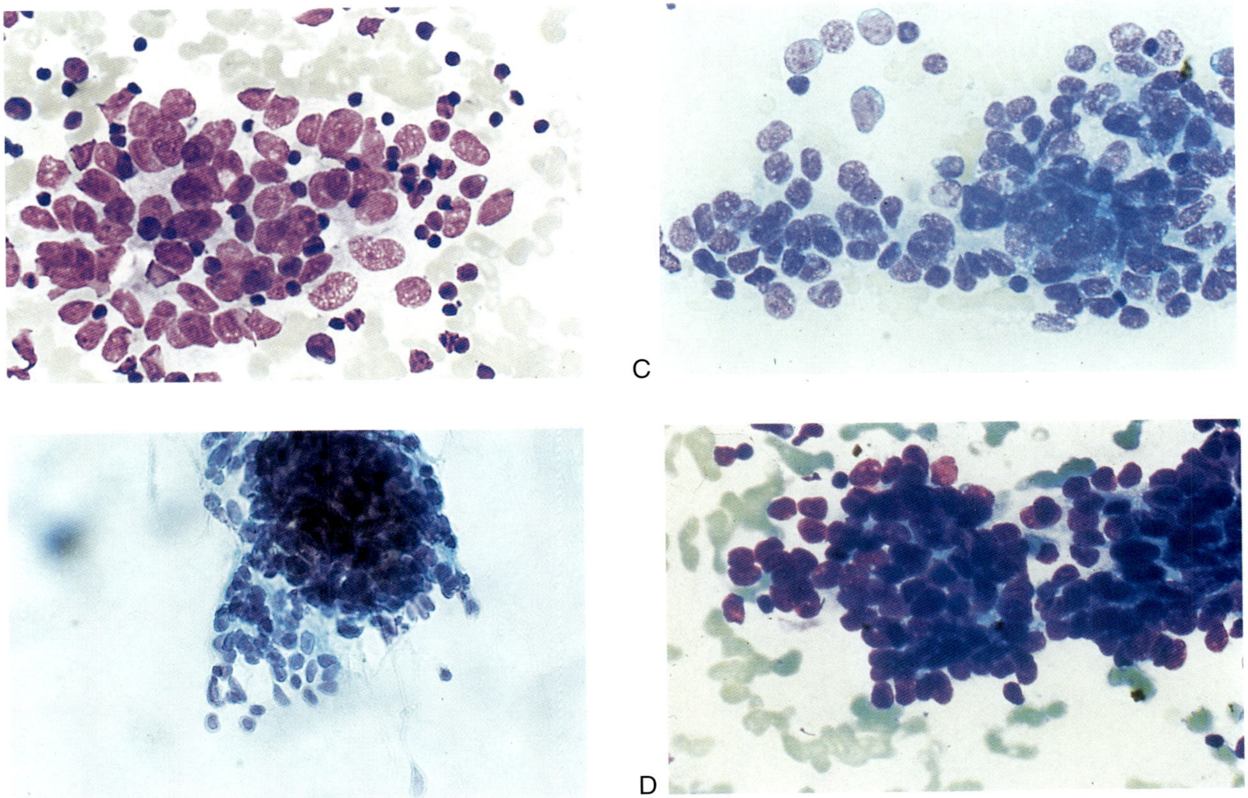

A C B D

Fig. 2-4. (**A–D**) In a given thymoma, the neoplastic cells are usually quite uniform among themselves. They possess solitary round to ovoid nuclei with thin nuclear membranes, evenly distributed pale stained chromatin, and small nucleoli. Typically, they do not manifest the features of obvious malignancy. (**C & D**) The thymoma cells have round nuclei with very high nuclear to cytoplasmic ratios. Superficially, this could cause confusion with a metastatic oat cell carcinoma. However, on closer inspection, the nuclear contours are smooth and round and the chromatin is finely reticulated (**A, C, D**, Diff-Quik; **B**, Papanicolaou).

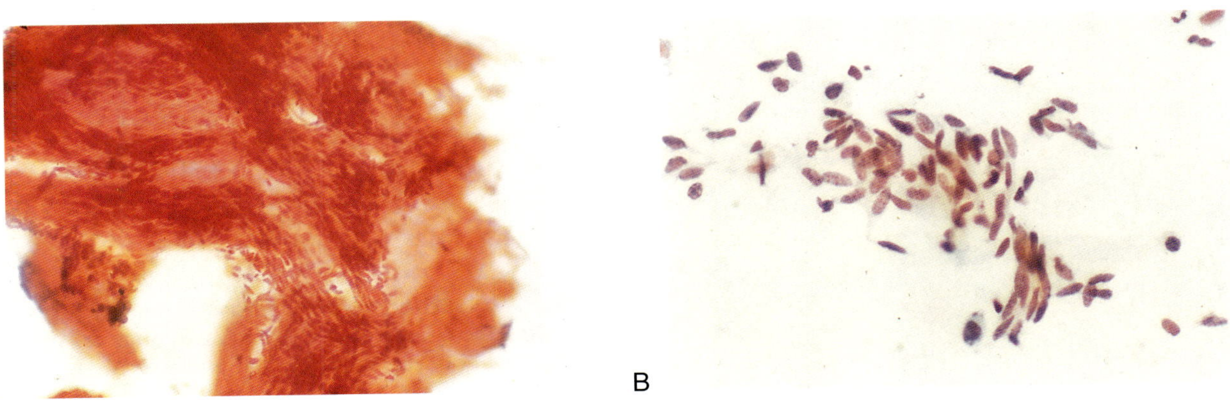

A B

Fig. 2-5. (**A & B**) In the scanning lens view (**A**), this spindle cell thymoma is characterized by interlacing fascicles of elongated tumor cells. Intercellular cohesion is well preserved. There is a distinct absence of associated lymphocytes. The chromatin is very finely granular and uniformly distributed throughout the elongated nuclei. Nucleoli are not apparent, and cytoplasm is inconspicuous (Papanicolaou). (Case courtesy of Dr. Celeste Powers, Syracuse, NY.)

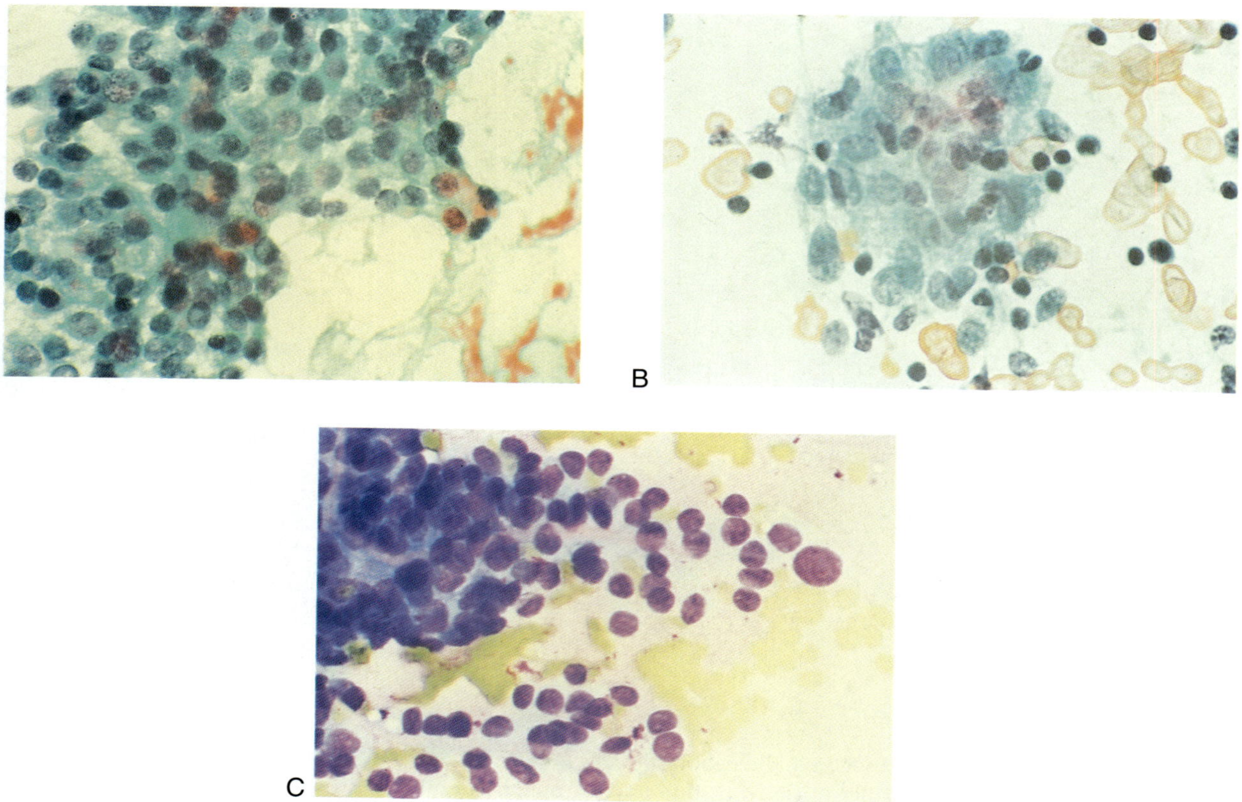

A B C

Fig. 2-6. (**A–C**) In malignant thymomas, the individual neoplastic cells are indistinguishable from those of a clinically benign thymoma. They have round or ovoid nuclei with delicate nuclear membranes, finely granular, pale staining chromatin, and small nucleoli. That is, the cells appear cytologically bland. The neoplastic cells may be intimately admixed with lymphocytes (**A & B**, Papanicolaou; **C**, Diff-Quik).

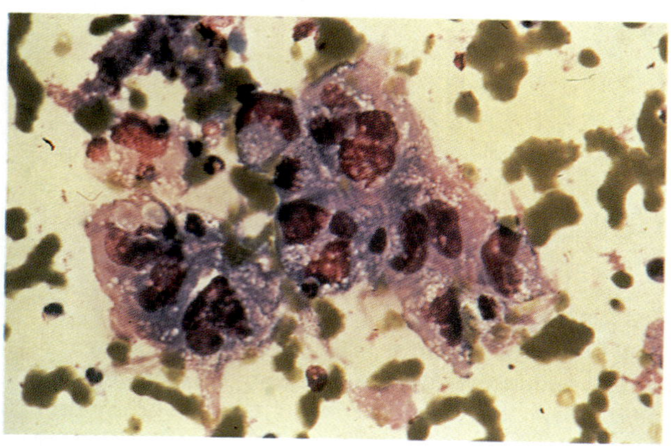

Fig. 2-7. Nuclear pleomorphism is readily apparent among these aspirated thymic carcinoma cells. Chromatin is more coarsely clumped and irregularly distributed; nucleoli are apparent in some nuclei. Cytoplasm is present in greater quantities than is typically seen in benign thymomas and shows a distinct vacuolization. The differential diagnosis is not that of a benign thymoma, but rather of a metastatic carcinoma or a germ cell neoplasm (Diff-Quik).

benign thymomas. In addition, hyperchromatic chromatin, huge nucleoli, and elevated nuclear to cytoplasmic ratios will be evident.

For most thymomas, the cytomorphologic differential diagnosis includes Hodgkin's disease, non-Hodgkin's malignant lymphomas, and germinoma.[6,31–38] Clinically, patients with thymoma tend to be older than those with these other entities. In the non-Hodgkin's malignant lymphomas, the lymphoid cells in a specimen are the malignant elements. As a result, they will appear rather monomorphic and cytologically atypical. In these other entities (e.g., Hodgkin's disease and germinoma), the vast majority of lymphocytes are small and mature-appearing. Smears of lymphomas should not include, to any extent, benign epithelial cells. Only infrequently will rare thymic epithelium, including Hassall's corpuscles, be sampled by the needle.

True intercellular cohesion among the neoplastic cells will be essentially nonexistent in both Hodgkin's disease and the non-Hodgkin's lymphomas and generally minimal in germinoma.[31–38] This same lack of cohesion may exist in aspirates of thymomas, but in general some degree of intercellular cohesion will be evident within the smears.[26] The distinction of Reed-Sternberg cells from thymoma cells should not be difficult.[31,32] The former have much larger and more irregular nuclei with prominent nucleoli and are usually greatly outnumbered by the associated benign lymphoid cells. In germinomas, the neoplastic nuclei will have much more prominent nucleoli than are typically evident in thymoma cells.[37,38]

Immunocytochemistry may be helpful in distinguishing thymoma cells from these other entities.[39,40] Only thymomas (Fig. 2-8) should possess neoplastic cells which are positive for epithelial markers (Table 2-2). The neoplastic cells of lymphomas and germinomas are generally negative for cytokeratin.[6] Ultrastructural examination of aspirated cells may also be quite useful in this distinction.[7,8,41] Well-developed desmosomes and bundles of cytoplasmic tonofilaments will be evident in aspirated thymoma cells. Although intercellular junctions may also be seen in germinoma, they will be far fewer in number and much more primitive in appearance. In addition, glycogen, and not tonofilaments, will predominate in the cytoplasm. Lymphocytes, both benign and malignant, will not manifest junctions or tonofilaments.

The differential diagnosis of thymic carcinomas includes primary mediastinal germ cell tumors,

Cytomorphologic Features of Hodgkin's Disease

- Low to moderate cellularity
- Polymorphous lymphoid cells
- Small lymphocytes predominate
- Classic Reed-Sternberg cells, mononuclear variants, polyploid cells in low numbers

Cytomorphologic Features of Lymphoblastic Lymphoma

- High cellularity
- Uniform noncohesive blasts twice the diameter of small lymphocytes
- Nuclei: variable contours, often convoluted; fine chromatin; inconspicuous nucleoli
- Very high nuclear to cytoplasmic ratio
- Lymphoglandular bodies

Cytomorphologic Features of Large Cell Lymphoma

- High cellularity
- Noncohesive uniform to variably sized large lymphocytes
- Variable nuclear contours
- Prominent nucleoli
- Variable nuclear to cytoplasmic ratio
- Lymphoglandular bodies
- Fibrosis

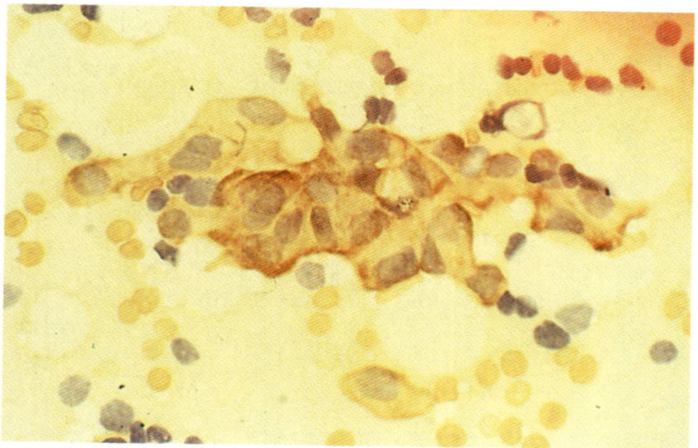

Fig. 2-8. In this direct smear preparation, thymoma cells are uniformly positive for cytokeratin.

Table 2-2. Differential Diagnoses of Mediastinal Aspirates Containing Lymphocytes

	Thymoma	*Non-Hodgkin's Lymphoma*	*Hodgkin's Disease*	*Germinoma*
Age group	Middle-aged adults	Children and young adults	Children and young adults	Young adults
Lymphoid cells	Small, mature	Atypical, monomorphic	Small, mature	Small, mature
Tumor cells	Bland nuclei	Atypical, monomorphic	Huge, sparse	Large
Ancillary studies	CK+ ICJ+ Tonofilaments	LCA+ LCL:B cell LL:T cell	$LeuM_1$+ Ki-1+ LCA–	CK- Primitive ICJ

Abbreviations: CK, cytokeratin; ICJ, intercellular junctions; LCA, leukocyte common antigen; LCL, large cell lymphoma; LL, lymphoblastic lymphoma.

especially embryonal carcinoma, and metastatic carcinomas, most frequently from the lung.[5,42] Based purely on cytomorphologic grounds, it may be impossible to distinguish these entities. Furthermore, electron microscopy and immunocytochemistry may not be helpful. In all these entities, the neoplastic cells should be immunocytochemically positive for cytokeratin and ultrastructurally demonstrate epithelial differentiation. In embryonal carcinoma, the malignant cells may be positive for α-fetoprotein and placental alkaline phosphatase, which does support a germ cell tumor. In this situation, clinical information can be of paramount importance.

MALIGNANT LYMPHOMAS

Although thymomas are the most common tumor in the anterosuperior mediastinum, in our experience, malignant lymphomas are seen more often in aspiration biopsies from this site.[5] Clinically, patients with malignant lymphomas may present with symptoms related to the effects of a mass lesion or invasion of adjacent structures.[43–46] Many patients, however, are asymptomatic at the time of discovery of the mass on routine chest radiography. The vast majority of all primary mediastinal lymphomas are accounted for by three major forms. Both Hodgkin's disease and the large cell variant of non-Hodgkin's lymphomas most often affect young adult women. By contrast, teenage boys are the group most often involved by lymphoblastic lymphoma. As both prognosis and therapy differ, the distinction among these three neoplasms is crucial. These tumors are the primary neoplasms most likely to be treated solely on the basis of an FNA biopsy diagnosis without a tissue confirmation. Accordingly, accurate cytologic diagnosis and classification are essential.

Hodgkin's Disease

According to Strickler and Kurtin,[46] Hodgkin's disease is the most frequent primary mediastinal lymphoma. The vast majority of these are of the nodular sclerosing variant. Not infrequently, supraclavicular lymphadenopathy and/or supradiaphragmatic serous effusions may be associated with the mediastinal mass. The overall prognosis is excellent with greater than 90 percent of all patients attaining a 5-year sfurvival.[46]

Frequently, the aspirates are not highly cellular, presumably reflecting the prominent fibrosis in the nodular sclerosing variant[31,32] (Fig. 2-9). The smears will be composed of polymorphous lymphoid populations with small mature-appearing lymphocytes predominating numerically, as shown in the box on page 59. The cytologist may need to search carefully for Reed-Sternberg cells and their cytologic variants to make a specific diagnosis of Hodgkin's disease. Classically, the Reed-Sternberg cells possess two large mirror-image nuclei with huge nucleoli (Fig. 2-10). In most cases, however, mononuclear forms are more numerous (Figs. 2-9 and 2-11). Polyploid cells are common; as described by Kardos et al.,[31] these consist of large, highly lobulated, dark-stained nuclei without any visible attached cytoplasm (Fig. 2-12). The contours of both nuclei and nucleoli are often quite irregular (Fig. 2-13). Compared to adjacent small lymphocytes, the chromatin of the neoplastic cells is much more open or immature in appearance (Fig. 2-14). Although the Reed-Sternberg cells may be present in relatively low numbers in the smears, the cells are generally easy to identify because of their large sizes, especially in air-dried preparations. At times, gigantic tumor cells are present (Fig. 2-15).

The only other primary mediastinal cancer that may yield cells of equivalent sizes is a malignant germ cell tumor with a choriocarcinomatous component.[38] In the latter situation, the malignant syncytiotrophoblast usually possess more than two nuclei, have abundant, often vacuolated cytoplasm, and are positive for human chorionic

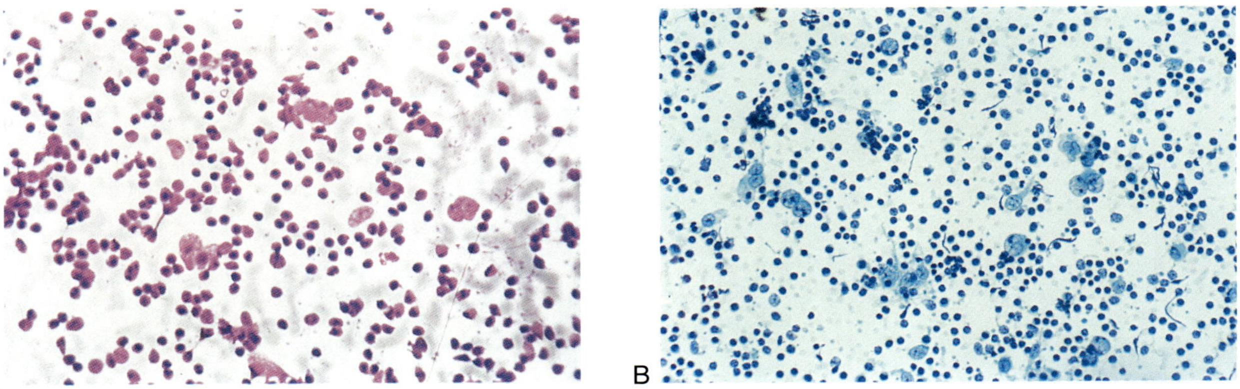

A B

Fig. 2-9. (**A & B**) These scanning lens views demonstrate the moderate level of cellularity that typifies aspirates of most examples of Hodgkin's disease. The predominate cell present is the small mature lymphocyte. Scattered throughout the smear are individually dispersed large neoplastic Reed-Sternberg cells. (**A**) A solitary classic binucleated Reed-Sternberg cell is present as are several mononuclear variants (Diff-Quik). (**B**) Both binucleated and mononucleated forms are present. Although their chromatin is relatively pale staining, the cells readily stand out due to their very large sizes (Papanicolaou).

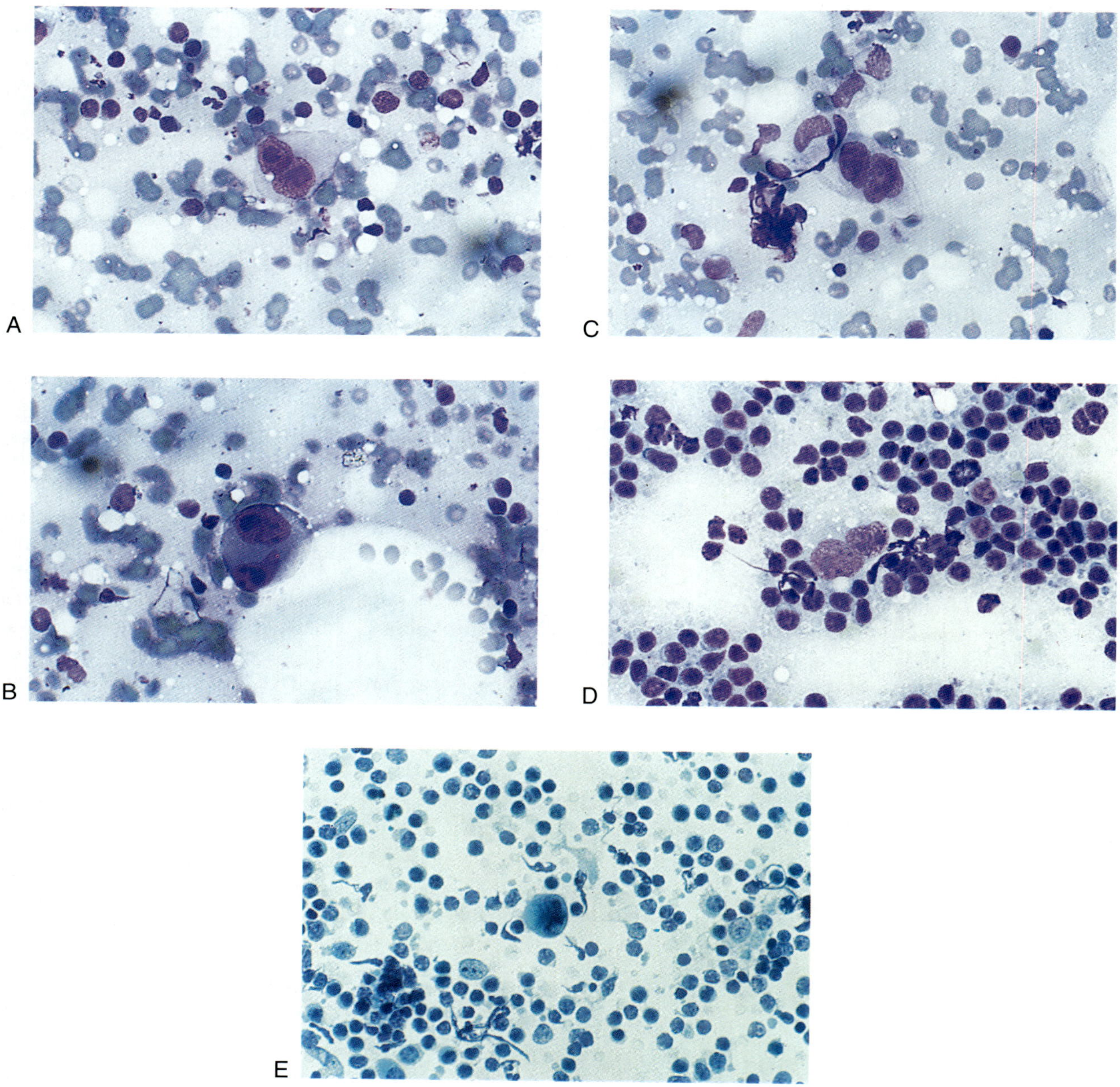

Fig. 2-10. (**A–E**) Classic Reed-Sternberg cells are characterized by two large nuclei with prominent to massive nucleoli. In most of these examples, the tumor cell cytoplasm is well preserved. However, it often appears to have been stripped away (**D**), leaving two closely associated huge naked nuclei (**A–D**, Diff-Quik; **E**, Papanicolaou).

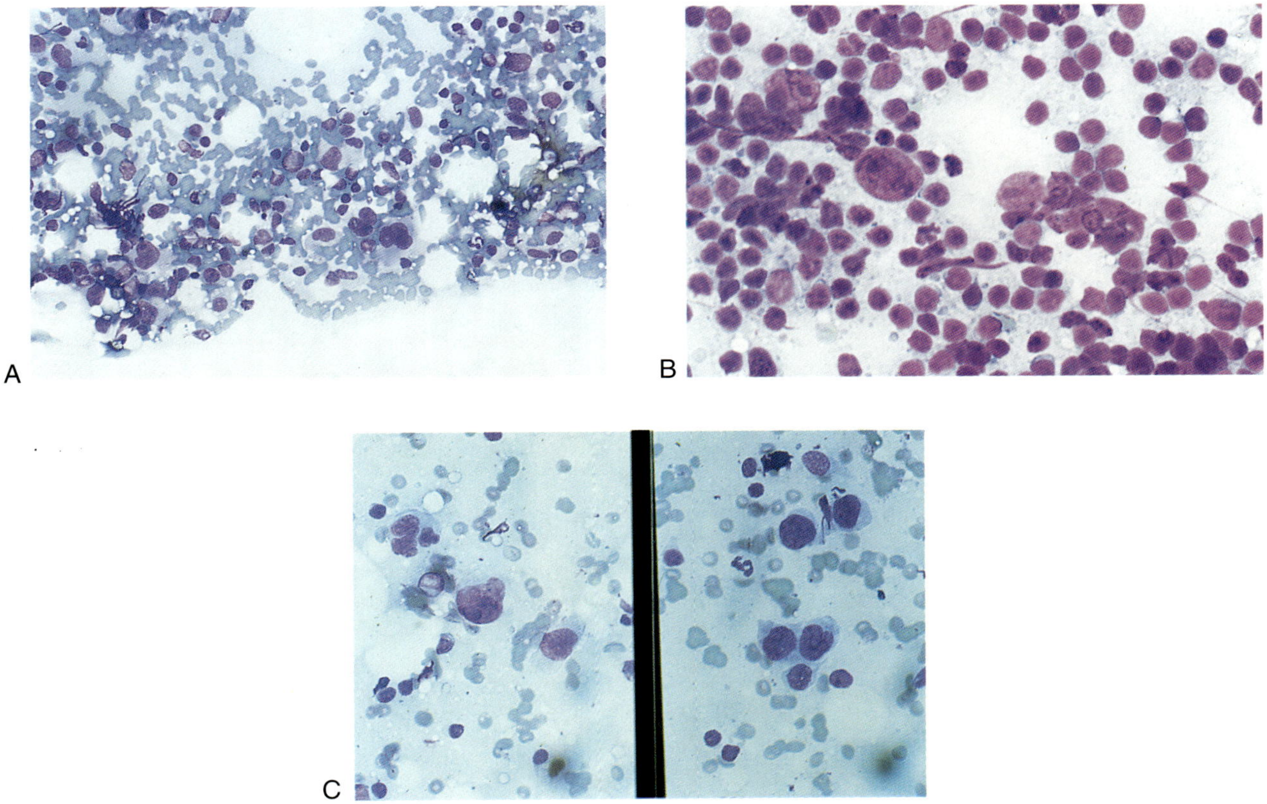

Fig. 2-11. (A–C) Compared to classic Reed-Sternberg cells, the mononuclear variants are usually present in larger numbers in aspiration smears. Similar to the classic forms, they have large irregularly contoured nuclei with prominent pleomorphic nucleoli. (A) Benign lymphocytes are rather sparse. (B) Benign lymphocytes manifest the usual predominance (Diff-Quik).

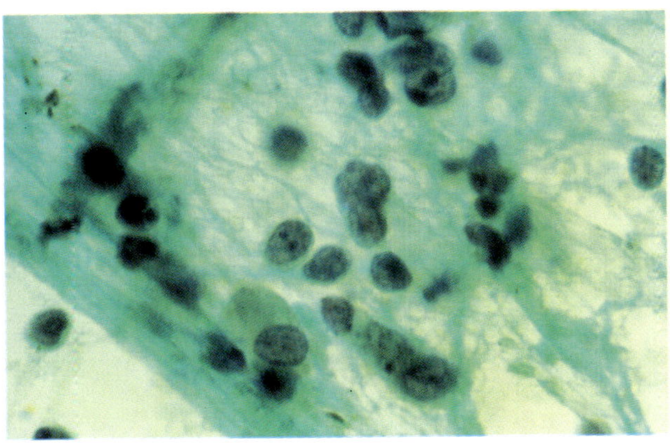

Fig. 2-12. Polyploid cells are characterized by darkly stained lobulated nuclei which appear to have been totally stripped of their cytoplasm. Although alcohol fixation reduces the sizes of these neoplastic cells, they still stand out compared to the adjacent small benign lymphocytes (Papanicolaou).

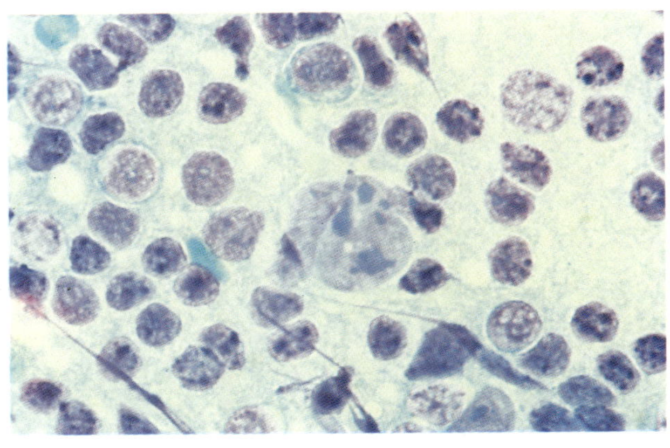

Fig. 2-13. The outlines of the neoplastic nuclei and nucleoli of Reed-Sternberg cells are uneven and jagged. This is in contrast to the smooth round or ovoid shapes that one expects in benign immunoblasts, which might be mistaken for Reed-Sternberg cells (Diff-Quik).

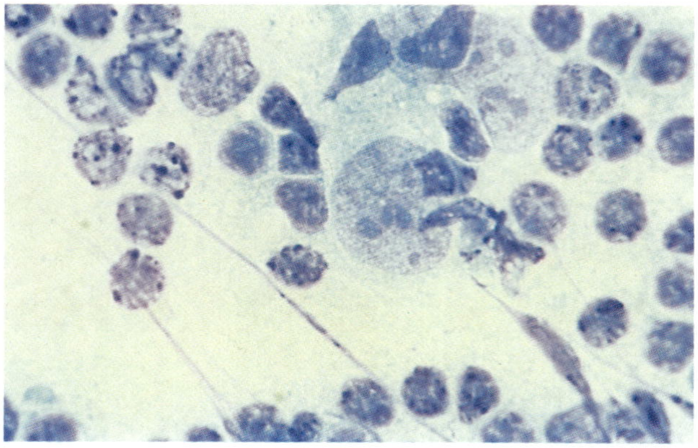

Fig. 2-14. Reed-Sternberg cells have very finely reticulated or granular (immature) chromatin, which contrasts sharply with the coarsely granular and blocky chromatin seen in small mature-appearing lymphoid cells. In this example, much of the cytoplasm appears to have been stripped away (Diff-Quik).

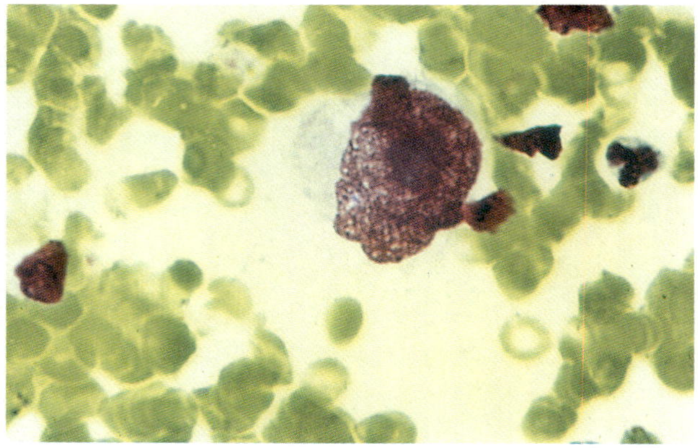

Fig. 2-15. Massive Reed-Sternberg cells and variants may be present in the smears. Except for metastatic giant cell carcinoma and choriocarcinoma, such huge cellular elements would not be expected in an aspirate of the mediastinum (Diff-Quik).

gonadotropin (HCG) by immunocytochemistry. Furthermore, the cells will be admixed with smaller mononucleated cytotrophoblasts. Giant cell carcinoma of the lung metastatic to mediastinal lymph nodes may also yield huge neoplastic cells.[5] Clinically, giant cell carcinoma typically occurs in older patients with a long history of cigarette smoking and a lung mass. As described in Chapter 1, pleomorphism will exceed anything seen in aspiration smears of Hodgkin's disease. In addition, these malignant cells will be positive for cytokeratin. Reed-Sternberg cells may be positive immunocytochemically for $LeuM_1$ and Ki-1, but some metastatic carcinomas may also be decorated by the former reaction (Fig. 2-16). Certain anaplastic lymphomas are also Ki-1 positive but, unlike Reed-Sternberg cells, are also decorated by leukocyte common antigen (LCA).

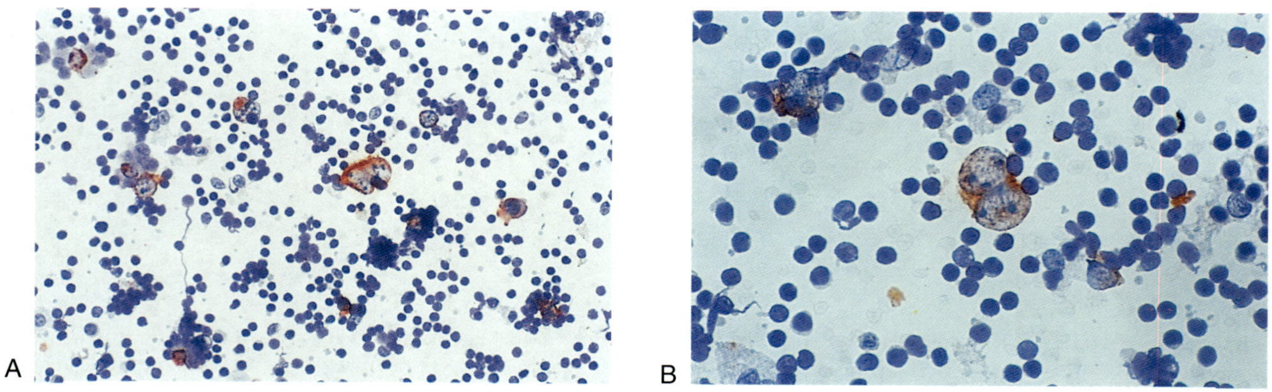

Fig. 2-16. (**A & B**) In this example of aspirated Hodgkin's disease, all the neoplastic cells are intensely immunoreactive for $LeuM_1$. The stain is confined to their cytoplasm. The small mature lymphocytes are nonreactive.

Non-Hodgkin's Malignant Lymphomas

Lymphoblastic lymphoma is a common cause of a mediastinal mass in children and young adults.[33,34,46,47] This histologically high-grade cancer may comprise nearly one-half of all childhood non-Hodgkin's lymphomas. It is predominantly a disease of adolescents and young adults and involves males much more often than it does females. Supердiaphragmatic lymphadenopathy may also be evident. Immunophenotypically, most of these tumors are T cell in type and are positive for terminal deoxynucleotidyl transferase (tdT). Lymphoblastic lymphoma of the mediastinum and T-cell acute lymphoblastic lymphoma are the same disease in which different clinical manifestations predominate. Not infrequently, lymphoblastic lymphoma may progress to acute leukemia. Prognosis is guarded with about a 55 percent 3-year survival rate.[46] Histologically, the thymus or lymph nodes, or both, are effaced by a diffuse growth of small, relatively uniform, malignant lymphocytes with very high nuclear to cytoplasmic ratios.

Aspiration smears are usually highly cellular and composed of dispersed monotonous-appearing lymphoblasts[32,34,36] (Fig 2-17). There should be no evidence of true intercellular cohesion, as shown in the box on page 59. Lymphoglandular bodies will be numerous in the smear background.[48] The malignant cells have diameters about twice that of small, mature-appearing lymphocytes. Nuclear contours may be smooth and round, but typically are quite irregular with prominent convolutions and indentations (Fig. 2-18). Chromatin is very finely reticulated and nucleoli are inconspicuous (Figs. 2-17 to 2-19). Only a very thin rim of cytoplasm surrounds each nucleus yielding mononucleated tumor cells with very high nuclear to cytoplasmic ratios. The smear may also contain relatively small numbers of tingible body

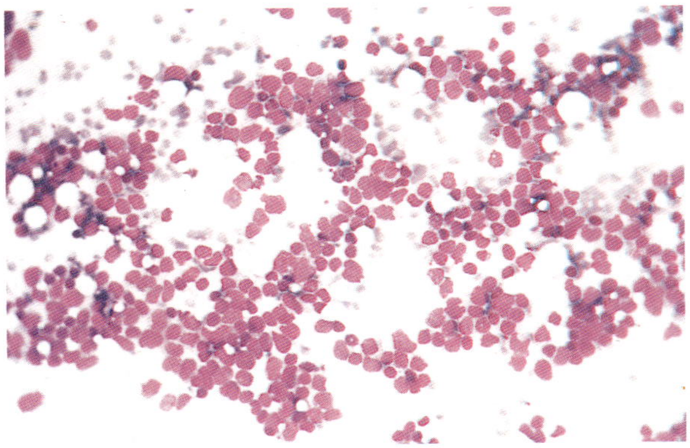

A

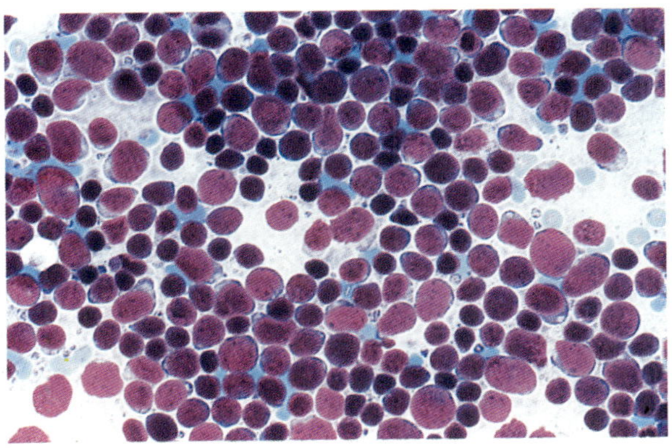

B

Fig. 2-17. (**A & B**) Smears of lymphoblastic lymphoma are moderately to highly cellular and consist of individually dispersed rather homogeneous-appearing small lymphoid cells. Although the cells may be closely packed within smears, there is no evidence of intercellular cohesion. Lymphoglandular bodies will be present in the background (Diff-Quik).

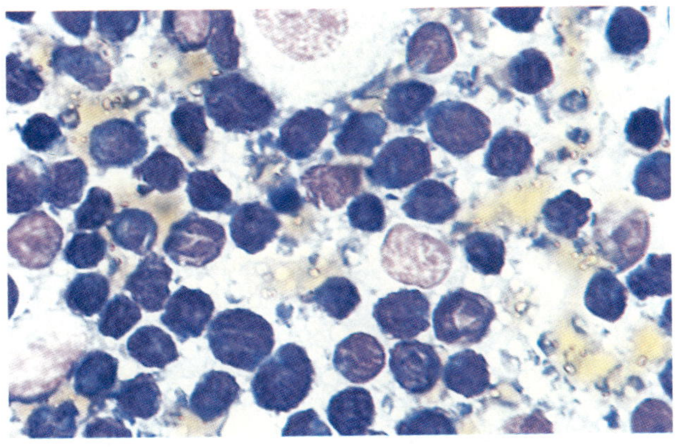

Fig. 2-18. Although the nuclei may appear round and smooth, most often the malignant lymphoblasts have highly contorted nuclei. In this example, nuclear indentations, cleavage points, and convolutions are seen. Note the very high nuclear to cytoplasmic ratios and the lack of nucleoli. Compared to the adjacent histiocytes, the tumor cells are quite small (Diff-Quik).

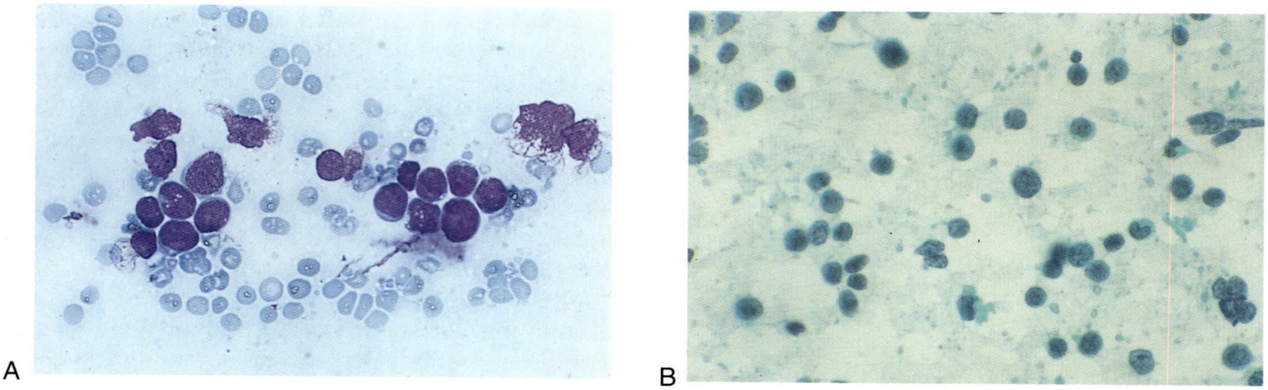

Fig. 2-19. (**A & B**) In lymphoblastic lymphoma, the tumor cells are quite uniform. (**A**) Each has a single irregular to round nucleus with finely reticulated chromatin and only a scanty rim of cytoplasm (Diff-Quik). (**B**) The chromatin can be seen to be very finely granular and uniformly distributed (Papanicolaou). Although irregular in shape, the nuclear membranes are rather delicate. Nucleoli are not apparent. Note the lymphoglandular bodies in the background.

macrophages. These histiocytic cells are characterized by larger ovoid nuclei and abundant cytoplasm that contains phagocytized cellular debris. Aliquots of aspirated cellular material may be evaluated to confirm the T-cell nature of the malignant cells, either by flow cytometry or immunocytochemistry (Table 2-2). Using either flow cytometry or immunocytochemistry, the nuclear localization of the tdT antigen can also be detected. In addition, T-cell receptor gene rearrangements can be determined on aspirated material.[36] Although lymphoblastic lymphomas may have a precursor B-cell immunophenotype, it is rare for this immunologic form to result in a mediastinal mass.

The cytomorphologic differential diagnosis of lymphoblastic lymphomas includes other small cell malignancies and benign lymphoid proliferations. Of the cancers, only lymphoblastic lymphoma should show a complete lack of intercellular cohesion and prominent lymphoglandular bodies. On a cell-for-cell basis, the nuclear contours will tend to be much more irregular in lymphoma than in any other malignancies as well. The other small cell malignancy of childhood that occurs with some frequency as a primary mediastinal lesion is neuroblastoma. Evidence of intercellular cohesion, nuclear molding, and rosette formation will be seen in the smears, as will the characteristic neuropil.[49] In addition, one may see evidence of neuroblastic maturation and ganglion cell differentiation in these specimens. The tumor cells of lymphoblastic lymphoma will be positive for LCA, tdT, and T-cell surface markers. These reactions should all be negative in neuroblastoma, which, by contrast, should be positive for neuron-specific enolase (NSE). In our experience, the single most frequent neoplasm sampled by FNA biopsy in the mediastinum is metastatic small cell undifferentiated carcinoma of bronchogenic origin.[5] The cytomorphologic features of this cancer have been well detailed in Chapter 1. Again, in contrast to lymphoblastic lymphoma, intercellular cohesion, including characteristic nuclear molding, will be evident. Although nuclear contours may appear angulated, the marked irregularities in lymphoblastic lymphoma are unexpected in small cell carcinoma. In addition, small cell carcinoma almost always occurs in older individuals with a long history of cigarette smoking. Although fleetingly rare, small cell carcinomas primary in the thymus may also occur in children.[15] We have never seen one of these in an FNA biopsy, but we would suspect that it would be indistinguishable cytomorphologically from the bronchogenic form.

Two specific forms of lymphoid hyperplasia need to be mentioned. The first of these is Castleman's disease, a peculiar benign lymphoid proliferation.[46] Aspiration smears contain numerous, generally small and rather mature-appearing lymphoid elements. It is doubtful that a specific diagnosis of Castleman's disease can be rendered solely on the basis of FNA cytomorphology. The second entity that deserves mentioning is thymic follicular hyperplasia.[8] This process is most commonly associated with myasthenia gravis. Usually, the thymus is not enlarged; thus an FNA biopsy would be unlikely, and we are unaware of a reported example. Histologically, normal-appearing thymus is encased by a florid proliferation of lymphoid follicles with prominent germinal centers.

These two benign lymphoid proliferations would also be in the differential diagnosis of large cell non-Hodgkin's malignant lymphomas. This type of lymphoma typically presents in young adult patients, often with a female predominance, as a large mass which may be associated with a cytologically positive effusion.[43–46] Large cell lymphomas comprise a morphologically heterogeneous group of neoplasms. The most common type is composed of large, noncleaved lymphocytes. Histologically, sclerosis may be seen throughout the entire neoplastic mass and may be so extensive as to mask the neoplastic cells. These neoplasms can be considered relatively aggressive clinically; the prognosis is guarded, with a 5-year survival rate of approximately 60 percent.[46]

Aspiration smears are usually quite cellular with numerous noncohesive lymphoid cells which are at least three times the diameter of a small lymphocyte[35,36] (Fig. 2-20). In most cases, the neoplastic elements are fairly uniform among themselves, but examples with extreme pleomorphism may be witnessed (see box, p. 59). Nuclear contours are quite variable; some tumor cells have almost perfectly round nuclei but, typically, nuclear irregularities will be well developed (Fig. 2-21). With the Papanicolaou stain, nuclei will demonstrate distinct and irregular contours, vesicular chromatin, and one or more prominent nucleoli. The latter may range from small and multiple to a single huge centrally positioned nucleolus. In Diff-Quik–stained smears, nuclear membranes are less distinctive with a tendency to have smoother outlines. Cytoplasm varies in both quantity and quality. With the Diff-Quik stain, it often is intensely basophilic and may contain a few small vacuoles (Fig. 2-22). Although variable, nuclear to cytoplasmic ratios are generally high. Lymphoglandular bodies will be numerous in the smear background. The malignant cells are positive for LCA (Fig. 2-23); most mediastinal large cell lymphomas are immunophenotypically of the B-cell type.

A

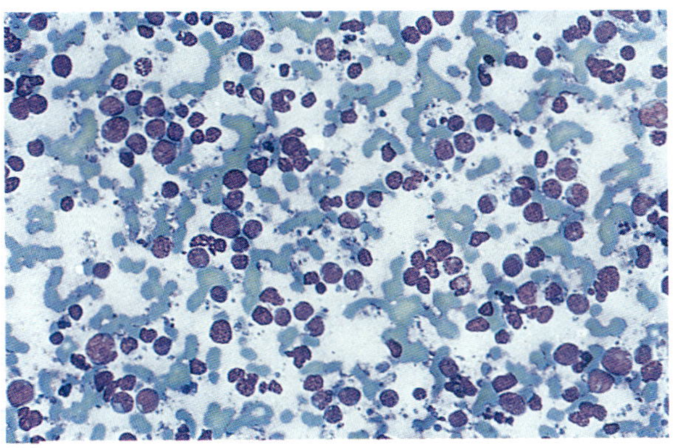

B

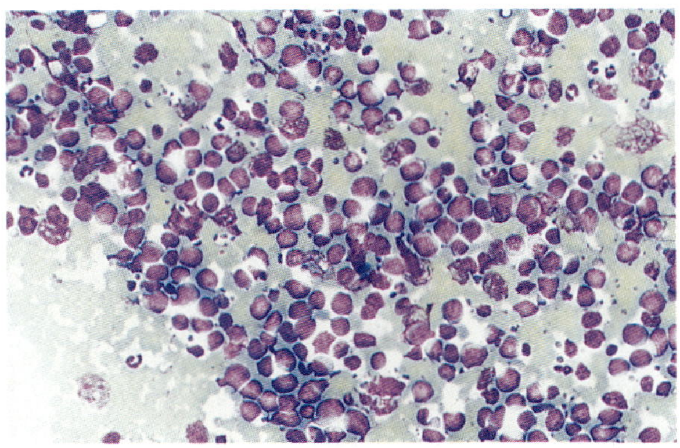

Fig. 2-20. (**A & B**) In large cell lymphomas not associated with prominent sclerosis, aspiration smears are usually highly cellular. Numerous individually dispersed and very loosely aggregated large malignant cells occupy the smear. In most examples, these tumor cells are rather uniform. Lymphoglandular bodies will be numerous, but when there is abundant blood in the background (**B**), these bodies may not be easily recognizable (**A & B**, Diff-Quik).

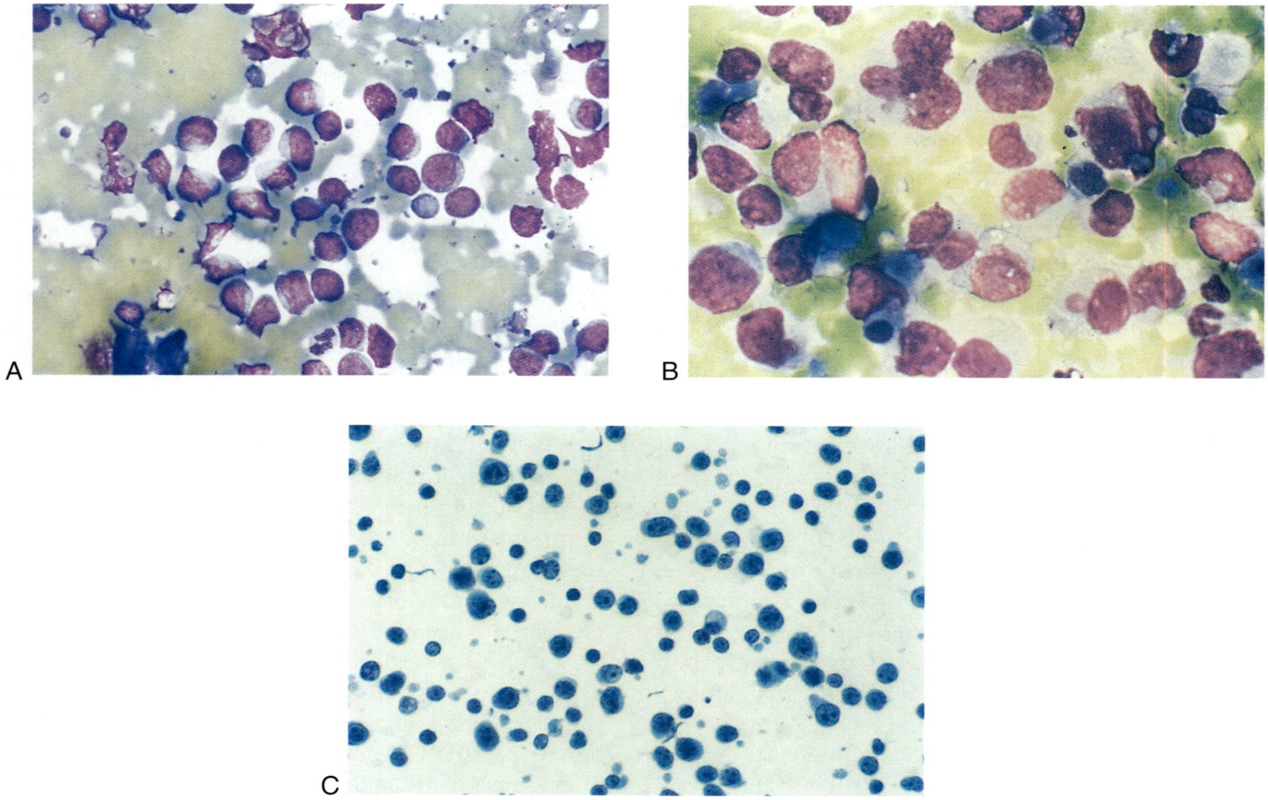

Fig. 2-21. (**A–C**) Nuclear contours among the malignant cells vary from round and smooth to quite contorted. Nuclear membrane irregularities are generally more evident with the Papanicolaou stain. Chromatin is immature in appearance and evenly distributed. Nucleoli are generally present, but in some examples they may be inconspicuous. For the most part, the nuclear to cytoplasmic ratios are very high. Note the complete absence of intercellular cohesion (**A & B**, Diff-Quik; **C**, Papanicolaou).

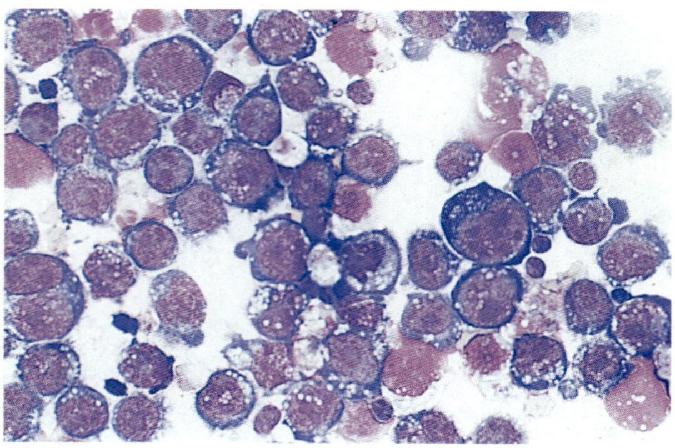

Fig. 2-22. In this large cell lymphoma, cytoplasm is scanty to moderate in volume, intensely basophilic, and often demonstrates multiple minute vacuoles. Nucleoli are also well developed in this example (Diff-Quik).

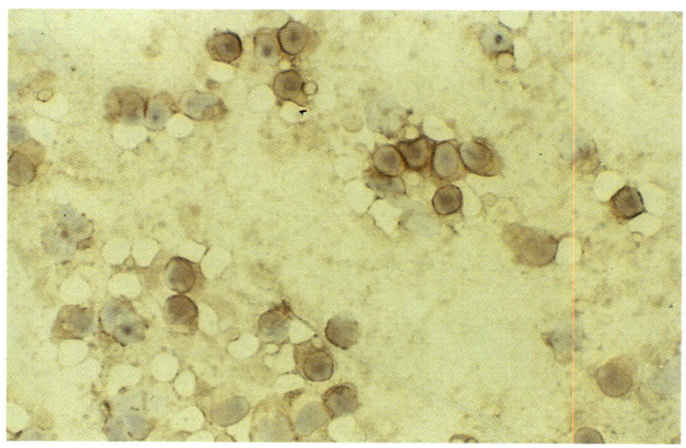

Fig. 2-23. The cytoplasm of the tumor cells are immunoreactive for leukocyte common antigen.

Although the aspiration biopsy diagnosis of large cell malignant lymphoma is generally straightforward, several potential diagnostic pitfalls need to be recognized. In the immunoblastic form of non-Hodgkin's lymphoma, the malignant cells may be quite large with huge nuclei and massively enlarged nucleoli. This could be cause for confusion with Reed-Sternberg cells, and consequently with the diagnosis of Hodgkin's disease. However, the number of tumor cells in aspirates of non-Hodgkin's lymphomas usually greatly exceeds the number of Reed-Sternberg cells sampled in mediastinal Hodgkin's disease. Furthermore, most of the lymphoid elements in the latter specimens are small and mature-appearing. Reed-Sternberg cells typically are positive for $LeuM_1$ and Ki-1 and are often nonreactive with LCA.

At times, areas of pseudocohesion may be seen in the smears of non-Hodgkin's lymphomas (Fig. 2-24). This refers to aggregates of lymphoid cells in clumps and small sheet-like arrangements that could lead to confusion with epithelial neoplasms, especially metastatic small cell carcinoma or thymic carcinoma. Although the cytoplasm of lymphomatous elements may contain small vacuoles, the intense basophilia so characteristic of the malignant lymphomas would be unusual in carcinoma. Furthermore, the aggregated neoplastic lymphoid cells usually form only a minority of all malignant elements. The usual picture of dispersed malignant cells and lymphoglandular bodies is found adjacent to foci of such pseudocohesion. The carcinomas will be positive for cytokeratin and/or neuron-specific enolase (NSE), which will never be seen in lymphoma. Epithelial differentiation will also not be recognized ultrastructurally.

As stated above, histologically, many large cell lymphomas of the mediastinum are associated with a prominent fibrotic component. This can cause diagnostic dilemmas. In some cases, aspirations may yield mostly minute fragments of collagen-containing tissue, rather than a predominant dissociated pattern of malignant lymphoid cells (Fig. 2-25). This fibrosis may also distort the appearance of the lymphoma cells, so that they have an elongated and somewhat more hyperchromatic-appearing nucleus[35] (Fig. 2-26). In this setting, the cytomorphologic differential diagnosis would include neoplasms composed predominantly of spindle-shaped cells (Fig. 2-27). Noteworthy among these are primary and metastatic soft tissue sarcomas, benign peripheral nerve sheath tumors, and the spindle cell variant of thymoma. This appearance has also been confused with metastatic melanoma.[35] Among these different

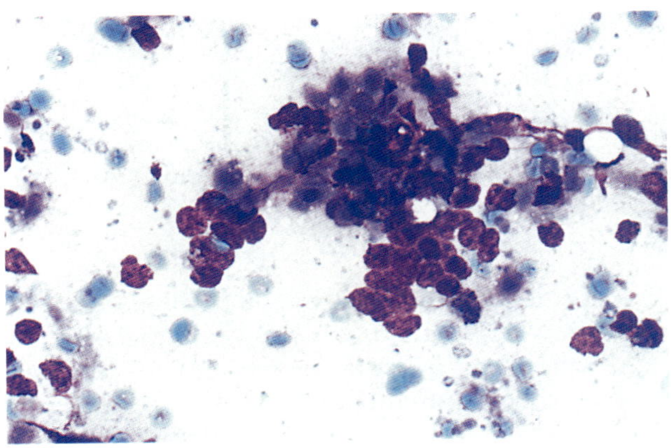

Fig. 2-24. An aggregated clump of tumor cells is present in this aspirate of a mediastinal large cell lymphoma. This is an example of pseudocohesion. The entire smear needs to be examined; most of the specimen showed the individually dispersed lymphoid elements with numerous lymphoglandular bodies (Diff-Quik).

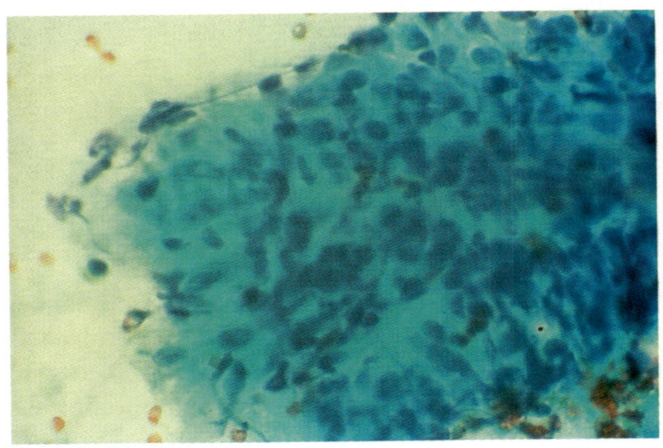

Fig. 2-25. Aspirate of a mediastinal large cell lymphoma with sclerosis. The specimen aspirated as chunks of collagen containing the malignant cells. Although some of the cells are round or ovoid, many of them have elongated or spindled contours (Papanicolaou).

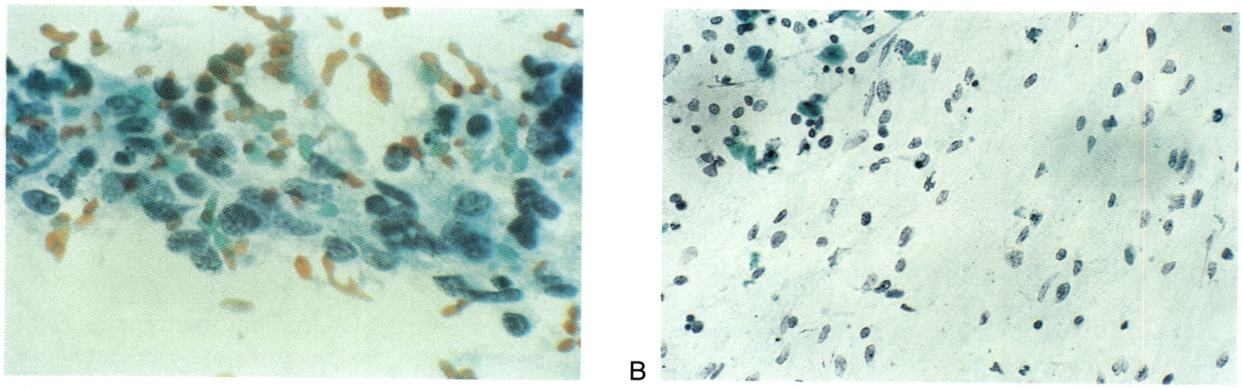

Fig. 2-26. (**A & B**) Many of the malignant lymphocytes in the mediastinal lymphoma have very elongated nuclear contours with indistinct cytoplasm. This creates a resemblance to an aspirate of a spindle cell sarcoma. (**B**) Most of the malignant lymphocytes have very elongated nuclear and cellular contours without being embedded within a collagenous matrix. This could be mistaken for an aspirate of a mesenchymal or nerve sheath neoplasm (**A & B**, Papanicolaou).

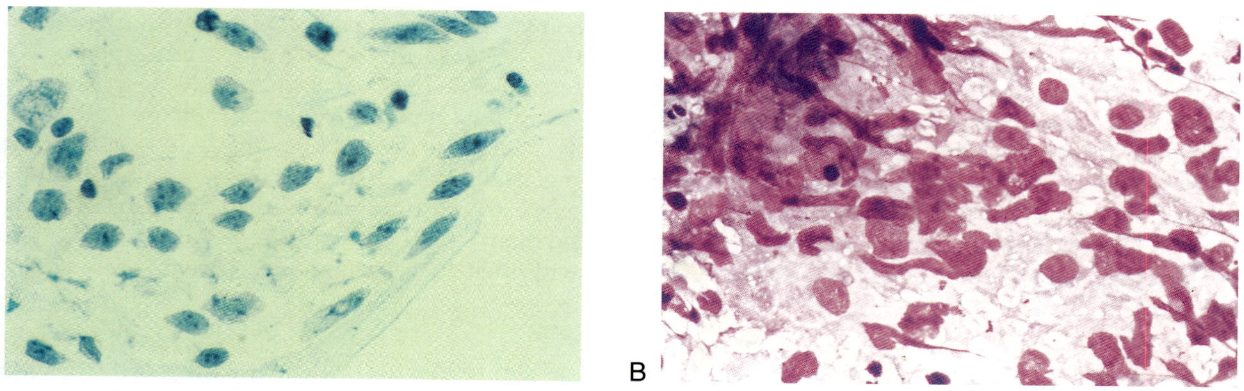

Fig. 2-27. (**A & B**) These aspirates are from two mediastinal large cell lymphomas which were not associated with well-developed sclerosis. A number of the lymphoma cells have rather elongated nuclear and cellular contours. Presumably, this resulted from an artifact induced during preparation of the smear. Note that many of the malignant cells have round contours and are easily recognizable as lymphoid (**A**, Papanicolaou; **B**, Diff-Quik).

considerations, only lymphoma cells should be positive for LCA. Both melanoma and peripheral nerve sheath tumors should be positive for S-100 protein, whereas cytokeratin will decorate the thymoma cells.

GERM CELL TUMORS

By far, the gonads are the most common sites of origin of germ cell neoplasms. Extragonadal germ cell tumors typically arise along the midline of the body and thus are clearly amenable to sampling by fine needle aspiration biopsy.[38] In adults, the mediastinum is the single most common site of origin of extragonadal germ cell neoplasms.[50–52] Men are affected much more frequently than women. In children, the mediastinum follows the gonads and the sacrococcygeal region as a site of origin for germ cell neoplasms. The majority of all primary mediastinal germ cell tumors are benign teratomas. In contrast to other mediastinal germ cell tumors, the male-to-female ratio for benign teratomas is approximately unity, with a relatively wide age range.

Many patients with teratomas are discovered to have an incidental asymptomatic mass on routine chest radiography. Most patients with nonteratoma malignant germ cell tumors are symptomatic due to the effects of a mass lesion and/or invasion into adjacent structures. Most patients with these neoplasms are males and most frequently are in their third decade of life. Histologically, these neoplasms fully resemble their counterparts in the ovary and testis. The most common malignant germ cell tumor is the germinoma.[50–52] However, embryonal carcinomas, yolk sac tumors, and choriocarcinomas may occur as well as mixed germ cell tumors. The prognosis for patients with these germ cell tumors depends on the specific histologic type of neoplasm and the patient's age. Regardless of age, surgical resection of a mature teratoma should be expected to be curative.[50] Immature teratomas generally act in a benign fashion in young children, but similar-appearing tumors in patients above the age of 14 years need to be considered as fully malignant.[50] Before the introduction of cisplatin combination chemotherapy, the nongerminomatous malignant germ cell tumors were basically uniformly fatal.

Germinomas generally yield moderately to highly cellular aspiration smears composed of uniform large malignant cells with relatively little maintenance of intercellular cohesion[37,38] (Fig. 2-28). Each tumor cell has a solitary large vesicular nucleus

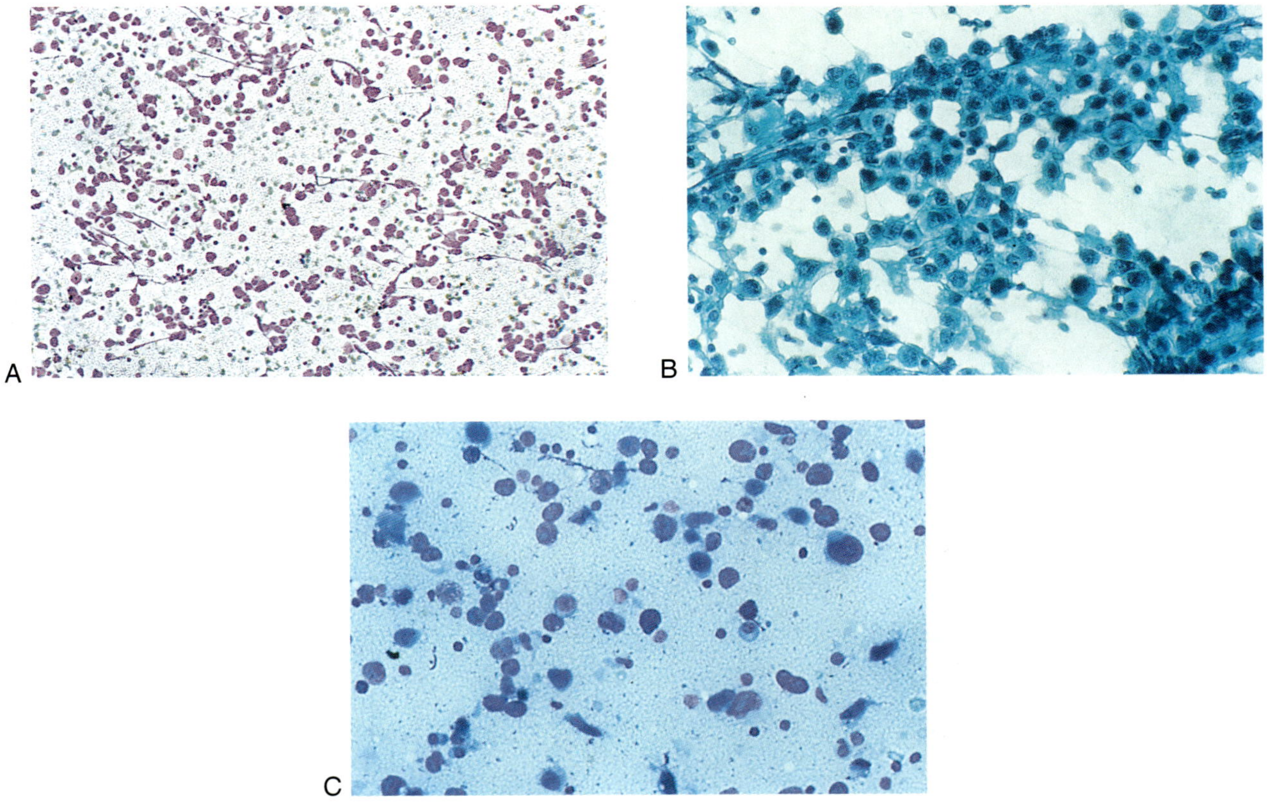

Fig. 2-28. (**A–C**) Low magnification views demonstrate the high level of cellularity in aspirates of germinomas. For the most part, intercellular cohesion is not well preserved, although in Fig. **B** loosely cohesive aggregates of cells are present. Germinoma cells are large, have a solitary vesicular nucleus, and one or more prominent nucleoli. Although the smears contain a large number of stripped nuclei, where cytoplasm remains intact, the nuclear to cytoplasmic ratios are high. Small numbers of lymphocytes are scattered among the tumor cells. (**C**) Some of these lymphocytes have a plasmacytoid appearance (**A & C**, Diff-Quik; **B**, Papanicolaou).

(Fig. 2-29) with a thick distinct membrane and one or more very prominent nucleoli, as listed on page 73. Cytoplasm may be scanty and naked nuclei are frequent as the cytoplasm is fragile and thus easily sheared away during the aspiration procedure. With the Diff-Quik stain, intact cytoplasm has a characteristic slate-gray color, whereas it appears homogeneous and green with the Papanicolaou stain (Fig. 29,30). The nuclear to cytoplasmic ratios appear high. Characteristically, the background of the smear is frothy, bubbly, or linear in appearance (Figs. 2-28, 2-29, and 2-31); this is referred to as "tigroid." Smear backgrounds often contain small, mature-appearing lymphocytes and possibly evidence of a granulomatous inflammatory reaction (epithelioid histiocytes).

Owing to the relative lack of intercellular cohesion, a major entity in the differential diagnosis is the large cell variant of non-Hodgkin's lymphoma. However, if one examines the smear carefully, evidence of true intercellular cohesion will be present, as small clusters and even sheets of tumor cells will be present in smears of germinomas. A tigroid background, rather than lymphoglandular bodies, is characteristic. A negative LCA reaction for the large tumor cells is expected in germinoma. Ultrastructurally, primitive intercellular junctions and

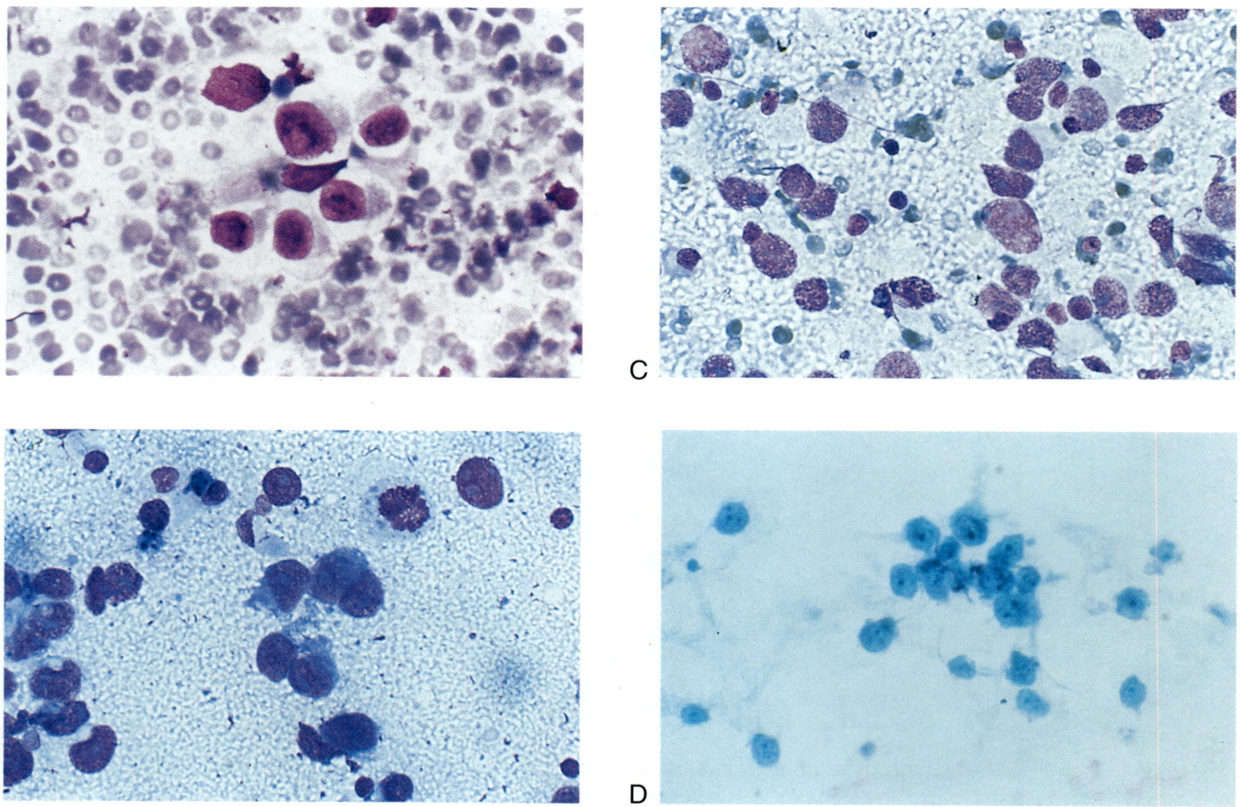

A C B D

Fig. 2-29. **(A–D)** The germinoma cells each possess a single large round nucleus with a heavy well-defined membrane, primitive chromatin, and one or more distinct nucleoli. With the Diff-Quik stain, intact cytoplasm takes on a gray coloration. With the Papanicolaou stain, it varies from faint green to optically clear. In all cases, the nuclear to cytoplasmic ratios are quite high (**A–C**, Diff-Quik; **D**, Papanicolaou).

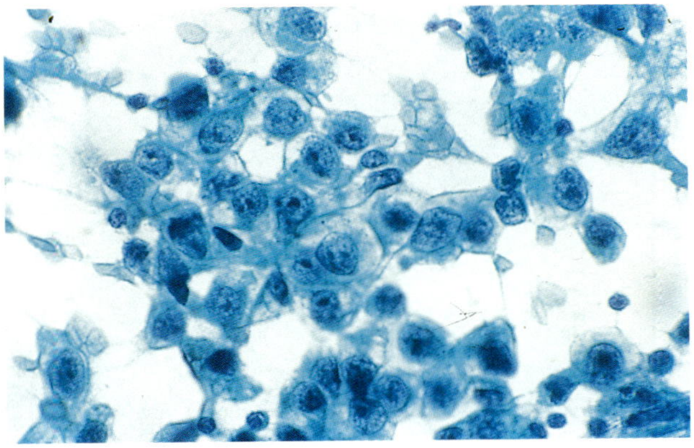

Fig. 2-30. In this field, intercellular cohesion is relatively well preserved. The intercellular borders are distinct, resulting in tumor cells with round to polygonal contours. Cytoplasm varies from scanty to moderate and is faintly cyanophilic. Each cell has a single darkly stained round nucleus with one or more prominent nucleoli. There is no evidence of an organoid arrangement of the tumor cells. A few lymphocytes are present in the background (Papanicolaou).

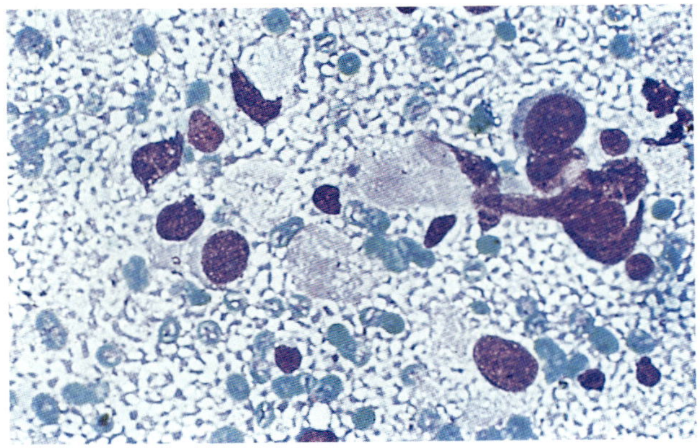

Fig. 2-31. In this air-dried smear, a well-developed tigroid background is evident. It consists of this finely reticulated lace-work of gray cytoplasm separated by clear staining areas. Presumably, this results from shearing of the fragile tumor cell cytoplasm. Such an appearance is not evident in alcohol-fixed specimens (Diff-Quik).

abundant cytoplasm glycogen are characteristic[7] (Fig. 2-31).

Because of the low levels of intercellular cohesion and the admixture of small lymphocytes in aspirates of germinomas, the differential diagnosis also includes thymomas and carcinomas metastatic to mediastinal lymph nodes. With thymomas, the nuclei do not possess the prominent, primitive nucleoli and greatly accentuated nuclear membranes characteristic of germinoma. In many metastatic carcinomas, the levels of both intercellular cohesion and pleomorphism are higher than one would expect with germinoma. Although not totally specific, the tigroid background would be unusual in carcinomas. Both thymomas and metastatic carcinoma will be positive for cytokeratin and will demonstrate epithelial differentiation ultrastructurally. Germinomas often do not react with cytokeratin but may have primitive junctions.

Cytomorphologic Features of Germinoma

- Moderate to high cellularity
- Uniform mononucleated tumor cells dispersed as single cells and small flat aggregates
- Cytoplasm: scant to moderate; delicate
- Distinct round to polygonal cell borders
- Nuclei: large; vesicular chromatin; distinct membranes; prominent nucleoli
- Background: frothy (tigroid); lymphocytes

Other pure nongerminomatous malignant germ cell neoplasms are uncommon as primary tumors of the mediastinum.[50–52] Yolk sac tumors are more frequent in children than in adults. They are also more common in the retroperitoneum than in the mediastinum. These aggressive tumors may be associated with elevations in serum α-fetoprotein levels. Although these tumors are aggressive clinically, good outcomes are possible with combined modalities of radical surgical resection and chemotherapy. Histologically, yolk sac tumors may show a very highly variable appearance from field to field. Aspiration biopsies may be highly cellular and composed of rather uniform, cohesive neoplastic cells, which are often closely associated with an extracellular, basement membrane-like material[38,53,54] (Fig. 2-32). Cell contours vary from polygo-

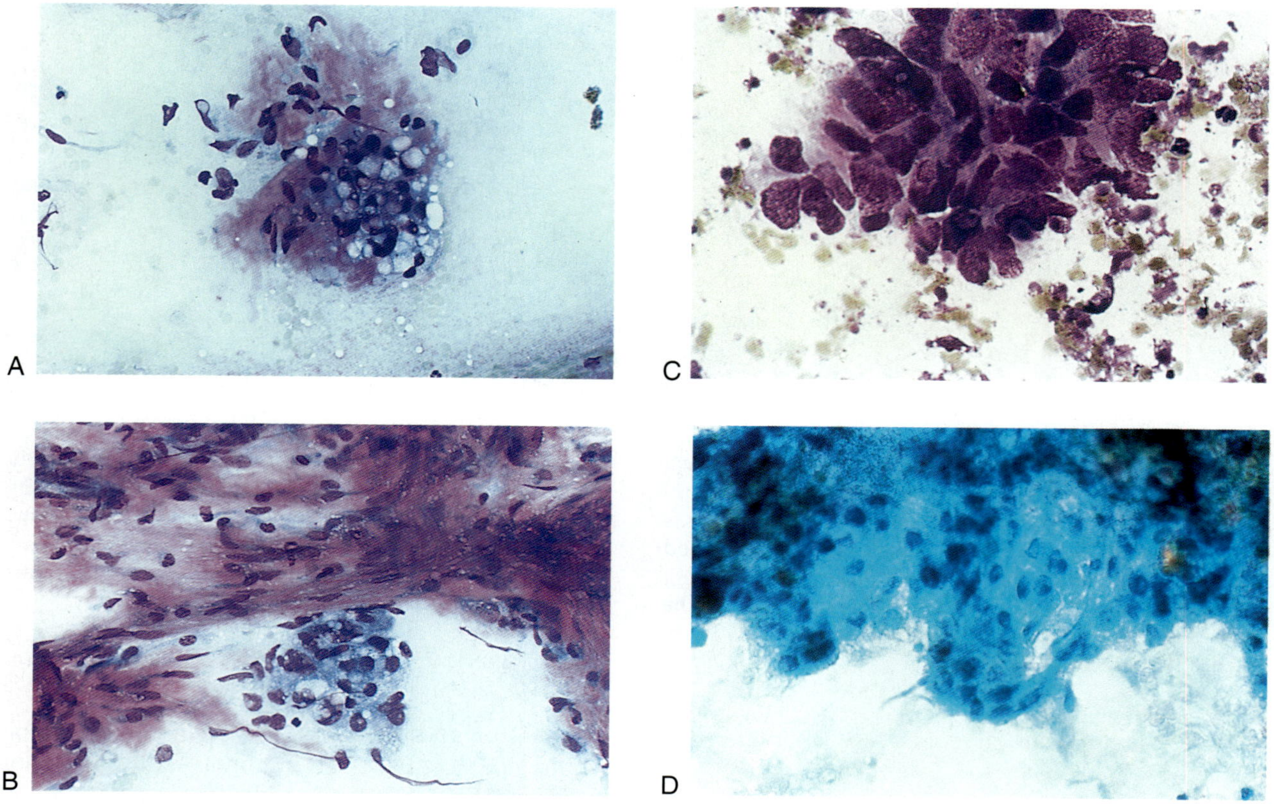

Fig. 2-32. (**A–D**) These aspiration biopsies of yolk sac (endodermal sinus) tumors are moderately cellular and composed of moderate-sized neoplastic cells with moderate volumes of cytoplasm. Cell borders are indistinct; in fact, a syncytial-like arrangement of the cells is present. Each cell has a single round to oval nucleus with one or more distinct nucleoli. Cytoplasmic vacuolization is apparent. Often, the tumor cells are closely associated or even embedded within matrix material. With the Diff-Quik stain, this matrix has a metachromatic quality. It is less distinctive with the Papanicolaou stain (**A–C**, Diff-Quik; **D**, Papanicolaou). (**A, B, D** courtesy of Dr. Paul Wakely, Jr., Richmond, VA.)

nal to somewhat elongated. Scanty cytoplasm results in high nuclear to cytoplasmic ratios; characteristically, the cytoplasm contains one or more sharply delineated vacuoles. Each cell possesses a single hyperchromatic nucleus that typically has a prominent nucleolus. With the Diff-Quik stain, the extracellular matrix material appears metachromatic. Both isolated and aggregated neoplastic cells may appear embedded within this matrix material. With the Papanicolaou stain, the matrix has a faint bluish green coloration. In addition to the prominent extracellular matrix material and the distinct cytoplasmic vacuoles, tumor cells may also be positive for α-fetoprotein by immunocytochemistry.

Pure embryonal carcinomas only rarely occur as a primary mediastinal neoplasm. These highly aggressive neoplasms usually result in death related to metastatic dissemination. In aspiration biopsies, they present as cellular smears consistent with a poorly differentiated or high-grade adenocarcinoma (Fig. 2-33). The tumor cells are arranged in syncytial and/or three-dimensional clumps with prominent nucleoli, moderate amphophilic cytoplasm, and high nuclear to cytoplasmic ratios[38] (Figs. 2-33 and 2-34). We are unaware of any reliable distinguishing cytomorphologic features that can distinguish with total certainty an embryonal carcinoma from a poorly differentiated adenocarci-

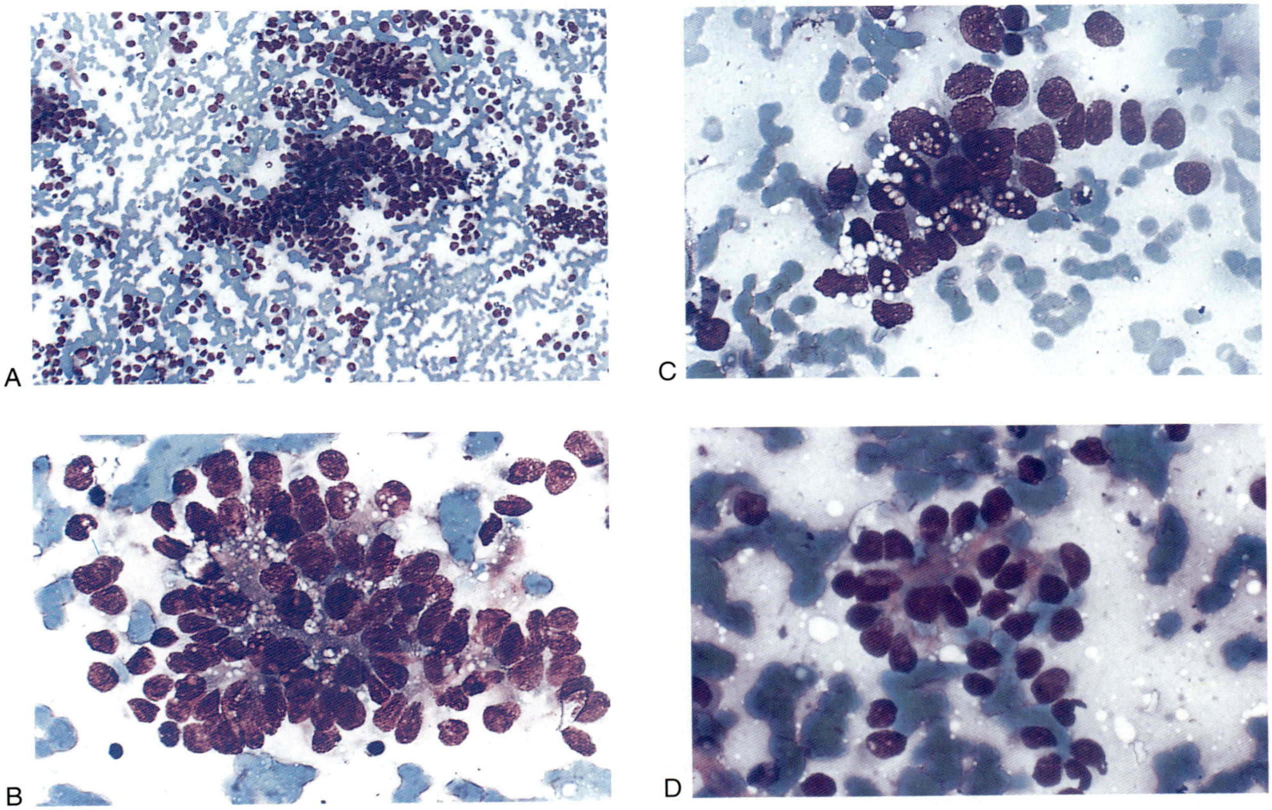

Fig. 2-33. Highly cellular aspirate of a pure embryonal carcinoma. Tumor cells are present in large and small cohesive aggregates. Some of the former have a branching papillary configuration. The epithelial quality of the cells is readily apparent at this low magnification. (**B**) Tumor cells are focally arranged in a manner suggestive of acinar formation. Each cell possesses a single large nucleus with one or more defined nucleoli and obviously vacuolated cytoplasm. These vacuoles appear to be sharply punched out and optically clear. Their exact nature is uncertain. In specimen represented in Fig. **D**, a truncated papillary configuration is suggested (**A–D**, Diff-Quik).

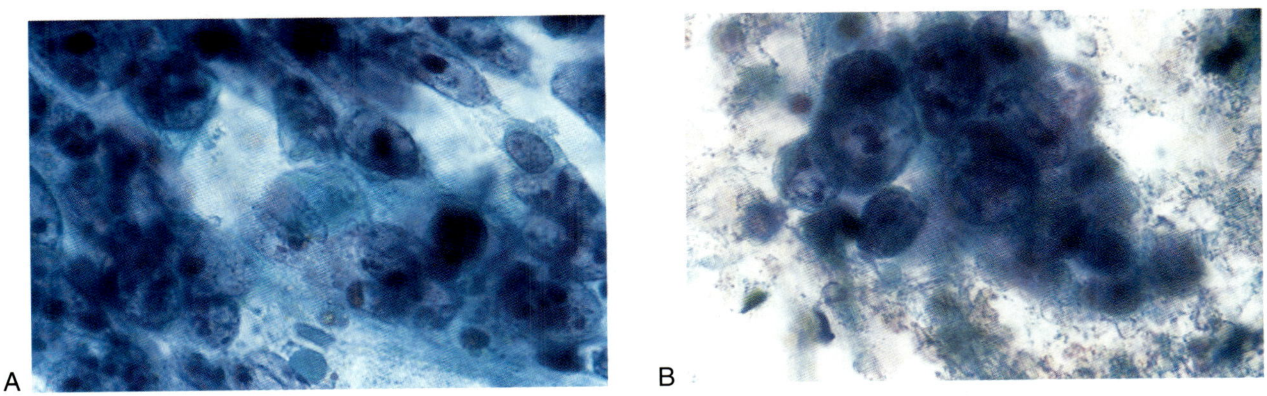

Fig. 2-34. (**A & B**) Each embryonal carcinoma cell has a large oval to irregular nucleus with a thick and uneven nuclear membrane, finely granular and dark-stained chromatin, and huge nucleoli. The nuclear to cytoplasmic ratios are high. No specific distinguishing features are recognizable (**A & B**, Papanicolaou).

noma metastatic to mediastinal nodes. Immunocytochemical reactions may help.

Pure nongestational choriocarcinoma primary in the mediastinum is an exceedingly uncommon event. Almost all reported cases have occurred in young adult men. More frequently, choriocarcinoma forms a component of a mixed malignant germ cell tumor. We have encountered only a single fine needle aspiration biopsy of a primary mediastinal choriocarcinoma.[38] Cellularity was scanty, perhaps reflecting the abundant necrosis in the mass (Fig. 2-35). The aspirate contained huge obviously malignant neoplastic elements. Both cytotrophoblasts and syncytiotrophoblasts were present as isolated cells. Nuclei were characterized by prominent membranes, vesicular to dense chromatin, and one or more well-developed nucleoli. With the Diff-Quik stain, cytoplasm was moderate to abundant in volume, had a dense amphophilic appearance, and contained multiple minute vacuoles. Choriocarcinoma is the only primary mediastinal neo-

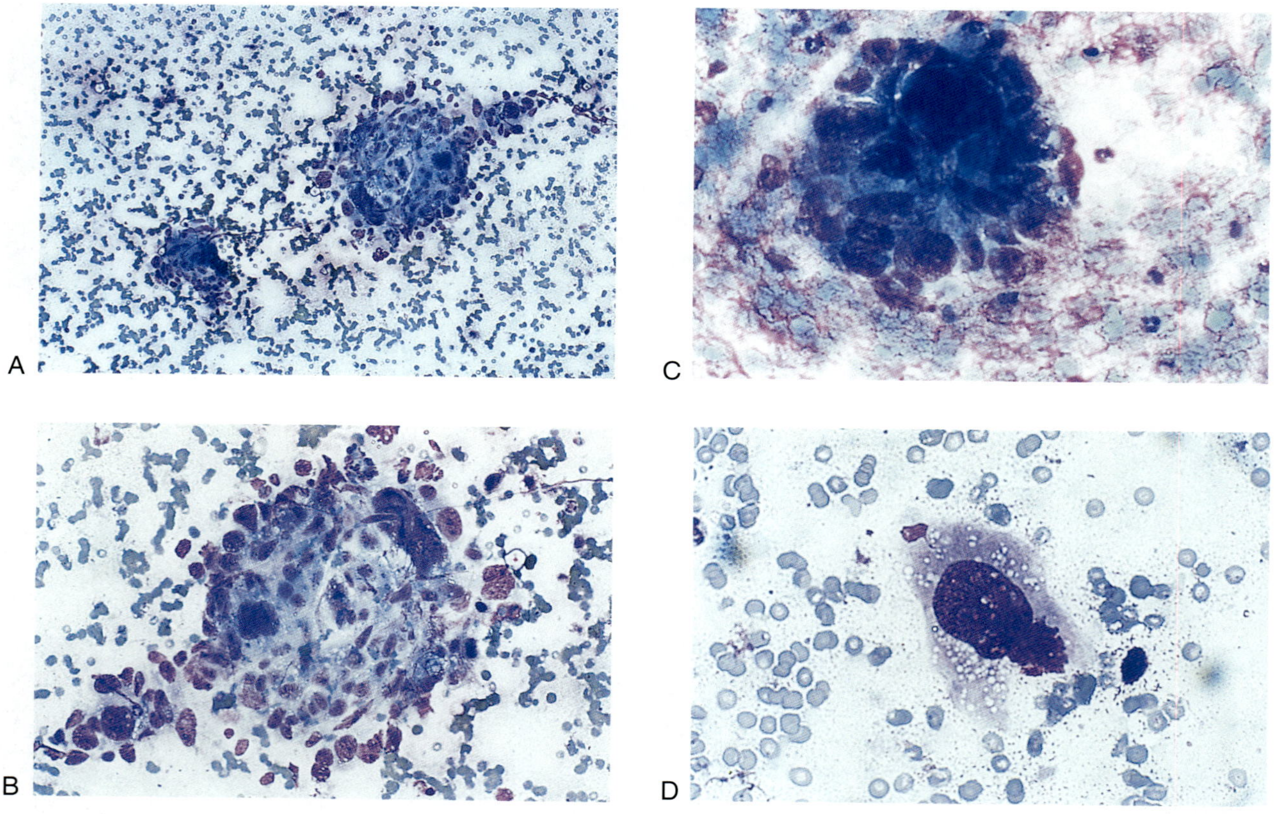

Fig. 2-35. (**A–D**) This is from the only aspirate we have ever seen of a primary mediastinal choriocarcinoma. (**A**) The scanning lens view demonstrates the relatively scanty cellularity in the specimen. The neoplastic cells are obviously malignant and are present in cohesive aggregates. Even at this very low magnification, the marked pleomorphism of the tumor cells is readily apparent. (**B**) Higher magnification of one of the aggregates. Most of the neoplastic cells are mononucleated and have high nuclear to cytoplasmic ratios and well defined nucleoli. Other tumor cells are much larger and contain multiple nuclei. Cytoplasm is faintly basophilic and vacuolated. (**C**) A three-dimensional spherical aggregate of malignant cells is apparent. There are not distinguishing cytomorphologic features. (**D**) A solitary tumor giant cell is present. It has abundant distinctly vacuolated dense cytoplasm and probably two (or one lobulated) nuclei with very dense chromatin (**A–D**, Diff-Quik).

plasm that would show such obvious anaplasia and pleomorphism. If one recognizes the presence of the dual neoplastic cell population, then the correct diagnosis is more apt to be made. As stated earlier, a major item in differential diagnosis would include large cell undifferentiated carcinoma of lung metastatic to the mediastinum. Patients with this latter neoplasm would tend to be older with a long cigarette-smoking history. In addition, choriocarcinomas invariably produce high levels of circulating human chorionic gonadotropin, which may result in endocrine effects such as gynecomastia.

The most common primary germ cell tumor of the mediastinum is the teratoma. Fortunately, most of these are histologically mature and thus clinically benign.[50] Although they occur most frequently in young adults, they do show a wide spectrum of age including presentation in infancy. Grossly, these tumors present as either cystic or a mixed cystic and solid mass that appears encapsulated. Histologically, the different germ cell layer components are mature in appearance. As in the ovary, the major component of benign teratomas is usually a mature stratified squamous epithelium with keratin production and often adnexal structures. We have never personally seen an example of an FNA biopsy of a benign mediastinal teratoma. One would expect to see mostly anucleated squamous cells, possibly admixed with a few viable mature squames. For benign cystic teratomas, squamous-lined non-neoplastic cystic lesions (e.g., a thymic cyst) might present the same exact cytomorphologic picture.

Only a small minority of all mediastinal teratomas are histologically immature; that is, they have one or more components that appear fetal in type or less differentiated. Most frequently, the major component in immature teratomas is neural. Solid foci of neuroblastic tissues and/or neuroepithelial canals are set within a neurofibrillary background.

Immature teratomas may yield cellular smears that, in our experience, most commonly are dominated by small neuroblastic elements (Fig. 2-36). The latter are characterized by small round hyperchromatic nuclei, small but conspicuous nucleoli, and a thin rim of cytoplasm. These elements are typically set within a filamentous matrix, which is more apparent with the Diff-Quik stain. Other immature tissues may also be sampled as well, for example, glandular epithelium and cartilage (Figs. 2-36 and 2-37). With aspiration biopsies of immature teratomas, a major component of the differential diagnosis would be primary neuroblastoma. The latter typically occurs in young patients and usually arises in the posterior mediastinum. They may be

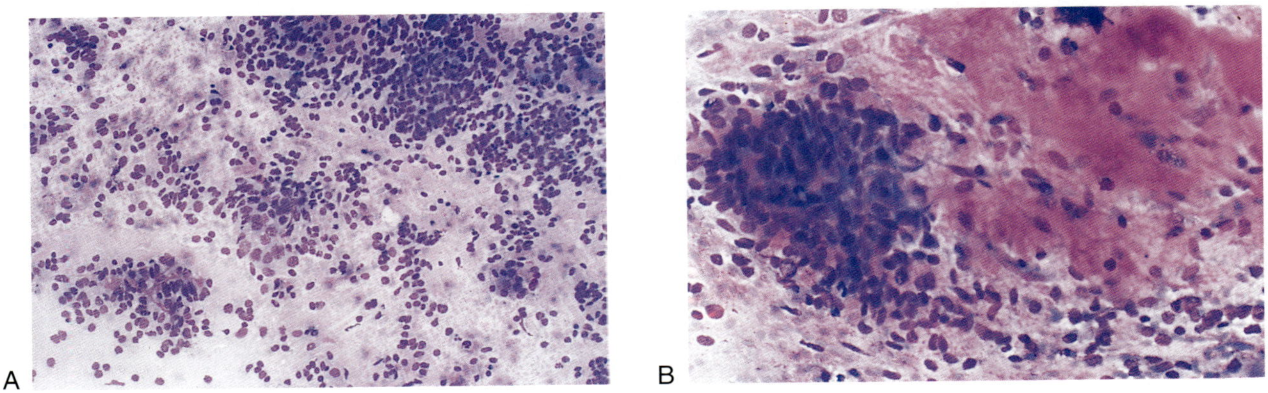

Fig. 2-36. (**A**) The scanning lens view demonstrates the high cellularity of an aspirated immature teratoma of the mediastinum. Most of the elements present in this field are probably derived from primitive neuroblastic elements. They are characterized by small round hyperchromatic nuclei and very high nuclear to cytoplasmic ratios. These cells are set within a loose eosinophilic matrix. Some of the cells present in this field are much larger; their exact nature is obscure. (**B**) Neuroblastic elements are present in the cohesive aggregate adjacent to dense metachromatic matrix material which may represent fetal-type cartilage (**A & B**, Diff-Quik).

associated with elevation in plasma catecholamine levels. Teratomas, on the other hand, almost always occur in the anterior mediastinum. The reader should recall that mediastinal germ cell tumors, as elsewhere, may consist of mixed types (Fig. 2-38).

Although quite rare, it needs to be mentioned that soft tissue sarcomas may arise in association with a malignant germ cell tumor of the mediastinum.[55,56] Most frequently, these are rhabdomyosarcomas, angiosarcomas, or spindle cell sarcomas. We are unaware of a single reported example of one of these rarities in the FNA biopsy literature. We are also unaware of a reported

A

B

C

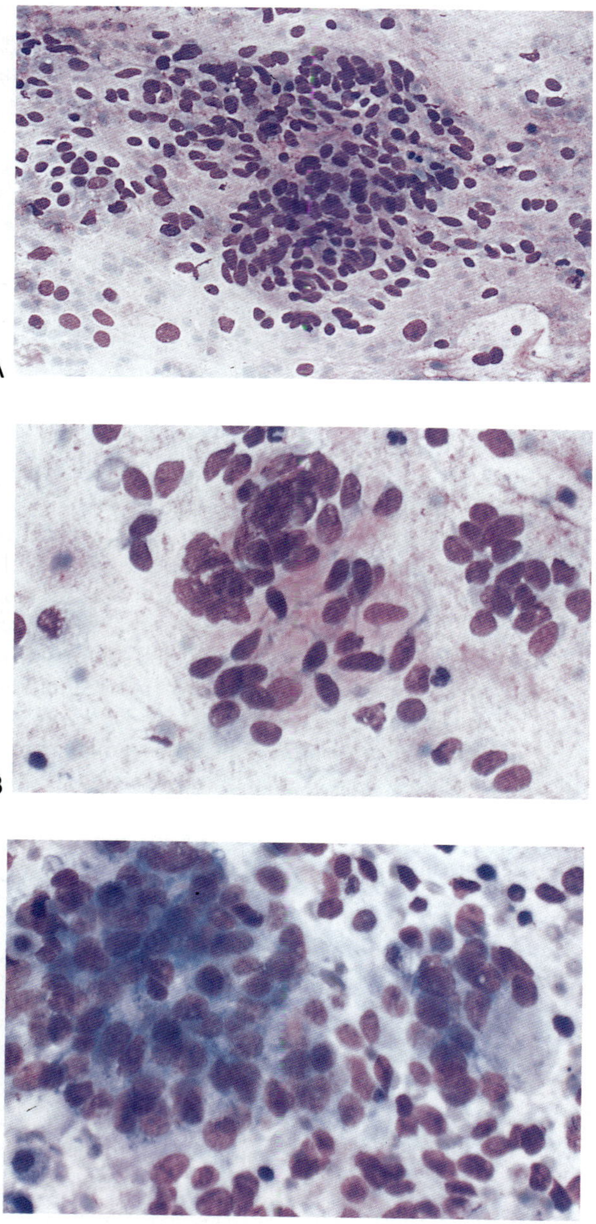

Fig. 2-37. It may be impossible to determine with certainty all the cell types present in an aspirate of an immature teratoma. (**A**) Many of these elements may represent immature neuroblasts. (**B**) The tumor cells have plump ovoid nuclei and bipolar cytoplasm. They are set within a pale matrix material. Although their nature is uncertain, they may very well represent a form of primitive mesenchyme. (**C**) Epithelial differentiation is probably represented by these cohesive three-dimensional aggregates of malignant-appearing cells (**A–C**, Diff-Quik).

A

B

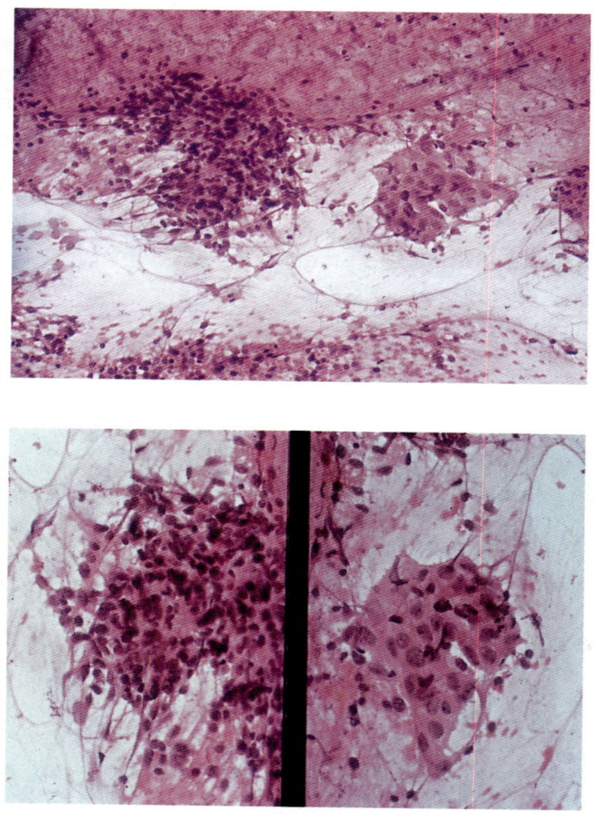

Fig. 2-38. (**A & B**) Aspirate of a mediastinal mixed malignant germ cell tumor. On the left are elements derived from an immature teratoma, specifically, primitive neuroblasts. On the right are larger cells with more vesicular nuclei and more abundant cytoplasm. These are derived from a yolk sac tumor component. (**B**) These elements are magnified (**A & B**, Papanicolaou).

example of another rare event, namely the occurrence of a low-grade spindle cell tumor (persistent yolk sac tumor) as a residual mass in a patient previously treated with chemotherapeutic agents for a malignant germ cell tumor.[57]

NEUROENDOCRINE NEOPLASMS

In the mediastinum, primary neuroendocrine tumors arise in the thymus.[58] As in other body sites, a full spectrum of histologic variants may be seen. Most of these represent conventional carcinoid tumors of the foregut type. Two-thirds of patients with carcinoid tumors of the thymus are either discovered incidentally as a mass lesion on chest radiography or secondary to a local mass effect.[58] Approximately one-third of patients, however, will present because of systemic endocrine manifestations, which is almost always Cushing's syndrome. Such manifestations are unexpected with malignant or atypical carcinoid tumors and with small cell (oat cell) carcinomas. Histologically, the malignant carcinoid tumors maintain an organoid pattern of growth similar to conventional carcinoid tumors, but show nuclear enlargement, hyperchromasia, more mitotic activity, and at times abundant necrosis.

Only rare reports of fine needle aspiration biopsies of primary mediastinal neuroendocrine neoplasms have been published.[59, 59a] Very recently, Sickel et al.[59] described three examples of atypical carcinoids of the anterior mediastinum in FNA biopsies. However, the correct diagnosis was initially made in only one. The two other cases were misdiagnosed as thymoma and small cell carcinoma. In all cases, the smears were highly cellular with small uniform cells with finely granular chromatin and small nucleoli. In the correctly diagnosed case, cytoplasm was voluminous and granular in nature. In the other two, it was scanty. We have seen an additional example of malignant carcinoid tumor of the thymus in a fine needle biopsy specimen. Cellularity was moderate and dominated by variably sized aggregates of small uniform-appearing tumor cells (Fig. 2-39). With the Papanicolaou stain, chromatin was finely granular and evenly distributed throughout these round to ovoid nuclei, which contained small nucleoli. The nuclear to cytoplasmic ratios were quite high, as most cells were surrounded by only a thin rim of cytoplasm. Within the aggregates, nuclear molding and rosette formation were recognized. One would expect the cytologic appearance of the conventional carcinoid tumors to resemble those seen in aspirates of other sites, notably the lung. Smears will be cellular and dominated by both aggregated and dissociated uniform small cells. With the Papanicolaou stain, chromatin is finely granular and uniformly distributed throughout the round to ovoid nuclei; nucleoli are inconspicuous. Cytoplasm varies from scanty to moderate. Characteristically, one may see plasmacytoid elements, that is, isolated tumor cells with eccentric round nuclei and basophilic cytoplasm; there may even be a clearing of the cytoplasm in a perinuclear location. These features are more readily appreciated with the Diff-Quik stain. Capillary blood vessels in the smears may be coated with these neoplastic elements.

Usually, the neuroendocrine nature of the aspirated tumor cells is readily apparent. Consequently, the major item in the differential diagnosis is a neuroendocrine neoplasm metastatic to the mediastinum from elsewhere, generally in the lung. Before ascribing a neuroendocrine tumor as primary in the mediastinum, a bronchogenic origin needs to be rigorously excluded (Fig. 2-40). Rarely, most of the neoplastic cells in a carcinoid tumor have a distinctly spindled contour (Table 2-3). The differential diagnosis in this setting includes spindle cell thymoma, peripheral nerve sheath tumors, and primary mesenchymal neoplasms. The combination of radiographic appearances, cytomorphology, immunocytochemistry, and ultrastructure should resolve this problem.

INTRATHORACIC THYROID

According to Finner and Hammond,[60] up to 5 percent of all mediastinal mass lesions are intrathoracic goiters. The vast majority of these arise as a downward extension of a cervical goiter.[60,61] Most occur within the anterosuperior mediastinum. Women are involved more frequently than are men and a large proportion occur in middle-aged adults. We have seen examples of mediastinal aspi-

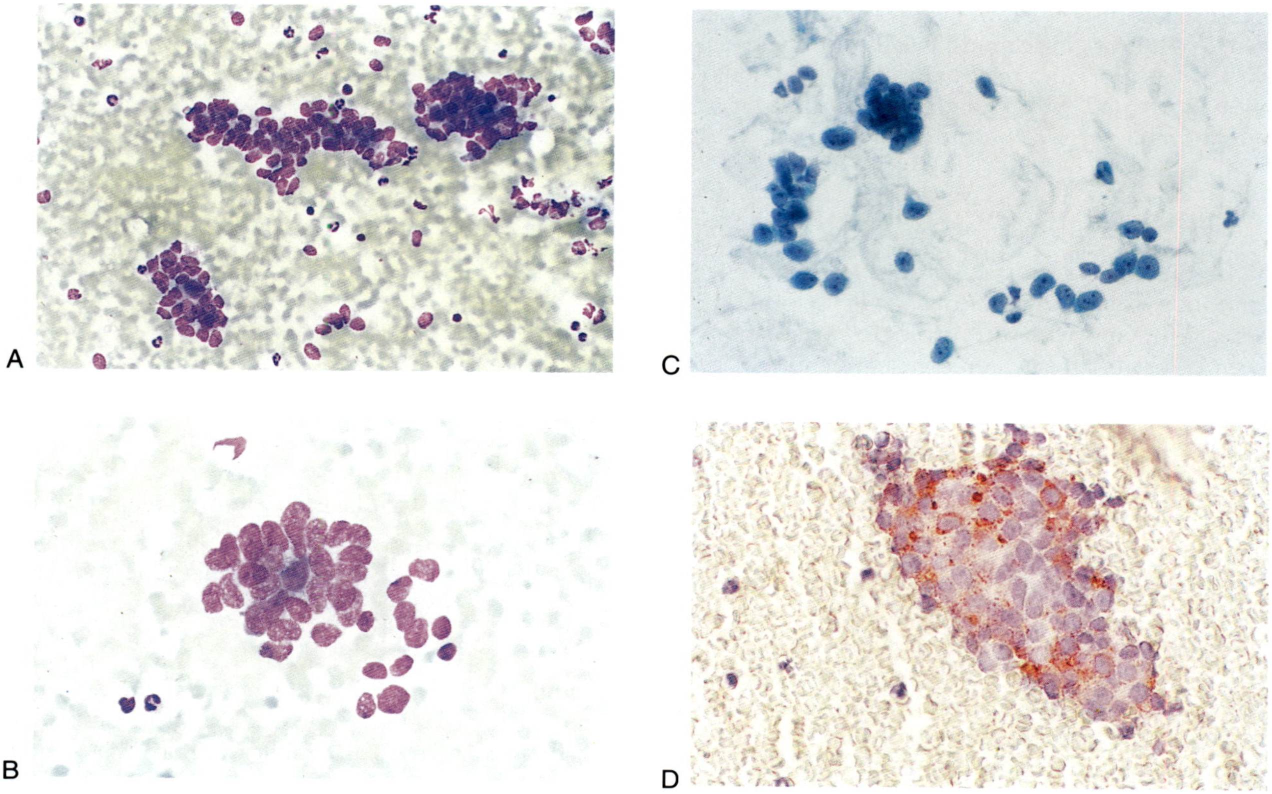

Fig. 2-39. (**A**) Cohesive aggregates of small uniform neoplastic cells are characterized the aspirate of this mediastinal malignant carcinoid tumor. Although most of these cells have very high nuclear to cytoplasmic ratios, a suggestion of acinar formation is present within the cellular clusters. (**B & C**) The chromatin is finely granular and evenly distributed throughout the nuclei. The latter vary from round to ovoid with generally smooth contours. Small but distinct nucleoli are evident in most nuclei. Although the tumor cells are closely opposed, molding is not well developed. (**D**) In this direct smear, some of the neoplastic cells are intensely immunoreactive for chromogranin (**A & B**, Diff-Quik; **C**, Papanicolaou).

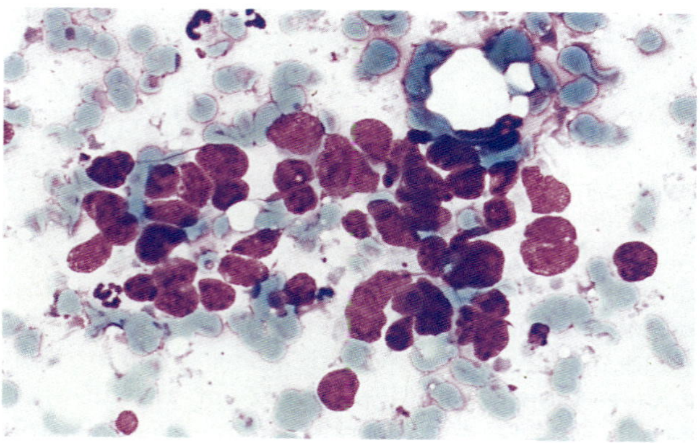

Fig. 2-40. Example of oat cell carcinoma of the lung metastatic to the mediastinum. In contrast to the carcinoid tumor, molding is better developed, nuclei are more sharply angulated, and nucleoli are less apparent (Diff-Quik).

Table 2-3. Differential Diagnosis of Primary Mediastinal Spindle Cell Tumors

	Thymoma	*Carcinoid Tumor*	*Peripheral Nerve Sheath Tumor*
Compartment	Anterosuperior	Anterosuperior	Posterior
Cell arrangement	Bundles, sheets	Sheets, rosettes Granular cytoplasm	Bundles Nuclear palisades
Nuclear contours	Smooth	Smooth	Wavy
Chromatin	Fine Pale staining	Coarser Darker	Variable Pale to dark
Background	Lymphocytes		Collagen
Immunocytochemistry	CK+	NSE+; CRG+: CK±	S-100 protein+ CK-

Abbreviations: CK, cytokeratin; NSE, neuron-specific enolase; CRG, chromogranin.

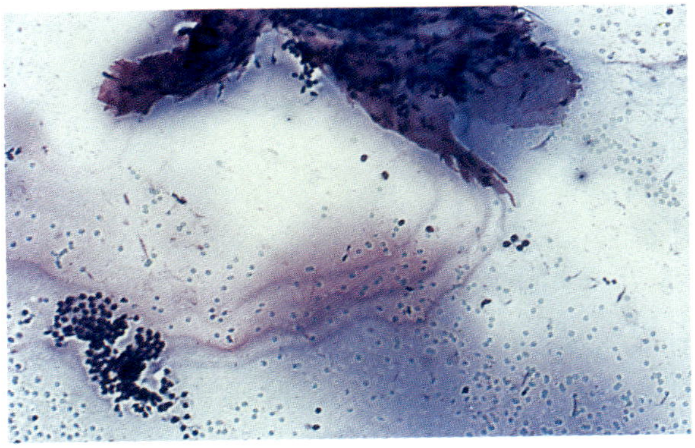

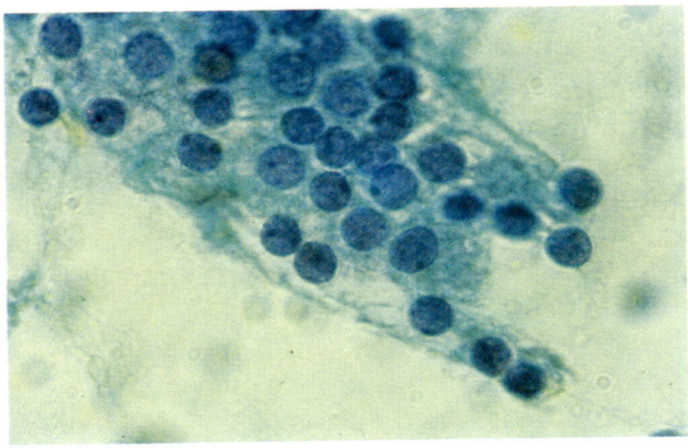

Fig. 2-41. (**A**) The overall cellularity of this aspirate is relatively low. In addition to a sheet of benign follicular epithelial cells, there is a large fragment of collagen-containing connective tissue and abundant colloid. The latter appears as a pale purple film (Diff-Quik). (**B**) The benign follicular epithelial cells are characterized by a solitary perfectly round nucleus with very finely granular and pale-staining chromatin and minute nucleoli. Although focally crowded there is no evidence of prominent nuclear overlapping, nuclear grooves, or pseudoinclusions (Papanicolaou).

ration biopsies with evidence of hyperplastic thyroid tissue. The smears contained sheets and small aggregates of small uniform follicular epithelial cells and colloid. Neoplasms, both benign and malignant, arising in mediastinal thyroid are distinctly uncommon. One would expect the cytomorphologic appearance of aspiration smears to resemble those procured from sampling of thyroid neoplasms in the neck (Fig. 2-41).

INTRATHORACIC PARATHYROID

Intrathoracic parathyroid tissue is not at all unusual.[58] A small but significant proportion of all parathyroid adenomas occur in the mediastinum. Many of these may descend from a cervical location, similar to the mechanism for many mediastinal goiters. However, neoplasms and hyperplasia may also arise in ectopic mediastinal parathyroid glands. Most will occur as small mass lesions in or near the thymus. Hypercalcemia, rather than a mass effect, brings the patient to medical attention. Consequently, aspiration biopsy of mediastinal parathyroid tissue is an unusual event. The mediastinal proliferations are usually noted during or subsequent to a neck exploration in patients with hypercalcemia. Cytomorphologically, smears should resemble those seen in specimens obtained from cervical parathyroid tissue.[62,63] Aggregated and isolated small homogeneous polygonal cells with perfectly round nuclei, finely granular euchromatic chromatin, frayed pale cytoplasm and high nuclear to cytoplasmic ratios will be evident. Aggregation about capillary vessels and follicle formation would be expected. Clustered stripped nuclei arranged in a morular pattern in a frothy lipid background can be seen. Immunochemical analysis of aspirated material for parathyroid hormone would confirm the diagnosis. We are unaware of a reported example of an aspirate of a mediastinal parathyroid carcinoma.

NEUROGENIC TUMORS

Neurogenic tumors comprise a large proportion of all neoplasms arising in the posterior mediastinum. There are basically two major categories: tumors of the peripheral nerve sheath and of the neuroectoderm. Of the two, the former are more common. The majority of these neoplasms are benign representing either schwannomas or neurofibromas. These tumors occur most commonly in young adults and most often are detected as an asymptomatic mass on routine chest radiography. However, malignant neoplasms of the peripheral nerve sheath may also originate in the posterior mediastinum. These are much more likely to present secondary to the effects of a mass lesion. A much greater proportion of neuroectodermal neoplasms are clinically malignant and present in the pediatric population. These include neuroblastoma, ganglioneuroblastoma, and ganglioneuroma.

In aspiration biopsies, neurofibromas and schwannomas are generally indistinguishable from each other.[64–66] The smears will be only moderately cellular at best and predominated by cohesive-appearing tissue fragments (Fig. 2-42). The most distinctive cytomorphologic feature is the elongated neoplastic nuclei. The nuclear length-to-width ratio may exceed that seen in any other spindle cell proliferation. Nuclei may end in sharply pointed tips. Chromatin is usually finely granular and evenly distributed; nucleoli are inconspicuous. Cytoplasmic volumes vary from scanty to somewhat moderate. Characteristically, it presents as very long, thin bipolar tails. Within the aggregates, it may be very difficult to discern the cytoplasm belonging to individual cells; rather, a blending of adjacent cytoplasm is suggested. Infrequently, a palisaded arrangement of the neoplastic nuclei, simulating Verocay bodies, may be seen in aspirates of schwannomas. Mitotic figures, if present at all, are exceedingly uncommon.

Aspirates of malignant peripheral nerve sheath tumors may have a higher level of cellularity and possess a much greater proportion of individually dispersed neoplastic cells.[64,67] Typically, the nuclei remain elongated and have euchromatic to hyperchromatic chromatin (Fig. 2-43). Mitotic figures may be visible in the smears. Cytoplasm is more variable than in the benign tumors and may actually be conspicuous. It should be emphasized, however, that it may be exceedingly difficult or even impossible to distinguish between a benign and a malignant neoplasm of peripheral nerve sheath origin in aspiration smears.[67] We have seen an example of an aspirate of a cellular schwannoma primary in the posterior mediastinum in which we favored, to a degree, a malignant tumor in the smears.[68] However, there is no evidence of recurrence or metastases four years following surgical resection of the mass.

The cytologic differential diagnosis of benign and malignant peripheral nerve sheath tumors incudes primary and metastatic soft tissue sarcomas, spindle cell thymoma, spindle cell carcinoid tumor, sclerosing malignant lymphoma, and granulomatous inflammation (Table 2-3). Based purely on cytomorphologic grounds, it may be impossible to distinguish a malignant peripheral nerve sheath tumor and possibly a benign one from other spindle cell sarcomas. In both thymomas and carcinoid tumors, the nuclear contours will be more plump and ovoid rather than the very elongated serpentine appearance typical of the nerve sheath tumors. Cytokeratin immunostaining may be helpful in that thymomas will be positive, carcinoids will probably be positive, and the nerve sheath tumors will be negative. Both thymoma and carcinoid will be negative for S-100 protein, which is generally positive in benign nerve sheath tumors and occasionally in malignant schwannomas. In malignant lymphomas, the nuclear contours will be much more irregular than one expects in nerve sheath tumors, and nucleoli may be much more prominent. In lymphoma, the neoplastic cells are positive for LCA; this will be negative in the other spindle cell tumors of the mediastinum. Ultrastructural examination of aspirated material may also be helpful in this differential diagnosis. Occasionally, granulomatous inflammation producing a mass lesion may be considered because

A 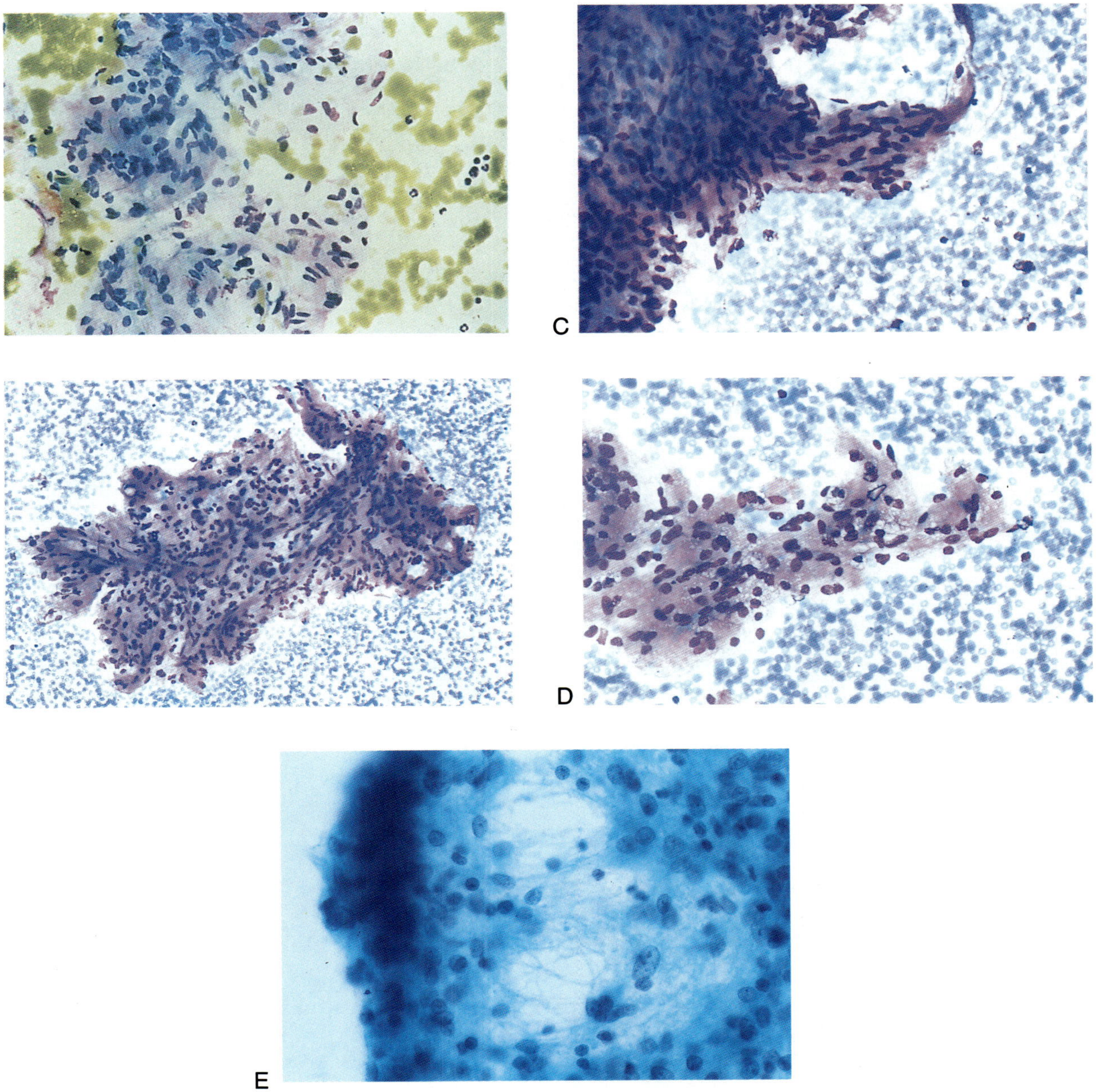 C

B D

E

Fig. 2-42. (A–E) Aspirates of benign nerve sheath tumors are characterized by cohesive-appearing aggregates of neoplastic cells embedded in a stromal matrix. (A–D) With the Diff-Quik stain, this matrix has a distinct eosinophilic quality. (C) For the most part, the neoplastic cells have solitary very elongated nuclei with condensed chromatin. Palisading may be evident when the nuclei line up parallel to one another (Diff-Quik). (D) However, rounded nuclei may be evident, perhaps representing Antoni B areas. Note the minute cytoplasmic vacuoles as well (Diff-Quik). (E) With the Papanicolaou stain, the filamentous nature of the elongated cytoplasm may be more apparent (Papanicolaou).

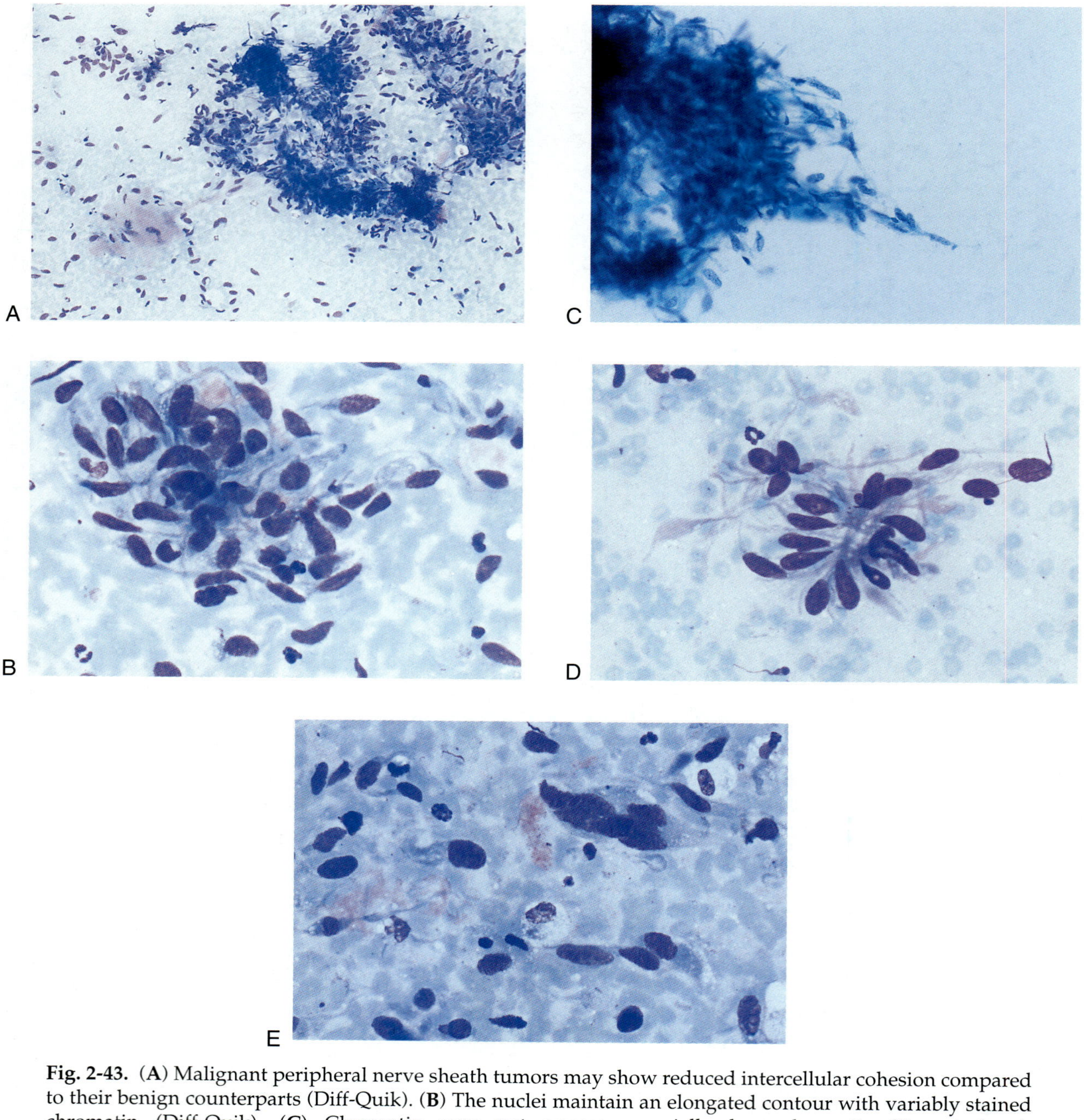

Fig. 2-43. (**A**) Malignant peripheral nerve sheath tumors may show reduced intercellular cohesion compared to their benign counterparts (Diff-Quik). (**B**) The nuclei maintain an elongated contour with variably stained chromatin (Diff-Quik). (**C**) Chromatin may not appear especially hyperchromatic (Papanicolaou). (**D**) Nuclear palisading may still be found in aspirates of these malignancies. Cytoplasm presents as delicate, elongated processes (Diff-Quik). (**E**) Pleomorphism may also be detected as variability in size and multinucleation (Diff-Quik).

of the elongated nature of the epithelioid nuclei. Smear cellularity will be considerably lower. Epithelioid cells have a characteristic appearance with bent-appearing ovoid nuclei, very bland chromatin, and greater volumes of cytoplasm. Both thymoma and carcinoid tumors arise predominantly in the anterosuperior mediastinal compartment, whereas the peripheral nerve sheath tumors are almost always restricted to the posterior mediastinum.

Tumors of sympathetic nerve may also arise in the posterior mediastinum. The cytomorphology of these tumors in aspiration biopsies is detailed in Chapter 6.

OTHER SOFT TISSUE SARCOMAS

Although uncommon, almost any form of soft tissue sarcoma may arise in the mediastinum. According to Mack,[69] the most common sarcomas in the mediastinum (including heart) are angiosarcoma, liposarcoma, and malignant fibrous histiocytoma. As stated earlier, sarcomas may arise in association with a mediastinal germ cell malignancy.[55,56] Therefore, if an aspiration smear from the anterior mediastinum contains sarcomatous elements, especially from a young adult man, the possibility of an associated germ cell tumor needs to be considered (Fig. 2-44). The cytomorphology of soft tissue sarcomas in aspiration biopsies is discussed in Chapter 7.

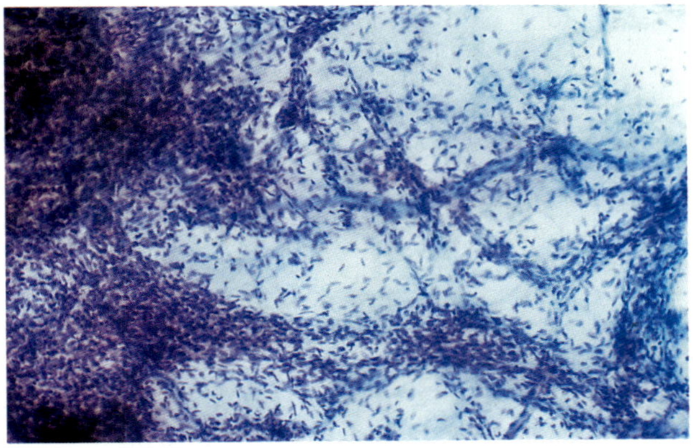

Fig. 2-44. Mediastinal aspirate of a myxoid liposarcoma. Numerous minute stellate and spindled mesenchymal cells are isolated within the pale-staining matrix material. The cellularity is moderate. The characteristic vasculature of this neoplasm is also apparent (Papanicolaou). (Case courtesy of Dr. Sherry Woodhouse, Lenoir, NC.)

METASTASES

In our experience, the most frequent indication for a mediastinal aspiration biopsy is to document the presence of a metastasis from a previously diagnosed malignancy. We recently participated in a multi-institutional evaluation of mediastinal aspiration biopsies. Most tumors sampled in this investigation were metastases.[5] Two-thirds of these were bronchogenic in origin, most of which were small cell anaplastic carcinomas. The remainder were diagnosed as adenocarcinomas (11 percent), squamous cell carcinomas (9 percent), large cell undifferentiated carcinomas (9 percent), and simply non-small cell carcinomas (19 percent). The second most frequent site of origin was carcinoma of the breast. Another very recent study also found pulmonary and breast carcinomas to be the major sites of origin for mediastinal metastases.[42] The cytologic features of many of these are discussed in Chapter 1.

CONCLUSION

The mediastinum may be the site of many different mass lesions that have a wide spectrum of therapeutic and prognostic implications. Many of these neoplasms produce distinctive cytomorphologic patterns and individual cellular features in aspiration smears. As elsewhere in the body, this information needs to be closely correlated with both relative clinical data and impressions from radiologic imaging studies. Frequently, this is sufficient to render a very accurate diagnosis. Ancillary diagnostic procedures serve to confirm impressions generated from the above information or may actually lead to a more specific diagnosis. In the series recently presented by Powers et al.,[5] the calculated diagnostic sensitivity and specificity for the detection of neoplasms was 87 percent and 88 percent, respectively. The sensitivity and specificity for distinguishing between benign and malignant diseases was 82 percent and 79 percent, respectively. Most diagnostic errors resulted from the misclassification of malignant tumors. With experience, aspiration biopsies of the mediastinum result in accurate, relevant diagnoses.

REFERENCES

1. Westcott JL: Percutaneous aspiration of hilar and mediastinal masses. Radiology 141:323, 1981
2. Sterett G, Whitaker D, Shilkin KB, Walters MN-I: The fine needle aspiration cytology of mediastinal lesions. Cancer 51:127, 1983
3. Sinner WN: Directed fine needle biopsy of anterior and middle mediastinal masses. Oncology 42:92, 1985
4. Herman SJ, Holub RV, Weisbrod GL, Chamberlain DW: Anterior mediastinal masses: Utility of transthoracic needle biopsy. Radiology 180:167, 1991
5. Powers CN, Silverman JF, Geisinger KR, Frable WJ: Fine-needle aspiration biopsy of the mediastinum: a multiinstitutional analysis. Am J Clin Pathol, accepted for publication
6. Geisinger KR: Differential diagnostic considerations and potential pitfalls in fine needle aspiration biopsies of the mediastinum. Diagn Cytopathol: 1995
7. Rosai J, Levine GD: Tumors of the Thymus. Armed Forces Institute of Pathology, Washington, DC, 1976, p 34
8. Levine GD, Rosai J: Thymic hyperplasia and neoplasia: a review of current concepts. Human Pathol 9:495, 1978
9. Marino M, Müller-Hermelink HK: Thymoma and thymic carcinoma. Relation of thymoma epithelial cells to the cortical and medullary differentiation of thymus. Virchows Arch A Pathol Anat Histopathol 407:119, 1985
10. Walker AN, Mills SE, Fechner RE: Thymomas and thymic carcinomas. Semin Diagn Pathol 7:250, 1990
11. Kornstein MJ: Controversies regarding the pathology of thymomas. Pathol Annu 27 (pt 2): 1, 1992
12. Kuo T-T, Lo S-K: Thymoma: a study of the pathologic classification of 71 cases with evolution of the Müller-Hermelink system. Hum Pathol 24:766, 1993
13. Morgenthaler TI, Brown LR, Colby TV et al: Thymoma. Mayo Clin Proc 68:1110, 1993
14. Park HS, Shin DM, Lee JS et al: Thymoma. A retrospective study of 87 cases. Cancer 73:2491, 1994
15. Ramon Y, Cajal S, Suster S: Primary thymic epithelial neoplasms in children. Am J Surg Pathol 15:466, 1991
16. Ho FCS, Fu KH, Lam SY et al: Evaluation of a histogenetic classification for thymic epithelial tumors. Histopathology 25:21, 1994
17. Quintanilla-Martinez L, Wilkins EW, Jr, Choi N et al: Thymoma: histologic subclassification is an independent prognostic factor. Cancer 74:606, 1994
18. Friedman HD, Inman DA, Hutchieson RE, Poiesz BJ: Concurrent invasive thymoma and T-cell lymphoblastic leukemia and lymphoma. A case report with necropsy findings and literature review of thymoma and associated hematologic neoplasm. Am J Clin Pathol 101:432, 1994
19. Kircher T, Schalke B, Buchwald J et al: Well-differentiated thymic carcinoma. An organotypical low-grade carcinoma with relationship to cortical thymoma. Am J Surg Pathol 16:1153, 1992
20. Snover DC, Levine GD, Rosai J: Thymic carcinoma: five distinctive histological variants. Am J Surg Pathol 6:451, 1982
21. Wick MR, Scheithayer BW, Weiland LH: Primary thymic carcinomas. Am J Surg Pathol 6:613, 1983
22. Pak HY, Yokota SB, Friedberg HA: Thymoma diagnosed by transthoracic fine needle aspiration. Acta Cytol 26:210, 1982
23. Sajjad SM, Lukeman JM, Llamas FT: Needle biopsy diagnosis of thymoma. A case report. Acta Cytol 26:503, 1982
24. Dahlgreen S, Sandstedt B, Sundstrom C: Fine needle aspiration cytology of thymic tumors. Acta Cytol 27:1, 1983
25. Tao LC, Pearson FG, Cooper JD et al: Cytopathology of thymoma. Acta Cytol 28:165, 1984
26. Tao L-C: Lung, Pleura and Mediastinum. Guides to Clinical Aspiration Biopsy. p. 212. Igaku-Shoin, New York, 1988
27. Riazmontazer N, Beydat C, Izadi B: Epithelial cytologic atypia in a fine needle aspirate of an invasive thymoma. A case report. Acta Cytol 36:387, 1992
28. Finley JL, Silverman JF, Strausbauch et al: Malignant thymic neoplasms: diagnosis by fine-needle aspiration biopsy with histologic, immunocytochemical, and ultrastructural confirmation. Diagn Cytopathol 2:118, 1986
29. Saad AJ, Schnadig VI, Davie K, Zaatari GS: Fine-needle aspiration cytology of thymic lesions, abstracted. Acta Cytol 38:859, 1994
30. Kuo T-T, Lo S-K: DNA flow cytometric study of thymic epithelial tumors with evaluation of its usefulness in the pathologic classification. Hum Pathol 24:746, 1993
31. Kardos TF, Vinson JH, Behm FG et al: Hodgkin's disease: diagnosis by fine needle aspiration biopsy: Analysis of cytologic criteria from a selected series. Am J Clin Pathol 86:286, 1986
32. Moriarity A, Banks E, Block T: Cytologic criteria for subclassification of Hodgkin's disease using fine needle aspiration. Diagn Cytopathol 5:122, 1989
33. Kardos TF, Sprague RI, Wakely PE, Jr, Frable WJ: Fine needle aspiration biopsy of lymphoblastic lymphoma and leukemia. A clinical, cytologic, and immunologic study. Cancer 60:2448, 1987

34. Jacobs JC, Katz RL, Shabb N et al: Fine needle aspiration of lymphoblastic lymphoma. A multiparameter approach. Acta Cytol 36:887, 1992
35. Silverman JF, Raab SS, Park HK: Fine-needle aspiration cytology of primary large cell lymphoma of the mediastinum: cytomorphologic findings with potential pitfalls in diagnosis. Diagn Cytopathol 9:209, 1993
36. Cartagena N, Jr, Katz RL, Hirsch-Ginsberg C et al: Accuracy of diagnosis of malignant lymphoma by combining fine-needle aspiration cytomorphology with immunocytochemistry and in some cases, Southern blotting of aspirated cells: a tissue-controlled study of 86 patients. Diagn Cytopathol 8:456, 1992
37. Akhtar M, Ali MA, Hug M, Barky M: Fine-needle aspiration biopsy of seminoma and dysgerminoma: cytologic, histologic and electron microscopic correlations. Diagn Cytopathol 6:99, 1990
38. Collins KA, Geisinger KR, Wakely P, Jr et al: Extragonadal germ cell tumors: a fine needle aspiration biopsy study. Diagn Cytopathol 12:223, 1995
39. Battifora H, Sun TT, Bahv RM et al: The use of antikeratin antiserum as a diagnostic tool: thymoma versus lymphoma. Hum Pathol 11:635, 1980
40. Chan WG, Zaatari GS, Tabei S et al: Thymoma: an immunohistochemical study. Am J Clin Pathol 82:160, 1984
41. Neill JS, Silverman JF: Electron microscopy of fineneedle aspiration biopsies of the mediastinum. Diagn Cytopathol 8:272, 1992
42. Miller MJ, Saigo P: Fine needle aspiration of mediastinal masses: an analysis of 42 cases, abstracted. Acta Cytol 38:860, 1994
43. Yousem SA, Weiss LM, Warnke RA: Primary mediastinal non-Hodgkin's lymphomas: a morphologic and immunologic study of 19 cases. Am J Clin Pathol 83:676, 1985
44. Perrone T, Frizzera G, Rosai J: Mediastinal diffuse large-cell lymphoma with sclerosis. A clinicopathologic study of 60 cases. Am J Surg Pathol 10:176, 1986
45. Lamarre L, Jacobson JO, Aisenberg AC, Harris NL: Primary large cell lymphoma of the mediastinum. A histologic and immunophenotypic study of 29 cases. Am J Surg Pathol 13:730, 1989
46. Strickler JG, Kurtin PJ: Mediastinal lymphoma. Semin Diagn Pathol 8:2, 1991
47. Sheibani K, Nathwani BN, Winberg CD et al: Antigenically defined subgroups of lymphoblastic lymphoma. Relationship to clinical presentation and biological behavior. Cancer 60:183, 1987
48. Francis IM, Das DK, Al-Rubah NAR, Gupta SK: Lymphoglandular bodies in lymphoid lesions and nonlymphoid round cell tumors: a quantitative assessment. Diagn Cytopathol 11:23, 1994
49. Silverman JF, Dabbs DJ, Ganick DJ et al: Fine needle aspiration cytology of neuroblastoma including peripheral neuroectodermal tumor with immunocytochemical and ultrastructural confirmation. Acta Cytol 32:367, 1988
50. Dehner LP: Germ cell tumors of the mediastinum. Semin Diagn Pathol 7:266, 1990
51. Hoffman OA, Gillespie DJ, Aughenbaugh GL, Grown LR: Primary mediastinal neoplasms (other than thymoma). Mayo Clin Proc 68:880, 1993
52. Dulmet EM, Maechiarini P, Suc B, Verleg JM: Germ cell tumors of the mediastinum. A 30-year experience. Cancer 72:1894, 1993
53. Akhtar M, Ali MA, Sackey K et al: Fine needle aspiration biopsy diagnosis of endodermal sinus tumor: histologic and ultrastructural correlations. Diagn Cytopathol 6:184, 1990
54. Yang, GCH: Demonstration of fibrils in the hyaline globules of yolk sac tumor with parietal differentiation in fine-needle aspiration smears. Diagn Cytopathol 10:216, 1994
55. Ulbright TM, Clark SA, Einhom LH: Angiosarcoma associated with germ cell tumors. Hum Pathol 16:268, 1985
56. Manivel C, Wick MR, Abenoza P, Rosai J: The occurrence of sarcomatous components in primary mediastinal germ cell tumors. Am J Surg Pathol 10:711, 1986
57. Ulbright TM, Michael H, Loehrer PJ, Donohue JP: Spindle cell tumors resected from male patients with germ cell tumors. A clinicopathologic study of 14 cases. Cancer 65:148, 1990
58. Wick MR, Rosai J: Neuroendocrine neoplasms of the mediastinum. Semin Diagn Pathol 8:35, 1991
59. Sickel JZ, Facik M, Feins RH: Atypical carcinoid tumor of thymic origin: a cytologic study of three cases, abstracted. Acta Cytol 38:858, 1994
59a. Wang D-Y, Kuo S-H, Chang D-B et al: Fine Needle aspiration cytology of thymic carcinoid tumor. Acta Cytol 39:423, 1995
60. Finner RL, Hammond EH: Pathology of the Mediastinum. p. 127. ASCP Press, Chicago, 1989
61. Wick MR: Mediastinal cysts and intrathoracic thyroid tumors. Semin Diagn Pathol 7:285, 1990
62. Abati A, Skarulis MC, Shawker T, Solomon D: Ultrasound-guided FNA of parathyroid lesions: a morphologic, immunocytochemical and biochemical approach, abstracted. Acta Cytol 38:847, 1994
63. Lerud KS, Frost AR, Tabbara SO, DelVecchio DM. Cytomorphology of cystic parathyroid lesions, abstracted. Acta Cytol 38:849, 1994

64. Hood IC, Qizilbash AH, Young JEM, Archibald SD: Needle aspiration cytology of a benign and a malignant schwannoma. Acta Cytol 28:157, 1984
65. Dahl I, Hagmar B, Idvall I: Benign solitary neurilemmoma (schwannoma). A correlative aspiration cytology and histologic study of 28 cases. Acta Pathol Microbiol Scand A 92:91, 1984
66. Zbieranowski I, Bedard Y: Fine needle aspiration of schwannomas. Acta Cytol 33:381, 1989
67. Vendraminelli R, Cavazzana AO, Poletti A et al: Fine-needle aspiration cytology of malignant nerve sheath tumors. Diagn Cytopathol 8:559, 1992
68. Woodruff JM, Goodwin TA, Erlandson RA: Cellular schwannoma. A variety sometimes mistaken for a malignant tumor. Am J Surg Pathol 5:733, 1981
69. Mack TM: Sarcomas and other malignancies of soft tissue, retroperitoneum, peritoneum, pleura, heart, mediastinum, and spleen. Cancer 75:211, 1995

3 Liver

The liver is one of the major targets of deep organ fine needle aspiration (FNA) biopsy. In our experience, it is second only to the lungs. The major indication for an FNA biopsy of the liver is the evaluation of a mass or focal defect. Most pathologists are more familiar with tissue core needle biopsies of the liver than with FNA specimens. Certainly, a tissue core biopsy is the specimen of choice whenever the clinical impression is that of a diffuse, non-neoplastic process, such as viral hepatitis, drug toxicity, and cirrhosis. The advantages of FNA biopsies become evident through the investigation of mass lesions, especially neoplasms. During a single needle pass, the mass can be sampled in several different planes; this is not the case with a tissue core biopsy. As with other body sites, we endorse immediate evaluation of aspirated material in the radiology unit to determine adequacy of the specimen, possibly provide an immediate interpretation, and request, in some situations, additional material for ancillary diagnostic procedures.[1] This is possible, as multiple needle punctures can be performed during a single aspiration biopsy session, which is generally not the case with tissue core specimens. In some patients with a bleeding tendency, FNA biopsy will be fully acceptable, whereas tissue core biopsies may very well be contraindicated. In fact, complications of aspiration biopsies of the liver are distinctly rare.

In all clinical scenarios, a systematic evaluation of aspiration biopsies is essential. An algorithmic approach lends itself very well to the evaluation of liver aspirates. In one scenario (Fig. 3-1), the first step is to determine whether the specimen is benign or malignant. As with most specimen types, this depends largely on nuclear features. If the aspirate is benign, inflammatory hepatic lesions must be distinguished from benign neoplasms and related proliferations, such as focal nodular hyperplasia. Benign growths in the liver may be composed predominantly of hepatocytes, ductal epithelial cells, or stromal elements. If the specimen is malignant, the major consideration is whether a primary liver cancer is present or the much more common metastases. In many situations, hepatocellular carcinoma can be distinguished from adenocarcinomas. It may be much more difficult to distinguish between a primary cholangiocarcinoma and a metastatic adenocarcinoma. Figure 3-2 suggests an alternative algorithmic approach. First, one determines whether the major cell type is epithelial or stromal. If it is the former, distinguishing hepatocellular from glandular lesions is key. From this point, the benign versus malignant decision is formulated.

Central to either of these approaches is the recognition of the spectrum of cytomorphologic appearances of benign hepatocytes (Figs. 3-3 to 3-8). These cells possess polygonal to round shapes with distinct cellular borders. Usually, cytoplasm is moderate in volume and relatively uniform from cell to cell and typically has a dense granular quality. With the Papanicolaou stain, minute red or deeply basophilic granules may be evident. Pigment granules such as bile or lipofuscin may be present, as well as lipid vacuoles. With the Diff-Quik stain, bile has a distinct green tint; with the Papanicolaou stain, it is golden brown (Figs. 3-6 and 3-7). This substance may be present external to the hepatocyte as well as intracytoplasmic. Most hepatocytes possess a solitary, centrally positioned nucleus. It often has a perfectly round contour, finely granular, evenly distributed

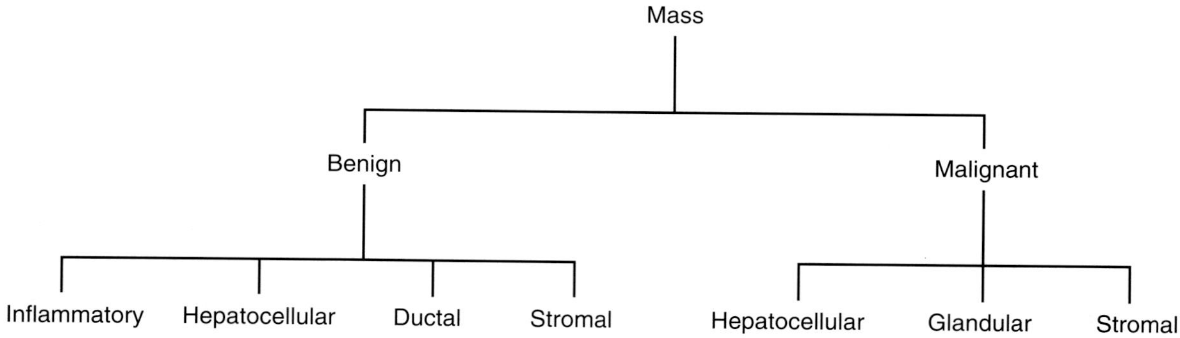

Fig. 3-1. The initial break point of this algorithm is the determination of whether the specimen is benign or malignant. This is based on traditional criteria such as smear cellularity, nuclear size and pleomorphism, and nuclear to cytoplasmic ratios. One must then decide whether the specimen represents an inflammatory lesion, a hepatocellular proliferation, a glandular neoplasm, or a mesenchymal neoplasm.

chromatin, and a single, round, fairly prominent nucleolus. The latter structure will be more evident with the Papanicolaou stain than with the Diff-Quik technique. Binucleated hepatocytes occur with some frequency; whether these cells have one or two nuclei, they usually are centrally placed. It is important to recognize that benign hepatocytes can show considerable variability in nuclear size (Fig. 3-5). However, benign hepatocytes have uniformly round and smooth nuclei and maintain relatively low nuclear to cytoplasmic ratios. The degree of intercellular cohesion among aspirated benign hepatocytes is variable. Several investigators have stated that the typical presentation is in small, flat aggregates composed of 15 or less hepatocytes (Fig. 3-9). However, these benign cells may also be present in broad sheets. Kupffer cells and vascular structures may also be seen within the sheets of benign hepatocytes;

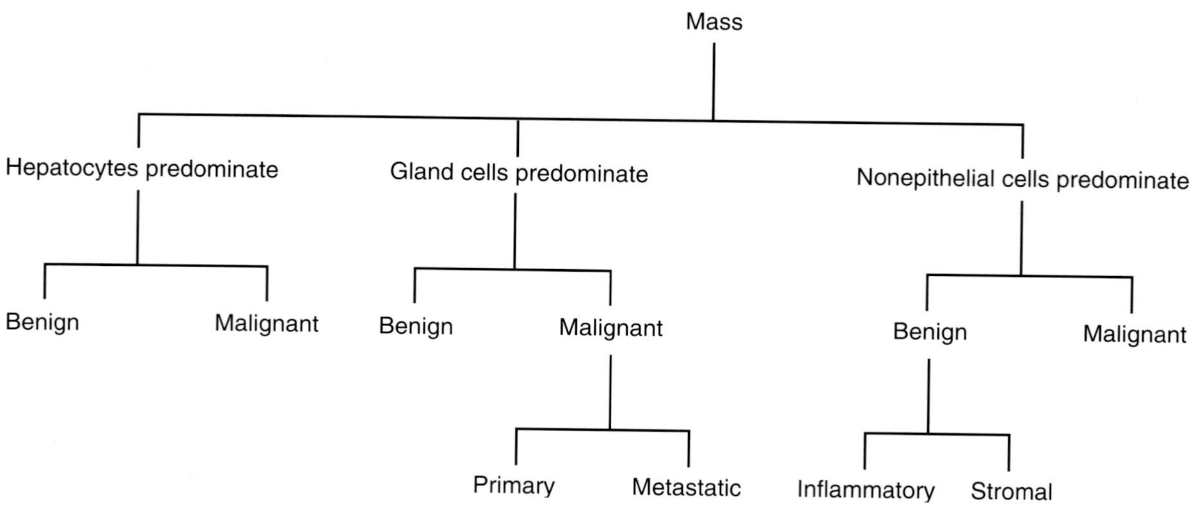

Fig. 3-2. In this approach, the initial decision involves the determination of the predominant cell type present. Major considerations include hepatocytic, glandular, or nonepithelial. Once the cell type is determined, the biological potential (benign or malignant) is assessed.

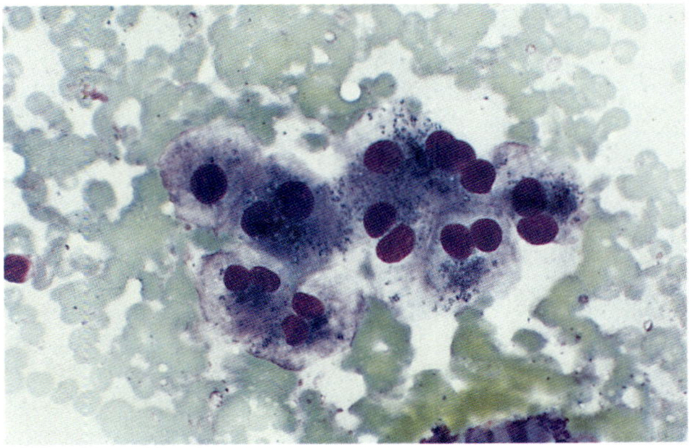

Fig. 3-3. In this Diff-Quik–stained aspirate, the benign hepatocytes demonstrate their characteristic polygonal contours, well-defined cell borders, and obvious cytoplasmic granularity. Binucleated hepatocytes occur with some frequency. Whether these cells possess one or two nuclei, they usually are round, with smooth contours and are centrally positioned.

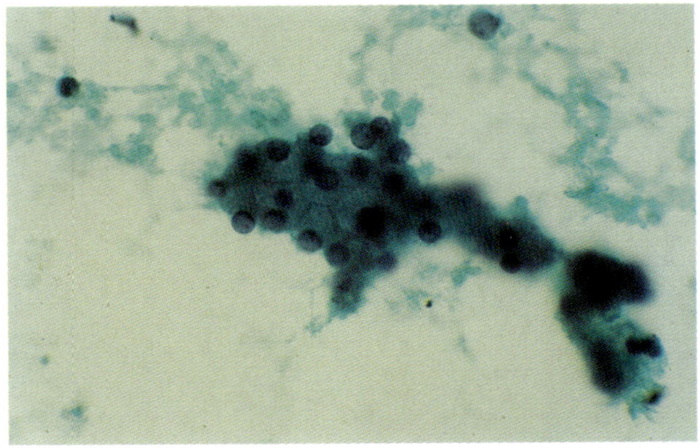

Fig. 3-4. With a Papanicolaou stain, the chromatin is very finely granular and evenly distributed throughout the round nuclei, which have delicate nuclear membranes. Small nucleoli may be more apparent than with the Romanowsky stains. Cytoplasm is opaque and granular in quality, and cell borders are readily apparent. The nuclear to cytoplasmic ratios are low.

these elements have elongated nuclei with pale, finely granular chromatin, inconspicuous nucleoli and indistinct, scanty cytoplasm. When a trabecular arrangement is evident, they are usually one or two cells thick.

A number of investigators have stated that the cell block preparation can be very helpful in liver aspirates, and we certainly agree.[2–6] Although characteristic arrangements of hepatocytes may be seen in direct smears, architectural patterns are better preserved and more easily recognized in cell blocks (Fig. 3-10). Furthermore, this material lends itself beautifully to immunocytochemistry, as multiple levels can be obtained for a panel of immunoreactions. Benign hepatocytes can be seen in aspirates of a number of focal conditions, for example, hepatic abscesses.

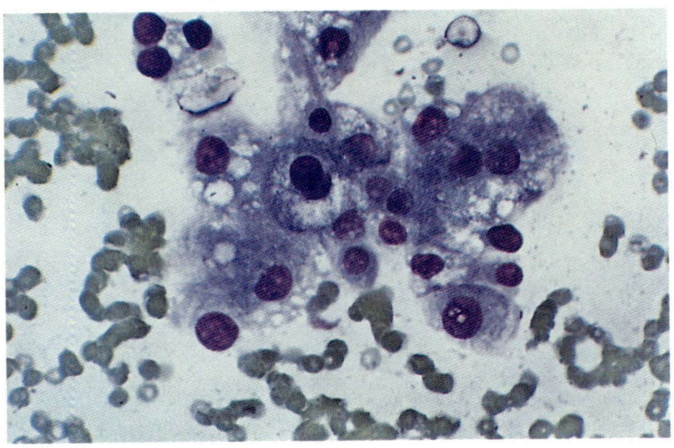

Fig. 3-5. Variability in nuclear sizes is fully acceptable among benign hepatocytes. They remain round and smooth with relatively low nuclear to cytoplasmic ratios. In this small, flat cluster, several of the hepatocytes manifest lipid vacuoles in their cytoplasm (Diff-Quik).

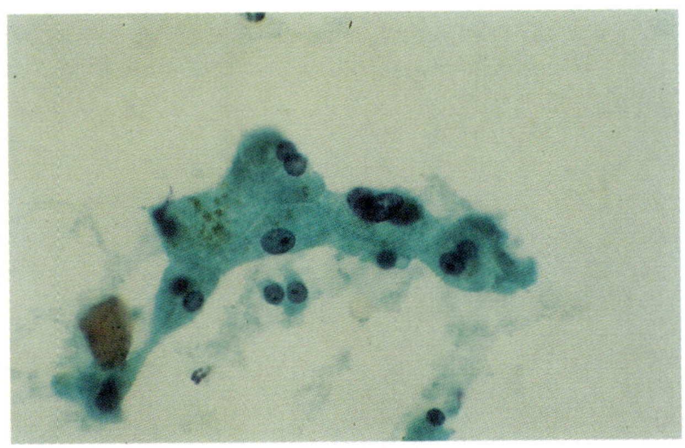

Fig. 3-6. With the Papanicolaou stain, bile is granular with a yellow to golden brown color. It may appear both within the cytoplasm of the hepatocytes as well as external to them. In this example, note also the variability in nuclear size and the presence of an occasional binucleated hepatocyte.

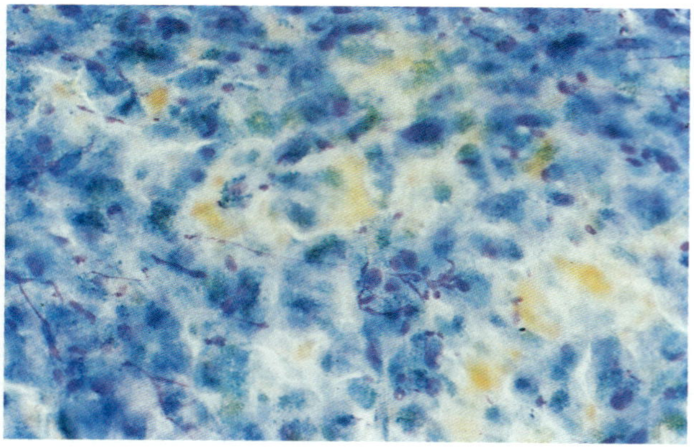

Fig. 3-7. With the Diff-Quik stain, bile has a distinct green color. It may appear somewhat less granular in texture. With this low magnification, it is possible to recognize the characteristic cord-like arrangement of benign hepatocytes.

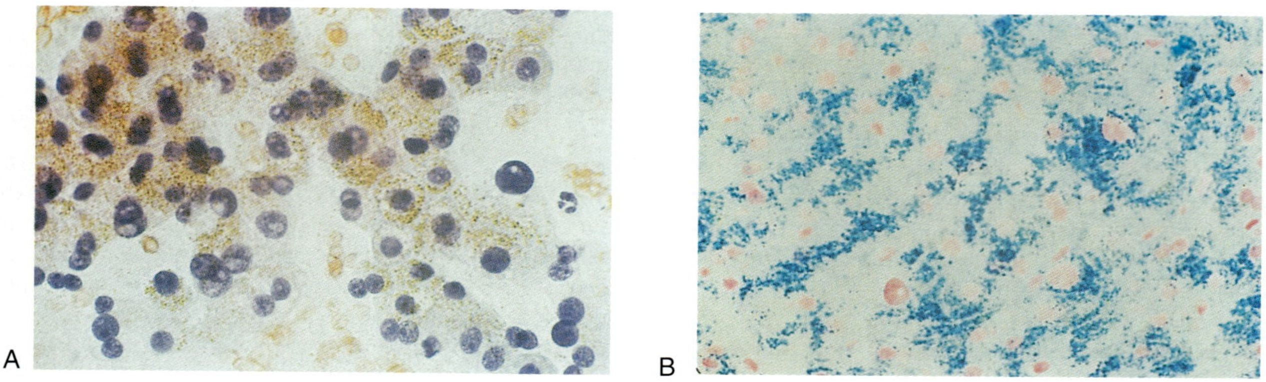

Fig. 3-8. (**A**) The cytoplasm in these hepatocytes is packed with hemosiderin granules. This iron-containing compound has a brown tint with the Papanicolaou stain. In this field, the intact hepatocytes have low nuclear to cytoplasmic ratios. A few stripped nuclei are also evident. (**B**) Prussian-stained cell block from the same specimen. The blue coloration confirms that the granular material contains iron.

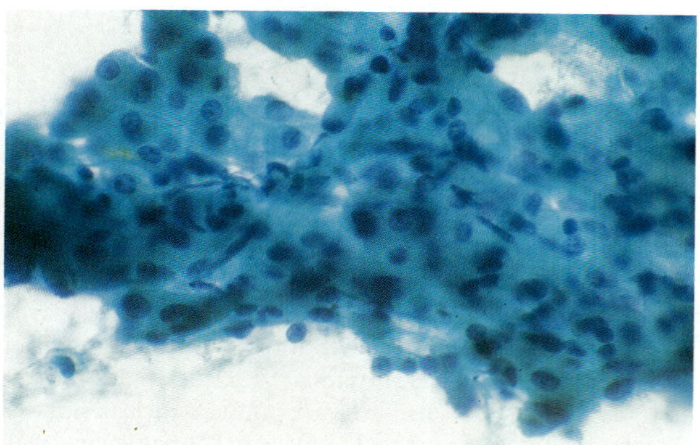

Fig. 3-9. Kupffer cells and vascular structures may be present within sheets of benign hepatocytes. These lining cells are characterized by elongated nuclei with benign-appearing chromatin, inapparent nucleoli, and very little discernible cytoplasm (Papanicolaou).

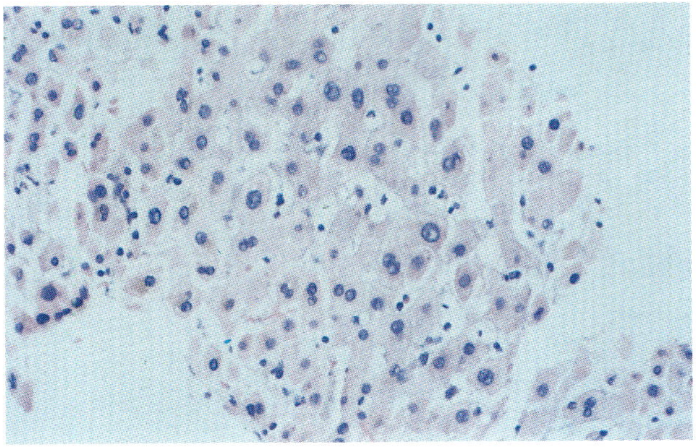

Fig. 3-10. In cell blocks, the expected features of hepatocytes are readily apparent. These include polygonal cell contours, well-defined cell borders, granular cytoplasm, and round central nuclei. In addition, the normal cord arrangement of the cells is more apparent than it is in direct smears. Although the nuclear to cytoplasmic ratios will vary, most of them appear low (H&E).

HEPATIC ABSCESSES

Most abscesses in the liver are bacterial in origin and are usually secondary to acute cholangitis, pylephlebitis, sepsis, or trauma. The typical clinical presentation is that of a patient with fever, right upper quadrant abdominal pain, and point tenderness. Radiographically, solitary or multiple mass lesions may be present that can mimic both primary and metastatic cancers.

Aspiration biopsies of pyogenic abscesses reveal loose sheets of neutrophilic leukocytes, abundant granular necrotic debris, and at times the causative organism (Figs. 3-11 to 3-13). Often, bacterial organisms are identifiable in Diff-Quik–stained material. However, Gram stain is preferable to identify and classify the bacterial population. In our experience, the most common fungus associated with hepatic abscesses is *Candida*. Both pseudohyphae and budding yeast may be present in the smears. In pyogenic abscesses, the neutrophils will greatly outnumber the benign hepatocytes. The latter may show both degenerative and regenerative changes. Degenerative alterations include cytoplasmic vacuolization, pyknosis, and fragmentation of the nuclear membrane. Prominent, large nucleoli are a hallmark of regeneration. Whenever a liver abscess is encountered, an additional aspirate should be performed with the material submitted for microbiologic culture.

We have seen a single example of a specimen in which the predominant inflammatory cell was the eosinophil, rather than the neutrophil (Fig. 3-14).

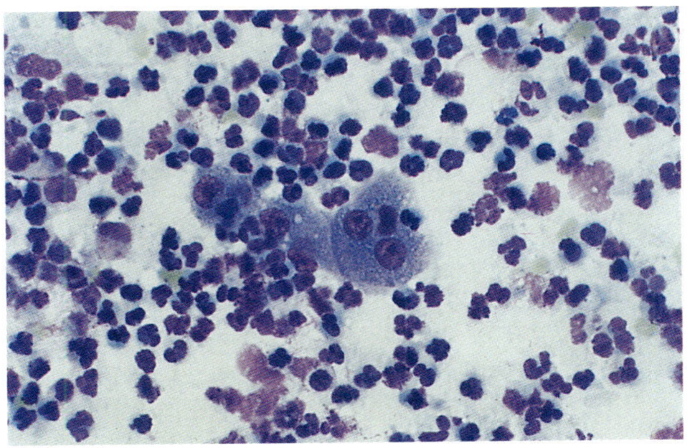

Fig. 3-11. In pyogenic abscesses, the segmented leukocytes greatly outnumber the benign hepatocytes. The latter cell type may demonstrate either regenerative or degenerative changes. As present in this field, the regenerative alterations include binucleation and large nucleoli (Diff-Quik).

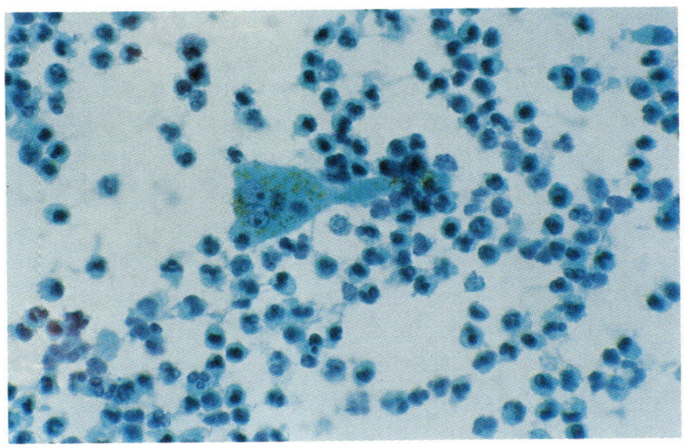

Fig. 3-12. Similar features may be seen with the Papanicolaou stain. Nucleoli may be more prominent, as may be the cytoplasmic granularity. In the face of such an inflammatory infiltrate, one should hesitate before rendering a malignant diagnosis in that benign inflammatory-related changes may result in prominent alterations.

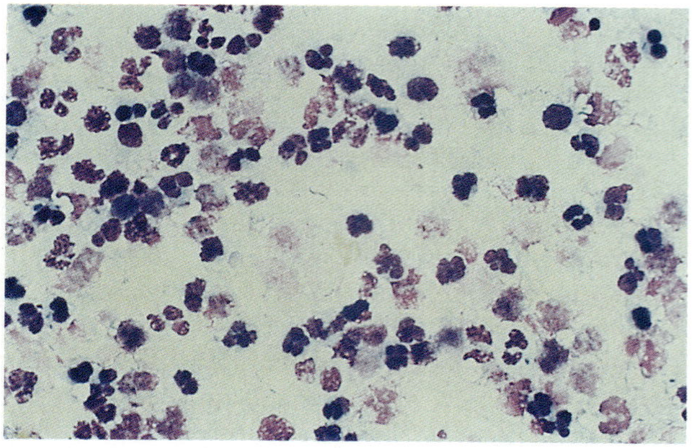

Fig. 3-13. Both neutrophils and hepatocytes are markedly degenerated in this example. In many of the cells, it is difficult to see any degree of cellular detail. Also present in this example are minute bacterial organisms in small chains and clusters (Diff-Quik).

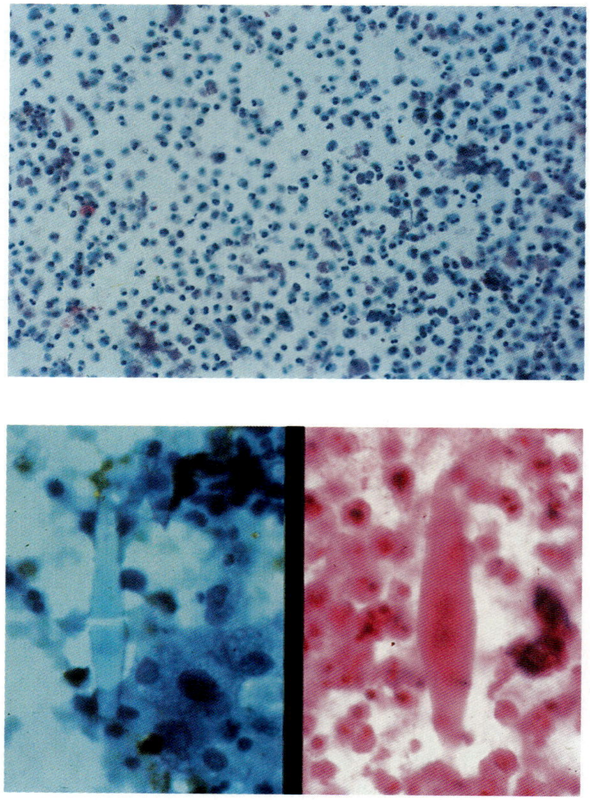

Fig. 3-14. (**A**) Aspirate from a middle-aged woman with fever and a solitary hepatic mass. Segmented leukocytes are numerous. However, these cells are eosinophils, rather than neutrophils (Diff-Quik). (**B**) Scattered about the specimen were numerous acellular structures consistent with Charcot-Leyden crystals. They have elongated polygonal contours with very sharply defined edges. With the Diff-Quik stain, they are pale blue in color. They can be accentuated with the periodic acid-Schiff stain, as exhibited in this example.

Scattered throughout the specimen were numerous acellular elongated structures consistent with Charcot-Leyden crystals. Although organisms were not recognizable within the smears, the patient (a middle-aged woman) became serologically positive for *Fasciola hepatica*. Tao[7] illustrated an example of hepatic clonorchiasis in aspiration specimens. In addition to numerous eosinophils and Charcot-Leyden crystals, the smears contained the ova of the fluke *Clonorchis sinensis*.

Amoebic abscesses are a relatively unusual clinical complication of amoebic enterocolitis. Patients present with fever and right upper quadrant pain. Most often, a single, large, irregular mass is present in the right hepatic lobe. In contrast to the pyogenic abscesses, segmented leukocytes are often not present in overwhelming numbers. Rather, the macrophage may be the predominant element (Fig. 3-15). In addition, the smears will contain granular debris and at times the trophozoites of *Entamoeba histolytica*.[8] The organism is characterized by rounded contours, abundant granular to vacuolated cytoplasm, and a single, small, round nucleus. The latter is characterized by a thick membrane with beading and a prominent karyosome. Phagocytized erythrocytes may be evident within the cytoplasm. The differential diagnosis of the trophozoites revolves about macrophages. The latter typically have a bean-shaped nucleus with finely granular chromatin, a delicate membrane, and small nucleoli. Frias-Hidvegi[9] illustrated an example of *Giardia lamblia* in a percutaneous aspiration biopsy of the biliary tree. The parasites were more readily identifiable in the Romanowsky-stained material than in the Papanicolaou-stained smears.

Although involvement of the liver with actinomycosis is most unusual, Shurbaji and colleagues[10] recently reported the cytomorphologic features of

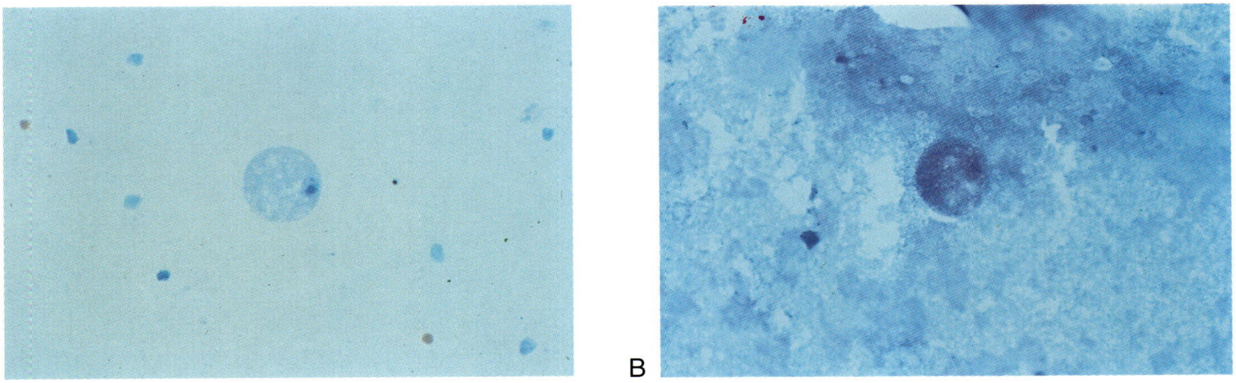

A B

Fig. 3-15. Liver aspirate containing trophozoites of *Entamoeba histolytica*. (**A**) They are characterized by rounded contours, abundant bubbly cytoplasm, and a single small round nucleus with a prominent karyosome. Although not evident in this example, phagocytized erythrocytes may also be recognized. A major differential diagnostic entity is benign histiocytes. In the latter, the nuclei will not be as perfectly round. Rather, they will be ovoid to elongated to banana shaped. Furthermore, nucleoli will not be nearly as prominent as the karyosome (Papanicolaou). (**B**) Similar features are evident with the Diff-Quik stain.

hepatic actinomycosis in a woman bearing an intrauterine contraceptive device. Scattered against a background of necrotic debris were irregular clumps of the organisms; gram-positive filaments with acute angle branching were demonstrated.

CYSTS

Cysts are a relatively frequent form of space-occupying mass lesion in the liver. They may be solitary or multiple and may attain rather large dimensions. Simplistically, hepatic cysts may be classified as congenital, parasitic, and neoplastic. Congenital cysts generally consist of spherical distentions of the biliary system within the parenchyma of the liver. They are lined with a single layer of cytologically bland glandular epithelial cells that vary from columnar to cuboidal to flat in shape. The surrounding benign hepatocytes may manifest atrophic, degenerative, or regenerative alterations. Aspirates of these benign cysts generally yield a large volume of clear fluid.[2,11,12] In the radiologic suite, it will probably be impossible to make direct smears of this fluid due to its watery nature; cellularity will therefore be scanty. This fluid is best processed in the cytology laboratory by one or more concentration techniques, for example, membrane filtration or cytocentrifugation. On a background of proteinaceous material, histiocyte-like elements will be present, as may be small aggregates of benign epithelial cells. The latter are derived from the cyst lining. The histiocyte-like cells, on the other hand, may represent true macrophages or dispersed, degenerated epithelial elements. In any case, the benign nature of this specimen is readily apparent.

Hydatid disease is caused by the tapeworm *Echinococcus granulosus*. When infective ova are inadvertently consumed by the human host, embryos develop within the proximal small intestine and may reach the liver through the blood. This may result in one or more echinococcal cysts, which usually develop in the right hepatic lobe. These cysts have a multilayered wall that includes an inner germinal layer from which brood capsules develop. Within the capsules, scoleces evolve; the latter are characterized by two rows of hooklets and suckers. Clinically, patients may present with right upper quadrant abdominal pain and/or findings consistent with biliary tract obstruction. Not infrequently, however, these masses are completely asymptomatic. If an echinococcal cyst is clinically suspected, FNA biopsy is probably contraindicated because of the potential risk of inducing an anaphylactic reaction.

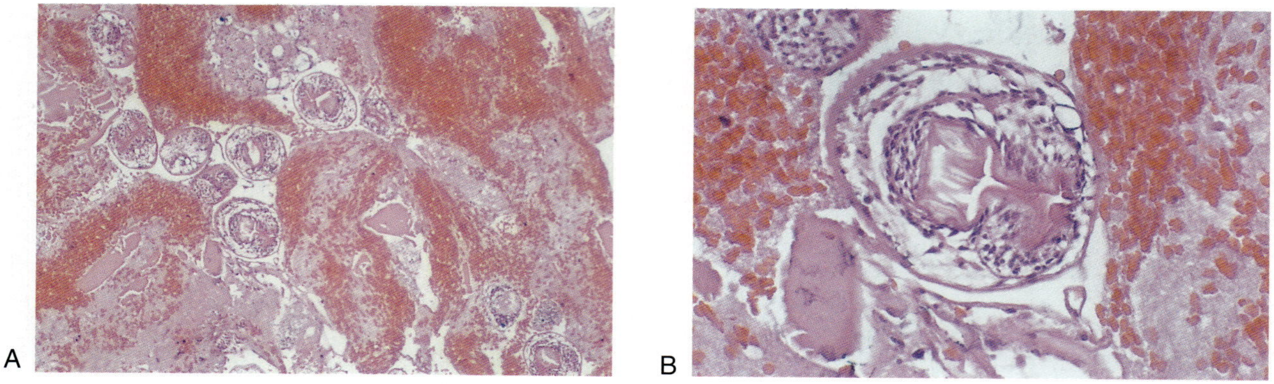

Fig. 3-16. Cell block preparation of a liver aspirate from a 15-year-old boy who was injured during a soccer game. This led to the discovery of his asymptomatic cystic liver mass. (**A**) In cross-sectional profile, numerous larvae of the *Echinococcus granulosus* are evident. Note the total absence of an associated inflammatory cell response. (**B**) At higher magnification, an inverted scolix, including hooklets, may be seen in detail (H&E). (Specimen courtesy of Dr. B. Naylor, Ann Arbor, MI.)

Aspiration biopsies of hydatid cysts, in contrast to pyogenic or amoebic abscesses, do not contain inflammatory cells to any extent (Fig. 3-16). Rather, acellular granular material predominates the smear background. A specific diagnosis may be rendered if one discovers scoleces that have numerous hooklets, some arranged in a ring-like pattern.[13] Theoretically, at least, one might also find a fragment of the laminated membrane from the cyst wall; this would appear as a uniformly thick band of somewhat refractile eosinophilic material.

CIRRHOSIS

Cirrhosis may be defined as a chronic condition in which the architecture of the entire liver is deranged by the combination of regenerating nodules of benign hepatocytes and interconnecting fibrous connective tissue. The latter varies from broad, irregular scar tissue to thin, barely perceptible fibrous bands. The regenerative nodules also vary in size and appearance; in general, the normal lobular architecture is disrupted within the nodules. In the United States, the long list of etiologic conditions is led by chronic alcoholism. In and of itself, cirrhosis is a major clinical problem with a relatively high mortality, with patients dying either from portal hypertension or hepatic failure. In addition, cirrhosis is the underlying soil for the development of the majority of hepatocellular carcinomas.

Aspiration biopsy of a typical case of cirrhosis is unlikely, but if one of the regenerative nodules predominates, this may simulate a malignant neoplasm leading to biopsy. Cellularity of these aspirates will vary from scanty to a high level (see below). Presumably, this reflects the influence of the fibrous connective tissue. A key feature is the variability in the appearance of the hepatocytes in these aspiration biopsies.[11,12,14–17] Some may appear totally unremarkable, whereas others show atypical features including nuclear enlargement, binucleation, anisonucleosis, and hyperchromasia (Figs. 3-17 to 3-19). In fact, the finding of this polymorphism within the same cell cluster is typical of cirrhosis and is helpful in distinguishing this benign entity from hepatocellular carcinoma. The level of cohesion among aspirated hepatocytes is also variable, even within the same

Cytomorphologic Features of Cirrhosis

- Variable cellularity and cohesion
- Within aggregates, hepatocytes are present in thin (1–3 cells thick) cords
- Variability in appearances of hepatocytes (polymorphism)
- Clusters of bile duct cells
- Chronic inflammatory cells in variable numbers

specimen. One may see a large proportion of individually dispersed, benign hepatocytes or a predominance of large aggregates. The latter will have a flat appearance in which the hepatocytes are arranged in rows or columns that are generally no more than two cells thick.[5,11,12] In the smears, one may also see numerous aggregates of benign bile duct epithelial cells and fragments of connective tissue.[12] The clusters of bile duct cells may appear flat or three-dimensional (Fig. 3-18). Chronic inflammatory cells may also be numerous in the smears. Although the hepatocytes may appear cytologically atypical, they generally maintain normal nuclear to cytoplasmic ratios, as described on page 96. Finally, as a true reflection of the underlying histopathology, the smears may contain dysplastic hepatocytes (Fig. 3-19), which can create diagnostic problems and confusion with hepatocellular carcinoma.[15] These dysplastic elements are discussed in a later section, as is the differential diagnosis with hepatocellular carcinoma.

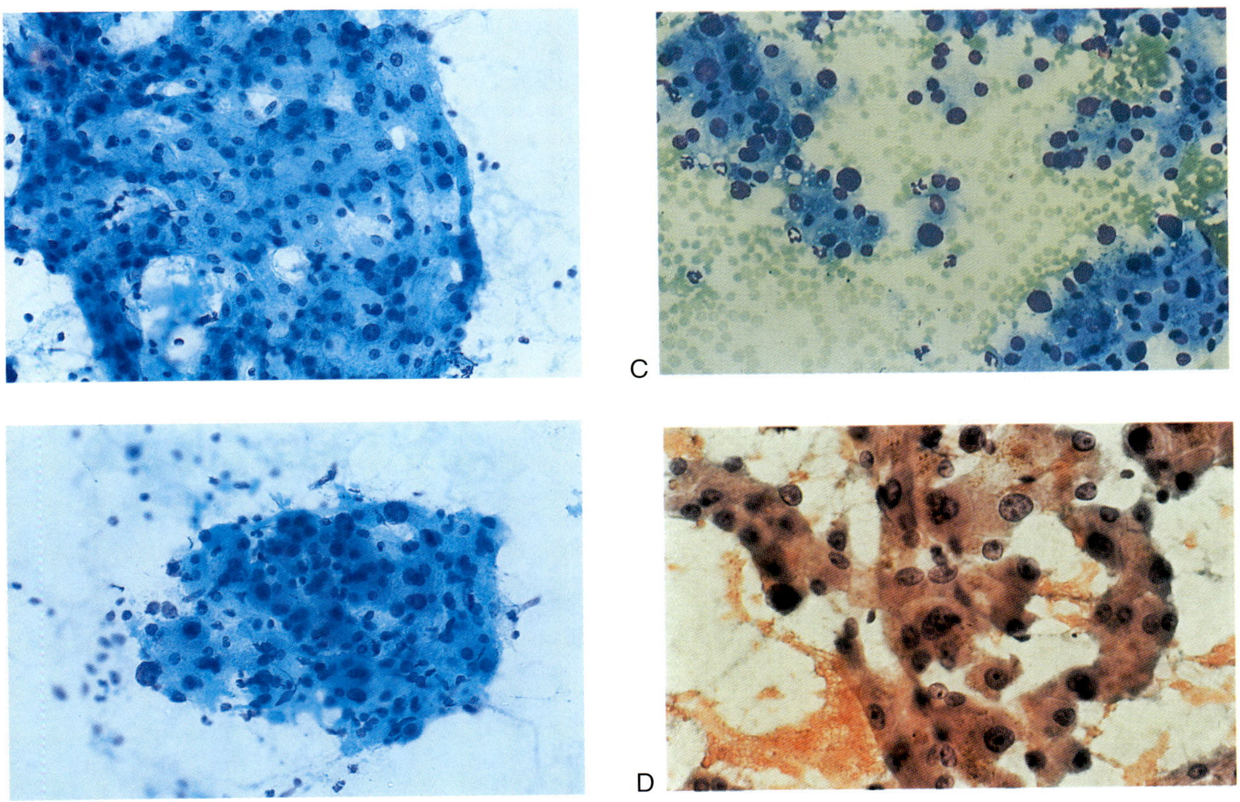
A B C D

Fig. 3-17. Aspiration biopsy of cirrhotic nodules yields a spectrum of hepatocytic appearances. (**A**) Most of the hepatocytes in this flat aggregate appear to be present in cord-like arrangements. The features of hepatocytes are readily apparent, including well-defined cell borders, polygonal cellular contours, and granular cytoplasm. The centrally positioned nuclei are round and manifest a mild degree of anisonucleosis. Spindle-shaped nuclei are also apparent and may represent endothelial cells, Kupffer cells, or fibroblasts. The amorphous extracellular material may be collagen (Papanicolaou). (**B**) A greater spectrum of nuclear sizes is present in this example. Overall, the nuclear to cytoplasmic ratios remain characteristically low (Papanicolaou). (**C**) A similar wide range in nuclear sizes is evident in this Diff-Quik–stained preparation. Several of the larger nuclei have lost some or all of their surrounding cytoplasm. (**D**) Nucleoli may also be well developed in hepatocytes in regenerating nodules. Note the absence of endothelial cells coating the external surfaces of these hepatocytic aggregates (Papanicolaou).

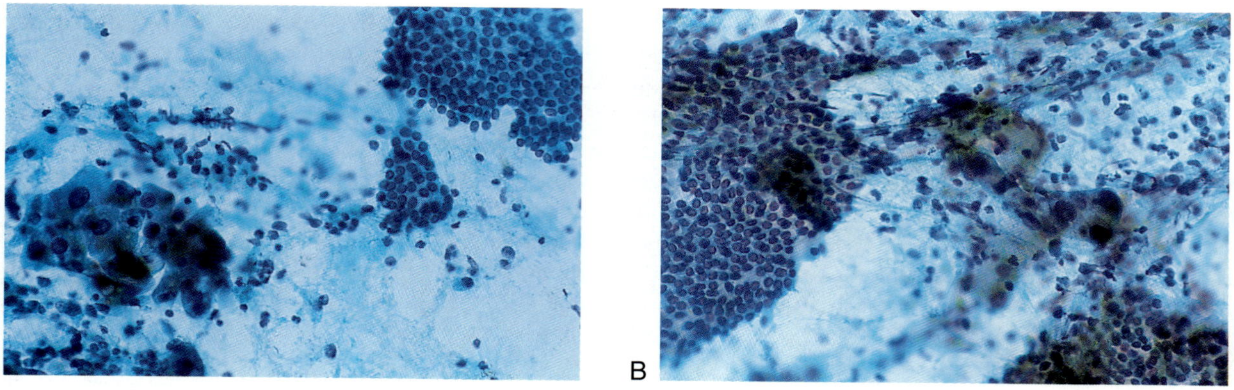

A B

Fig. 3-18. (**A**) Bile duct epithelium is characteristically present in aspirates of cirrhotic nodules. In this field, a small sheet of hepatocytes is adjacent to a large flat aggregate of ductular epithelium. The latter are characterized by small uniform round nuclei, scanty volumes of filmy cytoplasm, and a honeycomb arrangement (Papanicolaou). (**B**) A small aggregate of degenerating hepatocytes is surrounded by broad sheets of bile duct epithelium. The latter is easily distinguished from metastatic adenocarcinoma by the small benign-appearing nuclei and by the preservation of both intercellular cohesion and polarity (Papanicolaou).

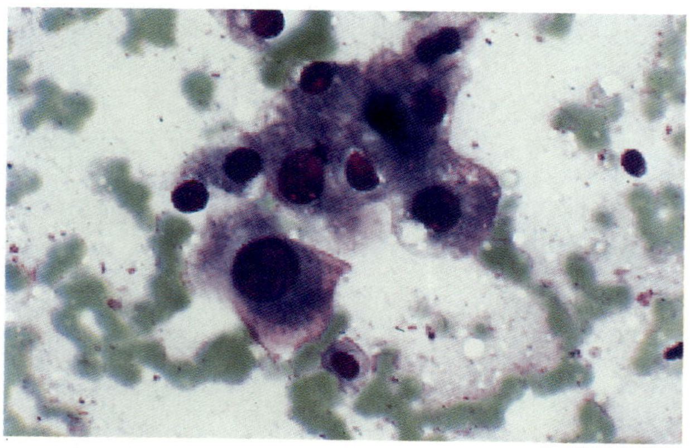

Fig. 3-19. This aggregate of hepatocytes was obtained from an aspirate of a cirrhotic nodule. It includes a single dysplastic hepatocyte characterized by a very enlarged nucleus, a huge macronucleolus, and a high nuclear to cytoplasmic ratio. However, note the persistence of a very smooth nuclear contour, as well as the company kept by this dysplastic cell. Several of the neighboring hepatocytes are entirely normal in appearance. Such an intimate polymorphic admixture would not be expected in an aspirate of hepatocellular carcinoma (Diff-Quik).

HEPATOCELLULAR CARCINOMA (HEPATOMA)

Hepatocellular carcinoma is the most common primary liver cell malignancy and is one of the most frequent cancers worldwide.[18,19] It is especially common in southeast Asia, China, and the sub-Saharan portions of Africa. By contrast, in the United States, it ranks as the 22nd most frequent form of malignancy, with less than 3,000 new cases diagnosed annually. An underlying cirrhosis is present in approximately 90 percent of all patients in whom this cancer develops, with major predisposing etiologies including infection with either hepatitis B or hepatitis C virus or alcohol abuse. Although much less frequent overall, a relatively large proportion of patients with hemochromatosis may go on to develop a hepatoma.

In the United States, hepatocellular carcinoma is predominantly a disease of older (greater than 60 years of age) men (85 percent of all patients) who present with abdominal pain, right upper quadrant abdominal mass, and weight loss. Most often, imaging studies will demonstrate a solitary large mass in the liver, but other radiologic patterns are also possible. The vast majority of patients will have an elevation in serum α-fetoprotein (AFP) levels. In the United States, the prognosis is dismal, with a median survival of less than 6 months.[20,21] As these tumors are essentially uniformly and rapidly fatal, the single most important prognostic factor is a small tumor size, making it amenable to a complete surgical resection. Accordingly, these neoplasms must be diagnosed in an early, generally asymptomatic stage. In our opinion, FNA biopsy is the procedure of choice when a dominant mass is seen in a liver. In this setting, aspiration biopsy possesses high levels of diagnostic sensitivity and specificity.[2,6,12,22,23]

Cytomorphologic Features of Hepatocellular Carcinoma

- High cellularity
- Polygonal cells with distinct borders and dense granular cytoplasm, possibly with bile
- One or two central round hyperchromatic nuclei with prominent nucleoli
- Uniformly high nuclear to cytoplasmic ratios
- Trabecular arrangements
- Fragments with sinusoids
- Naked atypical nuclei

The cytomorphologic features of hepatocellular carcinoma in FNA biopsy has been well described in detail by a number of publications.[2,5,6,11,12,22,24–36] As a general introduction, several correlates can be made between the histologic grade of hepatocellular carcinoma and the ease with which it is diagnosed in FNA biopsies. Moderately differentiated hepatocellular carcinomas are relatively easy to diagnose as the tumor cells will retain recognizable features of hepatocytic differentiation, yet may be obviously aberrant and malignant. On the other hand, it may be very difficult indeed to distinguish benign hepatocytes from some well-differentiated hepatocellular carcinomas. Conversely, high-grade hepatocellular carcinomas are easy to recognize as malignant, but they may be difficult to distinguish from metastases. Histologically, most hepatocellular carcinomas are well or moderately differentiated. Consequently, the malignant cells in the majority of aspirates will demonstrate features consistent with hepatocytic differentiation (Fig. 3-20), including polygonal contours, opaque granular cytoplasm, distinct cellular borders, and one or two centrally positioned nuclei, as noted in the box above. Nuclear membrane abnormalities in the form of uneven thickening and contour irregularities are fairly frequent in aspirates of hepatoma, although these alterations may be somewhat subtle (Fig. 3-20). Although subjective, hepatocellular carcinomas are said to have hyperchromatic chromatin and to have much more prominent nucleoli than are found in benign hepatocytes. The nucleoli may be irregular in contour and/or multiple. In addition to nucleoli, nuclear pseudoinclusions may also be evident (Fig. 3-21). The cytoplasm may include any number of the different types of inclusions, such as lipid vacuoles, Mallory bodies, other eosinophilic globules, and basophilic globules (Fig. 3-22). However, the only one specific for hepatocellular differentiation is bile (Fig. 3-23). The latter material may also be external to, and closely associated with, the malignant cells. For us, the single most important diagnostic criterion is a uniformly increased nuclear to cytoplasmic ratios among the malignant hepatocytes (Fig. 3-20). That is, essentially all the tumor cells will show an elevation of this ratio. Endothelial cell associations are also diagnostically significant. Trabeculae can be defined as aggregates of hepatocytes surrounded by endothelium (Fig. 3-24). Frequently, the trabeculae show a branching or an anastomosing pattern (Fig. 3-25). Such arrays are characteristic of hepatocellular carcinoma. Within larger fragments of neoplasm, sinusoidal capillaries can be seen separating columns of neoplastic hepatocytes (Fig. 3-26); these columns are often quite thick, being composed of four or more cell layers. Aspiration biopsy of hepatoma usually yields a very high level of cellularity in the smears, including both individually dispersed and many large aggregates and tissue fragments of malignant elements (Fig. 3-27). Some investigators have found that as the histologic grade increases, there is a progressive loss of intercellular cohesion, with a greater proportion of malignant cells occurring as isolated elements within the smears. In direct smear preparations, there may be numerous stripped (naked) malignant-appearing nuclei[30,32] (Fig. 3-28).

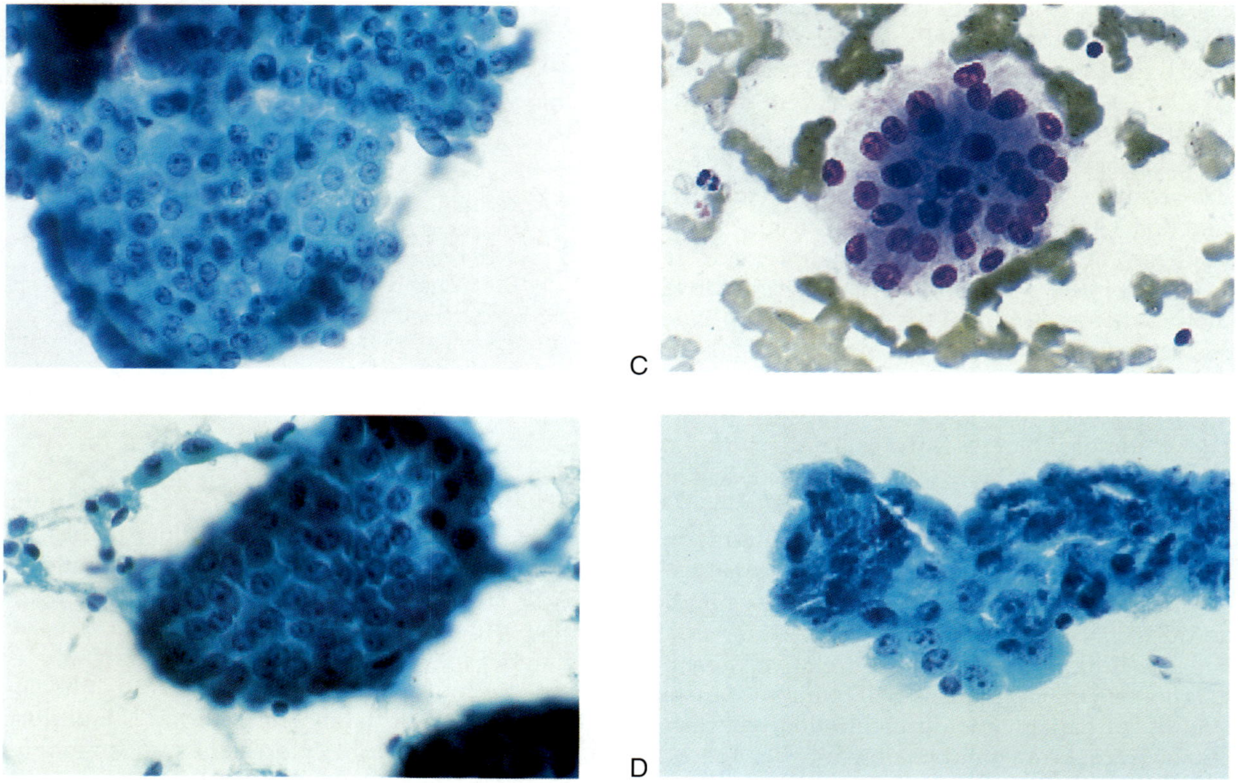

Fig. 3-20. In most hepatocellular carcinomas, aspirates will yield tumor cells recognizable as hepatocytic in type. (**A & B**) In these Papanicolaou-stained examples, the polygonal cell contours, well-defined cellular borders, opaque granular cytoplasm, and round central nuclei are all well demonstrated. The key to recognizing them as malignant is the uniformly high nuclear to cytoplasmic ratios. In both examples, nucleoli are also well developed. (**C**) In air-dried preparations, nuclear crowding may be even more apparent, although cytoplasmic granularity may be less evident (Diff-Quik). (**D**) Nuclear crowding may be very pronounced within aggregates of hepatoma when the nuclear to cytoplasmic ratios are very accentuated (Papanicolaou).

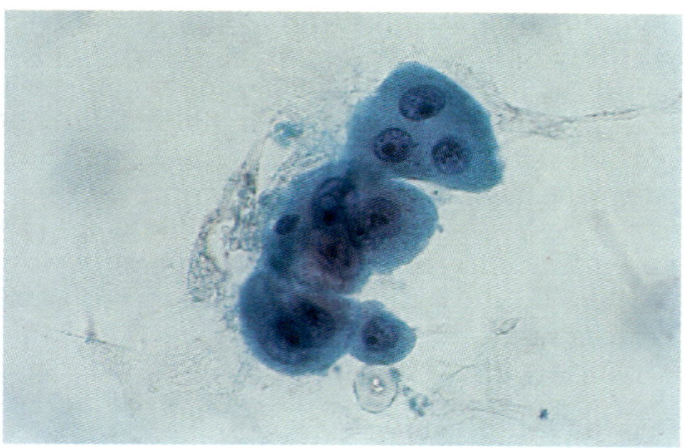

Fig. 3-21. Large nucleoli are evident in these hepatoma cells. In addition, one nucleus contains a cytoplasmic pseudoinclusion. It is characterized by a round contour, a distinct border (actually nuclear membrane), and a staining reaction similar to that of the cytoplasm (Papanicolaou).

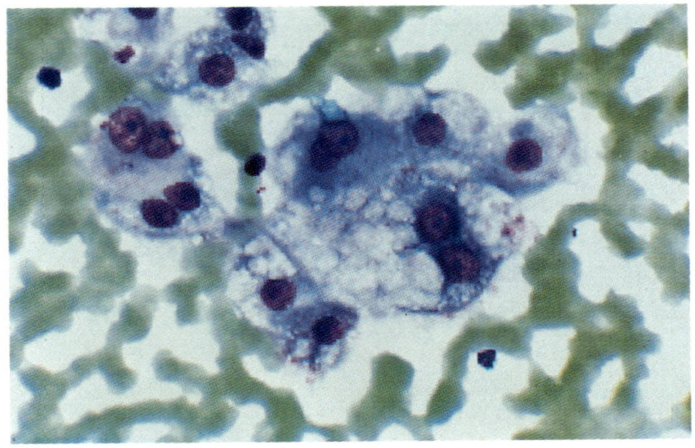

Fig. 3-22. Lipid vacuoles occupy much of the cytoplasm in several of these hepatoma cells. These should not be mistaken for cytoplasmic mucin vacuoles of an adenocarcinoma. Note that the nuclei of almost all tumor cells remain centrally positioned (Diff-Quik).

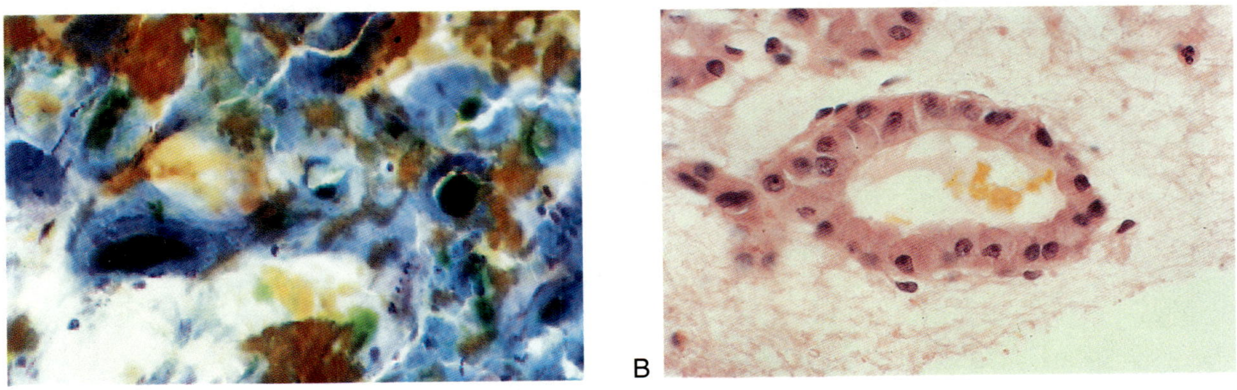

Fig. 3-23. (**A**) In this aspirate of hepatoma, bile is quite prominent. It is present both within the cytoplasm of the tumor cells, as well as external to them. Much of the latter appears to be present within lumens of pseudoacini, where it forms large plugs (Diff-Quik). (**B**) Cell block preparation from the same specimen. The lumen of the pseudoacinus contains bile. In addition, note the external surface is coated by endothelial cells and that many of the neoplastic elements have centrally positioned nuclei, in contrast to basal polarity expected in glandular neoplasms. All these features add up to a certain diagnosis of hepatocellular carcinoma (H&E).

As pointed out by Bottles and Cohen,[12] if aspirates of hepatocellular carcinomas are performed properly, the specimens should not contain any or very few benign hepatocytes. However, if the needle is moved outside the neoplastic mass, sampling the adjacent benign parenchyma, the smears will contain an admixture of both benign and malignant hepatocytes. In the experience of Nguyen,[11] this may not be unusual at all. In this scenario, the benign hepatocytes are not intimately admixed with the malignant elements;[5,12,25] rather, these two biologically different heptocellular components will occupy different areas of the slide.

Although the formation of trabeculae is the most characteristic common arrangement of malignant hepatocytes present in aspiration biopsies, pseudoacini may also be evident (Fig. 3-23). In some tumors, pseudoacinar arrangements will predominate, again reflecting the underlying histopathology of a minority of hepatomas. An apparent luminal space is surrounded by a palisade of malignant tumor cells, which thus resembles a glandular structure. One helpful clue may be bile within the lumens. Another clue that this is hepatocellular carcinoma and not an adenocarcinoma is the granular eosinophilic cytoplasm characteristic

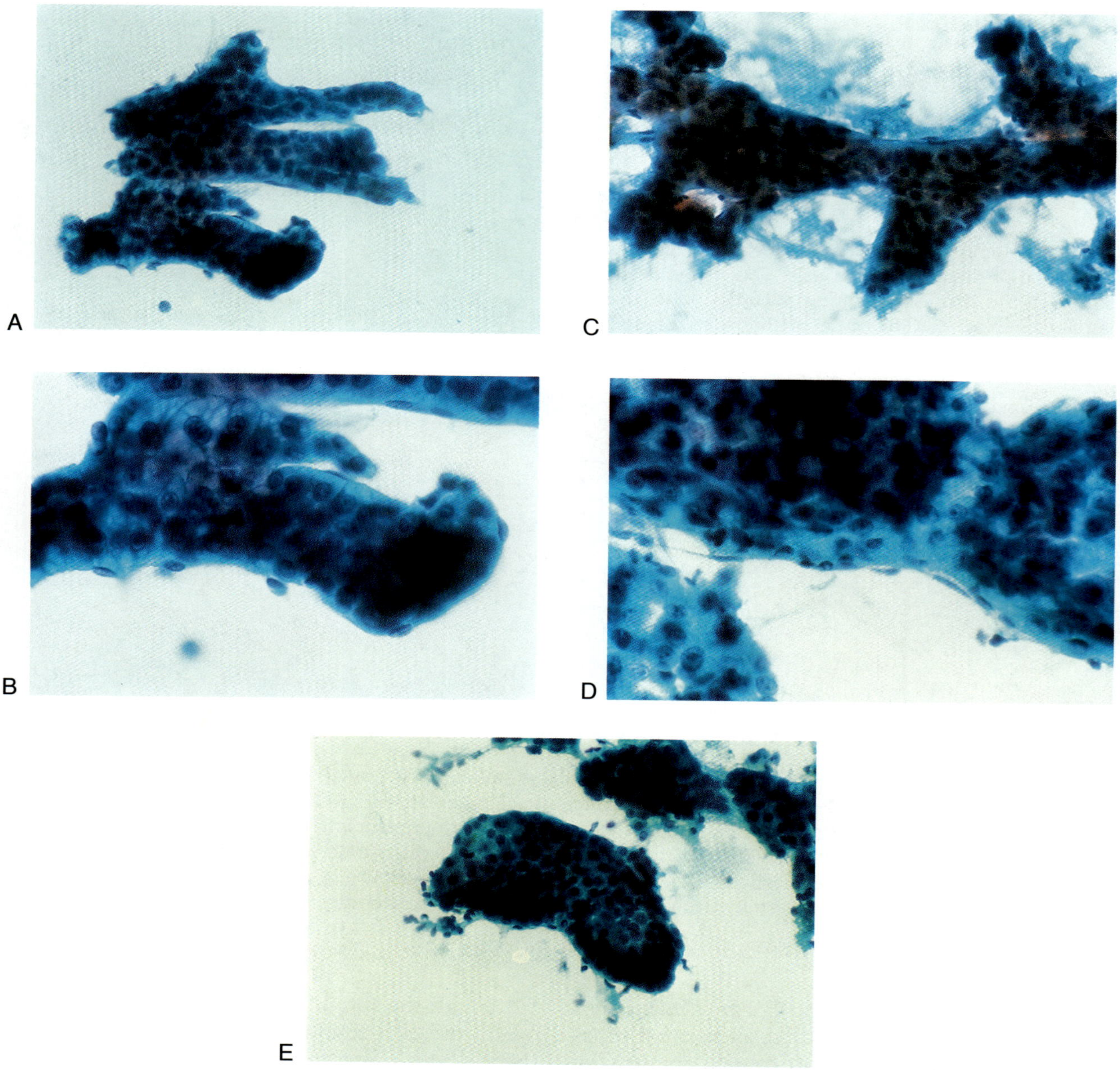

Fig. 3-24. (A) The rigid appearance of these trabeculae is evident. They are lined by, and separated from, each other by layers of endothelial cells. There is a suggestion of a branching pattern to these trabeculae as well (Papanicolaou). (B) Endothelial cells are characterized by solitary, small, oval to elongated nuclei with benign chromatin and very scanty cytoplasm. There may be an impression of endothelial cells falling off the surfaces of these trabeculae (Papanicolaou). (C) A continuous layer of endothelial cells line the external surface of these branched trabeculae. Note how these trabeculae are thick, composed of five or more layers of hepatoma cells (Papanicolaou). (D) These broad trabecular structures are lined by, and separated from, each other by continuous layers of delicate endothelial cells. Such a pattern would be most unexpected in a metastases or in a cholangiocarcinoma. In addition, note the high nuclear to cytoplasmic ratios and the well-developed nucleoli (Papanicolaou). (E) Although isolated, this broad trabecular structure is recognizable as such at low magnification by the external coating of endothelial cells (Papanicolaou).

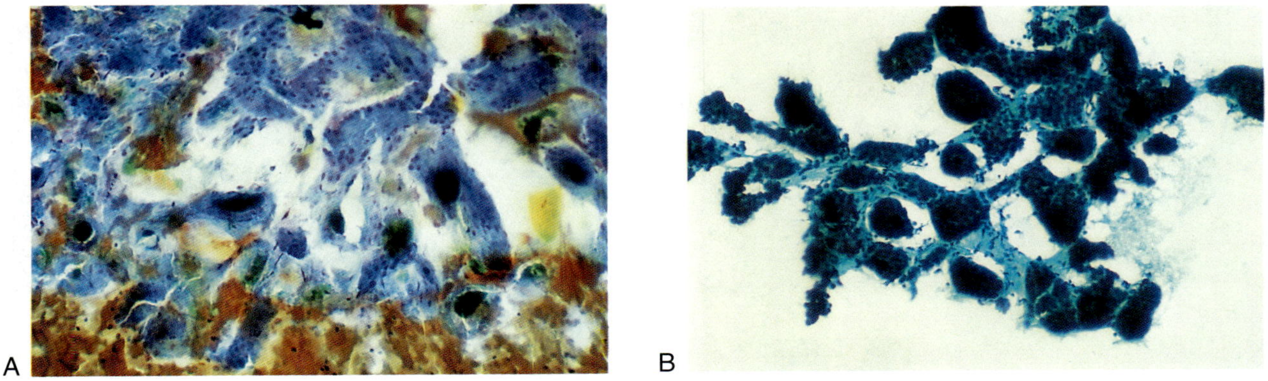

Fig. 3-25. At low magnifications, an anastomosing trabecular arrangement may be a helpful diagnostic clue to the diagnosis of hepatoma. (**A**) In this example, bile is prominent. The combination of this trabecular arrangement and bile is diagnostic of hepatocellular carcinoma (Diff-Quik). (**B**) In this example, the branched trabeculae appear to be oriented at right angles to one another; thus some of them resemble papillae (Papanicolaou).

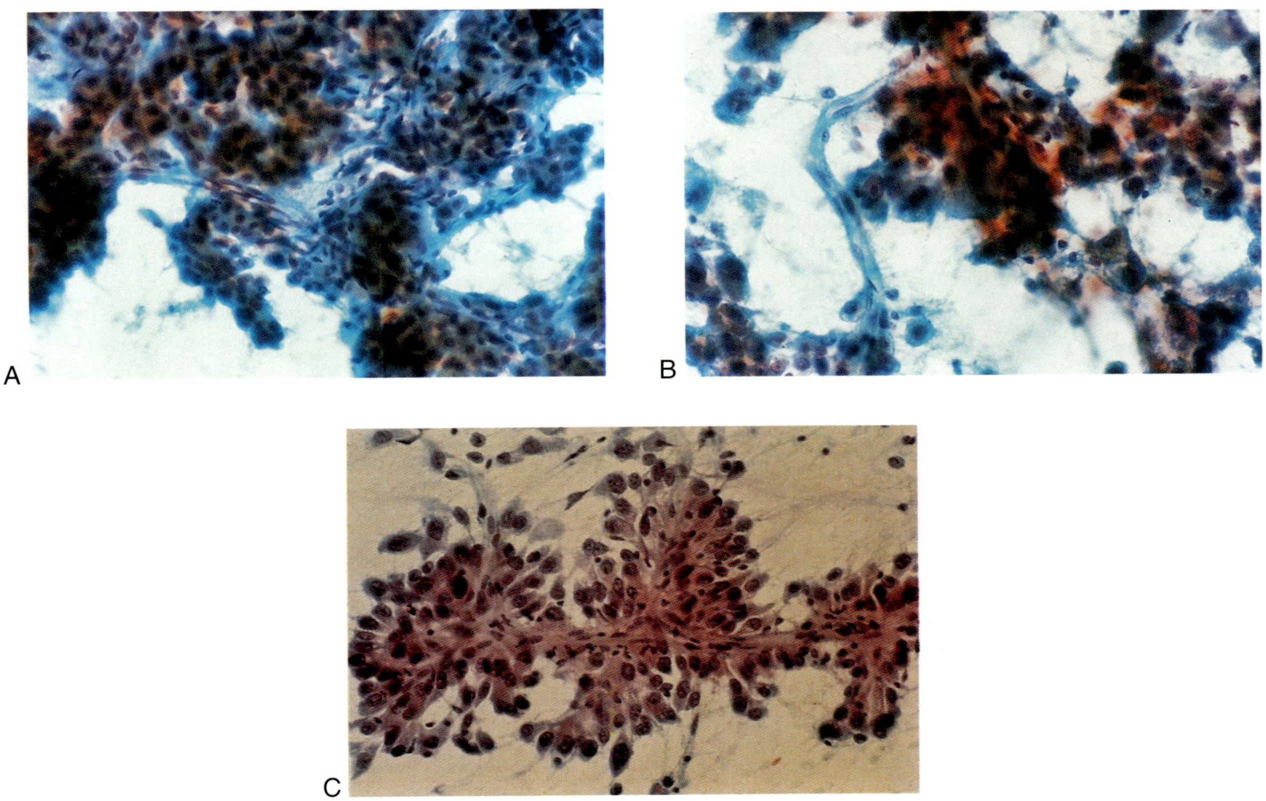

Fig. 3-26. (**A–C**) In addition to endothelial cells lining the external surfaces of trabeculae, capillary vessels course through fragments of hepatoma. At times, the vessels appear to be totally devoid of cells but in other instances are intricately coated by these neoplastic elements (Papanicolaou).

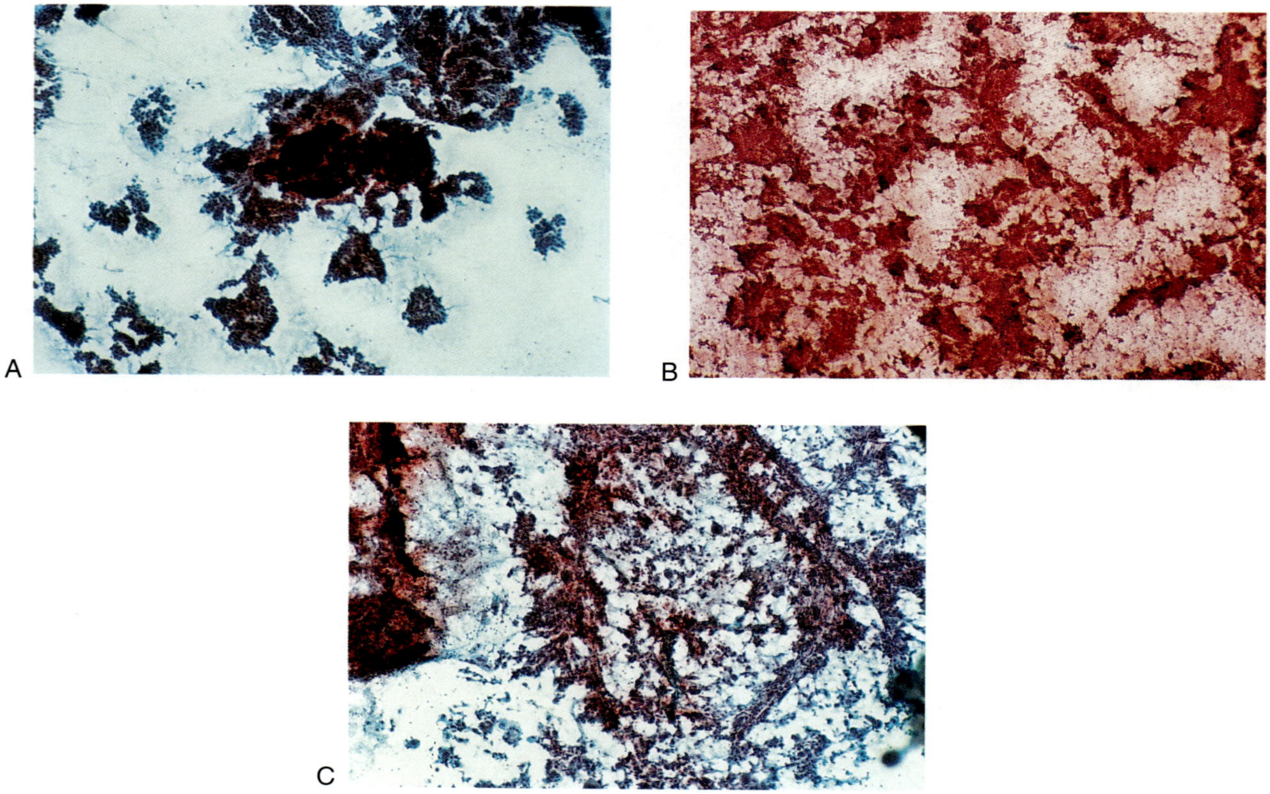

Fig. 3-27. Aspiration biopsies of hepatocellular carcinoma generally yield highly cellular smears. (**A**) As in this example, large fragments of neoplasm are frequently present within the specimen. Even at this low magnification, the presence focally of a trabecular arrangement of the hepatocytes is apparent. The smear background is distinctly clean (Papanicolaou). (**B**) In this example, smaller aggregates of tumor cells are evident without a well-developed anastomosing trabecular pattern (Papanicolaou). (**C**) Intercellular cohesion is greatly reduced in this specimen. The hepatoma cells are present both as isolated elements and in very small aggregates. In general, as the degree of differentiation declines, intercellular cohesion is progressively reduced (Papanicolaou).

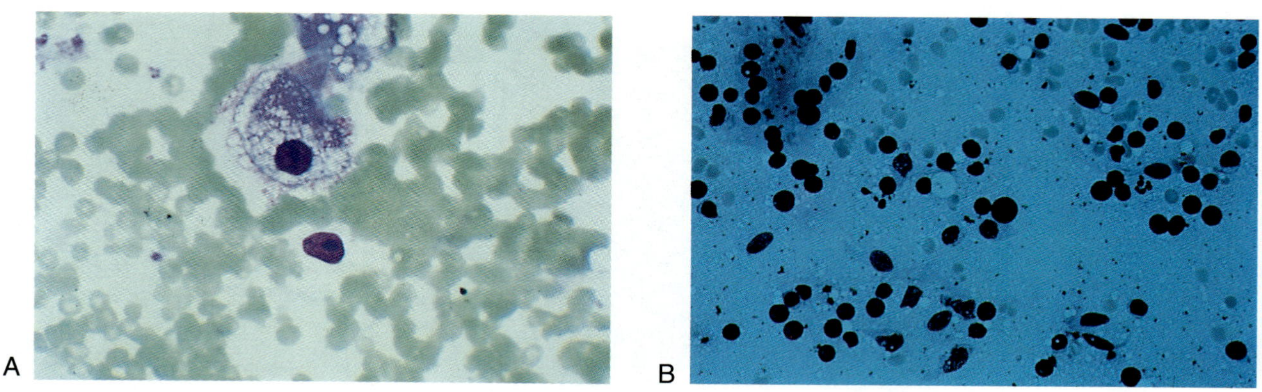

Fig. 3-28. Naked atypical hepatocytic nuclei are said to be characteristic of aspirated hepatocellular carcinomas. (**A**) One such stripped malignant-appearing nucleus is completely devoid of surrounding cytoplasm. It lies adjacent to a tumor cell with prominent cytoplasmic vacuolization (Diff-Quik). (**B**) The stripped atypical hepatocytic nuclei may be present in very large numbers (Diff-Quik).

of benign hepatocytes (Fig. 3-20). In addition, in many examples, the nuclei will be centrally positioned, again recapitulating normal hepatocytic appearances. Difficulties may result if the nuclei are peripherally (abluminally) oriented, as described by Wee and colleagues.[5] This may be compounded if cytoplasmic vacuoles (presumably lipid) are present within the cytoplasm of the malignant hepatocytes (Fig. 3-22). Another helpful clue for hepatoma is the presence of endothelial cells surrounding the circumference of the pseudoacini (Fig. 3-23B).

It is apparent that benign and malignant hepatocytes share many cytologic features in common, and except for a uniformly high nuclear to cytoplasmic ratio, we do not believe that a single other criterion is so absolute in this distinction.[31,36] However, no single cytomorphologic criterion is diagnostic of hepatocellular carcinoma.[26,31,35,36] Rather, it is a combination of morphologic alterations that permits one to make this diagnosis with a high level of certainty. For example, although malignant cells in smears tend to occur in larger aggregates and fragments than do benign hepatocytes, there certainly is overlap in this feature. The same can be said for the presence of trabeculae, especially relatively delicate ones.

As alluded to earlier, another problem in the distinction between benign and malignant hepatocytes is the occurrence of dysplastic cells, especially in cases of cirrhosis. On a cell-for-cell basis, dysplastic hepatocytes may be indistinguishable from malignant cells.[15,37] Importantly, it is the milieu in which the cells are found that distinguishes the two (Fig. 3-19). In smears from cirrhotic nodules, dysplastic hepatocytes will be relatively sparse and admixed with obviously normal, benign hepatocytes.[15] By contrast, in hepatocellular carcinoma one will see a relatively monomorphic population of hepatocytes. Furthermore, aberrations of nuclear contours are few to minimal in dysplastic hepatocytes, whereas they are fairly frequent in hepatoma.

That the distinction between benign and malignant hepatocytes is difficult has prompted several investigative groups to analyze long lists of cytologic features by statistical methods. Using regression analysis, Cohen and colleagues[32] found three statistically significant features for the presence of hepatocellular carcinoma. These included high nuclear to cytoplasmic ratios, the presence of a trabecular pattern, and naked atypical hepatocytic nuclei. We completely agree with their first feature, and partially with the second and third criteria. It is important to recognize that trabeculae may be present in aspirates of benign hepatocytes; this arrangement tends to be focal, however, and consists of thin columns of cells. If a trabecular pattern is present to any extent in the specimen, a malignant diagnosis is strongly favored. Although naked hepatocytic nuclei are more common in specimens from hepatomas, they may also be seen in aspirates of benign hepatocytic lesions. This reinforces the principle that one should never place major interpretive weight on nuclei that are devoid of their surrounding cytoplasm. Cohen et al.[32] described secondary criteria which did not reach significance in their statistical analysis. These included an irregular distribution of chromatin granules, uniformly prominent nucleoli, and the presence of multiple nucleoli. Features considered unhelpful included high smear cellularity, cellular pleomorphism, hyperchromatic nuclei, and an acinar arrangement of the cells. Recently, Sole et al.[36] esentially repeated the Cohen study.[32] They also found that uniformly high nuclear to cytoplasmic ratios are diagnostically important, but they did not discover that a trabecular arrangement nor naked hepatocytic nuclei were statistically of value.[34,36] These investigators did claim, however, that irregular chromatin distributions and loss of polarity within aggregates were crucial for the diagnosis of hepatoma.

We are unaware of any cytoplasmic antigens that can be used reliably to distinguish between benign and malignant hepatocytes. This is not surprising, in that most hepatocellular carcinomas show at least a modicum of hepatocellular differentiation. By contrast, ancillary diagnostic techniques that evaluate nuclear features may be helpful in this distinction.[38–43] For example, Ojanguren et al.[38] evaluated the efficacy of the immunocytochemical detection of proliferating cell nuclear antigen (PCNA) in FNA biopsies of the liver. In specimens derived from benign livers, immunoreactivity is either completely negative or minimal (less than 5 percent of nuclei staining). Conversely, hepatocellular carcinomas show increased staining for PCNA. Ploidy measurements, especially by image analysis, may prove to be of diagnostic value.[40–42] Aside from the possibility of dysplastic nuclei, aneuploidy should not be found in benign specimens; furthermore, high levels of polyploidy are much more frequent in malignant neoplasms.

The therapeutic and prognostic importance of distinguishing hepatocellular carcinoma from metastases is obvious. Several years ago, Bottles et al.[29] evaluated a series of hepatocellular carcinomas and metastases in aspiration biopsies for a large number of cytologic features. Subsequently, these features were subjected to a stepwise, logistic regression analysis. Three features emerged as statistically significant for the recognition of hepatoma. These included polygonally shaped tumor cells with centrally positioned nuclei, collections of neoplastic cells separated by sinusoidal capillaries, and bile intimately associated with neoplastic cells. Only the latter feature was completely specific for hepatocellular carcinoma (although quite insensitive).[29] Secondary diagnostic features included nuclear pseudinclusions and endothelial cells encasing clusters of neoplastic cells.[29] Cytologic features considered not to be helpful in this distinction included granular cytoplasm, cytoplasmic vacuoles and inclusions, large nucleoli, and multinucleated tumor giant cells. Although this was an excellent study, we do have one criticism of this analysis. Their metastatic category included a large number of neoplasms (lymphomas, sarcomas, and melanomas) that are typically not difficult to distinguish from hepatocellular carcinoma.[29] We believe this study needs to be repeated with only epithelial neoplasms in the metastatic group.

As stated above, poorly differentiated hepatocellular carcinomas comprise only a minority of all primary hepatic malignancies. We divide high-grade hepatomas basically into two types: tumors containing large, pleomorphic, and even bizarre malignant cells, and those composed almost exclusively of small malignant cells with very high nuclear to cytoplasmic ratios and little evidence of hepatocytic differentiation. In smears from the pleomorphic type, one can usually find tumor cells that are easily recognizable as hepatocytic in type (Fig. 3-29). Typically, they will be admixed with the much larger tumor giant cells.[24,33] The diagnosis of hepatocellular carcinoma will be straightforward if one recognizes that the smaller cellular component is also neoplastic and not simply reactive. In aspiration smears, the small cell type is composed of individually dispersed, rather uniform-appearing malignant cells that possess solitary nuclei and very high nuclear to cytoplasmic ratios (Fig. 3-30). The nuclei may be eccentrically placed within the scanty cyanophilic cytoplasm. Potential helpful clues are the presence of smooth round nuclei and granular dense cytoplasm. In this situation, it is especially important to search the entire smear carefully for malignant elements manifesting better evidence of hepatocytic differentiation.

As only a minority of all hepatocellular carcinomas are poorly differentiated, the malignant nature of tumor cells aspirated from hepatomas may not be immediately obvious in direct smears. This is one of the main reasons a very careful evaluation of architecture of the aspirated liver cells is so important. Certainly, to an extent, this can be achieved by careful examination of the cytologic smears. However, the overall arrangement of the cells and their interactions with each other, as well as other elements such as endothelium, is better achieved with histologic preparations (Fig. 3-31). Consequently, cell block preparations are frequently a very important component of FNA biopsy specimens of hepatic masses. The value of cell blocks and their complementary nature with the direct smears has been attested to by a number of investigators.[2–6,22,28,44] For example, in these preparations, it may be much easier to recognize thick (greater than three cells) trabeculae of hepatocytes lined by endothelium. In the study by Zainol and Sumithran,[6] nearly 80 percent of the hepatocellular carcinomas in their series could be absolutely diagnosed as hepatoma in the smear preparations. In the remaining 21 percent of tumors, the direct smears yielded only suspicious diagnoses. In all of these, the histologic attributes of the tumor visualized in the cell blocks permitted a diagnosis of hepatoma with confidence. By evaluating small tissue fragments obtained by FNA biopsy along with direct smears, Sangalli et al.[44] greatly increased the diagnostic yield of hepatocellular carcinomas. In addition to hepatomas, the use of cell blocks or bioptic tissue cores as a histologic tool may prominently increase the diagnostic yield in

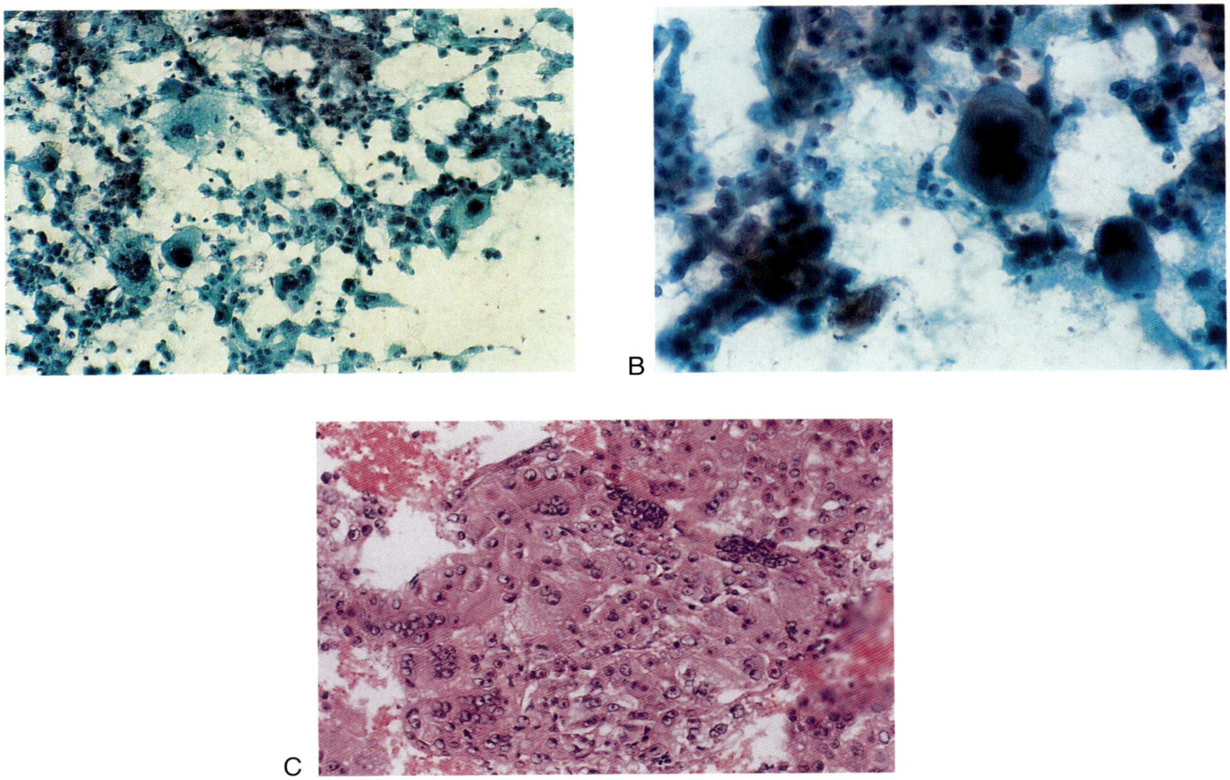

Fig. 3-29. (**A**) At low magnification, the marked pleomorphism of the neoplastic hepatocytes is readily apparent. Very simplistically, two basic tumor cells sizes are present. Most are relatively small, with very high nuclear to cytoplasmic ratios. They are admixed with much larger tumor giant cells; some of these are mononucleated, and others are multinucleated (Papanicolaou). (**B**) In both cell sizes, the neoplastic cytoplasm is dense and granular, in keeping with hepatocytic differentiation. On an individual basis, the nuclei in the tumor giant cells show the features expected of hepatocytic elements (Papanicolaou). (**C**) Pleomorphism of the neoplastic cells is apparent in this cell block. As seen in this poorly differentiated hepatoma, endothelial cell associations are less well developed in high-grade tumors (H&E).

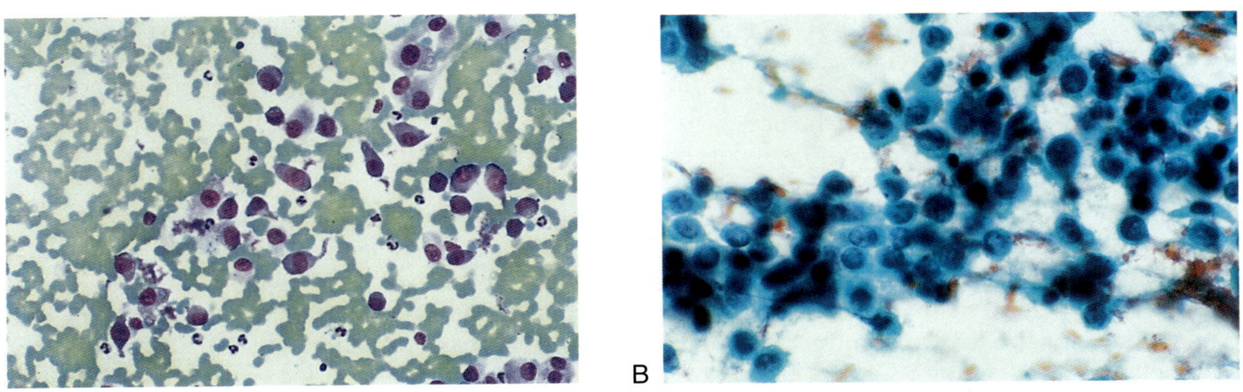

Fig. 3-30. (**A & B**) Small cell hepatomas are characterized by a general loss of intercellular cohesion. Consequently, smears contain numerous individually dispersed and loosely associated malignant cells. For the most part, the neoplastic cells will resemble one another closely with solitary nuclei and very high nuclear to cytoplasmic ratios. The thin rim of cytoplasm is dense and granular. Such specimens need to be distinguished from metastases of small cell malignant tumors (Diff-Quik). (**B**) Note the absence of sharply angulated nuclear contours and of nuclear molding; both would be expected in metastatic bronchogenic oat cell carcinoma, the most frequent metastatic small cell tumor seen in aspiration biopsies of the liver of adults (Papanicolaou).

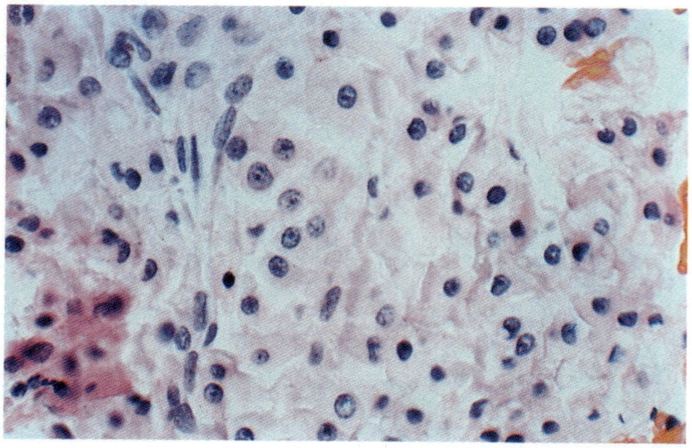

Fig. 3-31. The intimate relationship between neoplastic hepatocytes and reactive endothelial cells is readily apparent in this cell block preparation. Also note the uniformly high nuclear to cytoplasmic ratios of the tumor cells. Yet, the features expected of hepatocytes are present (H&E).

other focal hepatic lesions. This would include metastases and benign lesions, such as focal nodular hyperplasia (FNH).

In addition to conventional morphology, cell blocks lend themselves beautifully to ancillary diagnostic procedures, especially immunocytochemistry. A number of publications in the surgical pathology literature attest to the value of immunocytochemistry to distinguish among primary hepatocellular carcinomas, cholangiocarcinomas, and metastatic carcinomas.[45–50] Similarly, immunocytochemical reactions can be applied to cell blocks to enhance this distinction.[35,51–59] A number of antibodies have been proposed as helpful in distinguishing hepatocellular carcinomas from adenocarcinomas. The immunocytochemical reaction that appears to have the greatest strength for this distinction is polyclonal carcinoembryonic antigen (CEA).[55–57] In many metastatic adenocarcinomas, polyclonal CEA will produce cytoplasmic and membranous staining reactions among malignant cells. On the other hand, the cytoplasm of hepatocellular carcinoma is usually not decorated with this antibody. Rather, a characteristic outlining of the canalicular surfaces of the malignant hepatocytes occurs (Fig. 3-32). Monoclonal CEA appears to be of no value in the differential diagnosis. The cytoplasm of most hepatomas will react with low molecular-weight cytokeratins (e.g., CAM 5.2 and AE3). Adenocarcinomas may also react with these antibodies but in addition will stain with AE1, an antibody against higher molecular-weight cytokeratins. The combined high and low molecular-weight cytokeratin cocktail generally stains only bile duct epithelium, leaving hepatocytes unstained. Serum AFP levels are increased, often to very high levels, in most patients with hepatocellular carcinomas. Immunocytochemically, the AFP marker has a high level of diagnostic specificity, in that only a few other entities, such as certain malignant germ cell tumors, may, in general, be positive (Fig. 3-33). In our experience, however, many obvious hepatocellular carcinomas are not positive with this stain. In part, this may reflect technical difficulties with the antibody or the procedure. In fine needles biopsy specimens, the positivity rates for AFP in hepatomas has ranged from approximately 45 to 58 percent.[35,51,52] Similarly, immunostaining for the surface antigen of hepatitis B virus shows low sensitivity and high specificity.[51,54] Although we have no personal experience using this antibody in this setting, we suspect that it would be most unusual for a metastatic carcinoma to be positive with this reaction. Endothelial cell relationships are diagnostically important in aspirates of hepatocellular carcinoma. Often, they are obvious, as they outline trabeculae and separate masses of tumor cells. In difficult cases, the endothelium may be highlighted by immunochemical reactions. Both factor VIII-related antigen and vimentin should stain these elements rather intensely, whereas the neoplastic cells will be nonreactive.[53,58]

Less frequently, aspirated neoplastic cells are studied by electron microscopy.[22,27,54,60] Generally, this is attempted when it is recognized at the time of immediate interpretation that one is dealing with a relatively poorly differentiated neoplasm; at that juncture, an aliquot of the aspirate can be placed in glutaraldehyde. Ultrastructurally, the cytoplasm of hepatoma cells is generally organelle rich.[61] A characteristic feature is the presence of small intercellular lumens, which are formed by well-developed intercellular junctions and lined by blunt villi; these correspond to bile canaliculi. The presence of both

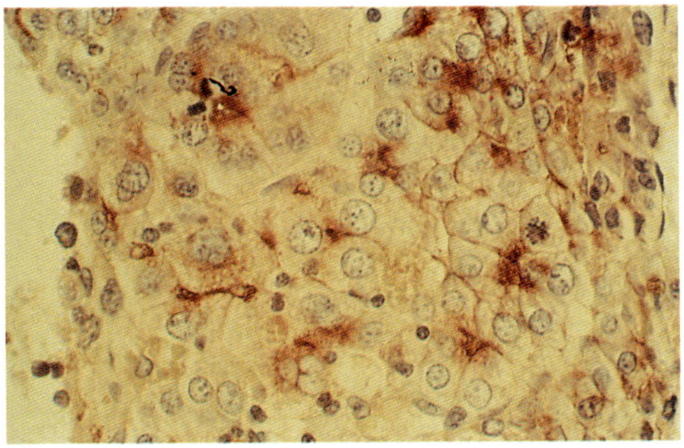

Fig. 3-32. This is a cell block reacted immunocytochemically with polyclonal carcinoembryonic antigen (CEA). The cannicular surfaces between malignant hepatocytes is highlighted by this antibody. This may be due to cross-reactivity with CEA-like macromolecules present on the cannicular surfaces.

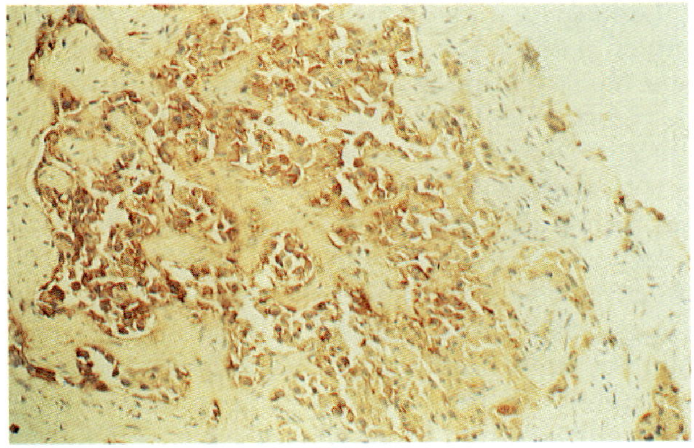

Fig. 3-33. In this cell block, the cytoplasm of most of the neoplastic cells is intensely immunoreactive for α-fetoprotein.

smooth and rough endoplasmic reticulum is also typical of hepatocellular differentiation. Both glycogen and lipid may be prominent. Bile may also be recognized (Fig. 3-34). Tumor cells will not contain mucous vacuoles, neurosecretory granules, or well-developed bundles of intermediate filaments. Metastatic adenocarcinomas and cholangiocarcinoma may demonstrate intracytoplasmic or large intercellular lumens lined with microvilli with a well-defined internal structure.

Several well-defined but rare morphologic variants of hepatocellular carcinoma need to be recognized. The best known of these is probably the fibrolamellar hepatoma.[62,63] Although it is quite uncommon, it accounts for a relatively large proportion of this malignancy in young adults. Unlike conventional hepatocellular carcinoma, the male-to-female ratio is approximately unity, it is not associated with an underlying preexisting cirrhosis, and it carries a generally better prognosis. Histologic hallmarks include the presence of very large neoplastic cells with voluminous granulated eosinophilic cytoplasm and prominent nucleoli, bundles of well-defined layers of collagen fibers, and cytoplasmic pale bodies. The obviously granulated cytoplasm, which is due to numerous mitochondria, confers an oncocytoid appearance.

Due to prognostic and possible therapeutic differences, it is important to distinguish the fibrolamellar variant from conventional hepatocellular carcinoma, as well as from other benign and malignant lesions. The cytomorphologic picture in FNA biopsies is rather characteristic and may permit a specific diagnosis (see box on p. 110). Smears tend to be moderately cellular at best.[64–66] The degree of intercellular

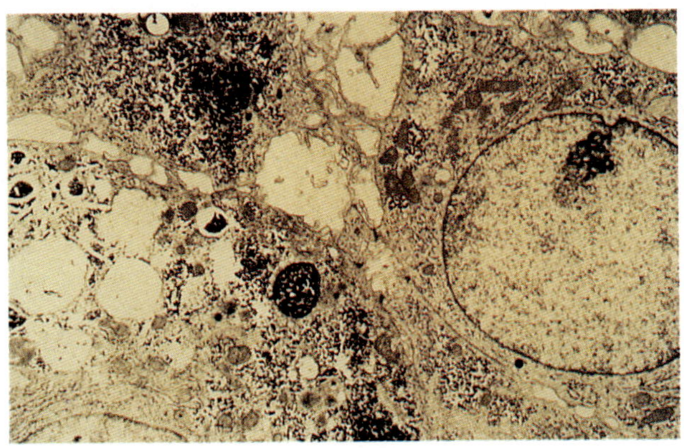

Fig. 3-34. In this electron micrograph, bile is characterized in the cytoplasm of malignant hepatocytes as dense irregular granules. An intercellular canniculus is suggested by a luminal space with primitive intercellular junctions and rudimentary microvilli.

cohesion is variable but is generally less well preserved than with conventional hepatoma (Figs. 3-35 and 3-36). In fact, in some specimens individually dispersed neoplastic cells predominate. The latter are characterized by abundant densely granulated cytoplasm that contains one or two perfectly round-appearing uniform nuclei. Typically, the latter possess a very prominent nucleolus. Characteristically, the nuclear to cytoplasmic ratios are very low. Cytoplasmic pale bodies may be recognized, but bile is uncommon. In addition, one may identify fragments of layered collagen. These large neoplastic cells are distinguished from benign hepatocytes by their uniformly very large sizes.[64–66]

Cytomorphologic Features of Fibrolamellar Carcinoma

- Moderate cellularity
- Large polygonal cells with voluminous granulated cytoplasm
- Pale cytoplasmic inclusions may be prominent
- One or two central round nuclei
- Low nuclear to cytoplasmic ratios
- Variable cohesion; individual cells may predominate
- Fragments of collagen

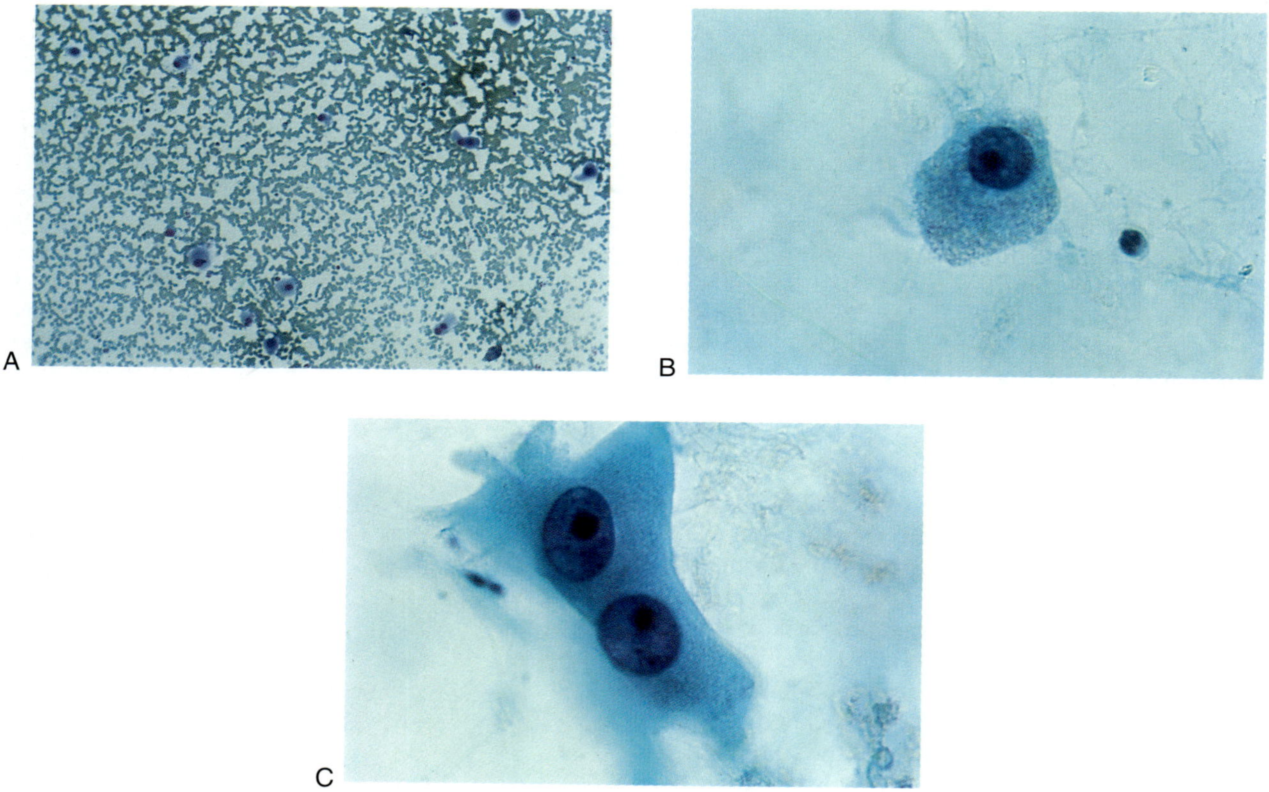

Fig. 3-35. In general, intercellular cohesion is less well preserved in the fibrolamellar variant of hepatocellular carcinoma. (**A**) Numerous individually dispersed neoplastic cells are seen in the scanning lens view. Cellularity is far lower than one would expect with a conventional hepatocellular carcinoma (Diff-Quik). (**B**) The typical isolated neoplastic cell is apparent. It is characterized by voluminous obviously granulated cytoplasm, a solitary nucleus with a smooth round contour, and a low nuclear to cytoplasmic ratio (Papanicolaou). (**C**) A binucleated tumor cell is present. Although it is quite large, note the preservation of the low nuclear to cytoplasmic ratio and the granularity of the cytoplasm (Papanicolaou).

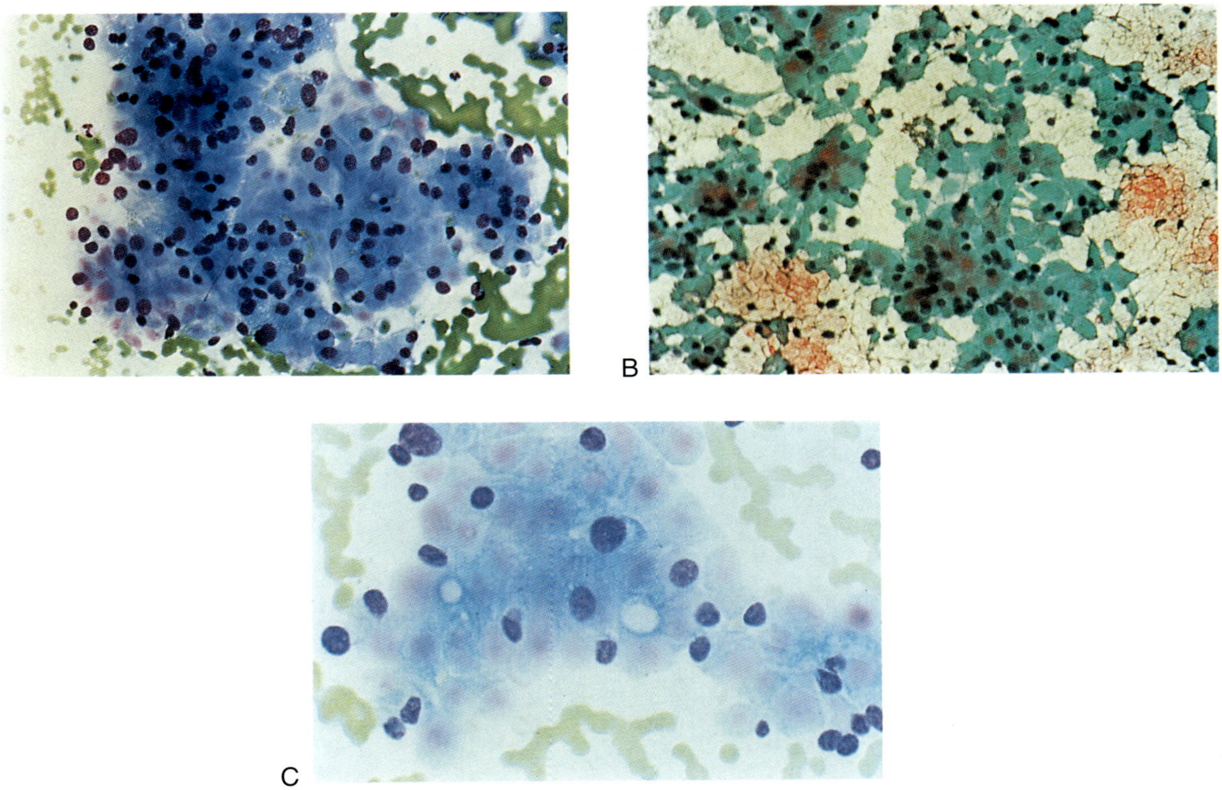

A B C

Fig. 3-36. Example of a fibrolamellar variant of hepatocellular carcinoma in which intercellular cohesion was better preserved than expected. (**A & B**) A superficial resemblance to benign hepatocytes is present. However, nuclei are focally very crowded. (**A**) Many of the hepatocytes appear to be intimately associated with round pale purple globules (Diff-Quik). (**B**) The structures have a glassy (homogeneous) cyanophilic quality (Papanicolaou). (**A & B**) Overall, the low nuclear to cytoplasmic ratios are obvious. (**C**) These pale purple structures represent cytoplasmic inclusions, which by immunocytochemistry were positive for fibrinogen (Diff-Quik).

Clear cell hepatocellular carcinoma does not have the therapeutic and prognostic implications of the fibrolamellar variant. That is, it appears to be more closely akin clinically to conventional hepatoma. The major reason for recognizing this morphologic variant is to avoid confusing it with clear cell neoplasms metastatic to the liver from other organs.[65–70] Much more frequently, clear cell carcinomas arise in the kidneys, ovaries, and adrenals. All may metastasize to the liver and thus need to be differentiated from a primary liver cell malignancy. The major morphologic attribute of this variant is the abundant optically clear cytoplasm[65–69] (Fig. 3-37). The latter may have a diffuse homogeneous appearance or may contain numerous minute vacuoles. Distinctive morphologic features in aspiration biopsies may include well-defined cellular borders and a relative reduction in intercellular cohesion. Otherwise, they closely resemble conventional hepatocellular carcinomas in smears. Donat et al.[67] claimed that immunostaining for α_1,-antitrypsin is helpful. Ultrastructurally, these tumors will demonstrate features in keeping with hepatocellular differentiation.[69]

Rarely, one may encounter a malignant neoplasm with morphologic and immunochemical features

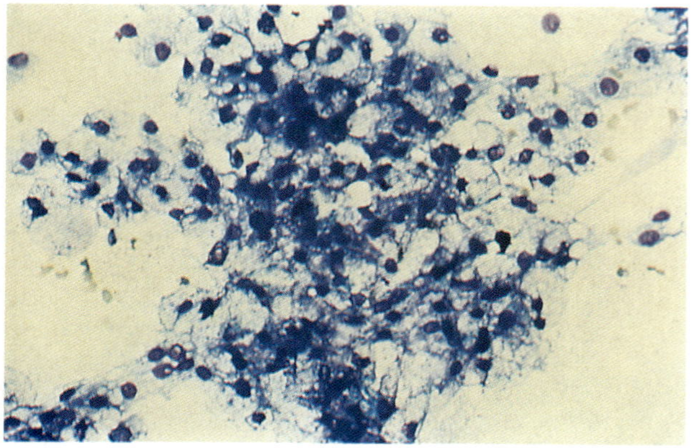

Fig. 3-37. In this clear cell hepatocellular carcinoma, the tumor cells uniformly possess abundant highly vacuolated cytoplasm and central nuclei. The nuclei are mostly small, but some pleomorphism is evident (Diff-Quik).

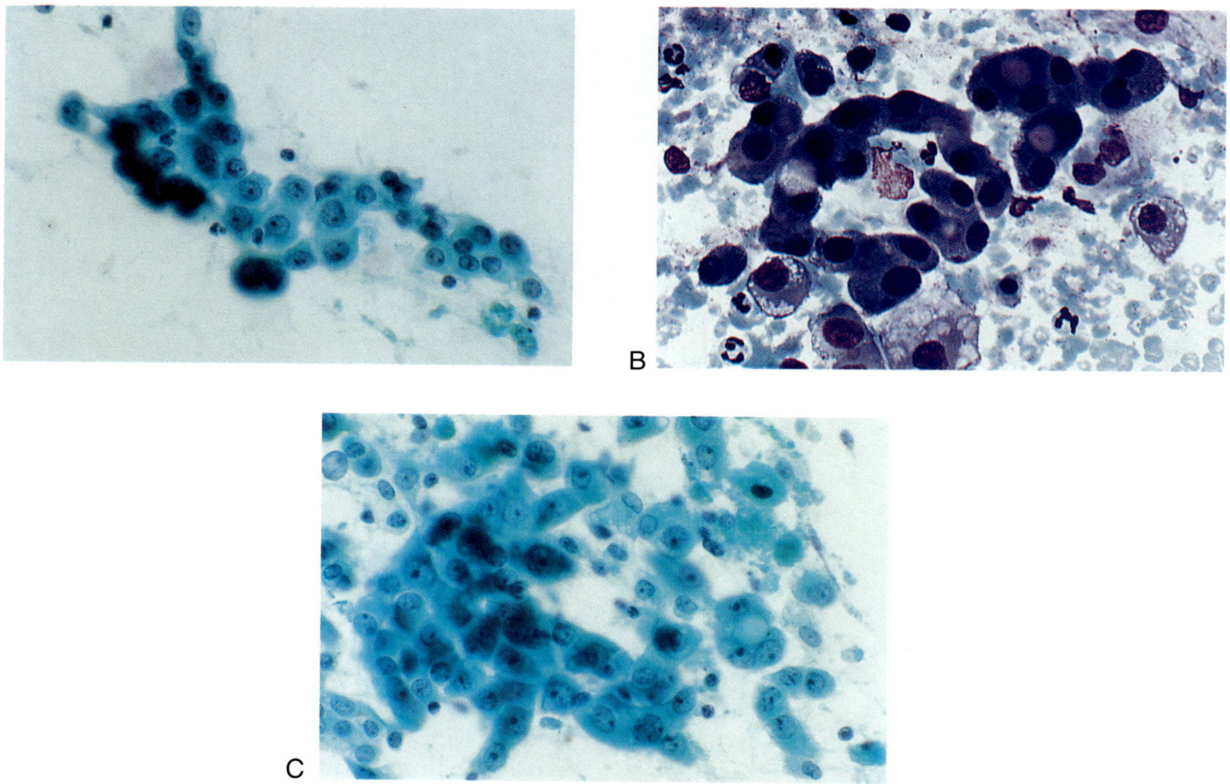

Fig. 3-38. (**A**) In this mixed carcinoma, the tumor cells in this field are consistent with malignant hepatocytes. They have polygonal contours, well-defined cellular borders, granular opaque cytoplasm, and round centrally positioned nuclei. An endothelial coating is not evident (Papanicolaou). (**B**) Several of the tumor cells in this field have large cytoplasmic vacuoles that displace the nucleus to one pole of the cell. Eccentric nuclear positions are also present in cells without such vacuoles. This represents glandular differentiation in a mixed carcinoma. These vacuoles were positive for mucicarmine in the cell block of this specimen (Diff-Quik). (**C**) In this field, the juxtaposition and possible transition between hepatocytic and glandular elements are evident (Papanicolaou).

consistent with both hepatocellular and glandular differentiation.[71] As a generalization, the clinical behavior of these primary combined hepatocellular cholangiocarcinomas parallels that of conventional hepatoma. Accordingly, the major reason for recognition of this variant is to prevent confusion with a pure adenocarcinoma, most likely metastatic in origin. Histologically, the pseudoacinar formation in hepatoma could be easily misconstrued as gland formation in adenocarcinomas. The finding of eosinophilic granular cytoplasm, centrally positioned nuclei, and luminal bile would be in keeping with liver cell differentiation. On the other hand, filmy vacuolated cytoplasm, basally oriented nuclei, ovoid to irregularly contoured nuclei, and luminal mucin point to adenocarcinoma. The immunocytochemical differentiation of hepatocellular carcinoma and adenocarcinoma have been previously described. In aspirates of the combined malignancy, one can expect to see evidence of both directions of differentiation. In our limited experience, this was this case.[59] One may see both cell types within the same microscopic field, and transitions between the two forms may be suggested (Fig. 3-38). By immunocytochemistry, the hepatocellular carcinoma demonstrates the features expected for liver cell differentiation, whereas the cholangiocarcinomatous component shows glandular differentiation, including positive cytoplasmic and luminal staining with mucicarmine.[51,59]

BENIGN HEPATOCELLULAR TUMORS

Both hepatocellular adenomas and FNH, although rather uncommon, must always be considered whenever an hepatic mass is being aspirated.[72] With rare exceptions, adenomas are restricted to the adult female population, as their development has been associated with the use of oral contraceptives. Although these may be asymptomatic, patients often present secondary to the effects of bleeding from the neoplasm. Quite infrequently, adenomas have been reported in children and in adult men in association with the use of exogenous anabolic steroids. The vast majority of adenomas occur as a solitary mass lesion. Although FNH predominates in adult women, it occurs relatively more frequently in men and children than does the adenoma. In most diagnosed cases, the patient is clinically asymptomatic, and a mass is discovered incidentally. Although the relationship is much weaker, there is a suggestion that the development of FNH may also be associated, at times, with the use of oral contraceptives. Histologically, adenomas are composed of normal-appearing hepatocytes present in delicate to densely arranged trabeculae that may yield the appearance of a diffuse growth pattern. Although thin-walled blood vessels are situated between trabeculae, neither bile ducts nor well-developed portal tracts are present. FNH presents as a multinodular proliferation of normal-appearing hepatocytes arranged in delicate trabeculae. Characteristically, the small nodules of hepatocytes are enveloped by fibrous septae that may contain numerous bile ducts and that radiate from a central to eccentric scar. On a cell-for-cell basis, the proliferating elements of both adenoma and FNH are indistinguishable from each other and from normal hepatocytes.

As might be expected, one cannot make the specific diagnoses of either FNH or adenoma on the basis of aspiration cytomorphology alone. Aspiration smears are usually cellular and contain numerous normal-appearing hepatocytes,[11] as described below. Most of the cells possess a solitary centrally positioned round nucleus with smooth outlines, finely granular chromatin, and a small nucleolus (Figs. 3-39 and 3-40). Cytoplasm is granular and may contain lipid vacuoles. Characteristically and importantly, the nuclear to cytoplasmic ratios

Cytomorphologic Features of Benign Hepatocellular Tumors

- Moderate to high cellularity
- Polygonal cells with distinct borders and dense granular cytoplasm
- One or two central round nuclei
- Variable but generally low nuclear to cytoplasmic ratios
- Fragments of connective tissue with bile ducts in focal nodular hyperplasia

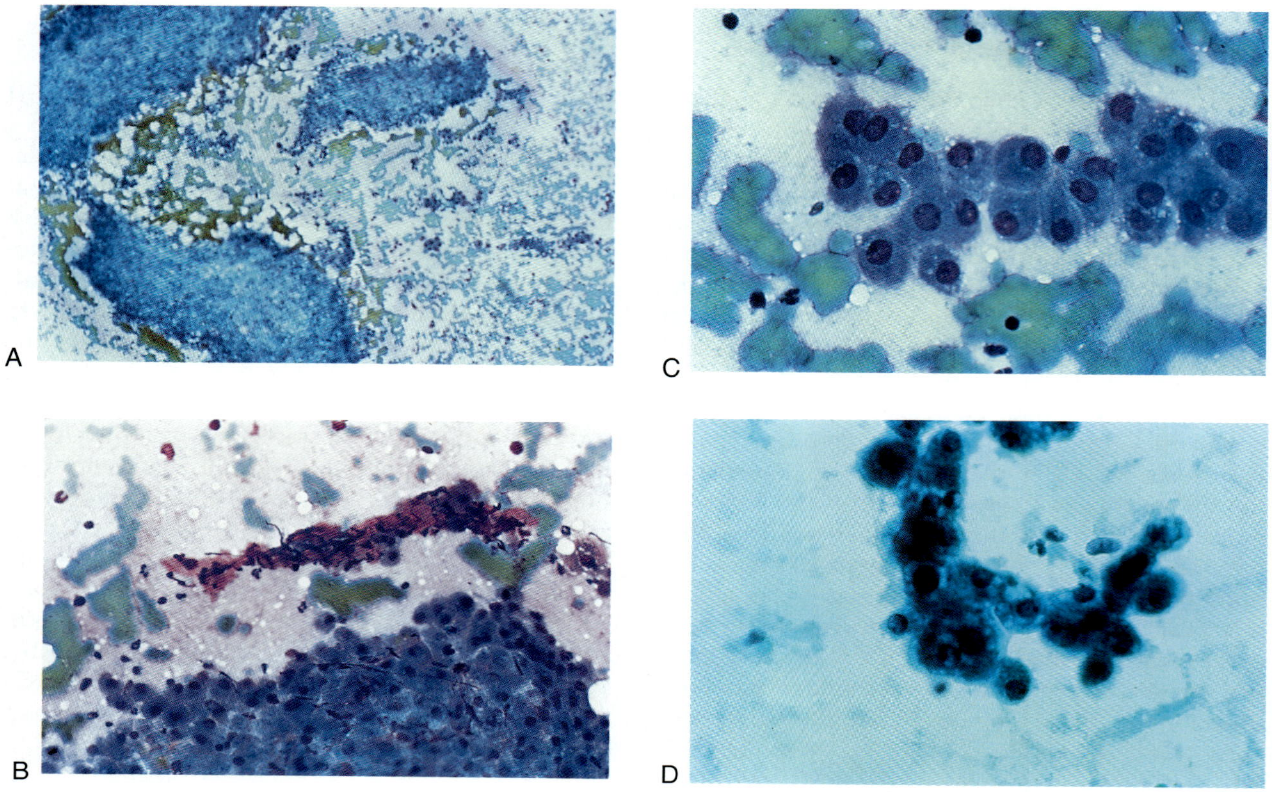

Fig. 3-39. Focal nodular hyperplasia. (**A**) In the scanning-lens view, marked cellularity of the aspirate is obvious. Most of the hepatocytes in this specimen are present in huge cohesive aggregates. Also present are small clusters and individual hepatocytes (Diff-Quik). (**B**) In addition to a large sheet of benign-appearing hepatocytes, a fragment of collagenized stroma is present. This may represent sampling of a connective tissue trabeculum separating nodules within the mass. However, bile duct structures are not evident within this small fragment (Diff-Quik). (**C & D**) The hepatocytes are indistinguishable from normal cells. They possess polygonal contours, sharply defined cellular borders, granular dense cytoplasm, and central round nuclei. Lipid vacuoles are present in the cytoplasm of a few of the cells. Nucleoli are not readily apparent. Most importantly, the nuclear to cytoplasmic ratios are uniformly low (**C**, Diff-Quik; **D**, Papanicolaou).

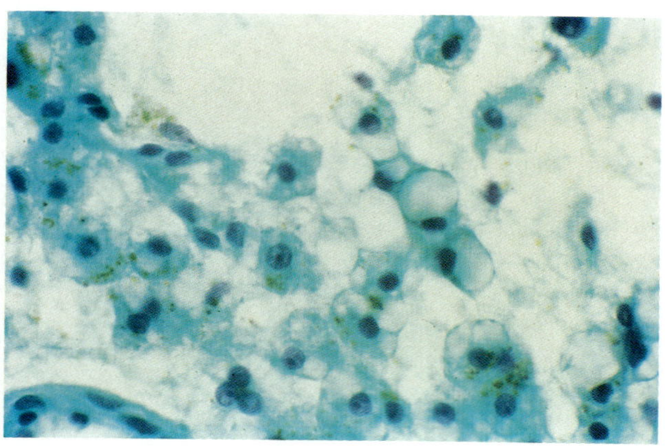

Fig. 3-40. In this hepatocellular adenoma, small aggregates and individual tumor cells predominate. Nuclei are small, round, and centrally positioned. In some, their cytoplasm contains vacuoles or bile, or both (Papanicolaou).

remain low. In our relatively limited experience, connective tissue septae containing bile ducts have not been a prominent component of aspirates of FNH. We are unaware of a specific cytomorphologic way to distinguish adenoma from focal nodular hyperplasia. In this setting, we make the diagnosis simply of "normal-appearing hepatocytes are present, consistent with either FNH or with adenoma." As the needle may sample parenchyma peripheral to the mass, the presence of small ductular elements in a smear does not exclude adenoma. These proliferations represent very good examples of the need for open communication lines between the radiologist and the pathologist. In fact, it is most helpful for the pathologist to witness personally (on the computed tomography [CT] monitor) the tip of the needle in the mass. When the needle placement is fairly certain, we are willing to make a diagnosis of adenoma or FNH.

The presence of numerous benign-appearing hepatocytes and bile ductular structures cannot be distinguished, based purely on morphologic grounds, from cirrhosis. Recently, Tao[37] called attention to the presence of dysplastic hepatocytes in smears from a small number of adenomas. This could cause further confusion with cirrhosis, as well as possibly with a well-differentiated hepatocellular carcinoma. The lack of uniformity of the atypical cells favors a benign interpretation.

METASTASES

With the exception of lymph nodes, the liver is perhaps the most common site in the body for metastatic tumor. Ninety-five percent of all malignant neoplasms involving the liver are metastases, far outshadowing hepatocellular carcinoma and other primary neoplasms. Statistically, the most frequent types of cancer to involve the liver are adenocarcinomas metastatic from the gastrointestinal tract, pancreas, breast, and lung, small cell carcinoma of the lung, and malignant melanoma. Infrequently, we have seen squamous cell carcinomas metastatic to the liver in FNA biopsies, predominantly from the esophagus and the lungs. It must be remembered, however, that essentially any form of neoplasm, including even exceptional primary brain tumors, may seed and grow in the liver.

Metastatic malignancies produce one to many separate nodules in hepatic parenchyma that can be easily detected by current imaging studies. Consequently, they readily lend themselves to sampling by fine needle aspiration biopsy. In our experience, the single most common neoplasm sampled by aspiration is metastatic colonic adenocarcinoma. Although certainly not specific, colorectal carcinomas may produce a characteristic cytomorphologic picture (Figs. 3-41 to 3-43). This consists of a triad of cohesive aggregates of tumor cells with a peripheral palisading of the nuclei, elongated (cigar-shaped) nuclei, and extensive dirty necrosis in the smear background. The latter refers to the presence of intact and degenerated neutrophils, necrotic tumor cells, and nondescript granular necrotic debris. Although often not well developed, cytoplasmic mucin vacuoles can occasionally be seen in metastatic colorectal adenocarcinomas. Occasionally, a predominance of signet-ring adenocarcinoma cells is present in aspirates; in this setting, carcinomas metastatic from the stomach and breast are most likely. More often than not, however, relatively specific cytomorphologic features are lacking (or at least not discernible), and a diagnosis simply of metastatic adenocarcinoma is rendered. One should not make an apology for a relatively nonspecific diagnosis. In our opinion, this is preferable to "sticking one's neck out too far," which could lead clinicians down the wrong path.

The differential diagnosis of metastatic adenocarcinoma includes poorly differentiated hepatocellular carcinoma, cholangiocarcinoma, and possibly bile duct adenoma. Earlier in this chapter, cytomorphologic features favoring hepatocellular carcinoma were detailed. In addition, the value of immunocytochemistry and electron microscopy were described. We have not seen an aspirate of a benign bile duct adenoma, but we would expect to see small numbers of cohesive ductal structures composed of cuboidal to columnar epithelial cells, similar to those seen in aspirates of cirrhotic nodules.[11]

Although not well discussed in the literature, we believe that one of the simplest ways to distinguish a high-grade hepatocellular carcinoma from a metastatic carcinoma is to look for two morphologically distinct cellular populations, namely, benign

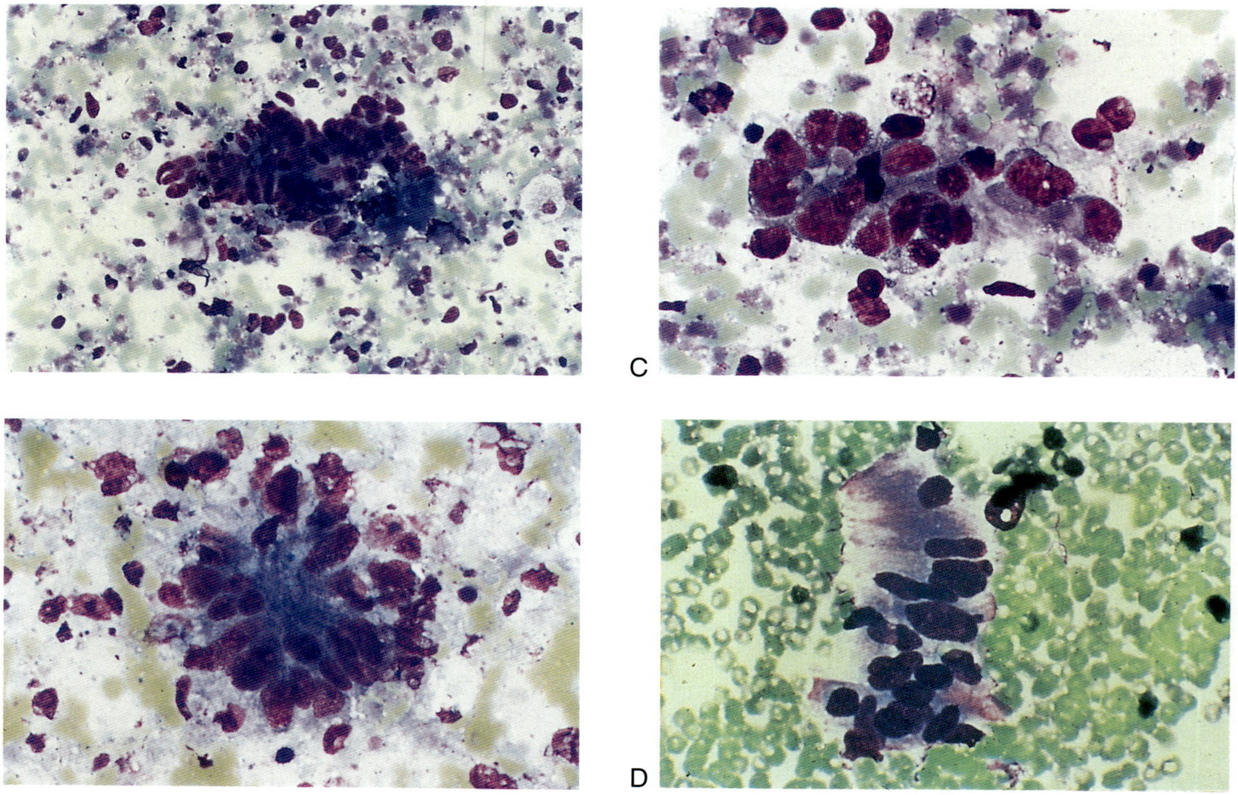

A C B D

Fig. 3-41. Adenocarcinoma of the large intestine metastatic to the liver. (**A**) The triad of aggregates of neoplastic cells with a peripheral palisading, elongated nuclear contours, and an extensive necrotic background is exemplified here (Diff-Quik). (**B**) Elongated nuclear shapes characteristic of metastatic colonic adenocarcinoma are well visualized in this field. Focally, a palisading of the nuclei is apparent. Cytoplasm is centrally concentrated. In the background are degenerating and necrotic tumor cells (Diff-Quik). (**C**) This carcinoma is less well differentiated. The nuclei are more rounded than is characteristic of this neoplasm. The nuclear to cytoplasmic ratios are very high. Minute cytoplasmic vacuoles are evident in some of the cells (Diff-Quik). (**D**) This small fragment of metastatic carcinoma is characterized by very tall columnar cellular configurations, abundant filmy cytoplasm, cigar-shaped nuclei, and a focal picket-fence arrangement of the nuclei (Diff-Quik). *(Figure continues.)*

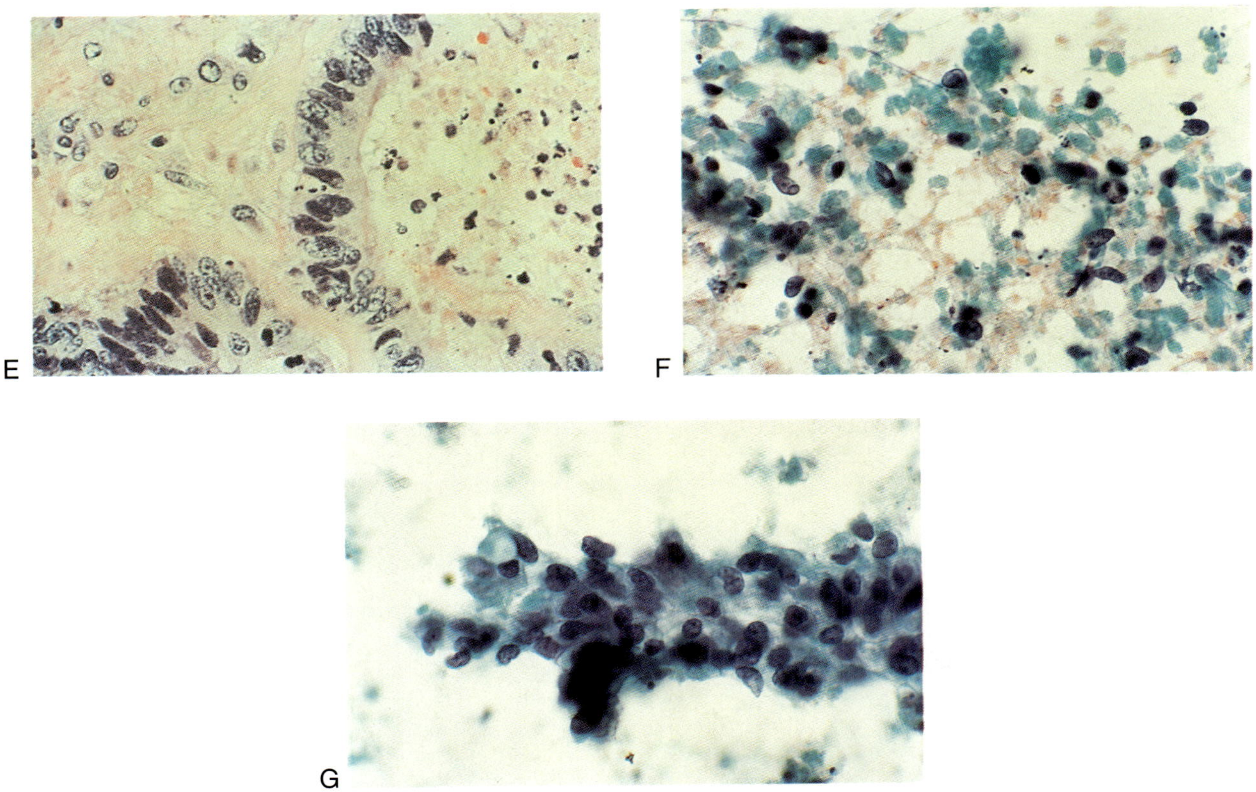

Fig. 3-41 *(Continued)*. (**E**) Cell block from the same specimen as in Fig. D. Similar cytologic features can be seen in this histologic preparation. In addition, abundant necrotic debris is evident (H&E). (**F**) The dirty necrotic background characteristic of this neoplasm consists of an admixture of necrotic tumor cells, degenerating neutrophils, and nondescript granular debris (Papanicolaou). (**G**) Occasionally, large cytoplasmic mucin vacuoles will be seen in colonic adenocarcinoma. Although helpful in the diagnosis of adenocarcinoma, this does not specifically point toward an intestinal primary (Papanicolaou).

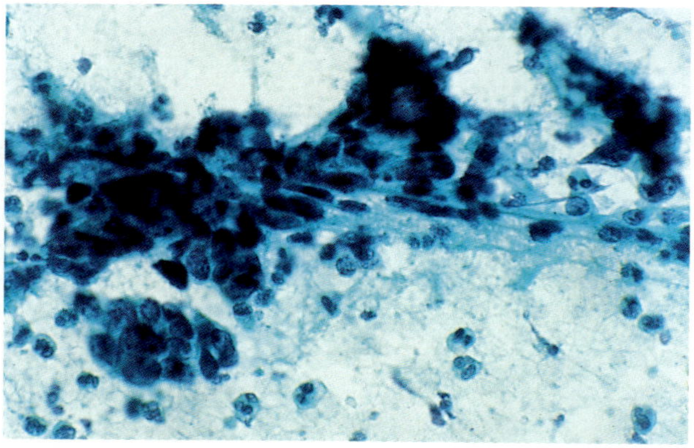

Fig. 3-42. In many metastatic carcinomas, including this colonic primary, large capillary-like blood vessels can be seen coursing through fragments of neoplasm. This highlights the lack of specificity for certain so-called "discriminating" cytomorphologic features. This emphasizes the need for careful scrutiny of the characteristics of the tumor cells encasing capillaries (Papanicolaou).

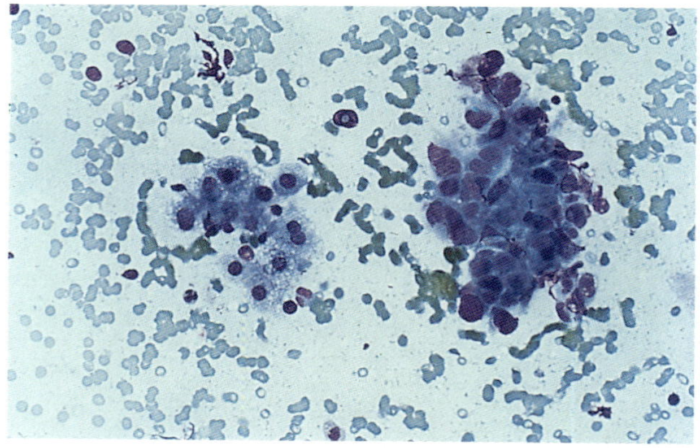

Fig. 3-43. Two cellular aggregates are physically closely related. Those on the left have polygonal contours, distinct borders, granular cytoplasm, and round central nuclei with low nuclear to cytoplasmic ratios. These are benign hepatocytes. The larger aggregate on the right is composed of cells with much larger and more variably contoured nuclei, possible mucin vacuoles, and higher nuclear to cytoplasmic ratios. The latter represents a metastatic colonic adenocarcinoma (Diff-Quik).

A

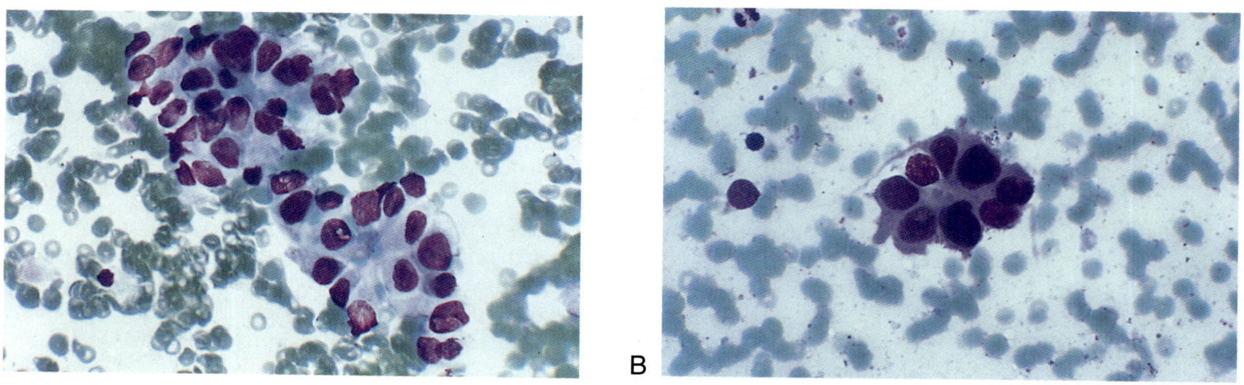

B

Fig. 3-44. **(A&B)** Two cholangiocarcinomas with very similar cytomorphologic features. Both consist of acinar arrangements of columnar epithelial cells. The latter are characterized by large basally polarized nuclei and centrally concentrated cytoplasm. Note the distinctly clean background; that is, it lacks the dirty necrosis characteristic of colonic carcinoma. However, there are no cytomorphologic features specific for cholangiocarcinoma (Papanicolaou).

hepatocytes and a secondary population of malignant cells (Fig. 3-43). Usually, but not always, the bimorphic pattern represents a metastasis.

CHOLANGIOCARCINOMAS

Cholangiocarcinoma is an uncommon primary adenocarcinoma of the liver that arises from intrahepatic bile ducts. Although quite infrequent, it is the second most common primary hepatic malignancy in adults; it is distinctly rare in children. The epidemiology of glandular carcinoma is different from that of hepatocellular carcinoma. Potential risk factors include congenital cystic liver disease, chronic inflammatory bowel disease, prior exposure to thorium dioxide, and parasitic infestations, particularly clonorchiasis. In contrast to hepatocellular carcinoma, there is no gender predilection or association with pre-existing cirrhosis or viral infection. The neoplastic cells are positive for cytoplasmic mucin, both high and low molecular-weight cytokeratin, and CEA. Similar to hepatocellular carcinoma, prognosis is dismal.

In aspiration biopsies, the cytomorphologic findings of cholangiocarcinoma are those of an adenocarcinoma so (Fig. 3-44) that it generally would be impossible to distinguish it from a metastasis, such as from the colon, stomach, or pancreas. Two potentially helpful morphologic features to suggest the possibility of cholangiocarcinoma include a relatively clean smear background (i.e., lacking dirty necrosis) and a relatively hypocellular smear, as noted below. The latter presumably reflects the prominent desmoplasia found so frequently in these neoplasms. Canalicular staining for CEA has not been reported in cholangiocarcinoma.[51] Only very rarely is this malignancy positive for AFP.[50,51]

Cytomorphologic Features of Cholangiocarcinoma

- Moderate cellularity with cohesive tumor cell aggregates, including gland formation
- Columnar cells with granular to bubbly cytoplasm
- Eccentric round nuclei with nucleoli
- Variable nuclear to cytoplasmic ratios
- Clean smear background

SMALL CELL TUMORS

A common morphologic pattern often seen in FNA biopsies of the liver is that of the small cell malignant tumor. A relatively long list of neoplasms can yield this picture. To some extent, these neoplasms can be divided on the basis of patient age. In the pediatric population, one must consider predominantly hepatoblastoma, neuroblastoma, and lymphoreticular malignancies. In adults, the list includes poorly differentiated hepatocellular carcinoma, small cell carcinoma of the lung, islet cell tumor, carcinoid tumor, lobular carcinoma of the breast, ovarian granulosa cell tumor, and malignant lymphoma. Many of these entities have been described in other portions of this book.

In our experience, the most frequent small cell malignancy seen in aspiration biopsies of the liver is that of metastatic small cell (oat cell) carcinoma from the lung. Helpful clinical hints include an adult patient with a long-standing cigarette-smoking history and a lung mass on chest radiography. Suggestive morphologic features include well-developed nuclear molding, sharply angulated nuclear contours, and the absence of obvious nucleoli (Fig. 3-45). The cytomorphology of pulmonary small cell carcinoma is discussed in detail in Chapter 1. Aspiration smears of both carcinoid and islet cell tumors will show morphologic features typical of well-differentiated neuroendocrine carcinomas, including the characteristic isolated plasmacytoid cells (Figs. 3-46 and 3-47). In all three of the above entities, ultrastructural examination may demonstrate dense-core neurosecretory-type cytoplasmic granules, and immunocytochemistry will reveal neuroendocrine differentiation. We have seen a few examples of one other neuroendocrine neoplasm in FNA biopsies of the liver, which may produce a characteristic cytomorphologic picture. Metastatic medullary carcinoma of the thyroid will consist of numerous small malignant cells with very high nuclear to cytoplasmic ratios and larger cells with abundant bright magenta cytoplasm (stained with Diff-Quik) (Fig. 3-48). The latter cells are characteristic of this thyroid cancer.[73]

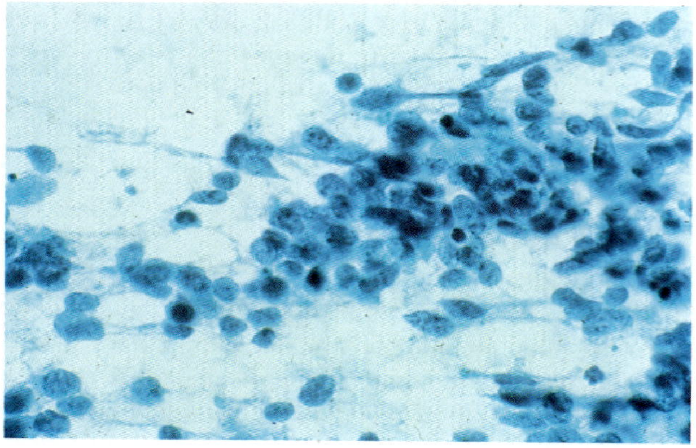

Fig. 3-45. Metastatic small cell carcinoma from the lung is easily recognizable as malignant. It is characterized by small rather uniform-appearing neoplastic cells, each with a single round to sharply angulated nucleus. The latter includes coarsely clumped hyperchromatic chromatin and inconspicuous nucleoli. Almost imperceptible rims of cytoplasm surround each nucleus and consequently in areas of intercellular cohesion nuclear molding will be well developed. Nuclear crush artifact is also characteristic. In our experience, of all primary lung carcinomas, oat cell carcinoma is the one most likely to be diagnosed in a liver aspirate prior to recognition of the primary neoplasm (Papanicolaou).

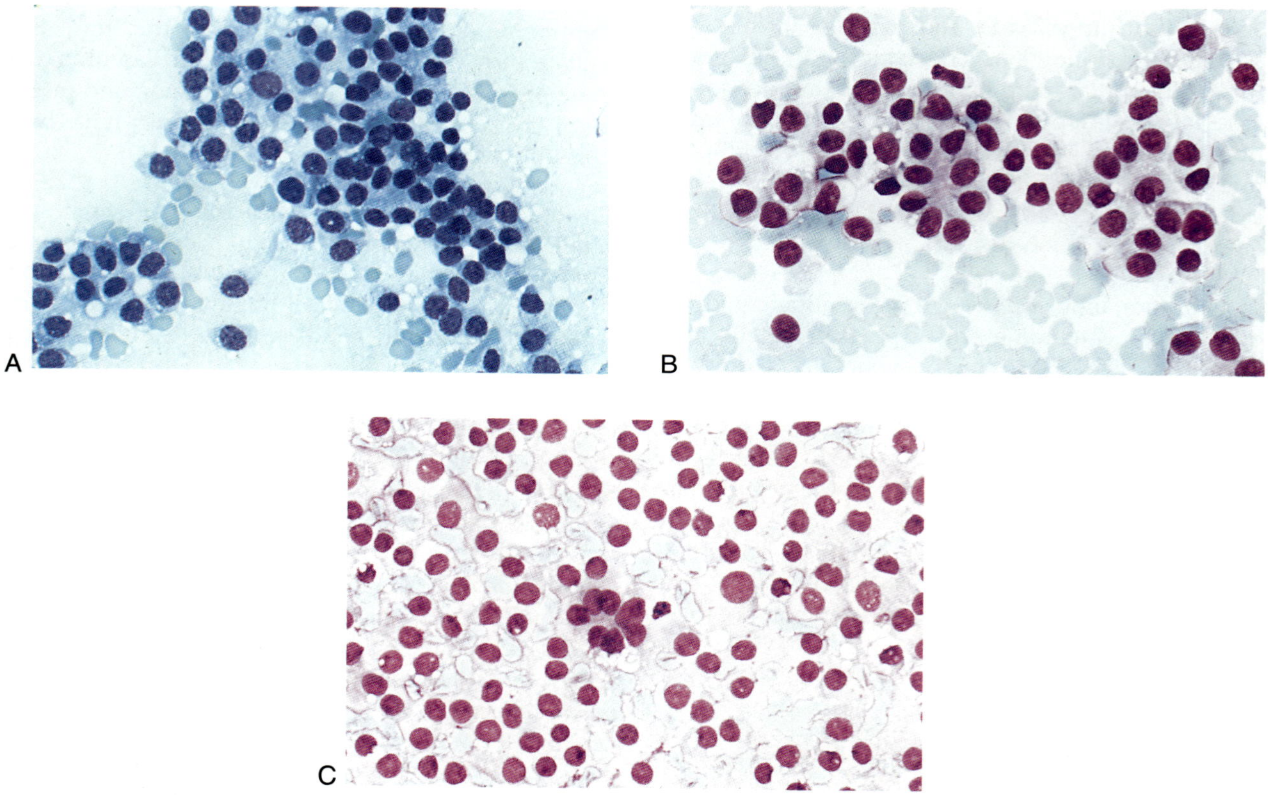

Fig. 3-46. Carcinoid tumor metastatic to the liver. (**A**) The primary site for this carcinoid tumor was the vermiform appendix. The smear is characterized by small flat sheets of uniform neoplastic cells. Each possesses a solitary round nucleus with finely reticulated and evenly distributed chromatin without obvious nucleoli. Scanty to moderate volumes of granulated basophilic cytoplasm are apparent. Within the larger sheet, a suggestion of acinar arrangement of the tumor cells is apparent (Diff-Quik). (**B & C**) This carcinoid tumor was primary in the large intestine. (**B**) The tumor cells are present in well-formed acini with the peripherally oriented homogeneous round nuclei surrounding central cytoplasm. A minority of the nuclei possess small nucleoli (Diff-Quik). (**C**) Intercellular cohesion is greatly reduced with the neoplastic elements presenting as isolated or dispersed cells (Diff-Quik).

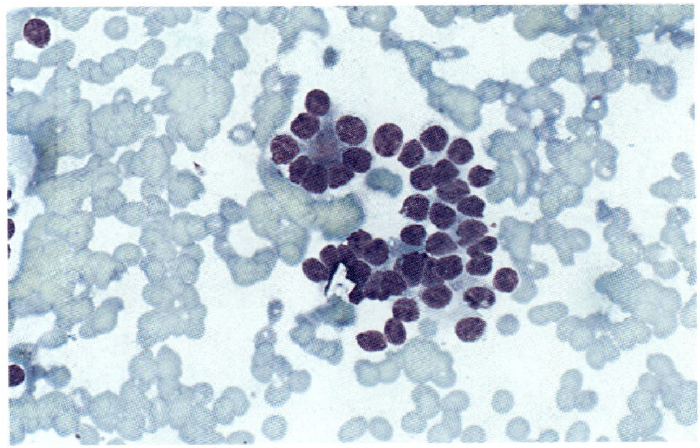

Fig. 3-47. This is an aspirate of an islet cell tumor metastatic form the pancreas. As in Fig. 3-46, these neoplastic cells have solitary round nuclei with evenly dispersed chromatin and scanty to moderate volumes of cytoplasm. Note the absence of nuclear molding. A single acinus is evident (Diff-Quik).

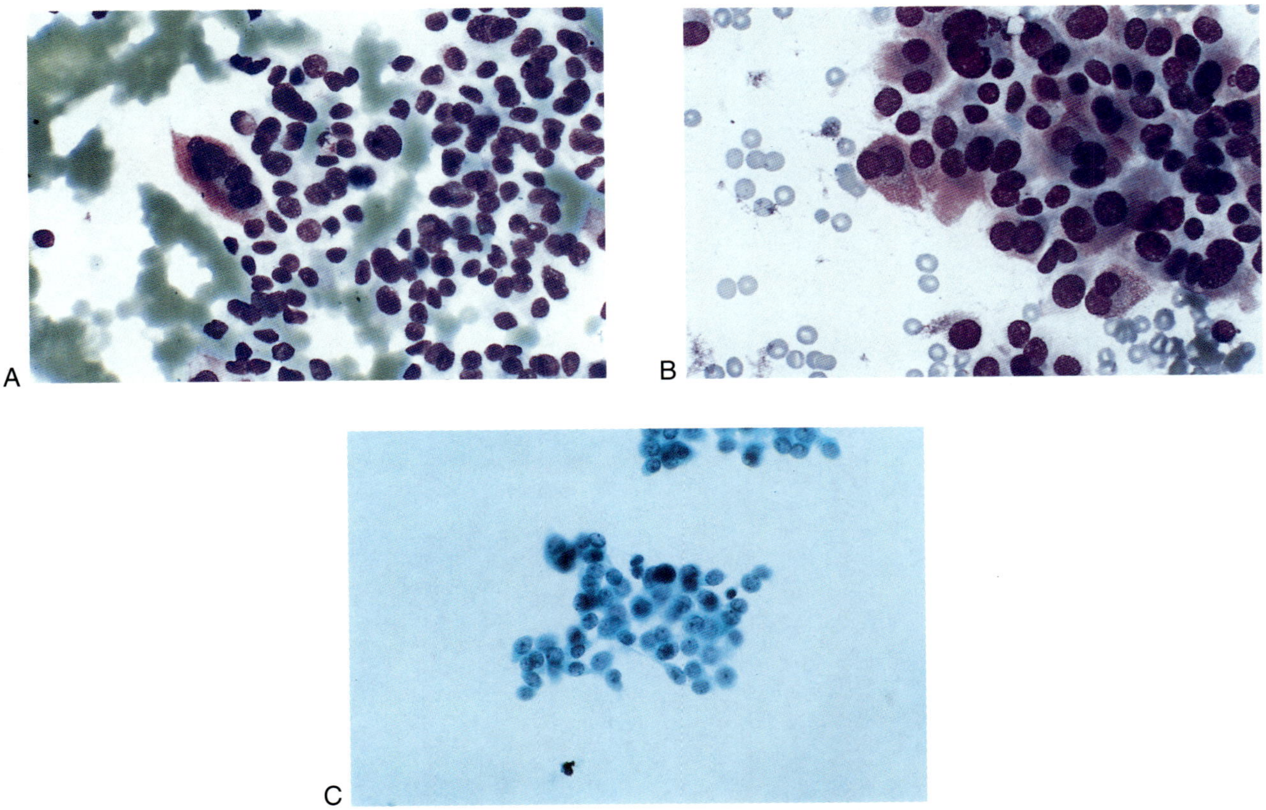

Fig. 3-48. (**A**) The predominant cell in some aspirates of metastatic medullary thyroid carcinoma is small and quite uniform. Each possesses a solitary hyperchromatic nucleus, inconspicuous nucleoli, and very high nuclear to cytoplasmic ratios. Generally, they are not arranged in any specific manner. (**A & B**) In addition, larger cells with moderate to abundant amounts of bright red to purple cytoplasm are present. The latter are a very helpful diagnostic clue. At times, this cytoplasm appears heavily granulated. Also evident are neoplastic cells containing two or more nuclei. In the binucleated cells, the latter are characteristically closely opposed to one another and eccentrically positioned (Diff-Quik). (**C**) In Papanicolaou-stained material, this striking feature is much less apparent. Rather, the cells assume an appearance basically indistinguishable from that of metastatic carcinoid and islet cell tumors.

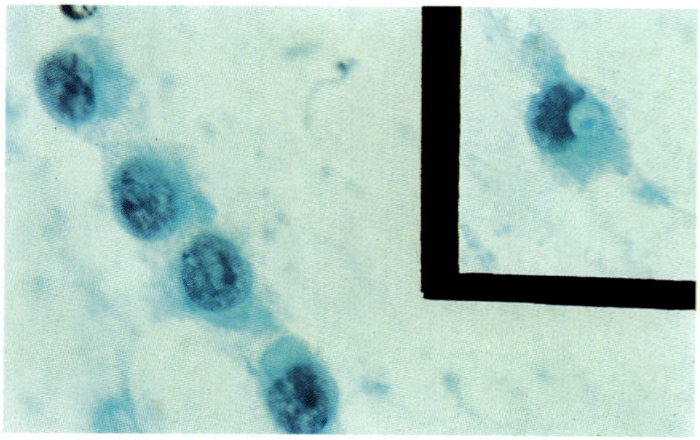

Fig. 3-49. In this metastatic lobular carcinoma from the breast, the tumor cells are small and uniform with rather high nuclear to cytoplasmic ratios. Each cell has a hyperchromatic nucleus with evenly distributed chromatin and small nucleoli. The cells are lined up, suggesting an Indian-file pattern. In the inset, the carcinoma cell demonstrates the characteristic cytoplasmic mucin vacuole. It is typified by a sharply demarcated border and central inspissated mucin. Although not totally specific, this appearance would be suggestive of breast carcinoma (Papanicolaou).

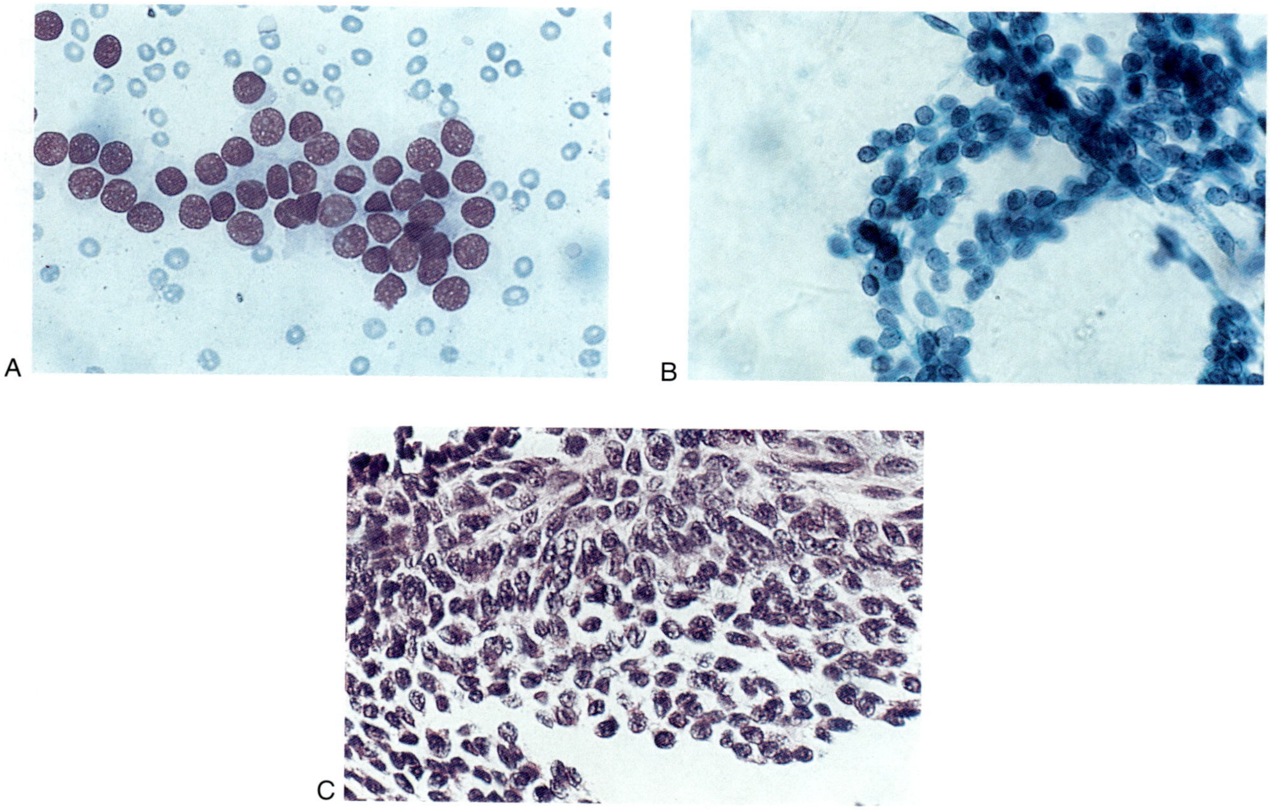

Fig. 3-50. The small neoplastic cells in this metastatic granulosa cell tumor are very uniform. (**A**) Each has a round or slightly ovoid nucleus, a delicate nuclear membrane, evenly distributed chromatin, and inconspicuous nucleoli. Characteristic but faint longitudinal grooves can be seen in a few of these neoplastic nuclei. Within this flat sheet-like arrangement, a pseudoacinar arrangement is suggested (Diff-Quik). (**B**) With the Papanicolaou stain, the homogeneous powdery consistency of the chromatin is readily apparent. Again, nucleoli are inconspicuous and longitudinal grooves can be seen in a few cells. The nuclear to cytoplasmic ratios are very high. However, nuclear molding is not apparent. (**C**) In this cell block of a liver aspirate, the morphologic appearance of the cells fully resembles those seen in the direct smears (H&E).

A primary thyroid carcinoma is not always known during sampling of the liver. Lobular adenocarcinoma of the breast may present in smears as variably cohesive small uniform neoplastic cells with solitary round nuclei (Fig. 3-49). The latter contain finely granular hyperchromatic chromatin, and possibly small nucleoli. An Indian-file arrangement without prominent nuclear molding is characteristic but is not always seen. Also typical is the presence of small cytoplasmic accumulations of mucin. More often than not, a history of prior breast cancer is known. If the primary tumor was positive for steroid receptors, immunocytochemical studies for estrogen and progesterone receptor proteins may help confirm a metastasis from the breast. Metastatic ovarian granulosa cell tumor yields smears containing highly uniform small cells with high nuclear to cytoplasmic ratios and a very finely granular, homogeneous chromatin arrangement (Fig. 3-50). Nucleoli are generally inconspicuous, but characteristic longitudinal nuclear grooves may be present. Rarely, tumors with a distinct basaloid appearance (e.g., adenoid cystic carcinoma and cloacogenic carcinoma) may be sampled (Fig. 3-51).

As discussed below, lymphoreticular malignancies present as individually dispersed malignant elements with high nuclear to cytoplasmic ratios, round to highly irregular nuclear contours, and generally readily apparent nucleoli[74] (Fig. 3-52). In addition, the smear background will contain numerous lymphoglandular bodies. Although exceedingly unusual, the possibility of sampling two distinct malignant neoplasms in the liver exists. For example, Cavanna et al.[75] recently reported the simultaneous sampling of a malignant lymphoma and hepatocellular carcinoma. In metastatic neuroblastoma, the smear background will often demonstrate a filamentous eosinophilic

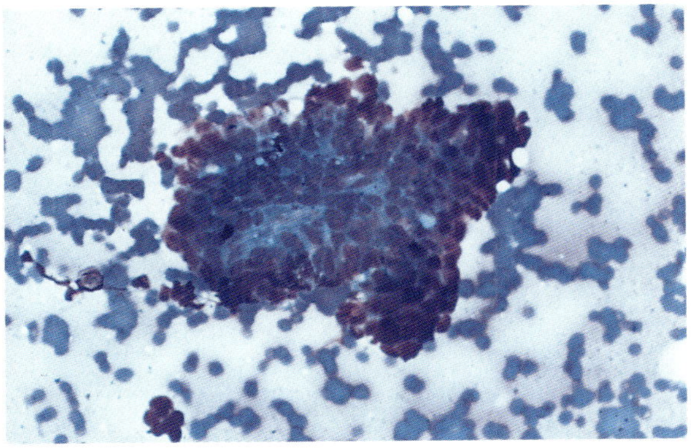

A

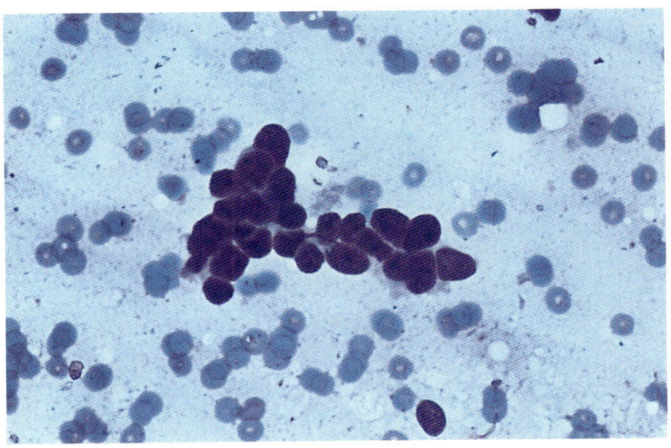

B

Fig. 3-51. (**A**) This aspirate of a cloacogenic carcinoma of the anus metastatic to the liver is characterized by cohesive aggregates of small uniform malignant cells. (**A**) At the edge of the aggregate, the ovoid nuclei are lined up, creating a palisading effect. In contrast to metastatic adenocarcinoma from the intestine, the smear background is clean (Diff-Quik). (**B**) Each neoplastic cell has a solitary dark-stained nucleus with one or two small nucleoli and very high nuclear to cytoplasmic ratios. Compression of adjacent nuclei is evident, somewhat simulating oat cell carcinoma (Diff-Quik).

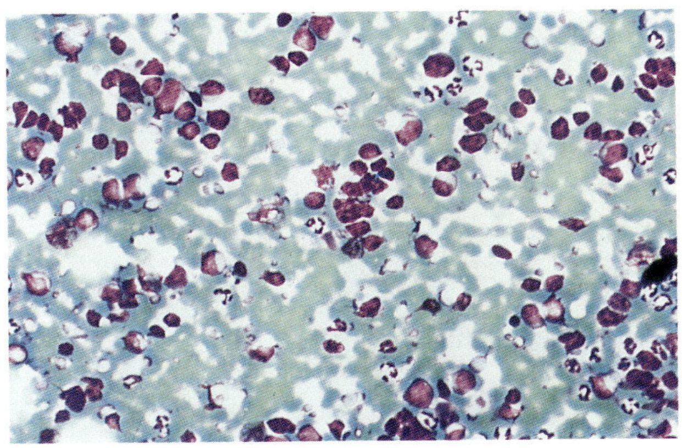

Fig. 3-52. Isolated malignant cells characterize this aspirated lymphoma. Small foci of pseudocohesion are represented. The neoplastic cells have distinct nucleoli and high nuclear to cytoplasmic ratios (Diff-Quik).

matrix material that corresponds to the neuropil. At least some of the neoplastic elements will appear embedded within this matrix. In addition to the small primitive elements, one may see an admixture of larger neoplastic ganglion cells. Less often, one would have to consider the possibility of a metastatic nephroblastoma. Triphasic differentiation should point in the correct diagnostic direction. Ancillary diagnostic studies may be invaluable in the evaluation of metasases from an unknown primary.

HEPATOBLASTOMA

Whereas hepatocellular carcinoma is overwhelmingly the most frequent primary liver malignancy in adults, in children hepatoblastoma is the most common primary cancer. Distinction between these two cancers is important and may be achieved on the basis of combined clinical and aspiration cytomorphologic features.[76–78] Almost all patients are less than 4 years of age. In almost all instances, hepatoblastoma is not associated with a preexisting cirrhosis. Histologically, the tumor cells resemble fetal and/or embryonal hepatocytes.[76] A mesenchymal-like neoplastic component may also be present and often includes material resembling osteoid. In addition extramedullary hematopoiesis is occasionally evident.

Aspiration smears are usually highly cellular and consist of uniform-appearing small malignant

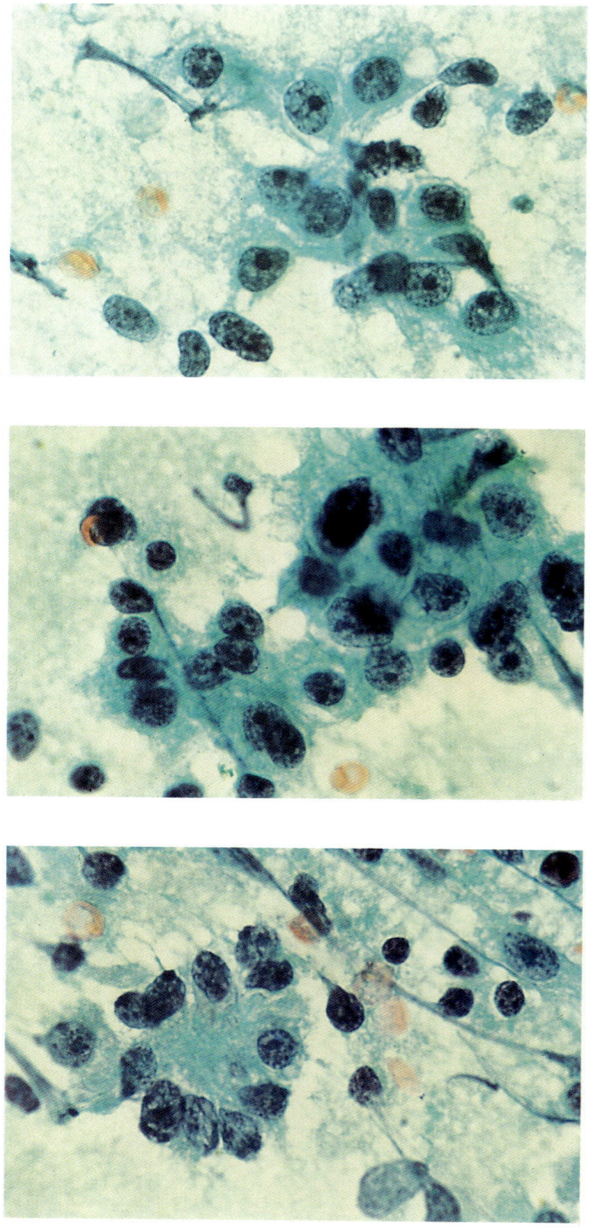

A B C

Fig. 3-53. (**A**) In hepatoblastoma, the neoplastic cells are highly uniform in appearance. Each possesses a solitary round nucleus with finely granular, uniformly distributed, and hyperchromatic chromatin. In this example, nucleoli are prominent (Papanicolaou). (**B**) Cell borders are distinct (Papanicolaou). (**C**) A pseudoacinar arrangement is apparent with peripherally oriented primitive nuclei surrounding central cytoplasm (Papanicolaou).

Cytomorphologic Features of Hepatoblastoma

- High cellularity and uniformity
- Small round to polygonal cells
- Single round hyperchromatic nuclei with or without prominent nucleoli
- Moderate to very high nuclear to cytoplasmic ratios
- Cells present in trabeculae, acini, cords and solitary
- Hematopoietic elements
- Mesenchymal elements (e.g., osteoid)

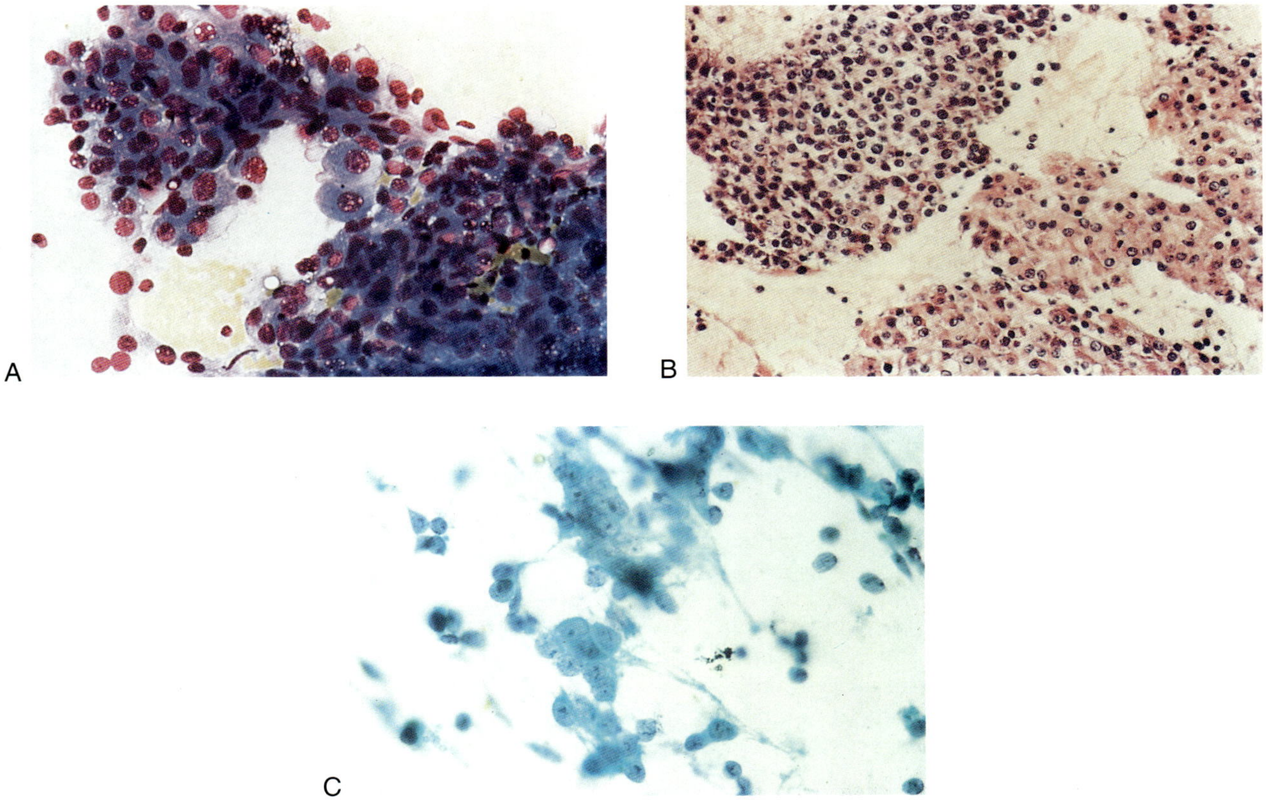

Fig. 3-54. (**A**) Hepatoblastoma cells manifesting fetal differentiation are present on the left, whereas those showing embryonal differentiation are present on the right. The latter are slightly smaller and have distinctly elevated nuclear to cytoplasmic ratios. In the fetal-type tumor cells, hepatocytic differentiation is readily apparent (Diff-Quik). (**B**) The differences between fetal and embryonal tumor cells are very apparent in this cell block (H&E). (**C**) Only rarely will one encounter a tumor composed almost exclusively of very primitive neoplastic cells. There is little or no evidence of hepatocytic differentiation. It would be very difficult or impossible to distinguish this specimen from many other metastatic small cell nonlymphoreticular malignancies of childhood (Papanicolaou).

cells arranged in trabeculae, acini, irregular aggregates, as well as individually,[79–84] (see box on p. 124). Each cell possesses a solitary nucleus which is round and centrally positioned within the cell (Fig. 3-53). Chromatin is finely granular, evenly distributed, and darkly stained. Nucleoli may or may not be prominent. Cytoplasm varies in quantity from scanty to moderate and is often distinctly granular in quality. Only very infrequently is bile associated with the malignant cells. One may attempt to distinguish tumor cells showing fetal and embryonal differentiation from each other (Fig. 3-54); in the latter, the nuclear to cytoplasmic ratios are exceedingly high due to scanty cytoplasm showing little or no evidence of differentiation in the form of granular eosinophilic cytoplasm. Rather, cytoplasm is basophilic and nongranulated. Similar to hepatocellular carci-

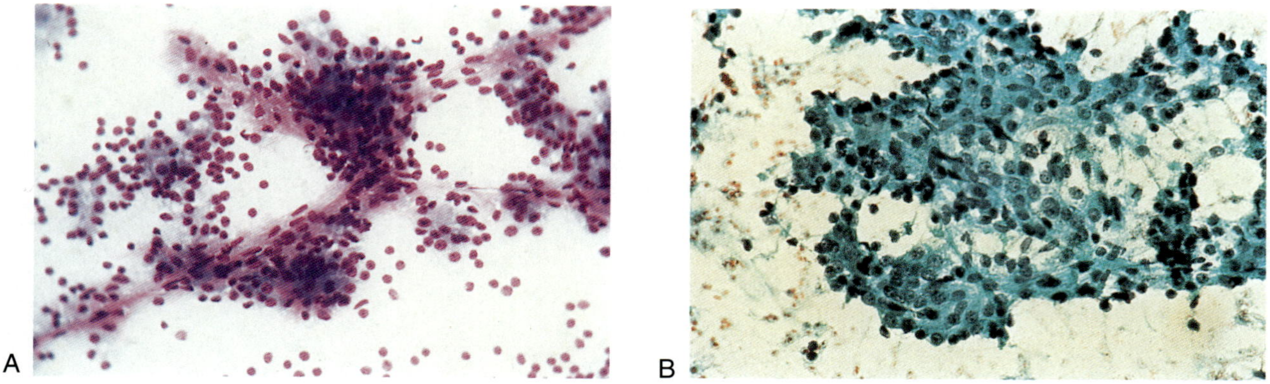

Fig. 3-55. (**A**) Trabecular arrangements of the neoplastic hepatoblasts are evident, as small uniform cells are arrayed about capillary-like vessels. The neoplastic cells are strikingly uniform. (Diff-Quik). (**B**) With the Papanicolaou stain, cytoplasmic granularity may be more apparent (Papanicolaou).

noma, the neoplastic cells may be arranged in characteristic anastomosing trabeculae (Fig. 3-55). Paralleling the histology, one occasionally may recognize primitive hematopoietic elements or osteoid, or both, in the smears.

MALIGNANT LYMPHOMA

Recently, we had the opportunity to review a relatively large series of FNA biopsies of malignant lymphoma involving the liver.[74] Of the 16 individuals, all were adults with a slight male predominance. Importantly, in nine of the patients, there was no prior diagnosis of malignant lymphoma. Furthermore, in many of the patients with a previous known diagnosis of malignant lymphoma, many of these were primary in extranodal sites. Most of the tumors were either large cell lymphomas or included a large cell component. The smears were generally quite cellular and dominated by the malignant lymphocytes (Fig. 3-52). Only a small number of benign hepatocytes were present in most of the specimens. We found ancillary diagnostic procedures to be quite helpful in confirming lymphoma. Most frequently, this consisted of immunocytochemistry, performed either directly on the smears or in cell block. In addition, this type of specimen would lend itself readily to immunophenotyping by flow cytometry for lymphoid cell surface markers. Not surprisingly, of the 11 tumors examined, 10 had a B-cell phenotype.[74] FNA biopsy may also sample hepatic involvement with myeloproliferative disorders.[85] A helpful hint is the presence of megakaryocytes and granulocytic precursous (Fig. 3-56). Diff-Quik stain is especially helpful in this regard.

STROMAL NEOPLASMS

The most common benign neoplasm of the liver is the hemangioma. In most instances, it is an incidental finding during surgery or radiologic imaging of the abdomen. Although most hemangiomas are small and solitary, they may attain a large size or occur as multiple lesions. Both grossly and radiographically, these have a typical appearance. Therefore, it is quite unlikely that these will be initially aspirated. We have seen a few examples of histologically confirmed hepatic hemangiomas that have been aspirated, almost always in an attempt to confirm the presence of a metastasis in a patient with a known primary neoplasm elsewhere (Fig. 3-57). As expected, these smears are dominated by blood.[2,11] One could easily "miss" the neoplastic cells as the endothelial cells are small and nondescript. They

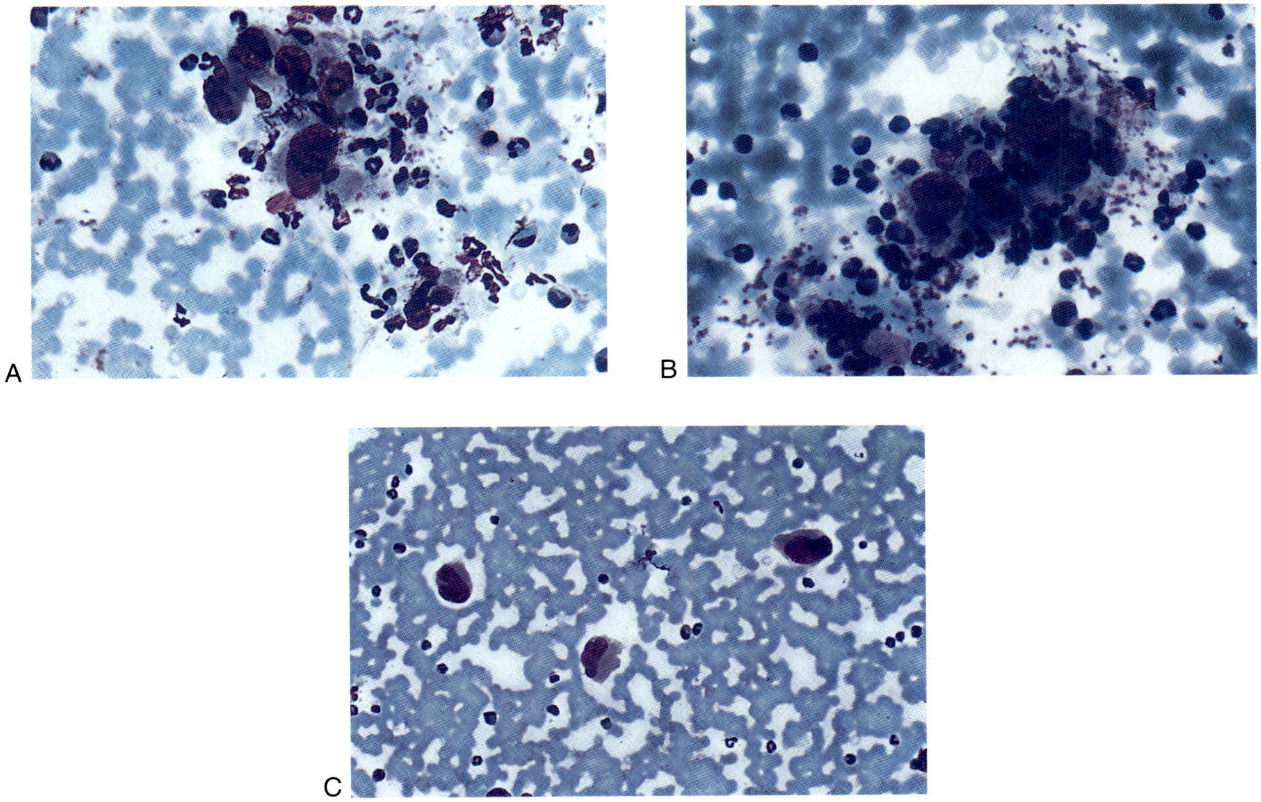

Fig. 3-56. (**A&B**) Small clusters of megakaryocytes are associated with elements of the granulocyte series in a liver aspirate from a patient with an unrecognized myeloproliferative disorder. Diagnostic clues reside in the recognition of the highly lobulated nuclei in the megakaryocytes and the presence of immature leukocytes in addition to the mature granulocytes (Diff-Quik). (**C**) When the megakaryocytes are individually dispersed, their true nature is more easily recognized (Diff-Quik).

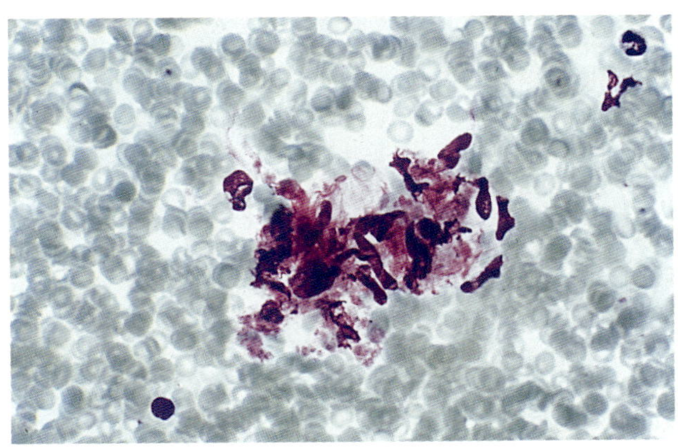

Fig. 3-57. As is expected in aspirates of hemangiomas, the picture is dominated by a background of blood. The smears may contain small clumps of endothelial cells. These cells have elongated or plump ovoid nuclei with bland chromatin and inapparent nucleoli. Cytoplasm is inconspicuous. In this example, the endothelial cells are associated with a fragment of collagen (Diff-Quik).

each possess a single ovoid to elongated nucleus and scanty cytoplasm. The chromatin is finely granular, uniformly distributed, and nucleoli are not evident. The tumor cells may be present singly within the blood or loosely aggregated in association with collagenous stroma. If one recognizes these elements, a correct diagnosis of hemangioma can be suspected, although the findings are not specific by themselves. Otherwise, a diagnosis of inadequate or insufficient will be mistakenly made.

Fortunately, primary sarcomas of the liver are distinctly rare. The most common type is angiosarcoma. Epidemiologically, it is well recognized that the development of this malignancy is associated with environmental exposure to thorium dioxide, arsenic, polyvinyl chloride, and anabolic steroids. Most patients are middle-aged or older, and men are affected more often than women. The most frequent clinical presentation is abdominal pain or the discovery of a palpable mass. These neoplasms are highly fatal, with most deaths related to either hepatic failure or massive hemorrhage. In addition, metastases occur with some frequency. We have seen only a few examples of angiosarcoma of the liver in fine needle aspiration biopsies, and only rarely have aspirates of this tumor been reported in the literature.[11,86] These smears may be highly cellular or diluted with blood. Tumor cells have spindled to polygonal contours and generally a solitary hyperchromatic nucleus (Fig. 3-58). Nuclear to cytoplasmic ratios are variable, as are nucleoli. We have not seen any specific cytoplasmic features. The tumor cells may be arranged in papillae, solid medullary patterns, individually, and in structures resembling blood vessels. The latter consist of tubular structures lined with malignant cells resembling those in the more solid areas. It is important to differentiate tumor cells lining the internal surface of vessels, compared to the much more common phenomenon of neoplastic elements coating the outside of benign, non-neoplastic blood vessels. If these structures are recognized, the specific diagnosis can be suggested.

In our experience and in that of others, the most common sarcoma sampled in FNA biopsies of the liver is leiomyosarcoma.[29,87] The vast majority of all leiomyosarcomas sampled in aspiration biopsies of the liver represent metastases, almost always from an infradiphragmatic primary neoplasm. One of the more recently described aspects of acquired immunodeficiency syndrome (AIDS) is the development of smooth muscle tumors in relatively unusual sites, including the liver.[87]

The cytomorphologic appearance of aspirated leiomyosarcomas generally mirrors those seen in aspirates of soft tissue tumors, as described in Chapter 7. Very briefly, smears generally have a moderate level of cellularity with varying degrees of intercellular cohesion.[87] Characteristically, there is a combination of individually dispersed neoplastic elements and fragments of tumor (Fig. 3-59). The latter are characterized by overlapping bundles of malignant spindle cells. The vast majority of the tumor cells have a solitary hyperchromatic ovoid to elongated nucleus with blunt ends and variable amounts of cytoplasm that tends to blend with that of its neighbors. Pleomorphism is usually not well developed. Infrequently, one may encounter an epithelioid leiomyosarcoma, usually metastatic from the gastrointestinal tract; smears will be dominated by solitary round to polygonal-shaped neoplastic cells with ovoid nuclei that may be centrally or eccentrically positioned (Fig. 3-60). Often, nucleoli are not well developed. In the absence of a known epithelioid leiomyosarcoma, the differential diagnosis in the smears would include metastatic adenocarcinomas, especially signet-ring cell types and lobular carcinoma of the breast, malignant melanomas, and possibly large cell lymphoma. This decision may be further complicated by clear-appearing cytoplasm. Lagrange[88] described a single instance of a metastatic myxoid leiomyosarcoma from the stomach in which there was a mixture of epithelioid and spindle tumor cells. Nguyen[89] also described the aspiration cytomorphology of epithelioid smooth muscle tumors.

One neoplasm that occurs overwhelmingly in the pediatric population is the undifferentiated or embryonal sarcoma.[90,91] These large neoplasms generally carry a fatal prognosis. Histologically,

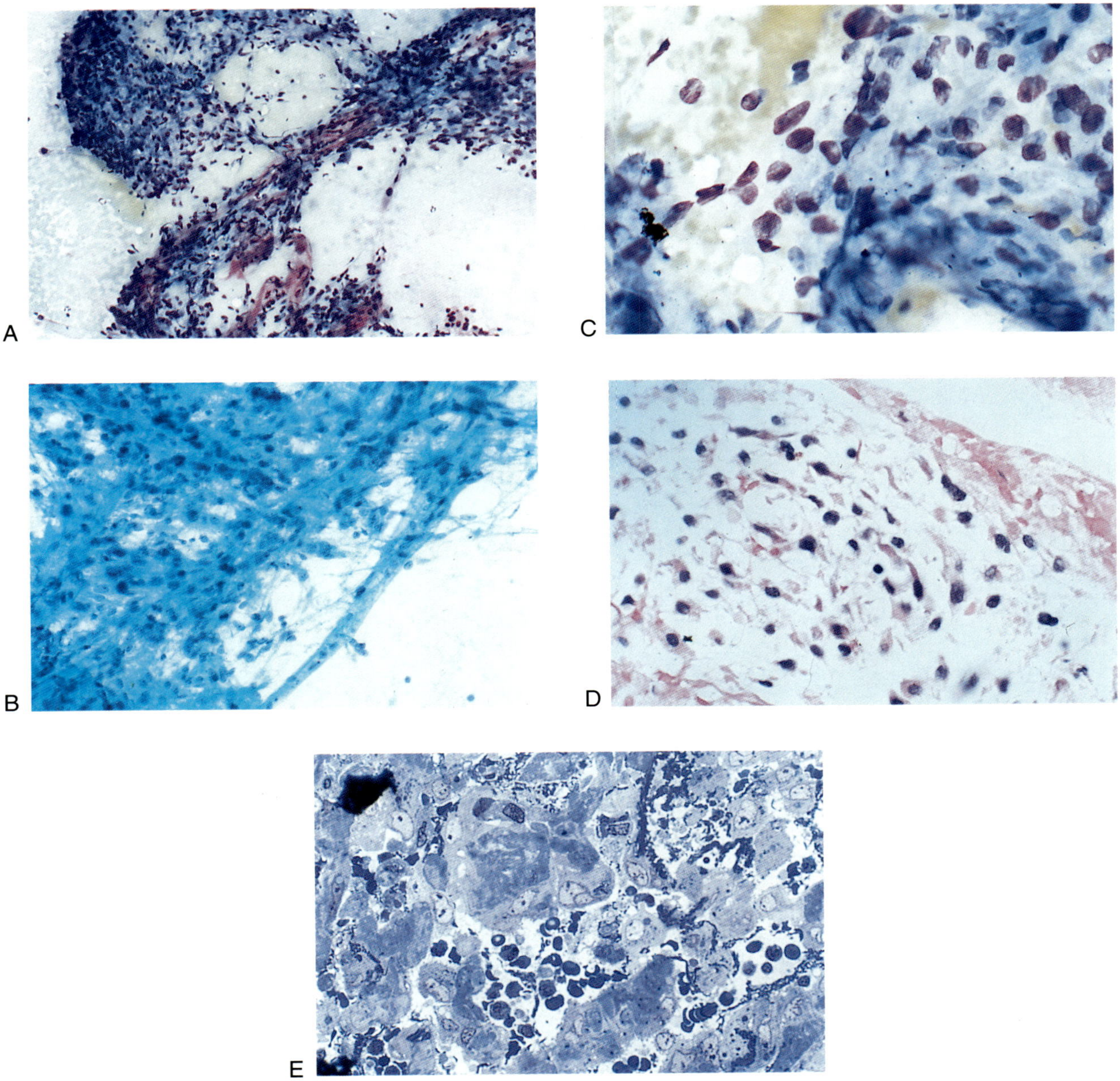

Fig. 3-58. (**A**) A key to the recognition of angiosarcoma in a liver aspirate is a high level of smear cellularity, yet an absence of recognizable epithelial qualities (Diff-Quik). (**B**) Many of the neoplastic cells are present in a solid or medullary arrangement. In one focus, however, the neoplastic cells appear to line a tubular or vascular structure (Papanicolaou). (**C**) The neoplastic cells have ovoid to elongated nuclei, an occasional nucleolus, and high nuclear to cytoplasmic ratios. Cell borders are indistinct (Diff-Quik). (**D**) The sarcomatous nature of the aspirated tumor cells is more readily apparent in this cell block. Each cell has an elongated hyperchromatic nucleus and elongated wispy cytoplasm (H&E). (**E**) In this thick section for electron microscopy, the neoplastic cells rest on a collagenous stroma and line irregular branching vascular structures (toluidine blue).

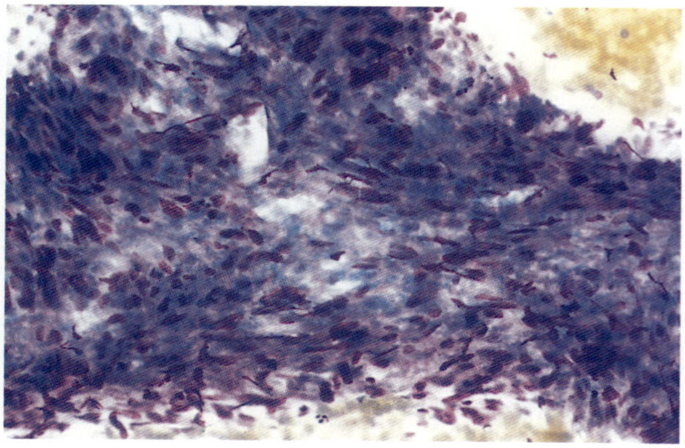

Fig. 3-59. Interlacing fascicles of elongated spindle-shaped malignant cells characterize this aspirated leiomyosarcoma. A moderate level of nuclear pleomorphism is apparent, as is a high level of cellularity (Diff-Quik).

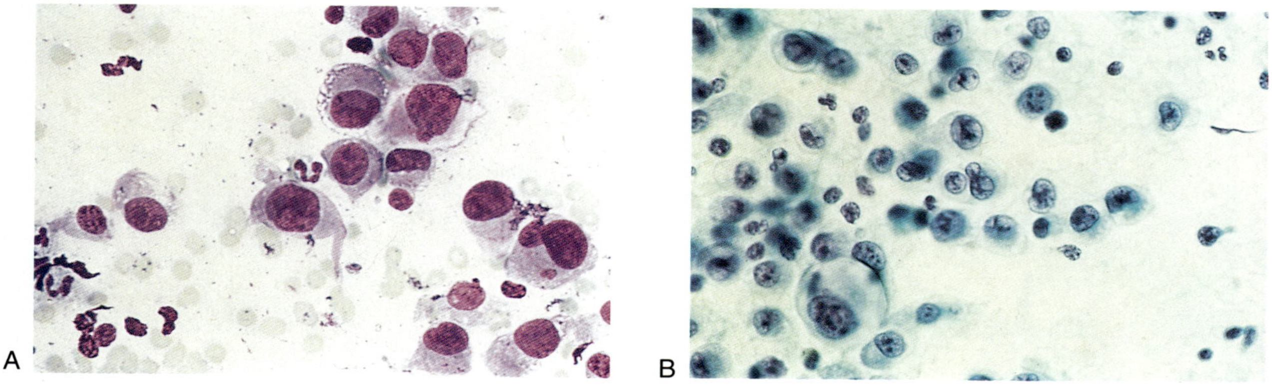

Fig. 3-60. (**A**) In this metastatic epithelioid leiomyosarcoma, individually dispersed polygonally shaped neoplastic cells are present. Each has a single eccentrically positioned round nucleus with a well-developed nucleolus and moderate volumes of cytoplasm. Intercellular borders are not highly distinct (Diff-Quik). (**B**) Similar features may be visualized with the Papanicolaou stain.

these neoplasms show alternating foci of hemorrhage and necrosis and viable tumor with a very well-developed myxoid background. Overall, tumor cellularity is not remarkably high in tissue sections. Pieterse et al.[92] described an example of embryonal sarcoma in an FNA biopsy. Smears were moderately cellular with individually dispersed neoplastic cells present within a myxoid matrix material. Cells varied from round to spindled to irregular in contour and generally possessed a solitary hyperchromatic nucleus and a high nuclear to cytoplasmic ratio. One could easily envision an inadequate specimen due to sampling of the large necrotic areas.

MISCELLANEOUS LESIONS

Amyloidosis is the pathologic deposition of proteinaceous material that histologically appears as a homogeneous and somewhat glassy-appearing eosinophilic substance. Typically, this material first appears in the interstitium with progressive accumulation and replacement of parenchyma. Amyloidosis frequently involves the liver as a diffuse enlargement, rather than as a relatively discrete mass lesion. Accordingly, tissue core biopsies are generally performed for the diagnosis of hepatic amyloidosis. Rare reports of amyloid in the liver have been documented by FNA biopsy.[93,94] In aspiration smears, amyloid appears as an acellular structureless material with rather sharply defined edges. In contrast to a matrix material seen in many mesenchymal neoplasms, cells are generally not evident within the amyloid itself. Rather, hepatocytes may be seen adjacent to the metachromatically staining amyloid with the Diff-Quik stain. This material may be confirmed histochemically. The most commonly used reaction is the Congo red stain. Although congophilia can be detected in air-dried smears, we prefer the use of the cell block preparation.

Steatosis usually involves the liver as a diffuse alteration secondary to any of a number of metabolic aberrations. As described by Brawer et al.,[95] infrequently fatty change occurs as a focal process in the liver that may yield the appearance of a mass lesion by various scanning modalities. Consequently, this could lead to an FNA biopsy. Smears will contain benign-appearing hepatocytes that contain one or more lipid vacuoles in their cytoplasm. As with benign hepatocytes in other conditions, the nuclear to cytoplasmic ratio is low but variable.[2,96] This is another example of the need for close interaction and communication between radiologists and pathologists.

Angiomyolipoma is a benign neoplasm composed of well-formed adipose tissue, smooth muscle cells, and blood vessels. As described in Chapter 5, this is a rare neoplasm in the kidney. As a primary hepatic neoplasm, angiomyolipoma is exceedingly unusual. Recently, Ma et al.[97] reported an example of a primary angiomyolipoma of the liver. The direct smears were characterized by a low level of cellularity and were considered nondiagnostic. However, the cell block preparation contained diagnostic material. This included thick-walled vascular structures, benign adipocytes, elements of extramedullary hematopoiesis, and arrangements of smooth muscle cells. For the most part, the latter elements were epithelioid in nature, rather than spindle shaped. These elements were positive for muscle specific actin and HMB-45.

Other rare entities presenting as hepatic mass lesions that may yield unusual cytologic findings include inflammatory pseudotumor, primary endodermal sinus tumor, and the familial hemophagocytic syndrome.[98–100]

REFERENCES

1. Silverman JF, Finley JL, O'Brien KF et al: Diagnostic accuracy and role of immediate interpretation of fine needle aspiration biopsy specimens from various sites. Acta Cytol 33:791, 1989
2. Whitlach S, Nuñez C, Pitlik DA: Fine needle aspiration biopsy of the liver. Acta Cytol 28:719, 1984
3. Axe SR, Erozan YS, Ermatinger SV: Fine-needle aspiration of the liver. A comparison of smear and rinse preparations in the detection of cancer. Am J Clin Pathol 86:281, 1986
4. Bell DA, Carr CP, Szyfelbein WM: Fine needle aspiration cytology of focal liver lesions. Results obtained with examination of both cytologic and histologic preparations. Acta Cytol 30:397, 1986
5. Wee A, Nilsson B, Chan-Wilde C, Yap I: Fine needle aspiration biopsy of hepatocellular carcinoma. Some unusual features. Acta Cytol 35:661, 1991
6. Zainol H, Sumithran E: Combined cytological and histological diagnosis of hepatocellular carcinoma in ultrasonically guided fine needed biopsy specimens. Histopathology 22:581, 1993
7. Tao L-C: Liver and pancreas. p. 829. In Bibbo M (ed): Comprehensive Cytopathology. WB Saunders, Philadelphia, 1991
8. Walsh TJ, Berkman W, Brown NL et al: Cytopathologic diagnosis of extracolonic amebiasis. Acta Cytol 27:671, 1983
9. Frias-Hidvegi D: Liver and Pancreas. Guides to Clinical Aspiration biopsy. Igaku-Shoin, New York, 1988
10. Shujrbaji MS, Gupta PK, Newman MM: Hepatic actinomycosis diagnosed by fine needle aspiration. A case report. Acta Cytol 31:751, 1987.
11. Nguyen G-K: Fine needle aspiration biopsy cytology of hepatic tumors in adults. Pathol Annu 21(pt 1):321, 1986
12. Bottles K, Cohen MB: An approach to fine-needle aspiration biopsy diagnosis of hepatic masses. Diagn Cytopathol 7:204, 1991
13. Agarwal PK, Husain N, Singh BH: Cytologic findings in aspirated hydatid fluid. Acta Cytol 33:652, 1989
14. Perry MD, Johnston WW: Needle biopsy of the liver for the diagnosis of nonneoplastic liver diseases. Acta Cytol 29:385, 1985.
15. Berman JJ, McNeill RE: Cirrhosis with atypia. A potential pitfall in the interpretation of liver aspirates. Acta Cytol 32:11, 1988
16. Russack V, Vass L, Gupta PK: Comparison of morphologic features of benign hepatocytes associated with nonmalignant and malignant liver lesions. Acta Cytol 37:153, 1993.

17. Abendroth CS, Grenko RT, Welch DR: Features distinguishing cirrhosis from hepatocellular carcinoma by fine needle aspiration, abstracted. Acta Cytol 38:800, 1994
18. Bisceglie AM, Rustgi VK, Hoofnagle JH et al: Hepatocellular carcinoma. Ann Intern Med 108:390, 1988
19. Colombo M: Hepatocellular carcinoma. J Hepatol 15:225, 1992
20. Falkson G, Canaan A, Schutt AJ et al: Prognostic factors for survival in hepatocellular carcinoma. Cancer Res 48:7314, 1988
21. Barbara L, Benzi G, Gaiani S et al: Natural history of small untreated hepatocellular carcinoma in cirrhosis: a multivariate analysis of prognostic factors of tumor growth rate and patient survival. Hepatology 16:132, 1992
22. Pinto MM, Avila NA, Heller CI, Criscuolo EM: Fine needle aspiration of the liver. Acta Cytol 32:15, 1988
23. Sbolli G, Fornari F, Civardi G et al: Role of ultrasound guided fine needle aspiration biopsy in the diagnosis of hepatocellular carcinoma. Gut 31:1303, 1990
24. Tao L-C, Ho CS, McLoughlin MJ WK et al: Cytologic diagnosis of hepatocellular carcinoma by fine-needle aspiration biopsy. Cancer 53:347, 1984
25. Greene C-A, Suen KC: Some cytologic features of hepatocellular carcinoma as seen in fine needle aspirates. Acta Cytol 28:173, 1984
26. Suen KC: Diagnosis of primary hepatic neoplasms by fine-needle aspiration cytology. Diagn Cytopathol 2:99, 1986
27. Ali MA, Akhtar M, Mattingly RC: Morphologic spectrum of hepatocellular carcinoma in fine needle aspiration biopsies. Acta Cytol 30:294, 1986
28. Noguchi S, Yamamoto R, Tatauta M et al: Cell features and patterns in fine-needle aspirates of hepatocellular carcinoma. Cancer 58:321, 1986
29. Bottles K, Cohen MB, Holly EA et al: A step-wise logistic regression analysis of hepatocellular carcinoma. An aspiration biopsy study. Cancer 62:558, 1988
30. Pedio G, Landolt U, Zöbeli L, Gut D: Fine needle aspiration of the liver. Significance of hepatocytic naked nuclei in the diagnosis of hepatocellular carcinoma. Acta Cytol 32:437, 1988
31. Kung ITM, Chan SK, Fung KH: Fine-needle aspiration in hepatocellular carcinoma: Combined cytologic and histologic approach. Cancer 67:673, 1991
32. Cohen MB, Haber MM, Holly EA et al: Cytologic criteria to distinguish hepatocellular carcinoma from nonneoplastic liver. Am J Clin Pathol 95:125, 1991
33. Wee A, Nilsson B, Tan LKA, Yap I: Fine needle aspiration biopsy of hepatocellular carcinoma. Diagnostic dilemma at the ends of the spectrum. Acta Cytol 38:347, 1994
34. Adeyanju MO, Dodd GD, Madariaga JR, Dekker A: Ultrasonically guided fine-needle aspiration biopsy of portal vein thrombosis: a cytomorphological study of 14 patients. Diagn Cytopathol 11:281, 1994
35. Goellner JR, Salomao DR: Hepatocellular carcinoma: needle biopsy findings in 74 cases, abstracted. Acta Cytol 38:800, 1994
36. Sole M, Calvet X, Cuberes T et al: Value and limitations of cytologic criteria for the diagnosis of hepatocellular carcinoma by fine needle aspiration biopsy. Acta Cytol 37:309, 1993
37. Tao L-C: Are oral contraceptive-associated liver cell adenomas premalignant? Acta Cytol 36:338, 1992
38. Ojanguren I, Ariza A, Llatjós M et al: Proliferating cell nuclear antigen expression in normal, regenerative, and neoplastic liver: a fine-needle aspiration cytology and biopsy study. Hum Pathol 24:905, 1993
39. Adachi E, Hashimoto H, Tsuneyoshi M: Proliferating cell nuclear antigen in hepatocellular carcinoma and small cell liver dysplasia. Cancer 72:2902, 1993
40. Russo A, Bazan V, Plaja S, Cajozzo M, Bazan P: Flow cytometric DNA analysis of hepatic tumours on ultrasound-guided fine-needle aspirates. J Surg Oncol 51:26, 1992
41. Cottier M, Jouffre C, Maubon I et al: Prospective flow cytometric DNA analysis of hepatocellular carcinoma specimens collected by ultrasound-guided fine needle aspiration. Cancer 74:599, 1994
42. Fuhr JE, Kattine AA, Nelson HS: Flow cytometry of liver fine needle aspirates. J Surg Oncol 56:153, 1994
43. Deprez C, Vangansbeke D, Fastrez R et al: Nuclear DNA content, proliferation index, and nuclear size determination in normal and cirrhotic liver, and in benign and malignant primary and metastatic hepatic tumors. Am J Clin Pathol 99:558, 1993
44. Sangalli G, Livraghi T, Giordano F: Fine needle biopsy of hepatocellular carcinoma: improvement in diagnosis by microhistology. Gastroenterology 96:524, 1989
45. Van Eyken PV, Sciot R, Paterson A et al: Cytokeratin expression in hepatocellular carcinoma: an immunohistochemical study. Hum Pathol 19:562, 1988
46. Hurlimann J, Gardiol D: Immunohistochemistry in the differential diagnosis of liver carcinomas. Am J Surg Pathol 15:3, 1991
47. Chedid A, Chejfec G, Eishorst M et al: Antigenic markers of hepatocellular carcinoma. Cancer 65:84, 1990

48. Ma CK, Zarbo RJ, Frierson HF, Lee MW: Comparative immunohistochemical study of primary and metastatic carcinomas of the liver. Am J Clin Pathol 99:551, 1993
49. Fucich LF, Cheles MK, Thung SN et al: Primary versus metastatic hepatic carcinoma. An immunohistochemical study of 34 cases. Arch Pathol Lab Med 118:927, 1994
50. Ganjei P, Nadji M, Albores-Saavedra, Morales AR: Histologic markers in primary and metastatic tumors of the liver. Cancer 62:1994, 1988
51. Bedrossian CWM, Davila RM, Merenda G: Immunocytochemical evaluation of liver fine-needle aspirations. Arch Pathol Lab Med 113:1225, 1989
52. Johnson DE, Powers CN, Rupp G, Frable WJ: Immunocytochemical staining of fine-needle aspiration biopsies of liver as a diagnostic tool for hepatocellular carcinoma. Mod Pathol 5:117, 1992
53. Rishi M, Kovatich A, Ehya H: Utility of polyclonal and monoclonal antibodies against carcinoembryonic antigen in hepatic fine-needle aspirates. Diagn Cytopathol 11:358, 1994
54. Guindi M, Yasdi HM, Gilliat MA: Fine needle aspiration biopsy of hepatocellular carcinoma. Value of immunocytochemical and ultrastructural studies. Acta Cytol 38:385, 1994
55. Wong MA, Yazdi HM: Hepatocellular carcinoma versus carcinoma metastatic to liver: value of stains for carcinoembryonic antigen and naphthylamidase in fine needle aspiration biopsy material. Acta Cytol 34:192, 1990
56. Wolber RA, Greene CA, Dapris BA: Polyclonal carcinoembryonic antigen staining in cytologic differential diagnosis of primary and metastatic malignancies. Acta Cytol 35:215, 1991
57. Carrozza MJ, Calafati SA, Edmonds PR: Immunocytochemical localization of polyclonal carcinoembryonic antigen in hepatocellular carcinoma. Acta Cytol 35:221, 1991
58. Domagala W, Lasota J, Weber K, Osborn M: Endothelial cells help in the diagnosis of primary versus metastatic carcinoma of the liver in fine needle aspirates. An immunofluorescence study with vimentin and endothelial cell-specific antibodies. Anal Quant Cytol Histol 11:8, 1989
59. Kilpatrick SE, Geisinger KR, Loggie BW, Hopkins MB III: Fine needle aspiration biopsy—cytomorphology of combined hepatocellular-cholangiocarcinoma. Acta Cytol 37:943, 1993
60. Neill JSA, Silverman JF: The role of ultrastructural examination in fine needle aspiration biopsy of liver neoplasms. Acta Cytol 36:621, 1992
61. Ordóñez NG, MacKay B: Ultrastructure of liver cell and bile duct carcinoma. Ultrastruct Pathol 5:201, 1983
62. Berman MA, Burhnam JA, Sheahan DG: Fibrolamellar carcinoma of the liver: an immunohistochemical study of nineteen cases and a review of the literature. Hum Pathol 19:784, 1988
63. Orsatti G, Greenberg PD, Rolfes DB et al: DNA ploidy of fibrolamellar hepatocellular carcinoma by image analysis. Hum Pathol 25:936, 1994
64. Suen KC, Magee JF, Halparin LS et al: Fine needle aspiration cytology of fibrolamellar hepatocellular carcinoma. Acta Cytol 29:867, 1985
65. Gupta SK, Das DK, Rajwanshi A, Bhusnurmath SR: Cytology of hepatocellular carcinoma. Diagn Cytopathol 2:290, 1986
66. Davenport RD: Cytologic diagnosis of fibrolamellar carcinoma of the liver by fine-needle aspiration. Diagn Cytopathol 6:275, 1990
67. Donat EE, Anderson V, Tao L-C: Cytodiagnosis of clear cell hepatocellular carcinoma. A case report. Acta Cytol 35:671, 1991
68. Gupta RK, AlAnsari AG, Fauck R: Aspiration cytodiagnosis of clear cell hepatocellular carcinoma in an elderly woman. A case report. Acta Cytol 38:467, 1994
69. Singh HK, Silverman JF, Geisinger KR: Fine needle aspiration cytology of clear cell carcinoma, abstracted. Acta Cytol 38:856, 1994
70. Jensen CS, Donnelly AD, Silverman JF et al: The role of fine needle aspiration cytology in the work-up of metastatic clear cell tumors, abstracted. Mod Pathol 8:41A, 1995
71. Goodman AD, Ishak KG, Langloss JM et al: Combined hepatocellular-cholangiocarcinoma. A histologic and immunohistochemical study. Cancer 55:124, 1985
72. Saul SH: Neoplasms of liver. p. 1162. In Sternberg SS (ed): Diagnostic Surgical Pathology. Raven Press, New York, 1989
73. Gomez-Fernandez C, Kraemer HJ, Ganjei P: Metastatic medullary thyroid carcinoma in liver diagnosis by aspiration cytology. Diagn Cytopathol 11:277, 1994
74. Collins KA, Geisinger KR, Raab SS, Silverman JF: Fine needle aspiration biopsy of hepatic lymphomas: cytomorphology and ancillary studies. Acta Cytol, in press
75. Cavanna L, Civardi G, Fomari F et al: Simultaneous relapse of liver cell carcinoma and non-Hodgkin's lymphoma in the liver. Report of a case with diagnosis by ultrasonically guided fine needle aspiration biopsy. Acta Ctyol 38:451, 1994

76. Weinberg AG, Finegold MJ: Primary hepatic tumors of childhood. Hum Pathol 14:512, 1983
77. Conran RM, Hitchcock CL, Waclawiw MA et al: Hepatoblastoma: the prognostic significance of histologic type. Pediatr Pathol 12:167, 1992
78. Schmidt D, Wischmeyer P, Leuschner I et al: DNA analysis in hepatoblastoma by flow and image cytometry. Cancer 72:2914, 1993
79. Bhatia A, Mehrotra P: Fine needle aspiration cytology in a case of hepatoblastoma. Acta Cytol 30:439, 1986
80. Dekmezian R, Sniege N, Popok S et al: Fine-needle aspiration cytology of pediatric patients with primary hepatic tumors: a comparative study of two hepatoblastomas and a liver cell carcinoma. Diagn Cytopathol 4:162, 1988
81. Wakely PE, Jr, Silverman JF, Geisinger KR, Frable WJ: Fine needle aspiration biopsy cytology of hepatoblastoma. Mod Pathol 3:668, 1990
82. Kaw YT, Hansen K: Fine needle aspiration cytology of undifferentiated small cell ("anaplastic") hepatoblastoma. A case report. Acta Cytol 37:216, 1993
83. Pérez JS, Pérez-Guillermo M, Bernal AB, Mercader JM: Hepatoblastoma. An attempt to apply histologic classification to aspirates obtained by fine needle aspiration cytology. Acta Cytol 38:175, 1994
84. Cangiarella J, Greco MA, Waisman J: Hepatoblastoma. Report of a case with cytologic, histologic, and ultrastructural findings. Acta Cytol 38:455, 1994
85. Raab SS, Silverman JF, McLeod DL, Geisinger KR: Fine-needle aspiration cytology of extramedullary hematopoiesis (myeloid metaplasia). Diagn Cytopathol 9:522, 1993
86. Abele JS, Miller TR: Cytology of well-differentiated and poorly differentiated hemangiosarcoma in fine needle aspirates. Acta Cytol 26:341, 1982
87. Smith MB, Silverman JF, Raab SS et al: Fine-needle aspiration cytology of hepatic leiomyosarcoma. Diagn Cytopathol 11:321, 1994
88. Lagrange W: Fine needle aspiration biopsy of myxoid variant of malignant leiomyoblastoma metastatic to the liver. Acta Cytol 32:443, 1988
89. Nguyen GK: Cytopathologic aspects of leiyomoblastoma in fine needle aspiration biopsy: report of two cases. Acta Cytol 27:173, 1983
90. Keating S, Taylor GP: Undifferentiated (embryonal) sarcoma of the liver: Ultrastructural and immunohistochemical similarities with malignant fibrous histiocytoma. Hum Pathol 16:693, 1985
91. Parham DM, Kelly DR, Donnelly WH et al: Immunohistochemical and ultrastructural spectrum of hepatic sarcomas of childhood: evidence of a common histogenesis. Mod Pathol 4:648, 1991
92. Pieterse AS, Smith M, Smith LA, Smith P: Embryonal (undifferentiated) sarcoma of the liver. Fine-needle aspiration cytology and ultrastructural findings. Arch Pathol Lab Med 109:677, 1985
93. Bose S, Kapila K, Verma K: Amyloidosis of the liver diagnosed by fine needle aspiration cytology. Acta Cytol 33:935, 1989
94. Srinivasan R, Nijhawan R, Gautam U, Bambery P: Potassium permanganate-resistant amyloid in fine-needle aspirate of the liver. Diagn Cytopathol 10:383, 1994
95. Brawer MK, Austin GE, Lewin KJ: Focal fatty change of the liver, a hitherto poorly described entity. Gastroenterology 78:247, 1980
96. Layfield LJ: Focal fatty change of the liver: cytologic findings in a radiographic mimic of metastases. Diagn Cytopathol 11:385, 1994
97. Ma T K-W, Tse M-K, Tsui M-S, Yuen K-T: Fine needle aspiration diagnosis of angiomyolipoma of the liver using a cell block with immunohistochemical study. A case report. Acta Cytol 38:257, 1994
98. Lupovitch A, Chen R, Mishra S: Inflammatory pseudotumor of the liver. Report of the fine needle aspiration cytologic findings in a case initially misdiagnosed as malignant. Acta Cytol 33:259, 1989
99. Wakely PE Jr, Krummel TM, Johnson DE: Yolk sac tumor of the liver. Mod Pathol 4:121, 1991
100. Silverman JF, Singh HK, Joshi VV et al: Cytomorphology of familial hemophagocytic syndrome. Diagn Cytopathol 9:404, 1993

4 Pancreas

Similar to the liver, the pancreas is a major abdominal organ involved in numerous aspects of digestion. In addition, many vital endocrine functions emanate from this organ. In contrast to hepatic parenchyma, however, the pancreas is much less accessible to physical examination and laboratory assessment. This is most unfortunate, in that monumentous inflammatory and neoplastic disorders may involve this retroperitoneal organ.

Fine needle aspiration (FNA) biopsy has had a major impact on the type of specimens interpreted by pathologists for the diagnosis of pancreatic diseases.[1–15] This results in part from the failings of tissue biopsy. First, frozen section and even permanent section interpretation of the pancreas have proved difficult.[16] Specifically, the distinction between pancreatitis and a well-differentiated adenocarcinoma can be difficult, especially in the intraoperative setting. Furthermore, sampling of the pancreas by either large-bore cutting needle or wedge section is associated with a high level of morbidity and even mortality, secondary to leakage of exocrine enzymes and bile leading to pancreatitis and peritonitis.[17] Although FNA biopsy of this organ is certainly not without its limitations, its relatively high diagnostic accuracy and safety overcome many of the negative aspects of a formal histologic examination.[13] As is detailed in this chapter, FNA examination has great value in the diagnosis of pancreatic neoplasms, in both the preoperative (percutaneous) and intraoperative settings. Accordingly, the proportion of these tumors diagnosed solely by aspiration cytology is increasing progressively in many medical institutions worldwide.

PANCREATITIS

As discussed in the excellent review article by Steinberg and Tenner,[18] acute pancreatitis has a myriad of etiologies. By far, the two most frequent are obstruction of the pancreatic ducts by gallstones and alcohol abuse. Acute pancreatitis has almost as many different clinical presentations. Yet, with an astute clinical evaluation and a judicious use of laboratory tests (e.g., serum amylase and lipase), a correct diagnosis can be made in the vast majority of patients. Furthermore, it would be unusual for acute pancreatitis to present as a mass lesion.[19] Consequently, FNA biopsy is usually not performed in patients with this inflammatory condition. Both benign and malignant pancreatic neoplasms, on the other hand, may obstruct the ductal system leading to release of pancreatic enzymes and the development of pancreatitis.[20] The aspirator may mistakenly place the needle in adjacent inflamed parenchyma rather than in the tumor. In this setting, aspiration smears will demonstrate the features of acute pancreatitis (Figs. 4-1 to 4-5). These may include a rather high level of cellularity (Figs. 4-1, 4-2, 4-5), largely due to numerous inflammatory elements, as listed on page 137. The latter, along with necrotic debris, provide an overall "dirty" appearance to the smear background.[1] Both intact and degenerated neutrophils will be present in large numbers and may be seen directly infiltrating epithelial cells (Fig. 4-3). Most of the latter will be acinar cells, with varying degrees of degenerative changes (Fig. 4-4). These cells will be present in small cohesive generally flat arrangements reminiscent of normal acini. The cells will have moderate volumes of granular cytoplasm and a single basally oriented small round nucleus. The chromatin will vary from the normal finely granular evenly distributed picture to complete pyknosis. In far fewer numbers, ductal cells may also be present in the smears and may demonstrate a normal honeycomb pattern or the changes of classic reparative atypia. Granular necrotic debris and foam cells complete the smear background. Aspirates of a pancreatic abscess will yield an indistinguishable cytomorphologic specimen with sheets of neutrophils. These abscesses may result from bacterial infection of a pseudocyst.

Patients with multiple episodes of well-documented acute pancreatitis may evolve into chronic pancreatitis. Many other patients insidiously develop chronic pancreatitis without any previ-

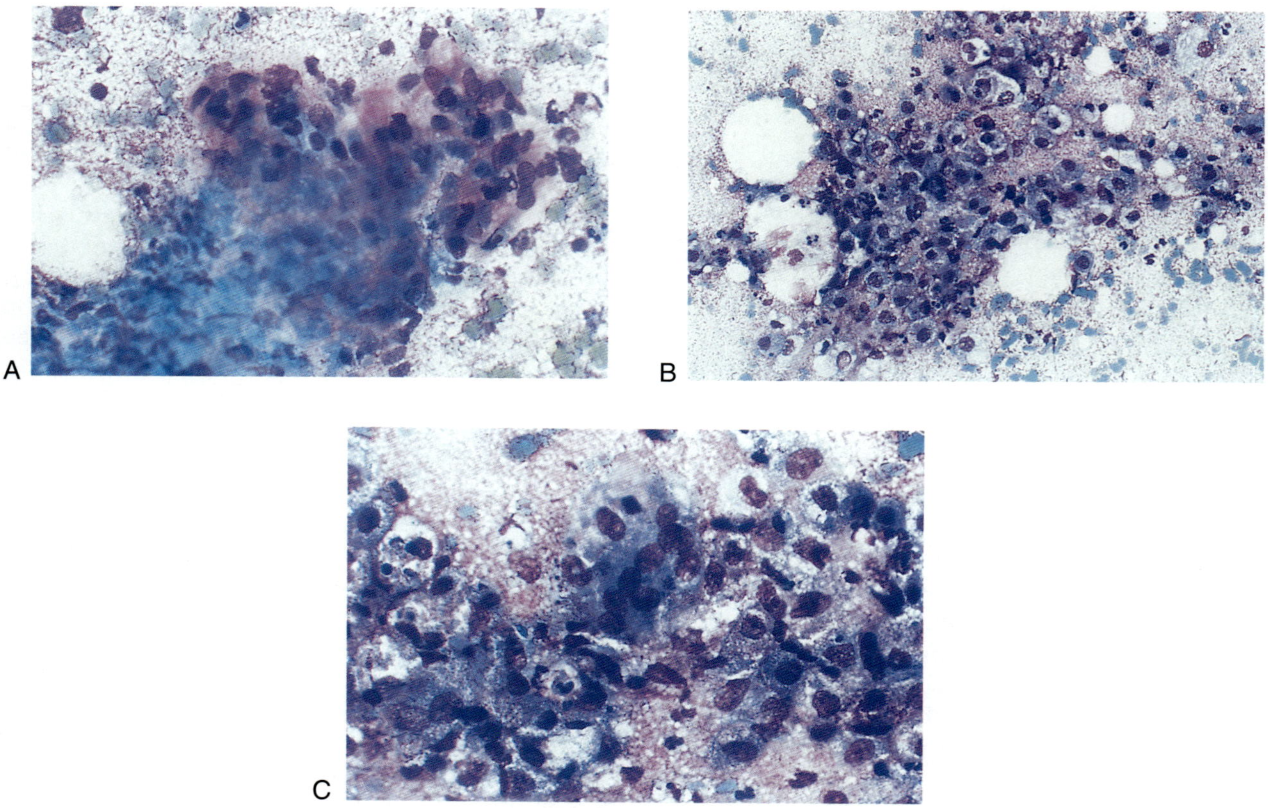

Fig. 4-1. Aspiration biopsies of acute pancreatitis may yield smears that are at least focally moderately to highly cellular. This may include a number of benign ductal epithelial cells. (**A**) Monolayer of small uniform benign epithelial cells set in a background of necrotic granular debris and a few neutrophils. The nuclei are larger than normal and a honeycomb pattern is not evident. However, polarity is relatively well maintained as crowding and molding are not apparent. Cohesion is also well preserved in that individually dispersed atypical epithelial cells are not present. (**B**) Intercellular cohesion disrupted with this rather loose arrangement of benign ductal cells. Fat globules and necrotic debris are present in the background. (**C**) Degenerative changes are apparent at higher magnification, in many of these epithelial cells; they include vacuolization of the cytoplasm and nuclei and cytoplasmic fragmentation. In addition, neutrophils are intimately admixed with these benign epithelial elements. A diagnosis of adenocarcinoma should not be made in the presence of such inflammatory and degenerative alterations. Note also that the epithelial nuclei are not much larger than the segmented leukocytes (**A–C**, Diff-Quik).

Cytomorphologic Features of Acute Pancreatitis

- Moderate to high cellularity (largely inflammatory) with an overall "dirty" background appearance
- Numerous neutrophils, often with degenerative features
- Acinar cells with variable degenerative changes
- Ductal cells in low numbers, possibly with reparative atypia
- Granular debris and foam cells

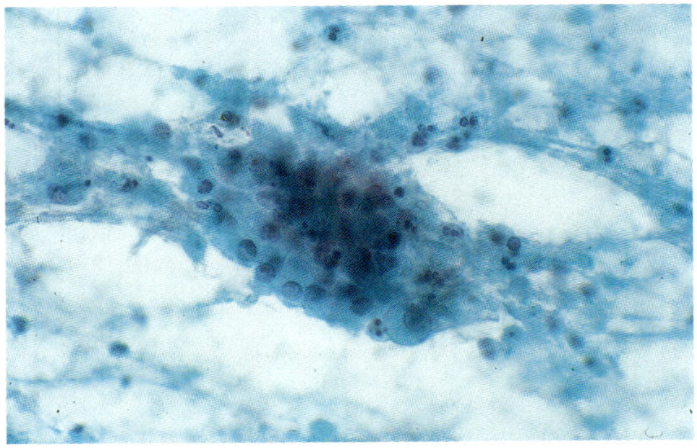

Fig. 4-3. In this small flat sheet of epithelial cells, the nuclei are enlarged and nucleoli are apparent. However, note the finely granular, evenly distributed, rather pale-staining chromatin, the relative preservation of polarity and intercellular cohesion, and the intimate association with neutrophils (Papanicolaou).

ously recognized episode of acute inflammation. Histologically, major alterations in chronic pancreatitis include a fibrous replacement of the exocrine parenchyma and, to a lesser extent, the islets, the presence of chronic inflammatory cells (often most densely concentrated about ducts), degenerative and regenerative alterations within ductal epithelial elements, and fat necrosis. As with acute disease, FNA biopsy of patients believed to have uncomplicated chronic pancreatitis is unusual. Due to the prominent fibrotic reaction in the tissue, smear cellularity is generally scanty (Fig. 4-6). The major component of the specimen may be ductal epithelial cells manifesting prominent reparative or inflammatory atypia, (see box on p. 138). For the most part, the ductal cells will be present in flat sheet-like arrangements, but these may

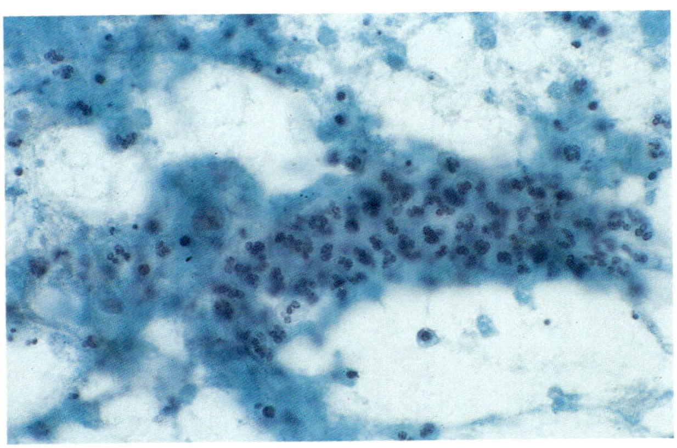

Fig. 4-2. Entire fields of the smears in acute pancreatitis may be occupied by neutrophilic leukocytes and granular debris. In the scanning lens mode, however, this may provide the initial impression of high smear cellularity (Papanicolaou).

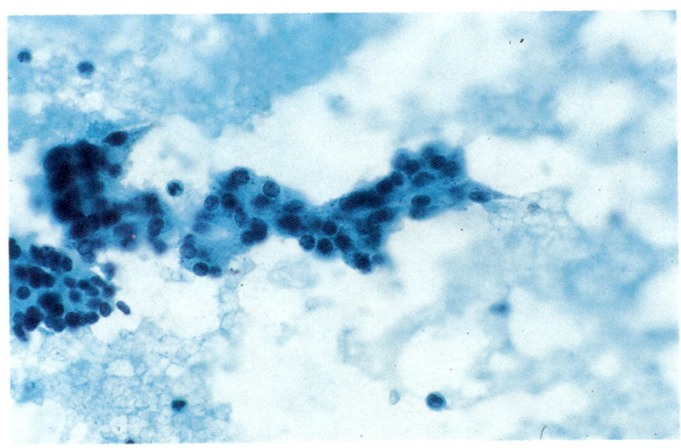

Fig. 4-4. Monolayer of small uniform benign acinar epithelial cells in a patient with acute pancreatitis. Both polarity and cohesion are well preserved. The nuclei are uniform and basally oriented and have a benign appearance. Note the complete absence of individually dispersed intact atypical epithelial cells (Papanicolaou).

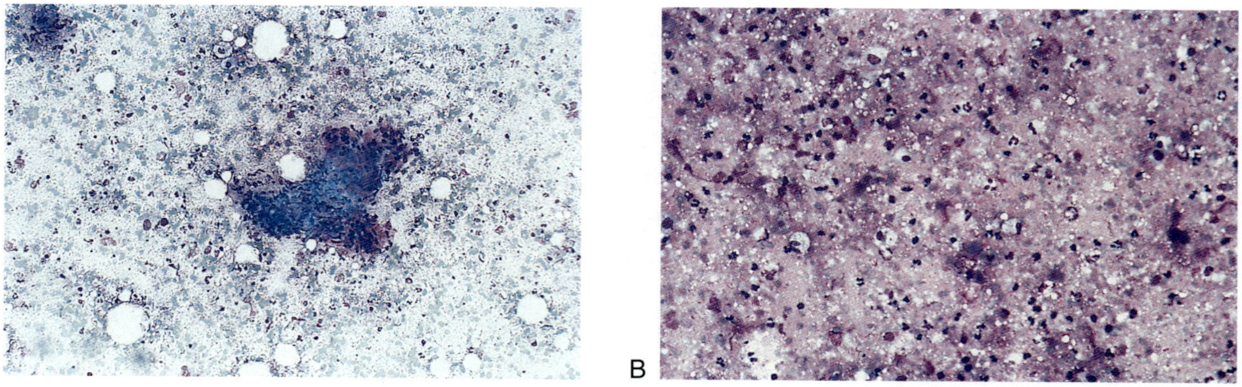

Fig. 4-5. In the setting of epithelial atypia and moderate cellularity, it is helpful to return to low magnification. It then becomes apparent that one is not dealing with a high level of epithelial cellularity. (**A**) In addition to scattered neutrophils, a few isolated enlarged epithelial nuclei are apparent. However, they are devoid of surrounding cytoplasm, and should therefore be diagnostically ignored. (**B**) Many fields in the slide will be totally devoid of epithelial elements. Only neutrophils, foam cells, and debris are present (**A & B**, Diff-Quik).

manifest altered polarity and reduced intercellular cohesion. In fact, individually dispersed atypical epithelial cells may be present, but only in very low numbers. These atypical cells have nuclei that are larger than normal and that often contain very well-developed nucleoli. Calcified debris secondary to saponification is characteristic (Fig. 4-7). At least from a cytopathologist's viewpoint, the major complication of chronic pancreatitis is the development of an inflammatory pseudocyst. As discussed later, the major consideration in aspirates of an inflamed pancreas is distinguishing the benign atypical alterations in epithelial cells from malignant neoplasm.[12,13]

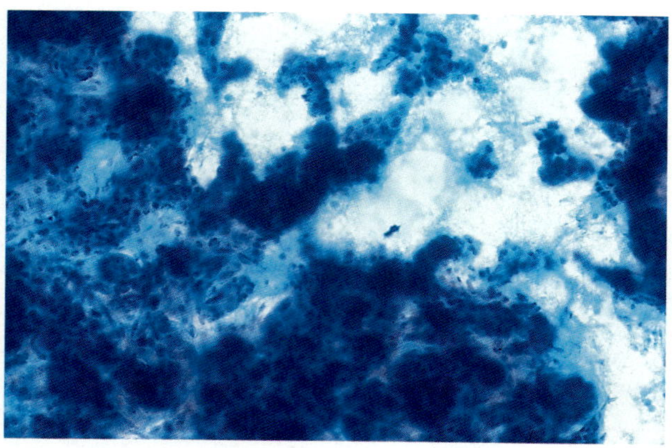

Fig. 4-6. Cellularity will be low in the majority of aspirates from patients with chronic pancreatitis. However, in this example, ductal and acinar epithelial cells are present in large numbers apparently entrapped within dense connective tissue. Intercellular cohesion appears to be well maintained (Papanicolaou).

Cytomorphologic Features of Chronic Pancreatitis

- Relatively low cellularity
- Sparse mixed inflammatory infiltrate
- Ductal cells with atypia
  - Flat sheets with altered polarity
  - Reduced cohesion
  - Large nuclei
  - Prominent nucleoli
- Granular and calcified debris

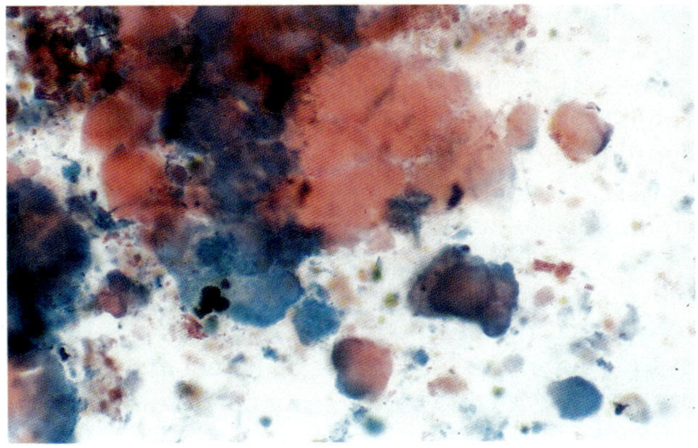

Fig. 4-7. No viable epithelial cells are evident in this field of fat necrosis. The "ghosts" of adipocytes are outlined by calcium salts (Papanicolaou).

CYSTS

Cystic lesions of the pancreas may be classified as congenital, infectious (parasitic), pseudocysts, and cystic tumors.[21–23] Concerning the last entry, almost any neoplasm of the pancreas may on occasion present clinically and radiographically as a cyst. As the name suggests, the cystic mucinous tumors of the pancreas are those which are most likely to present in this manner.

The most common of these cystic lesions is the pseudocyst. Most often, these result as a sequelae of pancreatitis, but may also occur secondary to upper abdominal trauma. In any case, the underlying pathogenesis consists of the release of pancreatic enzymes into the pancreas and surrounding tissues, with subsequent proteolytic and lipolytic damage. Pseudocysts do not have a true epithelial lining, but rather a luminal surface composed of granulation tissue with surrounding dense fibrosis. Most often, they occur within the lesser omental bursa.

Cytomorphologic Features of Pseudocyst

- Scanty cellularity
- Acute and chronic inflammatory cells, histiocytes
- Granulation tissue
- Debris

The cellularity of aspirates of pseudocysts is generally very low. Direct smears will be dominated by granular debris and lipid droplets.[2,6,8,15,24,25] If concentration procedures are employed, specimen cellularity will be enhanced. The predominant cells are inflammatory in nature and include neutrophils, lymphocytes, and plasma cells (Fig. 4-8). Young "active" fibroblasts from the surrounding granulation tissue may also be sampled. These will present as isolated cells and very small loosely arranged clusters of elongated to plump spindle-shaped fibroblasts. The fibroblasts may show the prominent nucleoli typical of cells derived from mesenchymal repair. In an attempt to

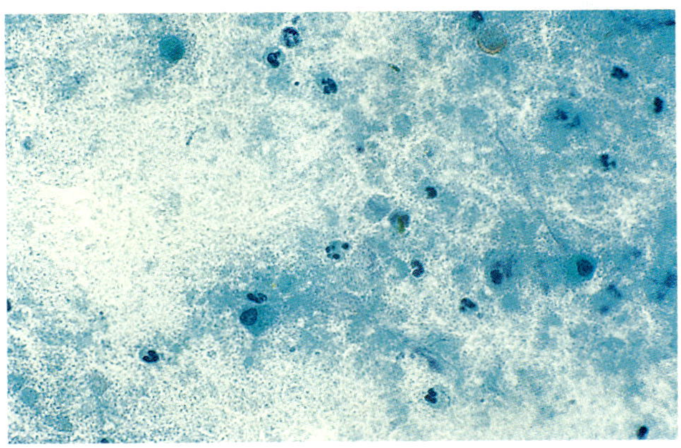

Fig. 4-8. Scant cellularity characterizes this pseudocyst aspirate. A few neutrophils and small mononuclear cells, presumably macrophages, are present on a granular background (Papanicolaou).

enhance the distinction between a benign pseudocyst and a cystic pancreatic malignancy, several investigators have suggested performing biochemical analysis of aspirated fluid in addition to cytomorphology.[26–31] Most helpful in this situation appear to be the measurement of amylase and carcinoembryonic antigen (CEA) levels. In pseudocysts, the fluid amylase level is very high, whereas the CEA level is low. The opposite pattern is recognized in pancreatic neoplasms, especially the cystic mucinous tumors. Both biochemical markers will be low to nondetectable in aspirates of pancreatic abscesses.

An unusual benign non-neoplastic condition, the lymphoepithelial cyst, which occurs most frequently in the parotid gland, has been described in the pancreas as well.[32,33] Recently, Cappellari[34] has depicted the aspiration cytomorphology of this rare cystic lesion. The specimen included numerous squamous epithelial cells, foam cells (histiocytes), lymphocytes, and cholesterol crystals (Fig. 4-9). Although a few superficial squamous cells were present, most were anucleated. The amylase level in the cyst fluid aspirate was within the upper range of normal. This cytomorphologic picture would probably be impossible to distinguish from an FNA biopsy of a dermoid cyst involving the pancreas.[35]

DUCTAL ADENOCARCINOMA

In the United States and many other parts of the world, malignant neoplasms of the pancreas are a common cause of death due to cancer.[36,37] In the United States, the incidence of pancreatic cancer has increased progressively over the past five decades, resulting in 2.5 percent of all new malignancies diagnosed each year. Adenocarcinoma of the pancreas is somewhat more common in men than in women and only rarely affects children. In fact, carcinoma of the pancreas before the age of 40 is distinctly rare. Black individuals are more often affected than are whites. Remarkable variations in the incidence of this malignancy are seen around the world. Malignancy of this organ is extremely aggressive, resulting in a dismal prognosis. As a result, cancer of the pancreas accounts for five percent of all cancer-related deaths and is the fifth most common cause of death from cancer in the United States. The 5-year survival rates are less than 5 percent, with one-half of all patients succumbing to their disease within 12 to 16 months. Most malignancies are ductal adenocarcinomas.[38]

Histologically, most ductal adenocarcinomas are well to moderately differentiated and are frequently associated with a prominent desmoplastic response. Several studies have strongly implicated

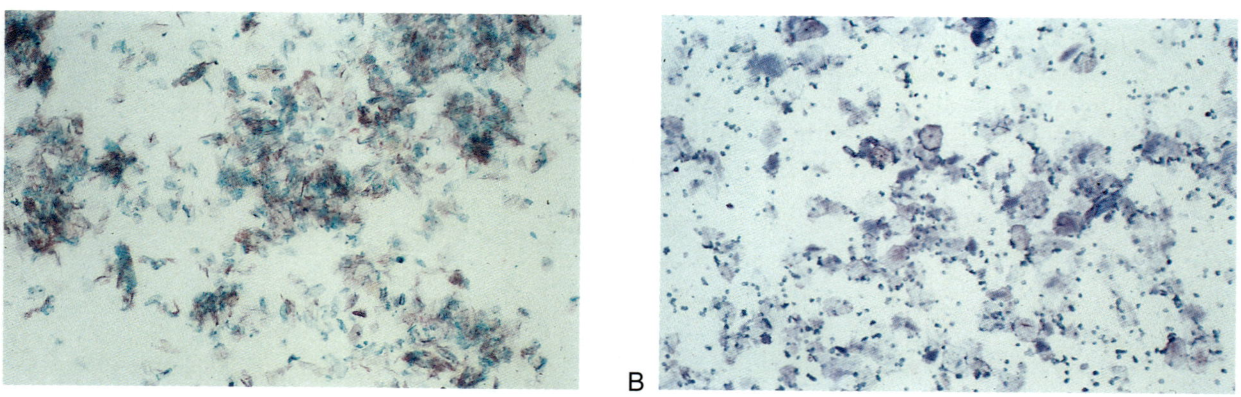

A B

Fig. 4-9. The smears of this pancreatic lymphoepithelial cyst are moderately cellular and are dominated by large anucleated squamous epithelial cells. (**A**) Only a minority of these squamous cells possess nuclei and in all of them they appear benign (Papanicolaou). (**B**) Numerous small mature-appearing lymphocytes are scattered along with the squamous cells (Diff-Quik).

two factors in their development: cigarette smoking and the consumption of a high-fat diet.[35] It remains controversial whether a history of chronic pancreatitis also predisposes to cancer. Factors that do not appear to be directly involved in the carcinogenesis of pancreatic cancer are caffeine and alcohol consumption or long-standing diabetes mellitus.[39]

The clinical presentation of pancreatic cancer depends in part on the exact site of origin of the tumor within the organ. Neoplasms in the head of the pancreas typically produce jaundice relatively early in the clinical course. In all cases, abdominal pain and weight loss are frequent. Ductal carcinomas are often discovered very late in the course in the vast majority of patients. At best, 10 percent of all carcinomas are considered amenable to total resection and hence are potentially curable. However, even in this fraction of patients, only 10 percent will survive for 5 years, with a median survival rate of just 20 months. Potential prognostic factors in patients with pancreatic carcinoma include mutations in the p53 tumor suppressor gene and DNA ploidy.[40–45]

The aspiration cytologic diagnosis of ductal adenocarcinoma is often straightforward.[1–15] One of the key features that we look for in direct smears is the presence of a high level of cellularity.[11] Although a high smear cellularity is neither totally specific nor sensitive, we would hesitate to diagnose carcinoma in its absence (Fig. 4-10). Although many of the malignant cells may be present in large sheets and smaller cellular aggregates, we require at least a few individually dispersed and intact abnormal cells before rendering a diagnosis of adenocarcinoma (Figs. 4-10 and 4-11). Thus, in addition to tumor cellularity, isolated abnormal cells with enlarged nuclei and surrounding cytoplasm are needed before an unequivocal diagnosis of carcinoma is made. As the histologic grade of the cancer increases, these dispersed cells are more apt to be present in large numbers. Compared to normal ductal epithelial cells, both nuclear and cytoplasmic volumes may be significantly increased (Fig. 4-12). However, the nuclear to cytoplasmic ratio is not necessarily elevated. The presence of large, irregular, and multiple nucleoli also favors a diagnosis of adenocarcinoma. Within the cohesive arrays, normal polarity is disrupted; however, in well-differentiated carcinomas, this alteration may be quite subtle (Fig. 4-13). As the histologic grade increases, crowding and molding become more evident, as will pleomorphism (Fig. 4-14). In occasional ductal adenocarcinomas, varying proportions of the tumor cells manifest distinct squamous differentiation (Fig. 4-15). This does not appreciably alter the clinical outcome.

Cytomorphologic Features of Ductal Adenocarcinoma

- High cellularity with reduced cohesion
- Altered polarity
- Nuclear and cellular enlargement
- Hyperchromatic, irregularly distributed chromatin
- Irregular nuclear contours, including grooves
- Prominent nucleoli

A number of publications have described cytomorphologic features of benign and malignant epithelial cells in aspirates. One of the more systematic comparisons between benign and malignant cells in aspiration smears was reported by Mitchell and Carney.[5] Seventeen different morphologic features were investigated. The cytologic characteristics which were most important in discriminating malignant from benign were loss of polarity within three-dimensional cellular aggregates, extreme enlargement of the nuclei, and nuclear membrane irregularities characterized as deep smooth convolutions ("popcorn" nuclei). Features not found to be helpful in this determination included hyperchromasia, coarse chromatin granularity, increased nuclear to cytoplasmic ratios, the presence of nucleoli, and small irregularities in the nuclear outline. The authors cautioned against diagnosing carcinoma when only small numbers of atypical-appearing epithelial cells are present in a smear.[5] As stated by these investigators, a "simple formula for determining the presence of pancreatic carcinoma" does not exist, at least for the well-differentiated neoplasms.[5]

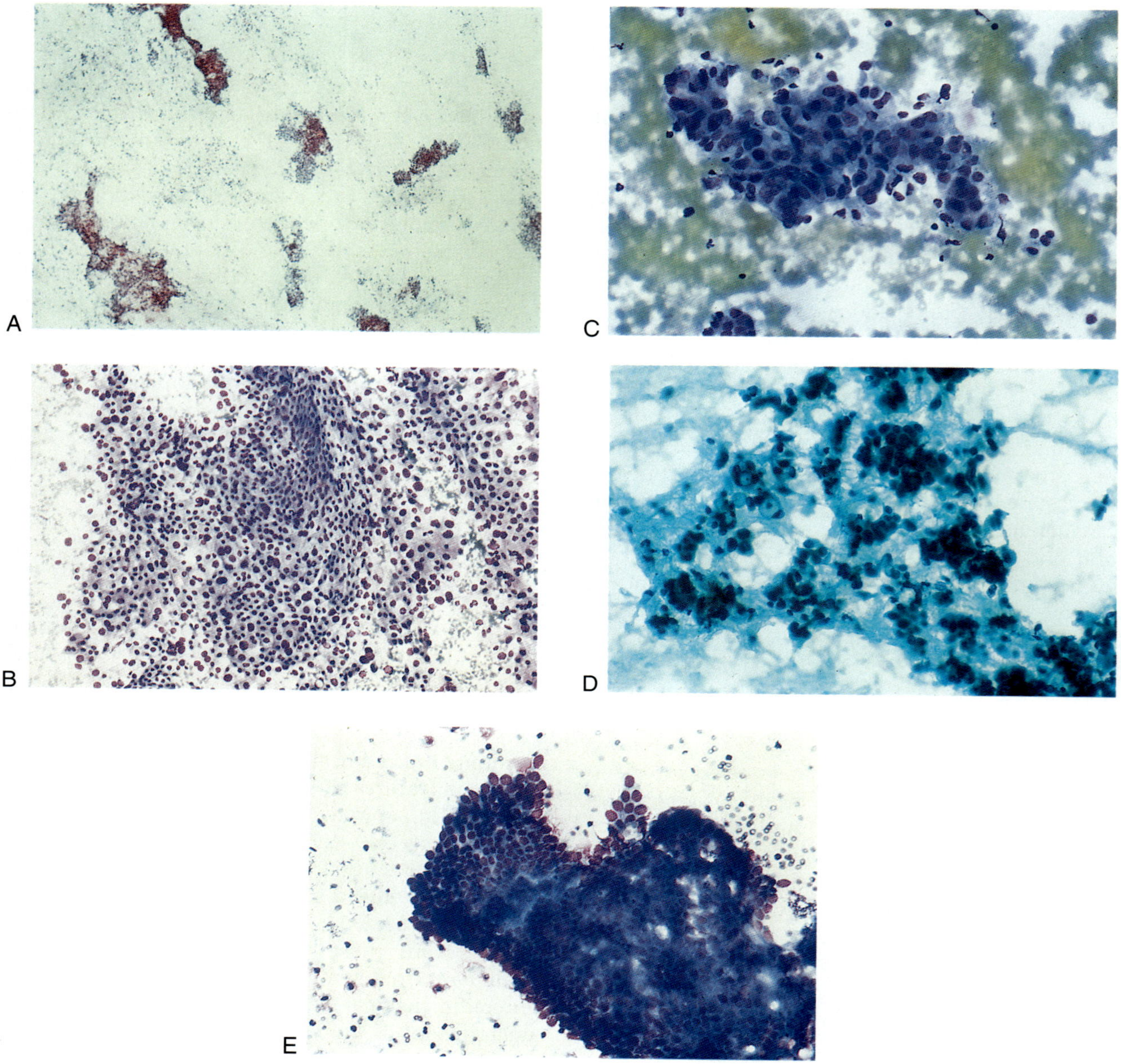

Fig. 4-10. High smear cellularity is an essential feature in the diagnosis of pancreatic adenocarcinoma. (**A**) Note the large number of irregularly shaped cellular aggregates and many isolated malignant cells. Some of the latter are closely associated with the aggregates of neoplastic cells, but others are widely dispersed (Papanicolaou). (**B**) Huge flat sheets of malignant cells may predominate the cytomorphologic picture. Even at this low magnification, a mild level of nuclear pleomorphism is apparent. Nuclear crowding is subtle in that polarity diverges only minimally from normal (Diff-Quik). (**C**) The neoplastic cells are present in typical three-dimensional aggregates. The latter are densely populated by crowded cells with large round nuclei. Again, nuclear pleomorphism, although not extensive, is easily appreciated (Diff-Quick). (**D**) High cellularity is again represented, but in this case the smears are dominated by relatively small three-dimensional spherical clusters. A few of the neoplastic cells have cytoplasmic mucin vacuoles. Individual tumor cells appear to be "escaping" from the cellular spheres (Diff-Quik). (**E**) Most of the field is dominated by a huge multilayered sheet of adenocarcinoma cells. For the most part, polarity and cohesion are maintained. However, individually dispersed intact tumor cells are also present in this field. Although subtle, the malignant nuclei are larger than normal (Papanicolaou).

A B

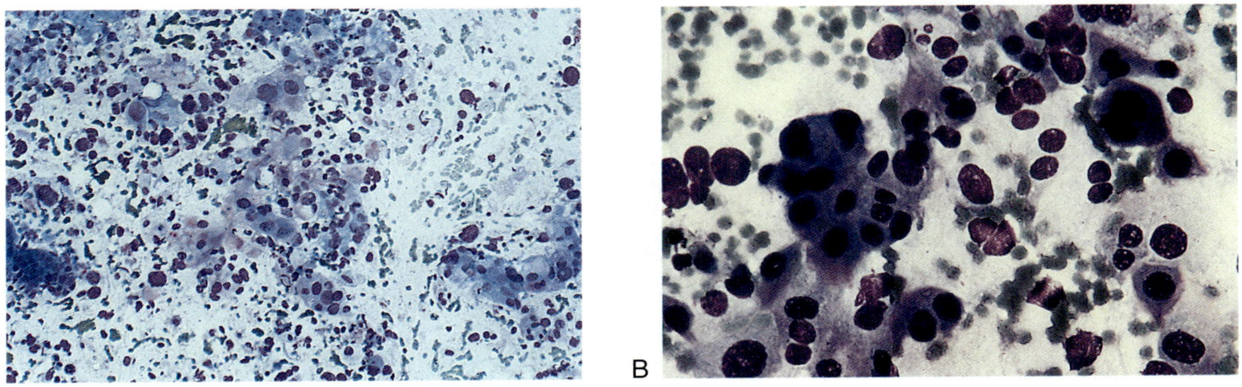

Fig. 4-11. **(A & B)** Two carcinomas in which a majority of the neoplastic cells show markedly reduced intercellular cohesion; that is, they present as single tumor cells and oligocellular clumps. Pleomorphism is apparent in both cases as well. Several nuclei that have been stripped of their cytoplasm are present. These should be diagnostically ignored. However, also present in this same field are individually dispersed malignant cells in which the cytoplasm surrounding the nuclei appears intact; these are diagnostically important (**A & B**, Diff-Quik).

In an attempt to enhance the discrimination between benign and malignant pancreatic aspirates, several investigative groups have used statistical analysis to compare cytomorphologic features. Cohen and colleagues[12] scrutinized 15 cytologic criteria in both Papanicolaou- and Romanowsky-stained smears. Using logistic regression analysis, three features emerged as key for the diagnosis of carcinoma. These included the presence of large nuclei (defined as at least two times the size of a normal nucleus), nuclear molding, and variability in nuclear size (defined as variation in nuclear dimensions by at least three times within the same epithelial aggregate). In their hands, when all three features were present in a specimen, diagnostic sensitivity and specificity were 98 percent and 100 percent, respectively, for the diagnosis of adenocarcinoma. Unfortunately, only about one-half of all specimens possessed all three features. Three other features were not found in any of the 29 benign specimens; these included the presence of nuclear grooves, nuclear pseudoinclusions, and macronucleoli.[12] Apparently, however, these features do not occur with a sufficiently high frequency in carcinomas to be of statistical value. Very recently, Robins et al.[13] examined 19 cytologic criteria by regression analysis. These workers also used both Romanowsky- and Papanicolaou-stained specimens. Three features emerged as major criteria: nuclear contour irregularities, irregular distribution of chromatin granules, and crowding and overlapping of nuclei within aggregates. In addition, four minor features were recognized as important predictors of malignancy. These included nuclear enlargement, mitotic figures, the presence of necrosis, and dispersed epithelial cells. When a specimen contained two (or three) of the major criteria or one major and three minor criteria, the diagnostic sensitivity and specificity were 90 percent and 100 percent, respectively. We agree with the statement of Robins and colleagues[13] that the scanning lens microscopic view is quite helpful in the diagnosis of pancreatic aspirates.

It is obvious that the major problem in diagnosing ductal adenocarcinoma by FNA biopsy is distinguishing well-differentiated neoplasms from the benign atypia that may be seen in specimens obtained from patients with pancreatitis. One must adhere to very strict criteria in order to avoid a false-positive diagnosis (Table 4-1). Smears from cases of acute pancreatitis contain cells in cohesive monolayer arrays. Only rare isolated epithelial cells with intact cytoplasm will be present in the smears. These elements may manifest the changes of inflammatory atypia, but definite malignant feature are lacking. Infrequently, three-dimensional aggregates of benign cells may be present. Within the

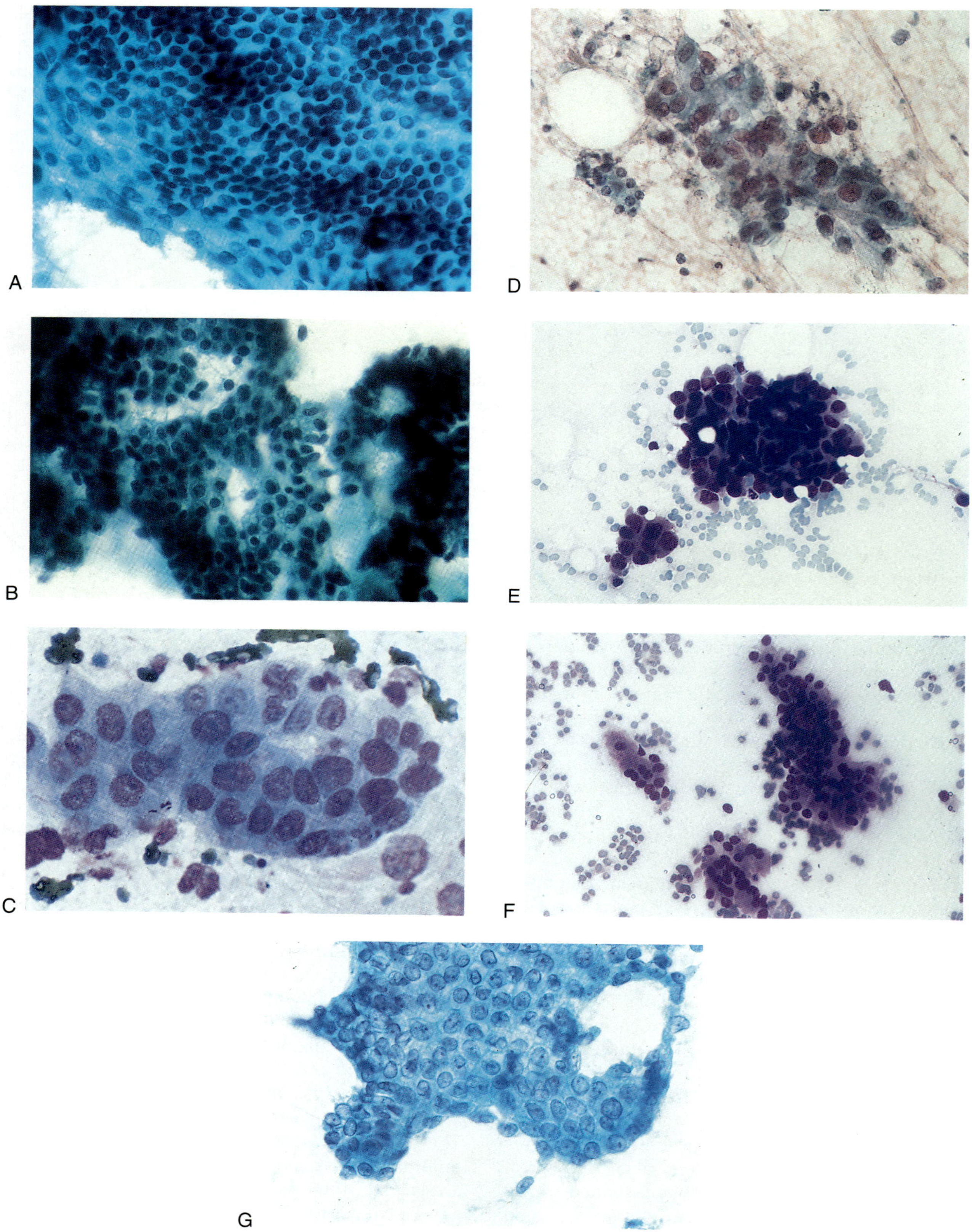
A
D
B
E
C
F
G

Fig. 4-12. As seen in these seven well-differentiated ductal adenocarcinomas, the malignant nuclei are appreciably larger than those found in benign duct elements. (**A**) Nuclear enlargement is best appreciated at the edge of the huge sheet. These cells also appear to possess the greatest volumes of cytoplasm. Intercellular borders are apparent. (**B**) Cell borders are even better demonstrated, with the nuclei exceedingly uniform. (**C**) Nuclear enlargement is better appreciated in air-dried smears. In this well-differentiated carcinoma, the nuclei are larger than normal, nucleoli are apparent, and polarity is slightly disrupted. (**D**) Nuclear enlargement in carcinoma can also be easily recognized when normal epithelial elements are also present in the smears. These malignant nuclei are at least three times the diameter of the adjacent benign epithelial nuclei. Additionally, slight irregularities in nuclear contours and altered polarity are apparent. (**F**) A papillary configuration for the neoplastic cells is suggested. (**G**) Nuclear grooves are easily appreciated (**A, B, D, G**, Papanicolaou; **C, E, F**, Diff-Quik).

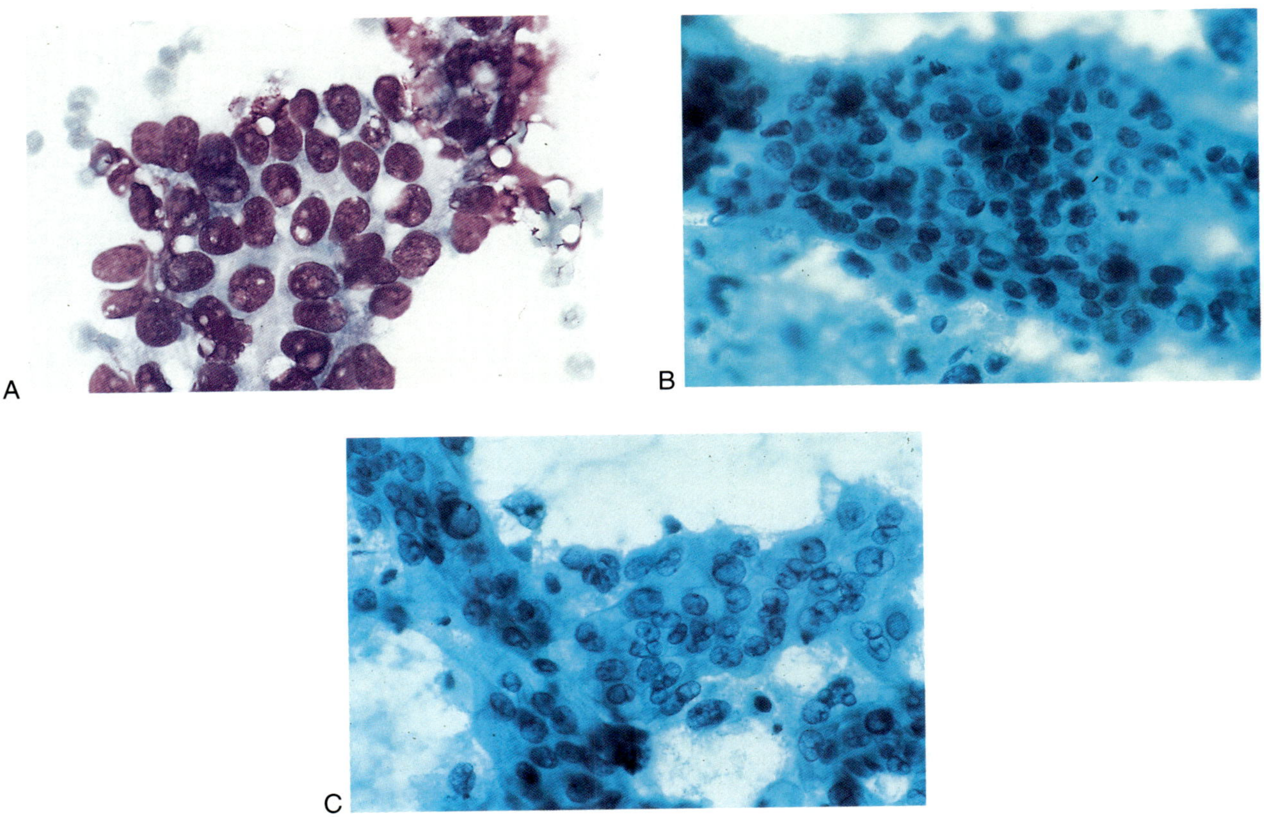

Fig. 4-13. (A–C) Within aggregates of ductal adenocarcinoma, normal polarity is usually visibly altered. In well-differentiated carcinomas, this will manifest as nuclear crowding and focal molding; this leaves zones of cytoplasm without any associated nuclei. In Fig. C, well-developed abnormalities of the nuclear contours are evident. This includes grooves and nuclei resembling popcorn kernels. Nucleoli are evident (**A**, Diff-Quik; **B & C**, Papanicolaou).

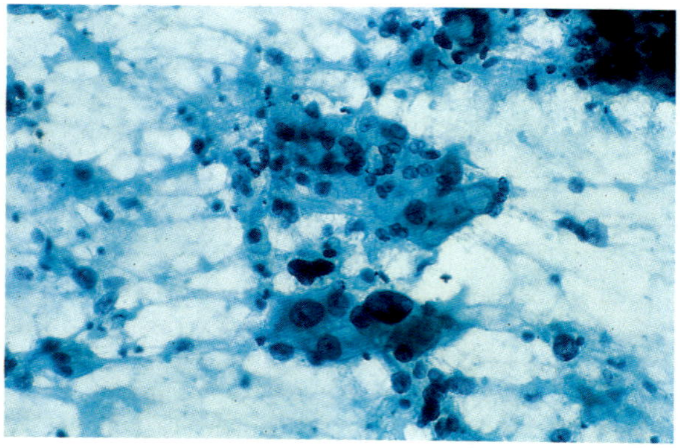
A

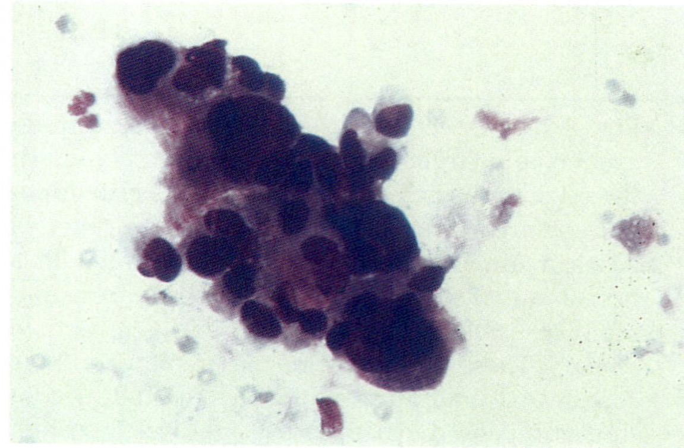
B

Fig. 4-14. Pleomorphism among the tumor cells and their nuclei is more easily appreciated as the histologic grade of the carcinoma diverges from normal. (**A**) Variability in nuclear dimensions is easy to appreciate by comparing the nuclei in adjacent neoplastic aggregates. Isolated malignant cells are also plentiful (Papanicolaou). (**B**) Nuclear pleomorphism is evident within the same cell cluster. Polarity is also disrupted (Diff-Quik).

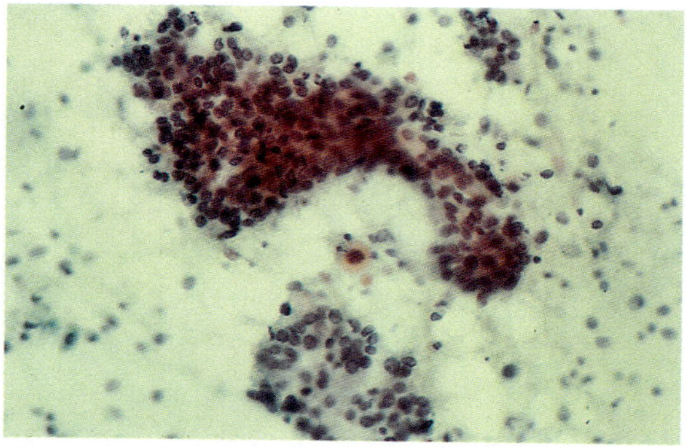
A

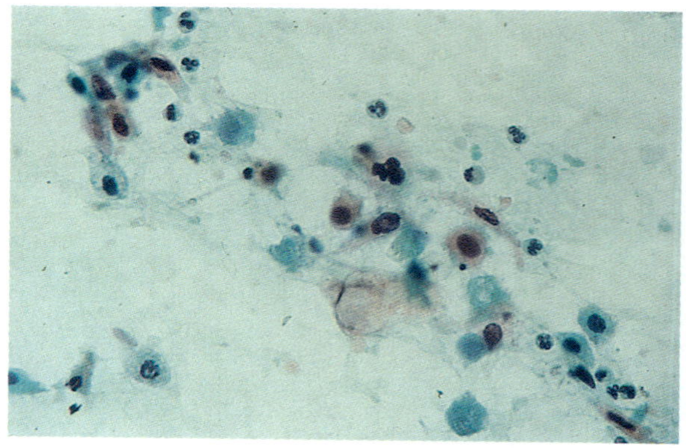
B

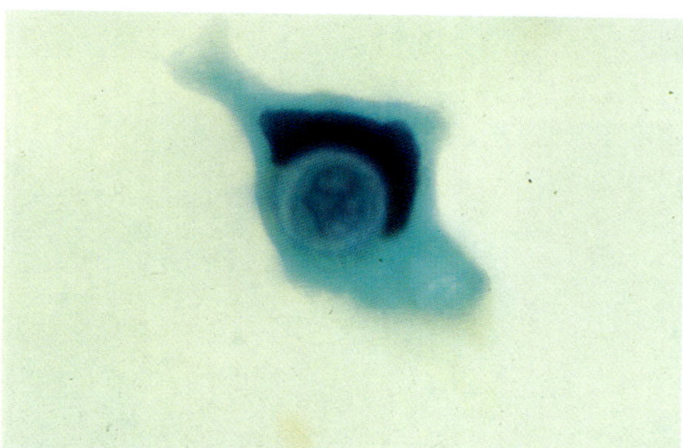
C

Fig. 4-15. In adenosquamous carcinomas, the malignant cells show divergent differentiation, with some of them manifesting glandular features and others squamous. (**A**) Almost all the malignant cells are present in huge sheets, reminiscent of typical ductal adenocarcinoma. However, in the very center of the field is a single isolated malignant cell with a central round very dark nucleus and a moderate volume of eosinophilic cytoplasm, a feature of squamous differentiation. (**B**) In a different field from the same specimen, many more of the neoplastic cells manifest squamous differentiation in the form of orangeophilic cytoplasm, elongated cellular contours, and pyknotic chromatin patterns. (**C**) This tumor cell has a large irregular nucleus with condensed chromatin and cytoplasm that is distinctly opaque, leading to a diagnosis of adenosquamous carcinoma (**A–C**, Papanicolaou).

Table 4-1. Comparison of Cytologic Features Between Benign and Malignant Aspirates

Feature	*Benign Inflammatory Atypia*	*Adenocarcinoma*
Cellularity	Low–moderate	Moderate–high
Epithelial aggregates	Flat, loose monolayers	Tight and three-dimensional
Polarity	Generally maintained	Altered: subtle to prominent molding
Isolated cells (reduced cohesion)	Rare	Noticeable
Nuclear enlargement	Present	Present
Nuclear contours	Smooth to slightly irregular	Smooth to very irregular
Macronucleoli	Rare When present, regular	Frequent
Nuclear to cytoplasmic ratio	Normal to slightly increased	Normal to slightly increased
Necrotic debris	May be present	May be present

cellular aggregates, however, crowding and molding will not be evident to any extent. Although these benign nuclei may be enlarged, the nuclear to cytoplasmic ratios are not greatly increased. Although nucleoli may be quite prominent, they usually are solitary and perfectly smooth. Within the sheets, distinct uniformity among the nuclei and nucleoli of neighboring cells will be apparent. When neutrophils are present to any extent in a specimen, we believe that the threshold for diagnosing adenocarcinoma should be noticeably elevated.

The area in which the pathologist is truly "put on the spot" for a rapid assessment of cytologic material is the intraoperative evaluation for pancreatic malignancy. In this situation, a false-positive diagnosis could result in an unnecessary Whipple operation, which carries significant risk of morbidity and mortality. It is our opinion that a false-negative interpretation in this situation has less important negative ramifications. Just how well do cytopathologists do in the operating room?[46–49] Edoute et al.[47] specifically examined publications that dealt with the intraoperative aspiration diagnosis of pancreatic malignancy. Including their own cases, their review included 274 patients in nine series. The diagnostic sensitivity of these specimens ranged from 73 percent to 100 percent. There were no false-positive cytologic interpretations; thus, the diagnostic specificity was 100 percent. Very recently, Blandamura and colleagues[48] evaluated their institution's experience with 92 intraoperative pancreatic aspirates. These investigators reported a diagnostic sensitivity and specificity of 100 percent each. They believed that the excellent sensitivity resulted from the presence of a pathologist in the operating suite.[48] Their series included 64 malignancies, most of which were adenocarcinomas. In addition to their ability to identify neoplasm, they correctly subtyped the malignancy in all but one instance (99 percent). Their single typing error involved a cytologic diagnosis of a papillary-cystic tumor that was histologically confirmed as a pancreatoblastoma.[48] In evaluation of intraoperative pancreatic FNA biopsies prepared with the Diff-Quik stained, Yang et al.[49] found nuclear size and variability in nuclear dimensions to be the most helpful features in distinguishing pancreatitis from ductal adenocarcinoma.

As a generalization, ancillary diagnostic procedures have only a limited role in the evaluation of the usual pancreatic ductal adenocarcinomas. Several studies have measured the DNA ploidy of pancreatic carcinomas.[42–45] Although most of these studies have used flow cytometry, image analysis has been employed as well. In ductal adenocarcinomas, a wide range of aneuploidy has been detected in aspirated specimens. Unfortunately, some ductal adenocarcinomas may appear diploid even by the more sensitive image analysis, and aneuploidy may be detected in some benign pancreatic aspirates. Thus, it is our opinion that ploidy analysis plays no role in distinguishing between benign and malignant aspiration specimens. Yet, several studies have demonstrated that patients whose carcinomas are diploid fair better than do those whose tumors are aneuploid. For example, in a recent multivariant analysis by Böttger and colleagues,[45] the DNA content was the second strongest prognostic indicator, with only the type of surgical procedure carrying more statistical significance. However, in some studies, this survival advantage is measured in only 2- to 4-month periods.

Several oncogenes and tumor suppressor genes have also been evaluated in pancreatic carcinoma. It has been demonstrated that most ductal adenocarcinomas possess a mutation in the K-ras oncogene. This mutation can be detected in aspiration biopsies when the signal is amplified by the polymerase chain reaction (PCR).[50] To date, benign pancreatic lesions have not shown this mutation. Thus, it has been proposed that K-ras analysis of FNA biopsies could enhance diagnostic sensitivity. Mutations in the p53 tumor suppressor gene have been identified by immunohistochemistry in a large proportion of invasive and in situ ductal carcinomas and may have prognostic implications. We are unaware of the application of p53 immunostaining to FNA biopsies pancreatic carcinoma.[51]

ISLET CELL NEOPLASMS

The diverse clinical manifestation of islet cell tumors has made them among the most fascinating of all neoplasms. This is largely related to their production and secretion of peptide hormones and to their resultant peculiar endocrine syndromes. However, many islet cell neoplasms, which represent approximately 6 percent of all primary tumors in the pancreas, may be hormonally "silent" or inactive and present clinically as a mass lesion similar to most other pancreatic tumors. Most patients have a solitary neoplastic mass, but a minority will have multiple, independent (synchronous or metacronous) tumors. Although the most common pancreatic tumor in children is the islet cell neoplasm, the vast majority of all patients with this tumor are adults.[14] In a recent epidemiologic review, it was found that these tumors are increasing to the greatest extent among black women.[37] It is important to remember that on purely morphologic grounds, both histologically and cytologically, it is probably impossible to predict accurately and consistently the biologic behavior of these neoplasms.[14] Fortunately, most are clinically benign, that is, they do not metastasize.

Aspiration smears are generally moderately to very cellular (Fig. 4-16) and are composed of relatively uniform-appearing, small- to medium-size, round to ovoid tumor cells,[14,52–58] as listed below. In most specimens, individually dispersed neoplastic cells predominate over loosely cohesive small and larger cellular aggregates (Fig. 4-17). The latter may include acinar and trabecular arrangements (Figs. 4-16 and 4-18). Occasionally, large sheets of neoplastic cells are also present (Figs. 4-16 and 4-19). Intercellular cohesion, in general, is less well preserved in aspirates of islet cell neoplasms than in specimens of ductal carcinomas (Fig. 4-20). Characteristically, the neoplastic nuclei appear perfectly round and typically are very eccentrically placed within the cell (Fig. 4-21). Binucleated tumor cells occur with some frequency. Nuclei contain finely stippled, evenly distributed chromatin ("salt and pepper") granules and possibly a small nucleolus. Cytoplasm varies from scanty to moderate in volume and is typically deeply basophilic and either granular or opaque in quality. This combination of isolated tumor cells with round eccentric nuclei set within basophilic cytoplasm yields the characteristic plasmacytoid appearance (especially with Romanowsky stains). In contrast to plasma cells, a perinuclear hof is not present. The cytoplasm is fragile and thus during the aspiration procedure it frequently gets stripped away from the nucleus.

Cytomorphologic Features of Islet Cell Neoplasms

- Moderately to very highly cellular with a monotonous population of rather small, uniform cells
- Low level of intercellular cohesion with cells singly and in small aggregates
  Vascularized fragments may be present
- Cytoplasm is fragile, scant to moderate, basophilic, granular, or dense
- Smooth round or ovoid nuclei with even, finely to moderately granular chromatin, often eccentric (plasmacytoid appearances); inconspicuous nucleoli
- Granular smear background

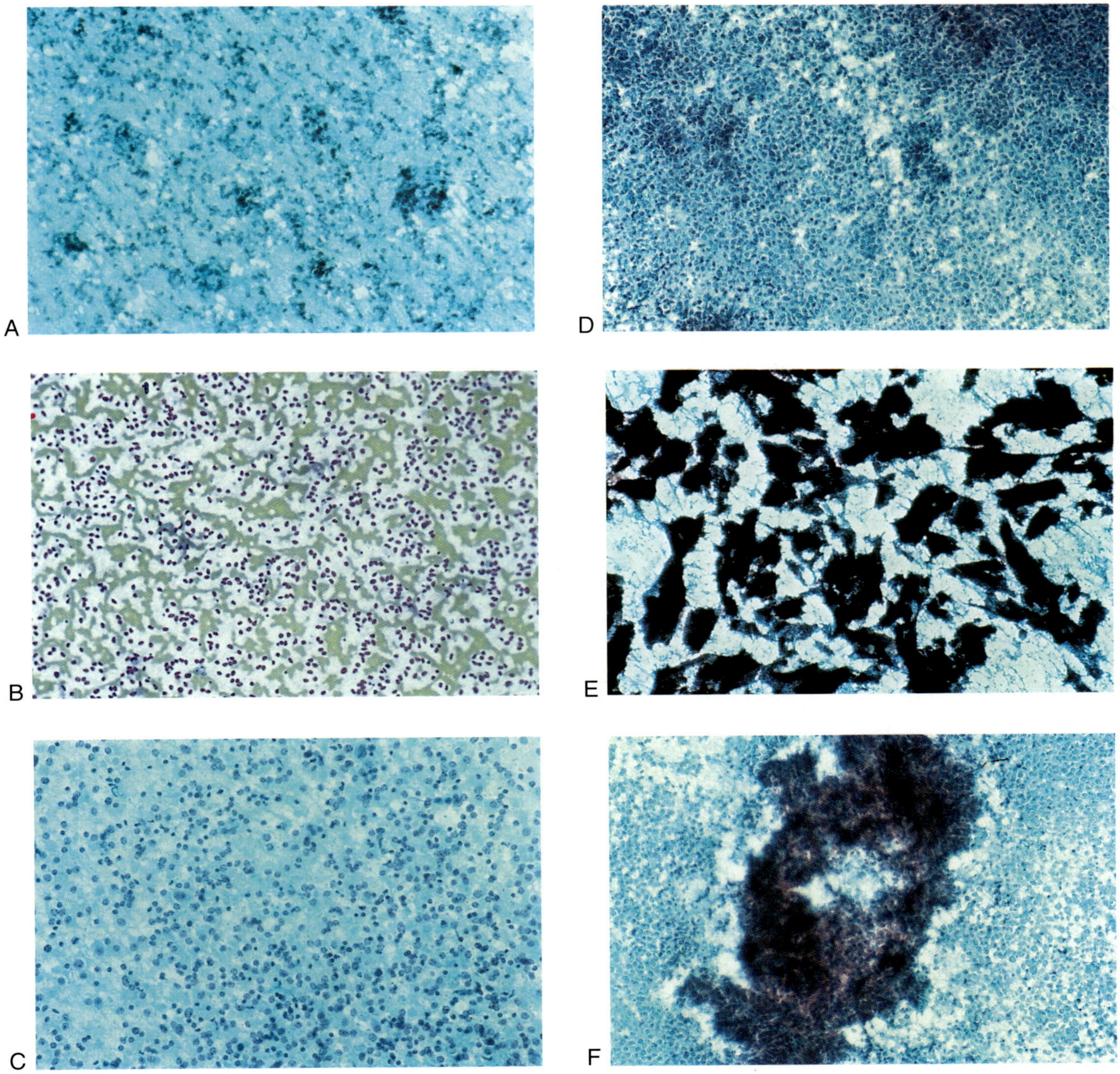

Fig. 4-16. Aspiration biopsies of islet cell neoplasms generally yield highly cellular smears. (**A**) The most common scanning lens view. There is a combination of small loosely cohesive aggregates and numerous individually dispersed tumor cells. Even at this low magnification, the uniformity of the cells is apparent. With the Papanicolaou stain, the smear background has a characteristic granular green tint. (**B**) The dispersed neoplastic cells have a somewhat linear arrangement suggesting that the cells were arranged in cords within the neoplasm. (**C & D**) The tumors presented as monolayers of cells with intact cyanophilic cytoplasm. The neoplastic cells appear monotonous. Occasionally, large three-dimensional aggregates of tumor cells will dominate the smears. (**E**) The tumor cells are arranged in a pattern suggestive of an anastomosing trabecular arrangement of the cells, the typical histology of many neuroendocrine neoplasms (**A, C–F**, Papanicolaou; **B**, Diff-Quik).

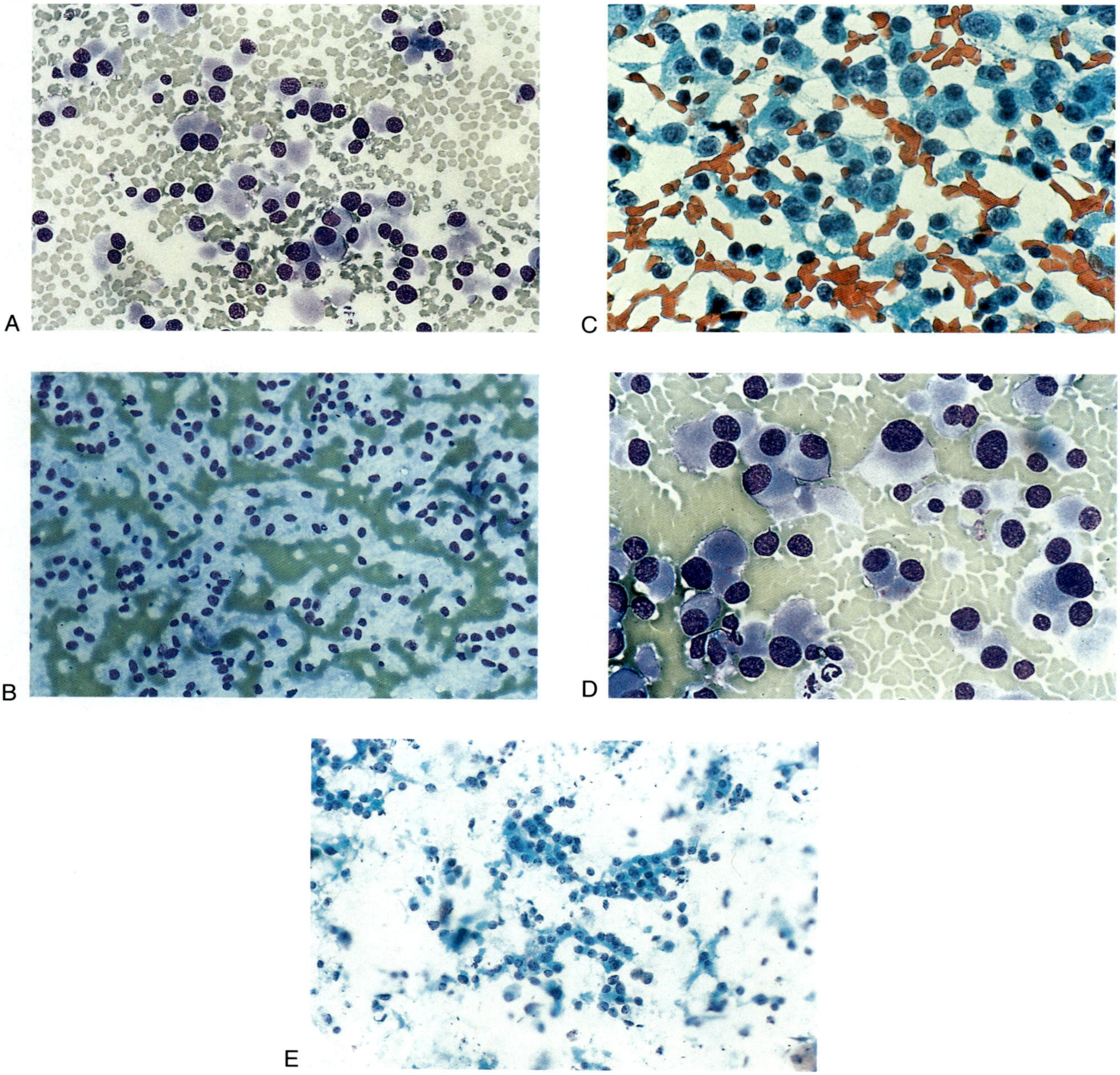

Fig. 4-17. Most aspirates of islet cell neoplasms include large numbers of individually dispersed and very loosely cohesive tumor cells. (**A–D**) The tumor cells are present as single cells and loose arrangements of two to four cells. Nuclei are round and characteristically very eccentrically positioned within the cytoplasm. (**B**) Many of the neoplastic cells appear to have been "stripped" of their cytoplasm. (**C**) The characteristic "salt and pepper" chromatin pattern and small nucleoli are readily apparent. Cytoplasm has a distinctly granular look. (**E**) Many of the tumor cells are present in small loosely arrayed sheets (**A, B, D**, Diff-Quik; **C & E**, Papanicolaou).

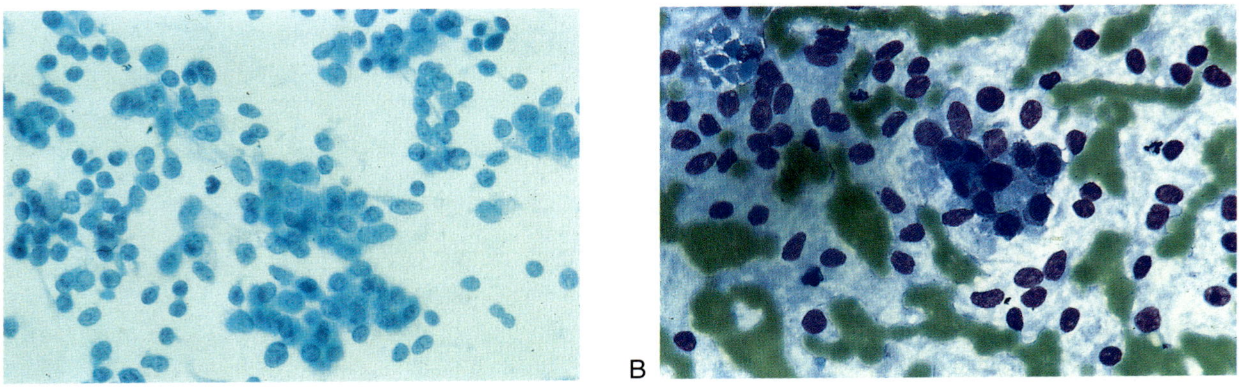

Fig. 4-18. (**A & B**) Small acinar structures are present in these examples. They are characterized by peripherally oriented small round nuclei and centrally concentrated cytoplasm (**A,** Papanicolaou; **B,** Diff-Quik).

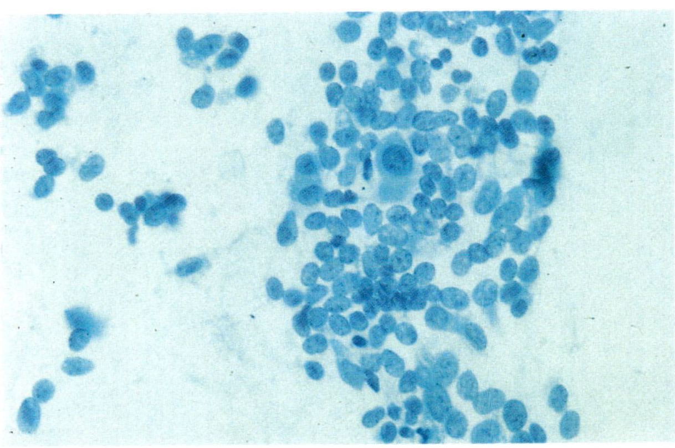

Fig. 4-19. In this example, many of the neoplastic cells are present in a loosely arrayed sheet. Although most of the tumor cells are strikingly uniform in appearance, in the center of this sheet is a tumor cell with a much larger nucleus and more abundant cytoplasm (Papanicolaou).

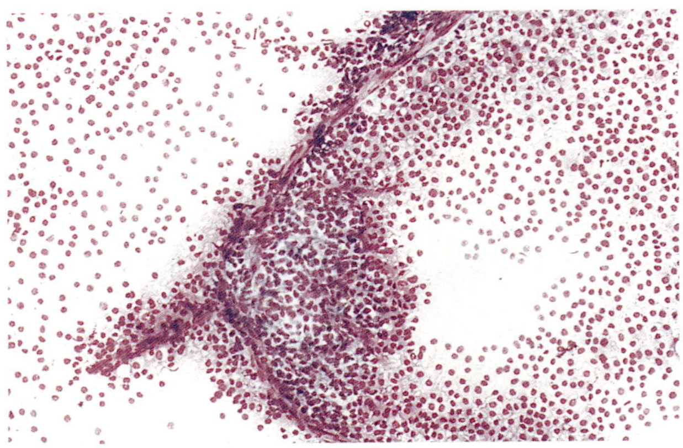

Fig. 4-20. Occasionally, aspirates of islet cell tumors yield highly vascularized tissue fragments. This refers to the presence of delicate and often branching capillaries, encased by the neoplastic cells. This finding would be unexpected in ductal adenocarcinoma (Diff-Quik).

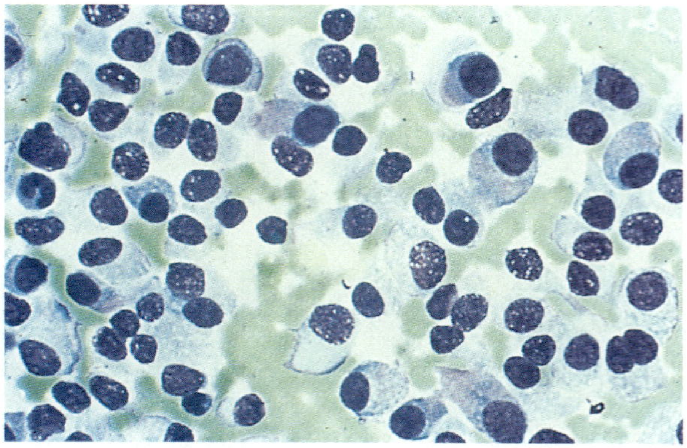
A

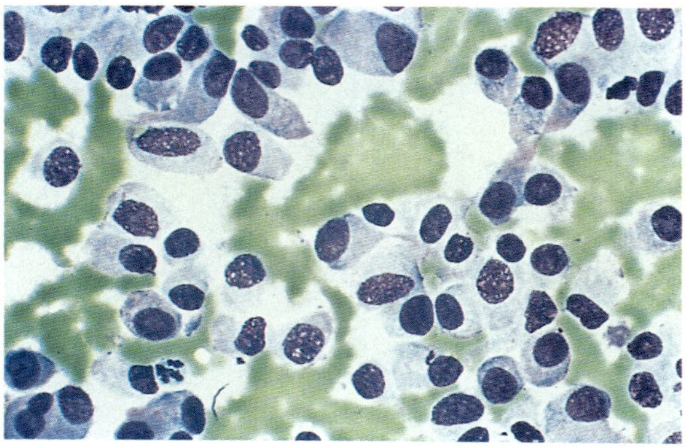
B

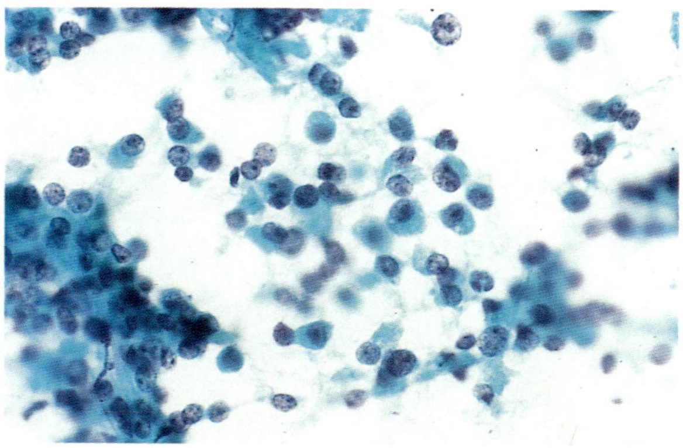
C

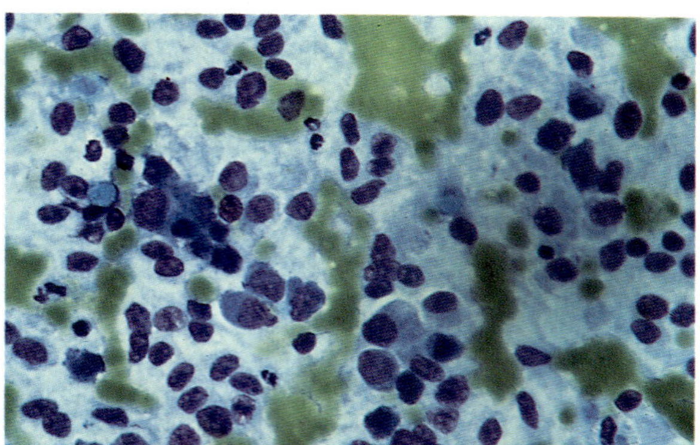
D

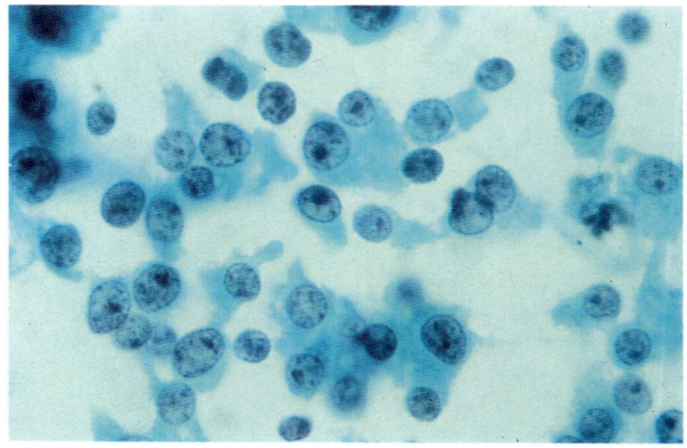
E

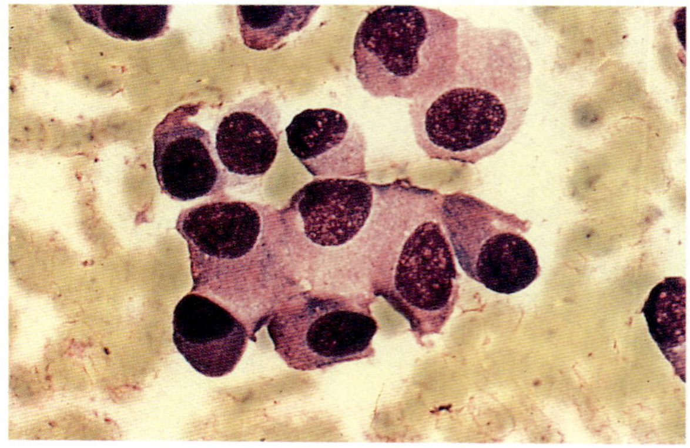
F

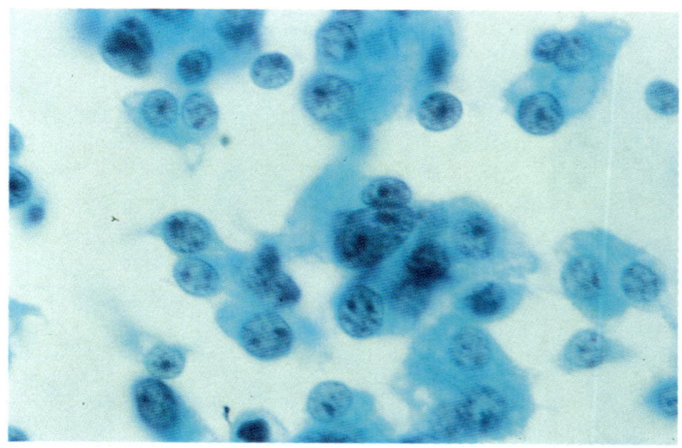
G

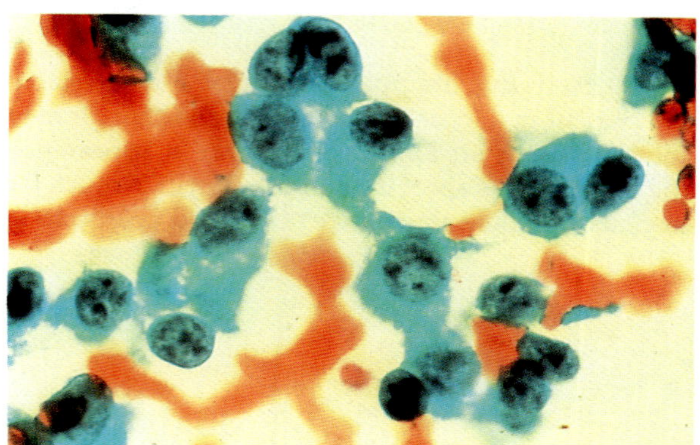
H

Fig. 4-21. (**A–H**) A typical tumor cell obtained from an islet cell neoplasm has one or two (and occasionally more) perfectly round-appearing nuclei, displaced to one pole of the cell. Chromatin is finely granular and uniformly distributed; small nucleoli may be apparent. As isolated elements, this creates the characteristic plasmacytoid appearance. Binculeated cells occur with some frequency (**A, B, D, F**, Diff-Quik; **C, E, G, H**, Papanicolaou).

This may result in numerous naked (stripped) nuclei in a granular smear background (Fig. 4-22). A true tumor diathesis is quite uncommon.

It is probably impossible to predict with certainty the clinical behavior of an individual patient's islet cell tumor based solely on cytomorphology. However, Tao[59] has suggested five cytomorphologic features that may be seen in aspiration biopsies that are predictive of aggressive behavior. These include the presence of huge nuclei, marked nuclear pleomorphism, multinucleated tumor giant cells, brisk mitotic activity, and a tumor diathesis. Tao claimed that when any four of these five features were present in an aspirate, the tumor was likely to act in a malignant fashion. Although this may be true, this conjecture remains to be proved.

The differential diagnosis of an islet cell neoplasm in FNA smears includes carcinoid tumor, metastatic neuroendocrine carcinoma, well-differentiated ductal carcinoma, papillary-cystic tumor, acinar cell carcinoma, malignant lymphoma, and plasmacytoma. We are unaware of any accurate way to distinguish an islet cell tumor from a carcinoid based purely on cytomorphologic grounds. A positive immunostain for serotonin or an associated carcinoid syndrome would favor a carcinoid tumor. This distinction may be important, in that it has been recently shown that carcinoid tumors of the pancreas have a better prognosis than that associated with islet cell carcinomas.[37] One must always consider the possibility of a metastatic neuroendocrine (small cell) carcinoma masquerading as a primary islet cell tumor.[60,61] On closer inspection, the metastatic cells will have more obviously malignant nuclear features, including coarsely granular chromatin and sharply angulated contours. The nuclear to cytoplasmic ratios are higher. There is obvious molding in cellular aggregates. Most often, these metastases arise in the lungs.[60] There is no way of distinguishing a metastatic oat cell carcinoma from the very rare primary small cell carcinoma of the pancreas.[62]

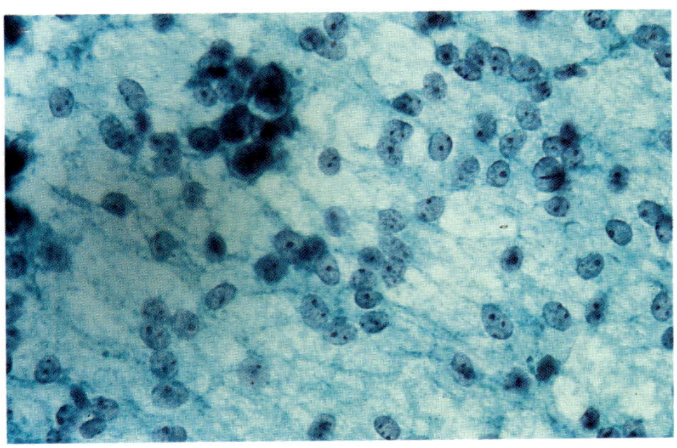

Fig. 4-22. The fragile cytoplasm is often sheared away from the neoplastic nucleus. This results in a background of basophilic granular material, which should not be mistaken for a true tumor diathesis (Papanicolaou).

A real diagnostic challenge may come in discriminating a very well-differentiated adenocarcinoma from an islet cell tumor. Both may have rather small round nuclei with finely granular chromatin. However, this distinction can usually be made. Features favoring adenocarcinoma include the presence of larger and more pleomorphic neoplastic cells; at least a minority of nuclei having obvious malignant features, including prominent nucleoli; and the preservation of intercellular cohesion.[14] Another neoplasm that may closely resemble islet cell tumors is acinar cell carcinoma.[58] The nuclei of the latter have clumped chromatin and well-developed nucleoli. Intercellular cohesion should also be more obvious in aspirates of acinar cell carcinomas, which may yield highly cellular specimens composed of uniform cells with round nuclei and granulated cytoplasm. Recently, Nguyen and colleagues[63,64] independently published two case reports demonstrating the overlapping cytologic features in these two neoplasms.

As a result of the numerous stripped nuclei that may be present in aspirates of some islet cell tumors, there may be a superficial resemblance to a malignant lymphoma.[14,60] However, on closer inspection it is easy to identify a malignant lymphoma. The nuclear contours will be much more irregular, nucleoli will be evident, cytoplasm will be intact, and the nuclear to cytoplasmic ratios are higher. In addition, lymphoglandular bodies will be present in the background, and there will be no evidence of true intercellular cohesion. Recently, Dodd et al.[65] reported two examples of extramedullary plasmacytomas involving the pancreas. This malignancy could be confused with an islet cell tumor displaying its characteristic plasmacytoid elements. However, the latter do not have a perinuclear hof or clumped, "clockface" chromatin. As discussed by Dodd and colleagues,[65] immunostaining will demonstrate light chain restriction and negative reactions for cytokeratin and neuron-specific enolase (NSE) in the plasmacytoma.

In cases in which the diagnosis is not clear cut by light microscopic examination, one can resort to ancillary diagnostic procedures.[14] If a portion of the aspirate is fixed in glutaraldehyde, electron microscopy can be performed directly on aspirated material. In essentially all cases, aspirated neoplastic islet cells will contain neurosecretory granules in their cytoplasm. Many of these will have a nonspecific dense-core appearance, whereas others may closely resemble granules associated with a specific peptide hormone. Immunocytochemistry may also be performed. Tumor cells should be positive for one or more neuroendocrine markers (e.g., NSE, chromogranin, and synaptosyn)[66] (Fig. 4-23). In addition, specific hormones may be identified in this manner.

A final consideration in this differential diagnosis includes a rare example of an aspirate of islet cell hyperplasia.[15,67,68] Potentially helpful microscopic clues distinguishing hyperplastic and neoplastic neuroendocrine cells include the smear cellularity, the arrangement of the cells, and the admixture of other elements. Compared to aspirates of islet cell neoplasms, specimens of hyperplastic endocrine cells tend to be much less cellular[67]; however, in the case reported by Gala et al.,[68] smear cellularity was remarkably high. In specimens with hyperplastic cells, the neuroendocrine elements should be mixed with benign ductal and acinar cells. This is not expected in aspirates of an islet cell tumor. Hyperplastic endocrine cells typically present in small aggregates with relatively well preserved cohesion. On the other hand, a much greater spectrum of cohesion is evident in the neoplasms, including numerous individually dispersed tumor cells.

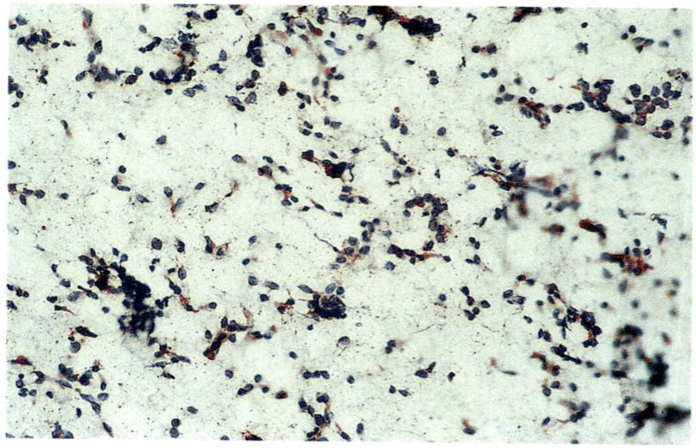

Fig. 4-23. Essentially all of the tumor cells in this direct smear are positive for chromogranin by immunocytochemistry.

PLEOMORPHIC GIANT CELL CARCINOMA

Pleomorphic giant cell or anaplastic carcinoma is an unusual pancreatic neoplasm, accounting for approximately 2 percent of all malignancies of this organ.[37,69,70] Although the overall clinical presentation of this tumor is similar to that of the usual conventional ductal adenocarcinoma, there are some differences. For example, pleomorphic carcinoma

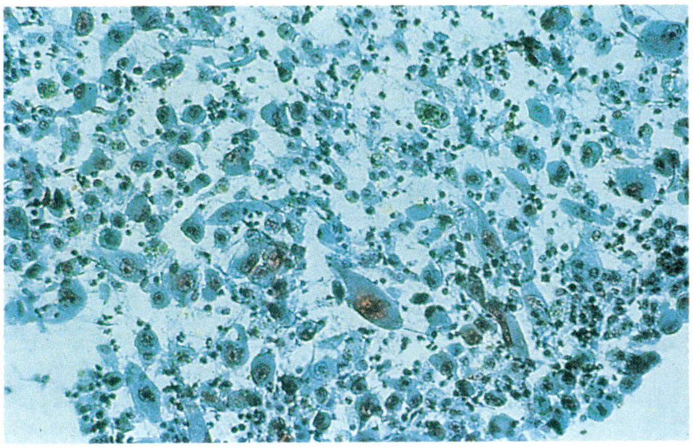

Fig. 4-24. In this aspirated pleomorphic giant cell carcinoma, the marked cellularity and the extreme variability among the tumor cells are both readily evident at low magnification. Intercellular cohesion is not apparent (Papanicolaou).

shows a stronger proclivity to arise in the body and tail of the organ. The male-to-female ratio is increased relative to conventional carcinoma. Finally, the prognosis is even worse, with almost all patients dying within the first 6 months of diagnosis.[37] This may be related to an increased tendency for hematogenous spread.

The cytomorphologic appearance of pleomorphic carcinoma in aspiration smears is characteristic[14,69,70] (Figs. 4-24 to 4-27). Uniformly, the smears

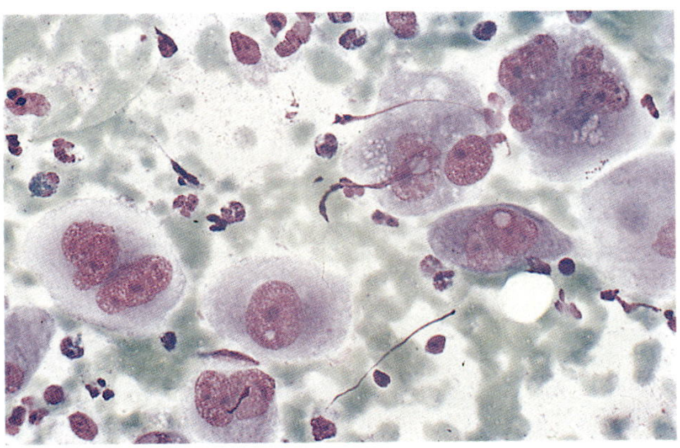

Fig. 4-25. These huge tumor giant cells dwarf the adjacent inflammatory elements. Several are multinucleated, and large nucleoli are a consistent finding (Diff-Quik).

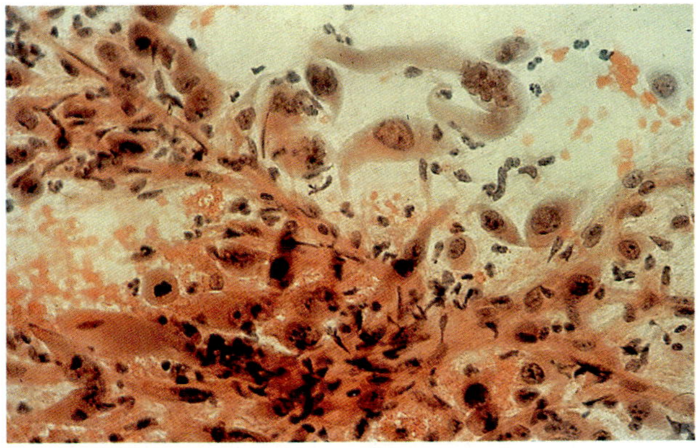

Fig. 4-26. In this field, most of the neoplastic cells of this pleomorphic giant cell carcinoma are mononucleated. They vary from round to polygonal to distinctly spindled in contour. Large irregular nuclei and nucleoli are prominent (Papanicolaou).

are extremely cellular with a striking dissociative pattern of individually dispersed obviously malignant cells, as listed on page 156. The sizes and shapes of the malignant elements are remarkably variable with round, polygonal, and distinctly spindled cells all represented (Figs. 4-24 to 4-26). As the name of the tumor implies, bizarre, monstrous tumor cells may be numerous and may be both mononucleated and multinucleated (Fig. 4-27). The

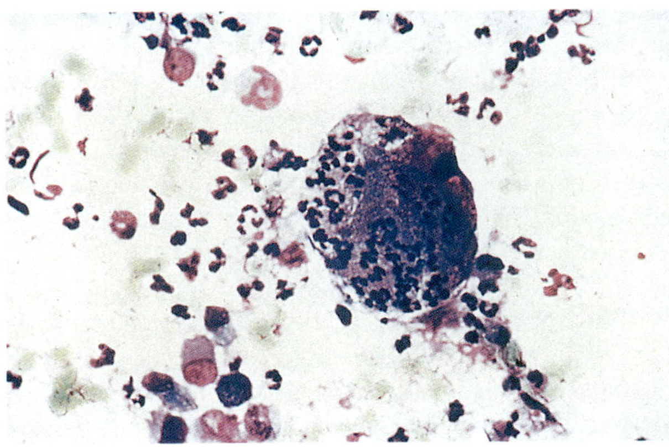

Fig. 4-27. The cytoplasm of this tumor giant cell contains numerous neutrophils, evidence of cytophagocytosis by the malignant elements. Note how much larger this tumor cell is compared to the few mononucleated malignant cells present (Diff-Quik).

Cytomorphologic Features of Pleomorphic Giant Cell Carcinoma

- Highly cellular with marked predominance of individual tumor cells
- Marked pleomorphism, including giant cells, both mono- and multinucleated
- Large irregular nuclei with coarsely granular hyperchromatic chromatin and one or more huge nucleoli
- Cytoplasm varies in quantity with polygonal to spindled contours
  Hyaline-like globules may be present in cytoplasm
  Phagocytosis of neutrophils
- Background of necrotic debris and many neutrophils

malignant nuclei are huge, lobulated, contorted, and obviously hyperchromatic and possess one to several massive nucleoli. These tumor giant cells are intimately admixed with the smaller yet obviously cancerous cells. Phagocytosis of neutrophils by the tumor giant cells is a very distinctive feature, as is a marked tumor diathesis (Fig. 4-27). In fact, this intermingling of inflammatory and neoplastic cells with necrotic debris is the expected picture. The cytoplasm of some of the neoplastic cells may contain rounded foci with a hyaline appearance. During the aspiration procedure, some of the cytoplasm may become sheared away from the nucleus, yielding naked round blobs in the smear background. In the minority of cases examined, a few three-dimensional spherical aggregates may be consistent with a poorly differentiated adenocarcinoma.

By immunocytochemistry, the tumor cells are positive for vimentin and are also usually positive for cytokeratin.[70] This coexpression of intermediate filaments is also characteristic of giant cell carcinomas elsewhere in the body such as in the lung and thyroid. Ultrastructurally, intermediate filaments also form a prominent cytoplasmic component. The rounded cytoplasmic globules with a glassy appearance correspond to whorled arrangements of filaments, which, at least in some cases, are vimetin.[70] Neither endocrine nor acinar differentiation will be recognized by electron microscopy. Specifically, neither zymogen nor neuroendocrine granules or intercellular junctions are evident. The differential diagnosis of pleomorphic giant cell carcinoma of the pancreas includes other large cell pleomorphic neoplasms that may occur in the retroperitoneum.[14] Foremost among these are pleomorphic soft tissue sarcomas (see Ch. 7 for a detailed discussion). On a cell-for-cell basis, it is probably impossible to distinguish a pleomorphic carcinoma from a high-grade sarcoma. Both present with very loosely arrayed and dissociated highly pleomorphic cells, including elongated elements. Distinction from a malignant fibrous histiocytoma would be most difficult. At the light microscopic level, the only potentially helpful clue may be the marked cytophagocytosis and neutrophilic background characteristic of the carcinomas. This aspect of the cytomorphologic picture would be unusual for pleomorphic sarcomas, as would the coexpression of intermediate filaments. Whenever a pleomorphic neoplasm with dispersed bizarre-appearing malignant cells is encountered in aspiration smears, the possibility of malignant melanoma must also be entertained. Occasionally, the neoplastic elements in melanoma will show prominent cytoplasmic pigment granules, as well as well-developed nuclear pseudoinclusions. More helpful would be immunocytochemistry in which the tumor cells should express one or more of the markers characteristic of melanoma (e.g., S-100 protein, and HMB-45). Ultrastructurally, aspirated tumor cells might demonstrate cytoplasmic melanosomes and premelanosomes. The possibility of a primary retroperitoneal or metastatic nonseminomatous germ cell tumor, especially choriocarcinoma, should be considered.[71] Usually, greater evidence of epithelial differentiation will be seen in the germ cell tumors. In choriocarcinoma, some of the neoplastic cells will be positive for human chorionic gonadotropin (HCG). Finally, the point at which one draws the line between a very poorly differentiated ductal carcinoma and a pleomorphic giant cell carcinoma is somewhat arbitrary.

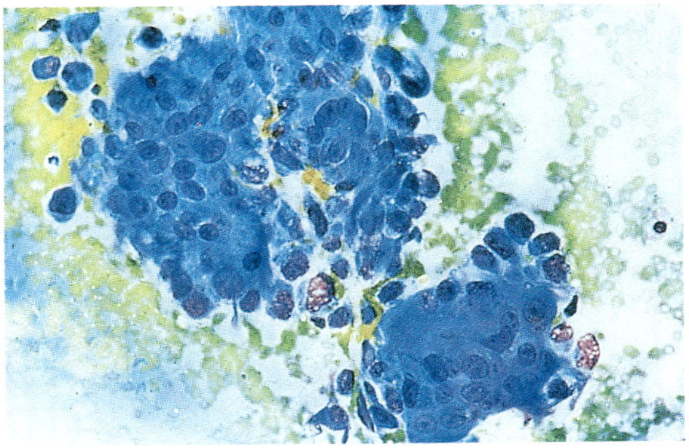

A

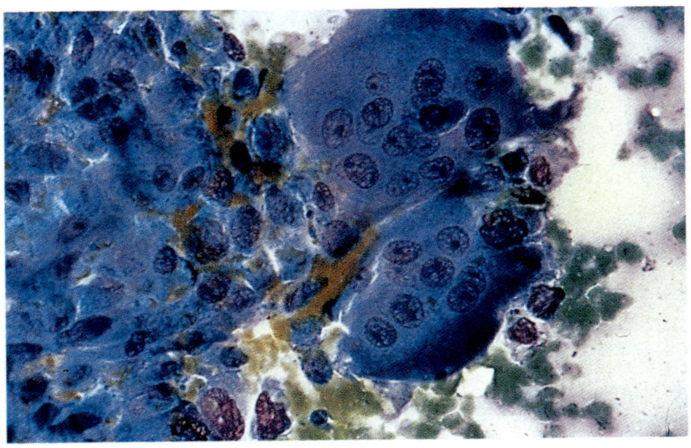

B

Fig. 4-28. (**A & B**) In this aspirated osteoclastic giant cell tumor, the mononucleated tumor cells have oval nuclei, evident nucleoli, and high nuclear to cytoplasmic ratios. Some are physically associated with the osteoclast-like giant cells that possess greater volumes of cytoplasm, but similar-appearing nuclei (Diff-Quik).

OSTEOCLASTIC GIANT CELL TUMOR

The osteoclastic giant cell tumor should not be confused with the pleomorphic giant cell carcinoma. The osteoclastic tumor is an exceedingly rare neoplasm that is histologically indistinguishable from the giant cell tumor of bone.[72,73] By using the small handful of reported cases as our database, the median age for patients is 60 years, with a possible slight female predominance.[74] Clinical signs and symptoms generally resemble those produced by conventional ductal adenocarcinoma. Prognosis remains poor, with only a 1-year survival from diagnosis.

Only rarely have the cytomorphologic features of this neoplasm in aspiration smears been detailed (see box on p. 158).[74–77] In the one example that we have personally examined, the smears were quite cellular and were dominated by rather small mononucleated neoplastic cells (Figs. 4-28 to 4-30), which, for the most part, were dissociated as isolated elements.[74] However, small loosely cohesive clusters of these cells were also present. Most of these mononucleated cells had an oval nucleus with a smooth delicate membrane, finely granular and uniformly distributed chromatin, and small but noticeable nucleoli. Some of the nuclei were distinctly hyperchromatic. Each cell possessed a thin rim of granular eosinophilic cytoplasm surrounding the nucleus. Scattered throughout the specimen were distinctive osteoclast-like tumor giant cells. Most of these had 10 to 20 centrally concentrated nuclei that often overlapped one another.

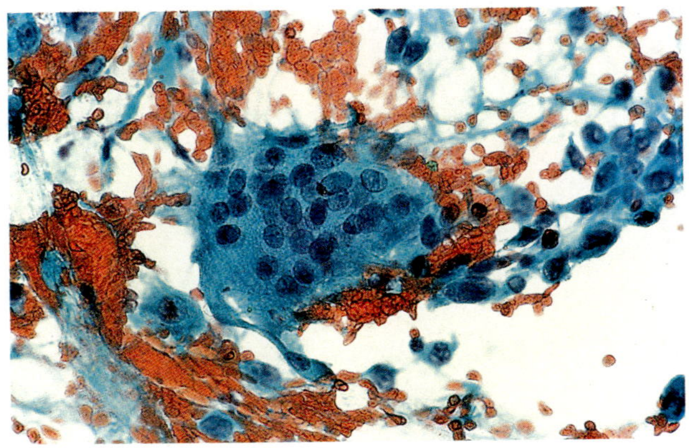

Fig. 4-29. The chromatin is finely granular, evenly distributed, and fairly darkly stained in this neoplastic osteoclast. Nuclear membranes are smooth and delicate, and small nucleoli can be seen. The mononucleated tumor cells are loosely cohesive (Papanicolaou). (From Geisinger and Silverman,[14] with permission.)

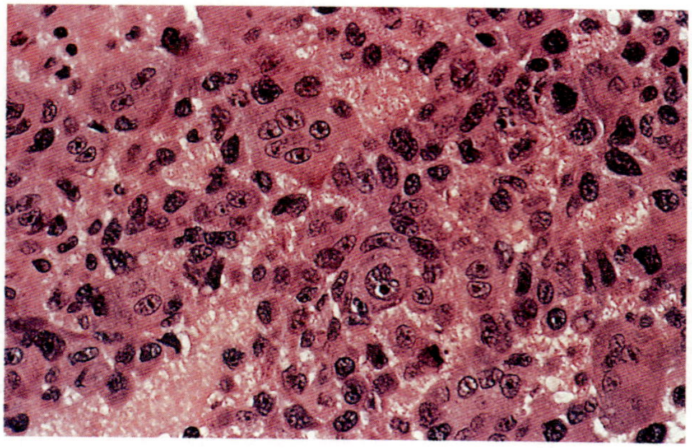

Fig. 4-30. Cell block in which the appearance of the osteoclastic tumor cells is indistinguishable from those of a giant cell tumor of bone. Nucleoli are more apparent than in the direct smears (H&E).

The giant cell nuclei were indistinguishable from those present in the smaller mononucleated cells. Moderate to voluminous cytoplasm encased these nuclei, creating sort of a syncytial appearance. Epithelial differentiation was not recognized in our specimen.[74] Both the multinucleated and the mononucleated neoplastic cells were intensely immunoreactive for vimentin but were negative for epithelial markers and CEA (Fig. 4-31).

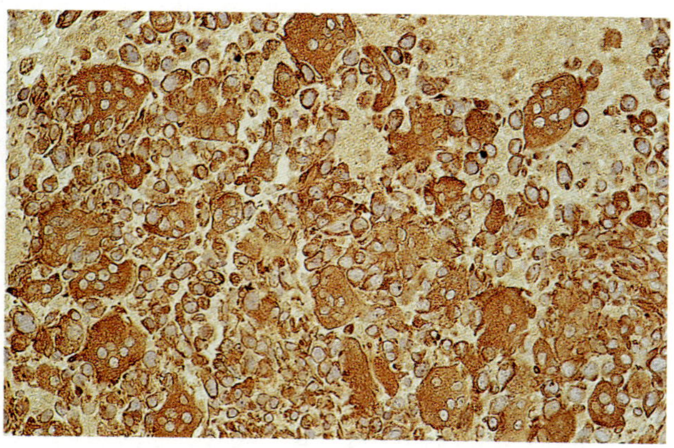

Fig. 4-31. Osteoclastic giant cell tumor in which the cytoplasm is uniformly decorated by vimentin in this immunocytochemical preparation.

Cytomorphologic Features of Osteoclastic Giant Cell Tumor

- Mononucleated polygonal to cuboidal cells, some aggregated, but mostly dissociated
- Uniform round to ovoid nuclei with fine, even chromatin and small but distinct nucleoli
- Eosinophilic, generally scanty cytoplasm
- Scattered multinucleated osteoclast-like giant cells with central nuclei that resemble those in smaller tumor cells, including well-defined nucleoli

PAPILLARY-CYSTIC TUMOR

The papillary-cystic tumor (or solid and papillary neoplasm) is a very uncommon neoplasm of unknown histogenesis that occurs almost exclusively in adolescent and young adult women.[78,79] The prognosis associated with these tumors is far superior to that of conventional ductal adenocarcinomas, as surgical resection is generally curative. However, if these tumors go unrecognized (and therefore untreated), they can be lethal. Clearly, all such tumors need to be considered at least potentially as low-grade malignancies. Only rarely have metastatic deposits been well documented, and only after a long clinical course.

Histologically, a major component of these neoplasms is papillary-like structures. Some investigators have referred to these as pseudopapillary, in that they are believed to result from degenerative changes in solid or medullary foci of the neoplasm, leaving the remaining neoplastic cells embedded in a delicate fibrovascular connective tissue scaffolding.[78] Further degenerative alterations result in cyst formation. In most described examples, the neoplastic cells are small and uniform, each possessing a single ovoid nucleus with pale delicate chromatin, inconspicuous nucleoli, and an occasional groove. Cytoplasmic and extracellular hyaline globules may also be present.

An early potentially very helpful diagnostic clue comes not from the aspiration smears, but rather from radiographic findings. Whether the modality is ultrasound, computed tomography (CT), or magnetic resonance imaging (MRI), the mass will appear at least partially cystic and demonstrate considerable heterogeneity in its composition.[78] This is another specific example of the valuable interaction between radiologists and pathologists. The cytologist should also incorporate into the diagnostic equation the fact that patients are almost always young females. A very high level of cellularity characterizes the scanning lens view of the aspiration specimens (Fig. 4-32). This will include individually dispersed neoplastic cells, small cellular aggregates, and numerous papillae.[78,80–89] The latter show both straight and branched configurations and are lined by one or more layers of radially oriented elements (Figs. 4-33 to 4-35). It would be most unusual to see such an abundance of papillary structures in aspirates of other primary pancreatic neoplasms. Each neoplastic cell contains a solitary nucleus that is homogeneous and pale stained and that has inconspicuous nucleoli (Figs. 4-36 and 4-37). The chromatin may be so finely granular as to have a powdery consistency, very similar to that associated with well-differentiated papillary adenocarcinomas of the thyroid. Very characteristically, the delicate nuclear membranes are thrown into longitudinal folds or grooves.[84] Only a thin rim of nondescript cytoplasm surrounds each nucleus, yielding very high nuclear to cytoplasmic ratios (Fig. 4-37). In many individual papillary structures, a slender fibrovascular core may be detected (Figs. 4-35 and 4-36).

Cytomorphologic Features of Papillary-Cystic Tumor

- Cellular smears with numerous individual cells, small aggregates, and papillae
- Straight and branched papillary structures with connective tissue cores and lined by one or more layers of neoplastic cells
- Uniform oval nuclei with very finely granular, euchromatic chromatin, and delicate membranes with grooves
- Very high nuclear to cytoplasmic ratio

Exfoliation of tumor cells from the papillary surfaces can also be imagined, resulting in the numerous isolated neoplastic elements.

As well reviewed by Pettinato et al.,[78] ancillary diagnostic testing procedures have been reported by a number of investigators. The neoplastic cells are consistently immunocytochemically positive for vimentin and negative for cytokeratin.[78,79,84,90] A number of investigators have reported the tumor cells being positive for NSE but negative for more specific neuroendocrine immunomarkers. In addition, only rarely have dense-core neurosecretory granules been demonstrated ultrastructurally. Immunoreactivity for α_1-antitrypsin is also frequently described, along with ultrastructural evi-

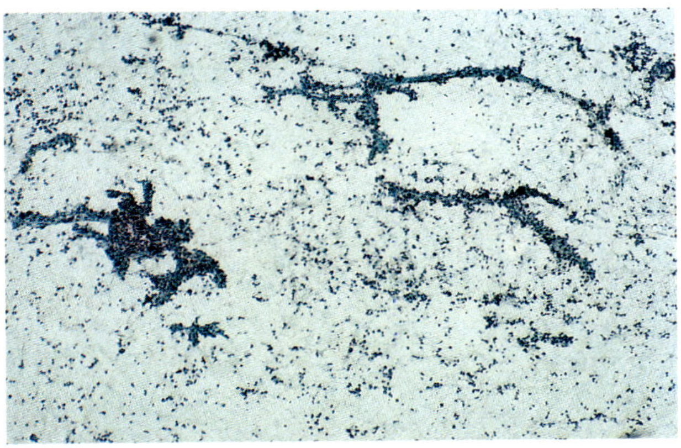

Fig. 4-32. The scanning lens view of this aspirated papillary-cystic tumor manifests high cellularity, many individual tumor cells, and branched papillary structures (Papanicolaou).

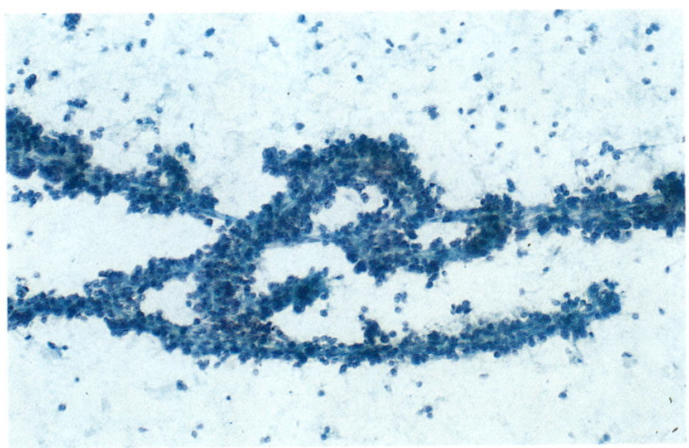

Fig. 4-33. One to several layers of small uniform neoplastic cells line delicate blood vessels, creating the appearance of papillae. The tumor cells have very high nuclear to cytoplasmic ratios (Papanicolaou).

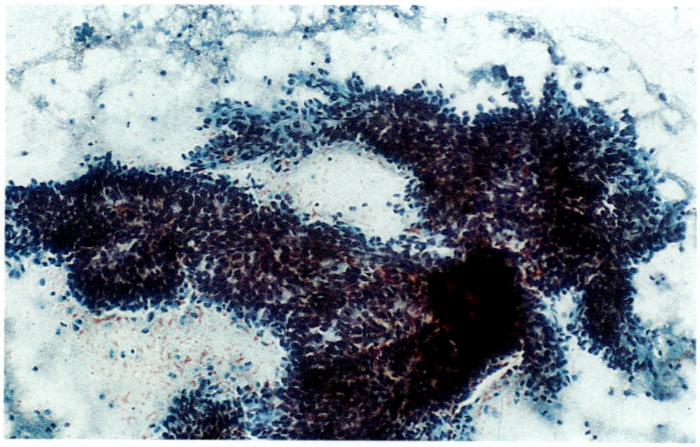

Fig. 4-34. Broad sheets appear to fuse with papillae, readily recapitulating the histologic appearance of papillary-cystic tumors (Papanicolaou).

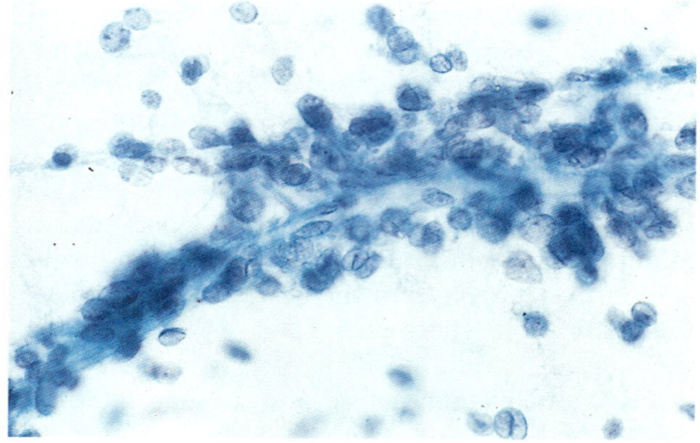

Fig. 4-36. The neoplastic nuclei in papillary-cystic tumors have pale chromatin, very inconspicuous nucleoli, high nuclear to cytoplasmic ratios, and delicate membranes; the latter may be folded to form grooves (Papanicolaou).

dence of zymogen-like granules. These features, however, are not specific for acinar differentiation. The results of analyses for both ploidy and steroid hormone receptors have also been mixed.

We have seen an example of a papillary-cystic tumor that recurred in a young woman following surgical and radiation therapy.[84] The aspirate of this mesenteric metastasis differed significantly from the characteristic picture described above. The nuclei of the tumor cells were much larger, round rather than ovoid, much more hyperchromatic, and noticeably more pleomorphic; they possessed discernable nucleoli, and longitudinal grooves were rare. The smears also contained absolutely bizarre-appearing neoplastic cells, some of which were multinucleated, quite reminiscent of the tumor giant cells of pleomorphic giant cell carcinoma. The number of identifiable papillary structures was remarkably reduced.[84]

A major item in the cytomorphologic differential diagnosis of the papillary-cystic tumor is an islet cell neoplasm.[14] Both tumor types may yield soli-

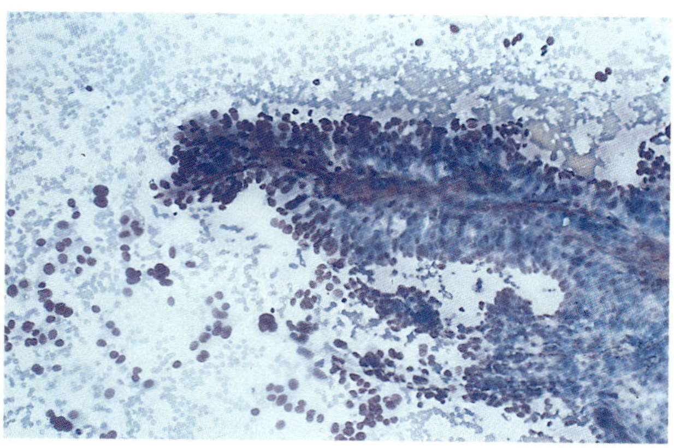

Fig. 4-35. Multiple layers of neoplastic elements radiate from a central capillary. Individual cells exfoliate from its surface (Diff-Quik).

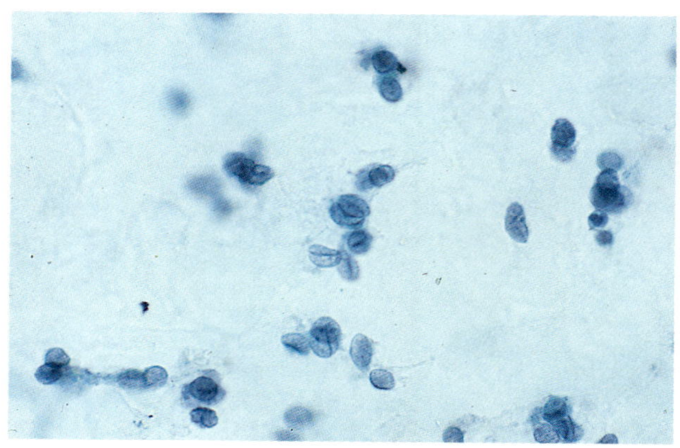

Fig. 4-37. Longitudinal nuclear grooves are a characteristic feature of papillary-cystic tumors, as are exceedingly high nuclear to cytoplasmic ratios (Papanicolaou).

tary relatively small and uniform neoplastic cells with inconspicuous nucleoli. This problem may be compounded by both clinical and immunocytochemical data; both neoplasms may occur in relatively young patients and may be positive for NSE. However, careful attention to the morphologic features should readily distinguish the two. The chromatin in islet cell tumors is more coarsely granular than one expects in the papillary-cystic tumor. Plasmacytoid appearances, so characteristic of the endocrine neoplasms, are not present in the papillary-cystic tumor. Islet cell tumors almost never show prominent papillary structures or longitudinal nuclear grooves. It would also be quite unusual for an islet cell tumor to manifest such prominent cystic change radiographically.

MUCINOUS CYSTIC TUMORS

Another uncommon low-grade cystic cancer with a marked proclivity to occur in women is the mucinous cystic tumor. Although these tumors may arise in any part of the organ, they occur most frequently in the body and tail of the pancreas. As with the papillary-cystic tumor, these neoplasms usually demonstrate a prominent cystic component radiographically. In the past, these neoplasms were divided histologically into benign cystadenomas and malignant cystadenocarcinomas. However, as was demonstrated by Compagno and Oertel, [91] all of these neoplasms need to be considered at least as potentially low-grade malignancies. Even though the bulk of many of these neoplasms histologically appear innocuous, if they are adequately sampled, a transition to cytologically malignant and architecturally more complex foci may be demonstrated.[21,91]

Cytomorphologic Features of Mucinous Cystic Tumors

- Abundant extracellular mucin in which tumor cells appear suspended in moderate numbers
- Cells occur singly, in flat sheets, and in three-dimensional aggregates
- Nuclei vary from small and bland to hyperchromatic and moderately pleomorphic
- Moderate volumes of clear to foamy cytoplasm

In contrast to most other primary pancreatic neoplasms, aspiration biopsies of mucinous cystic tumors generally yield only a moderate level of cellularity,[11,14,22–31,92,93] as listed above. The cytologic picture may be dominated by abundant extracellular mucinous material in which the neoplastic cells appear to be suspended (Figs. 4-38 to 4-41). Many of the neoplastic elements will be present in cohesive aggregates, which vary from small three-

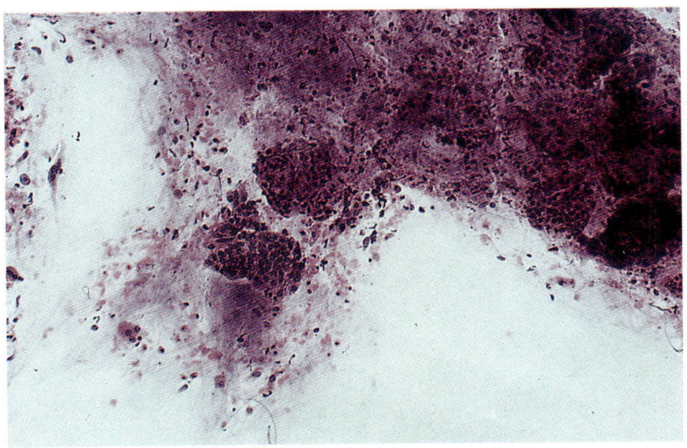

Fig. 4-38. Small monolayer-like aggregates of uniform tumor cells and extracellular mucin characterize this mucinous cystic tumor. Single cells, often degenerated, are also present; many of the latter appear suspended within the mucin (H&E). (Courtesy of Dr. S. Woodhouse, Lenoir, NC.)

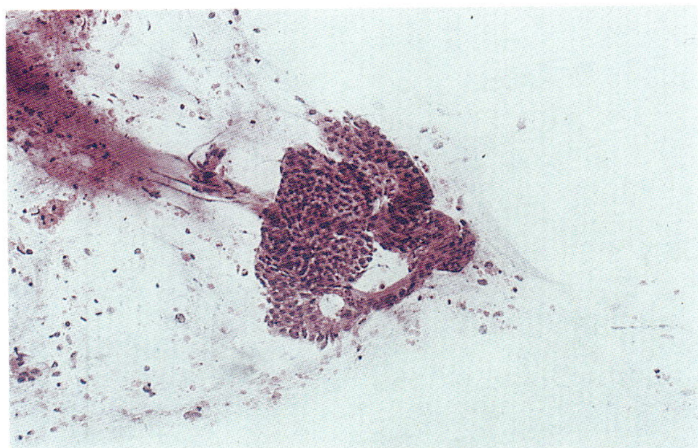

Fig. 4-39. In this mucinous cystic tumor, a larger sheet of tumor cells manifests a honeycomb arrangement and peripheral palisading. Mucin with a "stringy" appearance occupies a portion of the background (H&E). (Courtesy of Dr. S. Woodhouse, Lenoir, NC.)

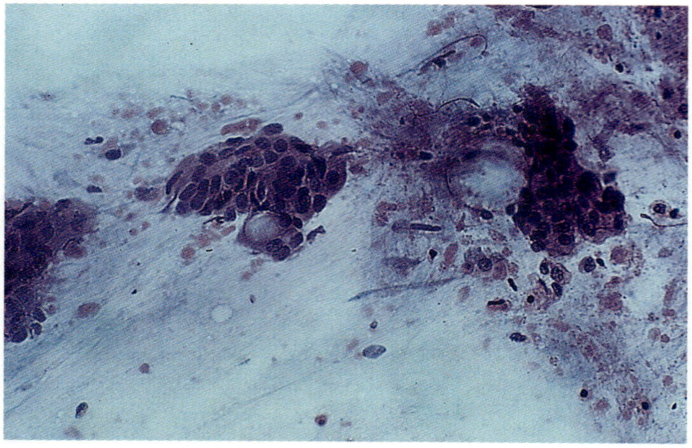

Fig. 4-40. The neoplastic cells of this mucinous cystic tumor have small oval uniform nuclei with finely granular, even chromatin. Mucin vacuoles are apparent, as are dispersed isolated tumor cells (H&E). (Courtesy of Dr. S. Woodhouse, Lenoir, NC.)

dimensional clusters to large flat sheets. In the latter, well-formed intercellular borders (honeycomb arrangement) are generally easily recognizable. Additionally, individually dispersed neoplastic cells may be recognized within the mucinous material. The latter typically demonstrates a very stringy appearance, in contrast to granular necrotic debris. The nuclear appearances vary from bland and benign to definitely malignant. Generally, the benign-appearing nuclei predominate and are characterized by round or slightly ovoid contours, delicate membranes, very finely granular chromatin, small nucleoli, and a striking uniformity (Figs. 4-40 and 4-41). Variably thick and irregular nuclear membranes, foci of parachromatin, and prominent nucleoli characterize the malignant elements; the malignant cells tend to occur in smaller and more three-dimensional aggregates that may manifest a syncytial pattern. Cytoplasm tends to be moderate to abundant in volume and foamy, bubbly, or clear in quality. Signet ring-like cells may also be present. Ancillary diagnostic procedures have relatively little to offer in the diagnosis of this tumor. A mucin stain (e.g., mucicarmine) will confirm the impression generated by the recognition of the above morphologic features (Fig. 4-42).

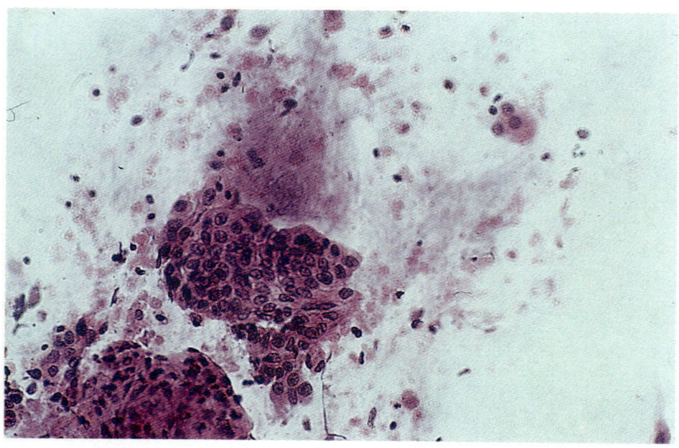

Fig. 4-41. Delicate nuclear membranes, pale chromatin, and inconspicuous nucleoli are evident in this aspirated mucinous cystic tumor (H&E). (Courtesy of Dr. S. Woodhouse, Lenoir, NC.)

A

B

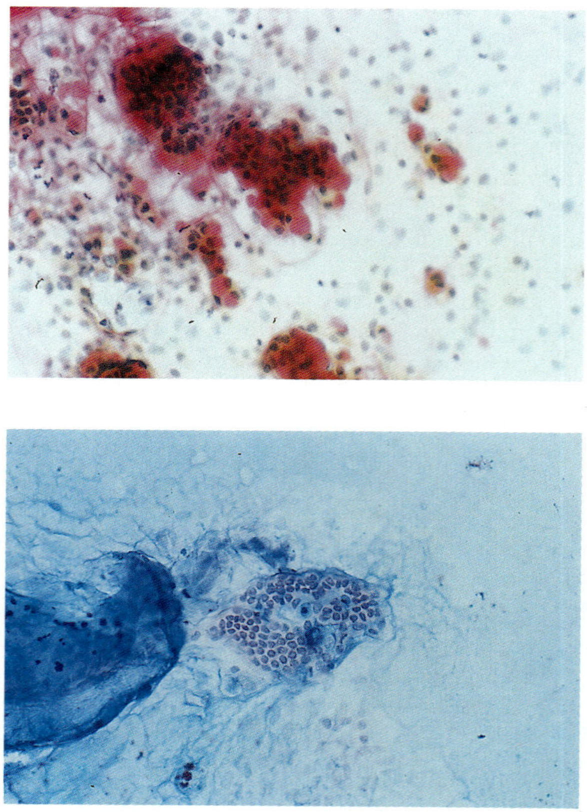

Fig. 4-42. (**A**) Cytoplasmic mucin is highlighted by the mucicarmine stain, revealing low nuclear to cytoplasmic ratios. (**B**) The extracellular mucin is stained with the Alcian blue reaction. Within the flat sheet of tumor cells, nuclei are homogeneous and evenly distributed.

The differential diagnosis of cystic mucinous tumors includes the much more common benign inflammatory pseudocysts.[14] As stated above, in aspiration smears of pseudocysts, the specimens are dominated by abundant granular necrotic debris, rather than stringy-appearing mucus. In general, neutrophils will be present in much greater numbers in pseudocyst specimens than in aspirates of the mucinous tumors. Conversely, epithelial cells may not be present at all in aspirates of pseudocysts; if present, they represent sampling of parenchyma from outside of the pseudocyst wall. However, such epithelial cells may demonstrate prominent inflammatory atypia. Intracytoplasmic mucin should not be a prominent feature. The differential diagnosis also includes ductal adenocarcinomas with extensive mucinous differentiation or cystic degeneration. Typically, aspirates of the ductal carcinomas will be much more highly cellular and contain a greater proportion of cells with obviously malignant nuclear features. The differential diagnosis must also include a recently described pancreatic neoplasm, the intraductal mucin-producing tumor.[94–99] Major features of these neoplasms include an exclusive or at least predominate intraductal (noninvasive) proliferation of mucin-producing columnar epithelial cells, marked mucus production with resultant distention of the main ductal system, and a rather good overall prognosis. Based purely on cytomorphology, we do not know any feature that could reliably distinguish this neoplasm from the mucinous cystic tumors. We are unaware of a report of one of these neoplasms with the FNA cytomorphologic features that have been detailed; in part, this may result from the fact that these patients present with a picture of chronic pancreatitis, rather than of neoplasm.

In addition to cellular morphology, aspirated material from the mucinous cystic tumors may be analyzed biochemically. Several investigators have found that most of these neoplasms contain exceedingly high levels of CEA[26–31]; these tumors are negative for amylase. Importantly, this is the reverse of the pattern expected for inflammatory pseudocysts. Lewandrowski and colleagues[30] have also measured the relative viscosity of cyst contents, demonstrating high values due to the high mucus content of these neoplasms. These workers have also investigated the potential heterogeneity of these different components in multiple loculi of a solitary tumor.[29] Although variability was seen among the different chambers, high diagnostic levels were observed in almost all cases. Pancreatic abscesses show low levels for all these tests.

ACINAR CELL CARCINOMA

Acinar cell carcinoma accounts for no more than 1 percent of all pancreatic malignancies.[37,38,100,101] Unfortunately, it is associated with a prognosis at least as dismal as that of the much more common ductal adenocarcinoma. Although many patients will present with a picture indistinguishable from that produced by conventional carcinomas, occasional patients have prominent arthropathies and widespread subcutaneous fat necrosis, consequent to the secretion of pancreatic proteolytic enzymes, including elastase by the tumor cells.[102] Histologically, the neoplastic elements are arranged in solid sheets, trabeculae, and well-formed acini. Individually, the neoplastic cells are characterized by voluminous densely granular cytoplasm and a solitary round nucleus, often with a very well-developed nucleolus.

Aspiration biopsies of acinar cell carcinomas produce highly cellular smears with relatively well-preserved intercellular cohesion,[14,58,64,90,102] as listed below. Although there may certainly be individually dispersed neoplastic cells, many of the malignant elements will be present in variably formed lobulated acinar structures (Fig. 4-43). These acini may occur singly or may appear to be

Cytomorphologic Features of Acinar Cell Carcinoma

- Cellular smear with many cellular aggregates, including lobulated acini with altered polarity
- Smooth to slightly irregular round nuclei with hyperchromatic chromatin and prominent nucleoli
- Moderate to abundant obviously granulated eosinophilic cytoplasm

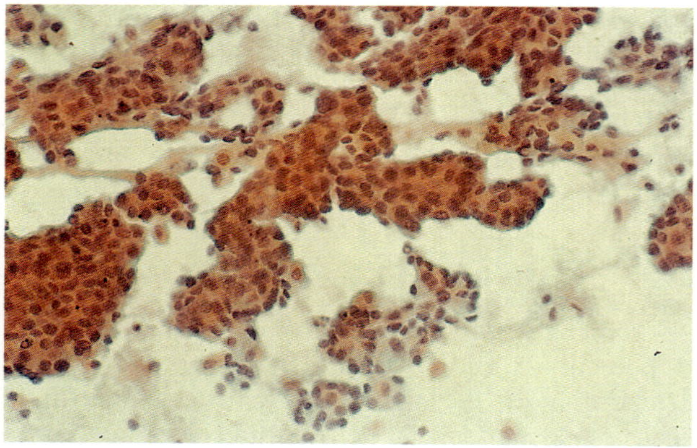

Fig. 4-43. Most of the tumor cells in this aspirated acinar cell carcinoma are located in anastomosing acini. The hyperchromatic nuclei tend to be peripherally displaced. Cytoplasm is moderate in volume, eosinophilic, and granular (Papanicolaou). (Courtesy of Dr. G. Thomas, Richmond, VA.)

anastomosed with one another. In contrast to normal pancreatic acini, polarity is altered with overlapped and displaced nuclei. For the most part, the neoplastic cells are uniform, each possessing moderate to abundant cytoplasm and a single centrally positioned enlarged nucleus (Figs. 4-43 and 4-44). The cytoplasm is distinctly granular and typically eosinophilic. The nucleus is round, has variably clumped and stained chromatin, and usually has a very prominent nucleolus. A simple stain for the periodic acid-Schiff (PAS) reaction will accentuate the cytoplasmic granularity, which is due to numerous zymogen granules. If an aliquot of the aspirate is placed in glutaraldehyde, electron microscopic examination will demonstrate the presence of these granules in large numbers as well.

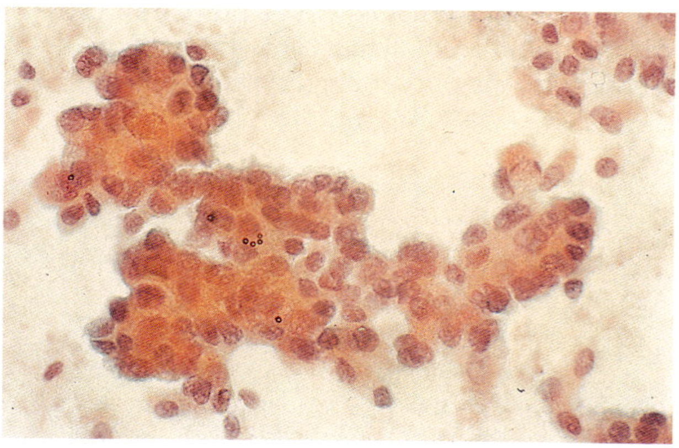

Fig. 4-44. Nucleoli are apparent in many of these round, darkly stained nuclei. In acinar cell carcinoma, cohesion is relatively well maintained (Papanicolaou). (Courtesy of Dr. G. Thomas, Richmond, VA.)

In aspiration smears, the clinically most important and cytologically most difficult differential diagnosis is with islet cell neoplasms. Both tumors may present with rather uniform neoplastic cells characterized by granulated cytoplasm and round nuclei.[14,58,64,102] Features favoring an islet cell neoplasm over an acinar cell carcinoma include a finely granular evenly distributed ("salt and pepper") chromatin arrangement, eccentrically positioned nuclei, an absence of large nucleoli, and marked reduction in intercellular cohesion. Concerning this last feature, the scanning lens view may demonstrate expected differences: the lobular arrangement of acinar cell carcinoma versus the dispersed pattern of islet cell neoplasms. Additionally, ancillary procedures will demonstrate the presence of distinct neuroendocrine differentiation.[51,56] One must also consider the possibility of ductal adenocarcinoma. As a generalization, nuclear pleomorphism will be much more prominent in the adenocarcinomas. A final consideration includes normal acini; in normal structures, basal nuclear polarity is maintained, the cells have lower nuclear to cytoplasmic ratios, and large nucleoli are exceptional.

PANCREATOBLASTOMA

Pancreatoblastoma is an exceedingly rare malignancy that occurs almost exclusively in children of all ages.[90,103–110] Although this neoplasm may have a relatively favorable clinical course, some patients succumb to metastatic disease or extensive local recurrence, or both. These tumors may arise in any portion of the pancreas and are usually quite large at diagnosis. Their histogenesis remains uncertain, and it may actually vary from tumor to tumor or be mixed within the same neoplasm. The histologic appearance of pancreatoblastoma is variable from case to case.

We reported a single FNA biopsy case of a pancreatoblastoma.[14,110] The patient was a 4-year-old boy who presented with a large retroperitoneal mass and acholic stools. The aspiration smear contained two major cellular components. The more primitive element consisted of small neoplastic cells with round, hyperchromatic nuclei and very scanty cytoplasm, resulting in very high nuclear to cytoplasmic ratios (Fig. 4-45). For the most part, they occurred as dissociated solitary elements. The nuclei had finely granular, evenly dispersed, and darkly stained chromatin with small nucleoli. The larger component possessed greater volumes of distinctly granular to foamy cytoplasm and manifested a much higher level of intercellular cohesion (Fig. 4-46). Some of these were arranged in acinar or ductal structures. The smears also contained large fragments of connective tissue in which benign-appearing, spindle-shaped mesenchymal cells were present (Fig. 4-46); additionally, some of the fragments appeared to be coated by an external layer of neoplastic epithelial cells. The neoplastic cells had contours that included spindled and elongated to pyramidal. In this example, the neoplastic cells were positive immunocytochemically for cytokeratin, CEA, NSE, and α_1-antitrypsin. Ultrastructurally, some of the aspirated tumor cells possessed distinct granules in their cytoplasm. In some instances, they were large resembling zymogen granules, whereas in others they were much smaller recapitulating neuroendocrine structures. Rare cells were hybrids in that their cytoplasm contained both types of granules.

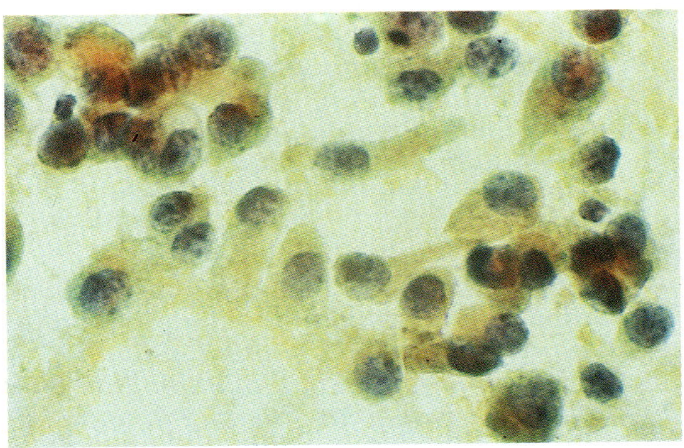

Fig. 4-45. The primitive component of pancreatoblastoma is composed of dispersed, small, obviously malignant cells with scanty cytoplasm and solitary hyperchromatic nuclei; the latter are often eccentrically positioned (Papanicolaou).

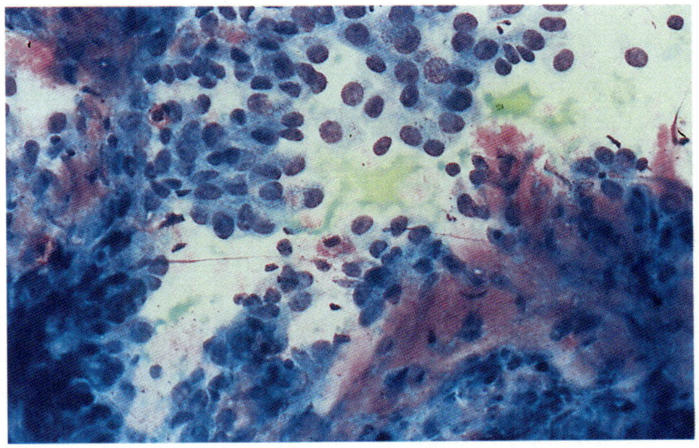

Fig. 4-46. The epithelial component of pancreatoblastoma consists of larger cells with greater volumes of cytoplasm. The latter may appear foamy or granular. An acinar-like arrangement of the tumor cells may be evident. In this Diff-Quik–stained preparation, the mesenchymal stroma readily stands out.

MICROCYSTIC CYSTADENOMA

Synonyms for this neoplasm include serous cystadenoma and glycogen-rich cystadenoma. These neoplasms tend to occur in older individuals as multiloculated fluid-filled cystic structures which are invariably benign.[111–113] Surgical resection is curative.

Aspiration biopsies of microcystic adenomas have been only rarely reported.[14,22,23,92,114,115] Specimens are generally not cellular and are composed of small uniform-appearing epithelial cells each possessing a single round or slightly oval nucleus. The chromatin will be finely granular, evenly distributed and rather darkly stained; nucleoli are inconspicuous. The nuclear to cytoplasmic ratios are high. The cells are present in cohesive flat sheets and individually dispersed. This arrangement may resemble aspirated benign mesothelium. The cytoplasm contains abundant glycogen, which may be demonstrated by a simple PAS stain.

The differential diagnosis of microcystic adenoma includes "contamination" of an abdominal aspirate by benign mesothelial elements. The differential diagnosis also includes pseudocysts, mucinous cystic tumors, and papillary-cystic tumor. Features in favor of a microcystic adenoma include a clean smear background lacking inflammatory cells and abundant extracellular mucin, the lack of well-formed papillary structures and no longitudinal nuclear grooves.

METASTASES

Neoplasms metastatic to the pancreas and peripancreatic lymph nodes may clinically simulate primary pancreatic cancers. Aspiration biopsy of these metastases may lead to a confusing cytologic picture, especially if the patient does not have a previously diagnosed malignancy. Two recently reported series have documented the relatively frequent occurrence of this type of specimen as seen by cytopathologists.[60,61] Our series evaluated 176 malignant diagnoses in a total of 304 pancreatic aspiration biopsies; 11 percent of these were metastases.[60] Of these, the most frequent were the non-Hodgkin's lymphomas and small cell carcinoma, usually of pulmonary origin. Other entities included Hodgkin's disease, squamous cell carcinoma from various sites, and hepatocellular carcinoma. Importantly, in six of the patients, the pancreatic mass was the initial presentation of the tumor; thus, the FNA biopsy was the first morphologic evidence of an occult extrapancreatic cancer.[60] These six tumors included five lymphomas, which probably involved peripancreatic nodes and a small cell carcinoma from the lung. Interestingly (and coincidentally) Carson et al.[61] also found that metastases represented 11 percent of their 117 pancreatic aspirates. Furthermore, in 30 percent of their aspirates, the primary diagnosis was also not previously recognized. However, in contrast to our study, Carson et al.[61] reported that the vast majority of the metastases were epithelial, with only 2 lymphoreticular malignancies in their series.

Recognition of cytomorphologic features unexpected for pancreatic carcinoma should suggest the possibility of an extrapancreatic cancer.[60] Ancillary diagnostic procedures such as electron microscopy and immunocytochemistry may be helpful in selected examples. It is essential for the cytopathologist to maintain a high level of suspicion in order to avoid a misdiagnosis of a primary pancreatic cancer which could lead to inappropriate therapy, such as a Whipple's procedure. Generally, these patients are best treated with systemic chemotherapy.

REFERENCES

1. Hastrup J, Thommesen P, Frederiksen P: Pancreatitis and pancreatic carcinoma diagnosed by perioperative fine-needle aspiration biopsy. Acta Cytol 21:731, 1978
2. Tao LC, Ho CS, McLoughlin MJ, McHattie J: Percutaneous fine needle aspiration biopsy of the pancreas: cytodiagnosis of pancreatic carcinoma. Acta Cytol 22:215, 1978
3. An-Foraker SH, Fong-Mui KK: Cytodiagnosis of lesions of the pancreas and related areas. Acta Cytol 26:814, 1982
4. Alpem GA, Dekker A: Fine needle aspiration of the pancreas: an analysis of its use in 52 patients. Acta Cytol 29:873, 1985
5. Mitchell ML, Carney CN: Cytologic criteria for the diagnosis of pancreatic carcinoma. Am J Clin Pathol 83:171, 1985
6. Fekete PS, Nunez C, Pitlik DA: Fine needle aspiration biopsy of the pancreas. Diagn Cytopathol 2:301, 1986
7. Hajdu EO, Kumari-Subaiya S, Phillips G: Ultrasonically guided percutaneous aspiration biopsy of the pancreas. Semin Diagn Pathol 3:166, 1986
8. Bret PM, Nicholet V, Labadie M: Percutaneous fine-needle aspiration biopsy of the pancreas. Diagn Cytopathol 2:221, 1986
9. Pinto MM, Avila NA, Criscuola EM: Fine-needle aspiration of the pancreas. Acta Cytol 32:39, 1988
10. Mitchell ML, Bittner CA, Willis JS, Parker FP: Fine needle aspiration cytology of the pancreas: a retrospective study of 73 cases. Acta Cytol 32:447, 1988
11. Oertel JE, Heffess CS, Oertel YC: Pancreas. p. 1057. In Sternberg SS (ed): Diagnostic Surgical Pathology. Raven Press, New York, 1989
12. Cohen MB, Egerter DP, Holly EA et al: Pancreatic carcinoma: regression analysis to identify improved cytologic criteria. Diagn Cytopathol 7:341, 1991
13. Robins DB, Katz RL, Evans DB et al: Fine-needle aspiration of the pancreas. In quest of accuracy. Acta Cytol 39:1, 1995
14. Geisinger KR, Silverman JF: Fine-needle aspiration cytology of uncommon primary pancreatic neoplasms: a personal experience and review of the literature. p. 23. In Schmidt WA (ed): Cytopathology Annual. Williams & Wilkins, Baltimore, 1992

15. Nguyen G-K: Percutaneous fine-needle aspiration cytology of the pancreas. Pathol Annu 22:163, 1985
16. Hyland C, Kheir SM, Kashlan MB: Frozen section diagnosis of pancreatic carcinoma: a prospective study of 64 biopsies. Am J Surg Pathol 5:179, 1981
17. Lightwood R, Reber HA, Way LW: The risk and accuracy of pancreatic biopsy. Am J Surg 132:189, 1976
18. Steinberg W, Tenner S: Acute pancreatitis. N Engl J Med 330:1198, 1994
19. Jorda M, Essenfeld H, Garcia E, Ganjei P: The value of fine-needle aspiration cytology in the diagnosis of inflammatory pancreatic masses. Diagn Cytopathol 8:65, 1992
20. Lin A, Feller ER: Pancreatic carcinoma as a cause of unexplained pancreatitis: report of ten cases. Ann Intern Med 113:166, 1990
21. Warshaw AL, Compton CC, Lewandrowski K et al: Cystic tumors of the pancreas. Ann Surg 212:432, 1990
22. Young NA, Villani MA, Khoury P, Naryshkin S: Differential diagnosis of cystic neoplasms of the pancreas by fine-needle aspiration. Arch Pathol Lab Med 115:571, 1991
23. Lauciricia R, Schwartz MR, Ramzy I: Fine-needle aspiration of pancreatic cystic epithelial neoplasms. Acta Cytol 36:881, 1992
24. Emmert GM, Bentra C: Fine-needle aspiration biopsy of mucinous cystic neoplasm of the pancreas: a case study. Diagn Cytopathol 2:69, 1986
25. Dodd LG, Farrell TA, Layfield LJ: Mucinous cystic tumor of the pancreas: an analysis of FNA characteristics with an emphasis on the spectrum of malignancy associated features. Diagn Cytopathol 12:113, 1995
26. Tatsuta M, Yamamoto R, Yanamamura H et al: Cytological examination and CEA measurement in aspirated pancreatic material collected by percutaneous fine-needle aspiration biopsy under ultrasonic guidance for the diagnosis of pancreatic carcinoma. Cancer 52:693, 1983
27. Pinto MM, Kaye AD, Brogan DA, Criscuolo EH: Diagnosis of cystic lesions of the pancreas: a biochemical and cytologic analysis of material obtained utilizing radiographic or intraoperative technique. Diagn Cytopathol 2:40, 1986
28. Pinto MM, Meriano FV: Diagnosis of cystic pancreatic lesions by cytologic examination and carcinoembryonic antigen and amylase assays of cyst contents. Acta Cytol 35:456, 1991
29. Lewandrowski KB, Warshaw AL, Compton CC et al: Variability in cyst fluid carcinoembryonic antigen level, fluid viscosity, amylase content, and cytologic findings among multiple loculi of a pancreatic mucinous cystic neoplasm. Am J Clin Pathol 100:425, 1993
30. Lewandrowski K, Southern J, Pins M et al: Cyst fluid analysis in the differential diagnosis of pancreatic cysts: a comparison of pseudocysts, serous cystadenomas, mucinous cystic neoplasms, and mucinous cystadenocarcinoma. Ann Surg 217:41, 1993
31. Yu HC, Shetty J: Mucinous cystic neoplasm of the pancreas with high carcinoembryonic antigen. Arch Pathol Lab Med 109:375, 1985
32. Weidner N, Geisinger KR, Sterling RT et al: Benign lymphoepithelial cysts of the parotid gland: a histologic, cytologic, and ultrastructural study. Am J Clin Pathol 85:395, 1986
33. Mockli GC, Stein RM: Cystic lymphoepithelial lesion of the pancreas. Arch Pathol Lab Med 114:885, 1990
34. Cappellari JO: Fine needle aspiration cytology of a pancreatic lymphoepithelial cyst. Diagn Cytopathol 9:77, 1993
35. Markowsky V, Russin VL: Fine-needle aspiration of dermoid cyst of the pancreas: a case report. Diagn Cytopathol 9:66, 1993
36. Warshaw AL, Fernández-Del Castillo C: Pancreatic carcinoma. N Engl J Med 326:455, 1992
37. Carriaga MT, Henson DE: Liver, gallbladder, extrahepatic bile ducts, and pancreas. Cancer 75:171, 1995
38. Cubilla AL, Fitzgerald PJ: Classification of pancreatic cancer (nonendocrine). Mayo Clin Proc 54:449, 1979
39. Gullo L, Pezzilli R, Morselli-Labate AM: Diabetes and the risk of pancreatic cancer. N Engl J Med 331:81, 1994
40. Scarpa A, Capelli P, Mukai K et al: Pancreatic adenocarcinomas frequently show p53 gene mutations. Am J Pathol 142:1534, 1993
41. DiGiuseppe JA, Hruban RH, Goodman SN et al: Overexpression of p53 protein in adenocarcinoma of the pancreas. Am J Clin Pathol 101:684, 1994
42. Allison DC, Bose KK, Hruban RH et al: What's new in general surgery. Pancreatic cancer cell DNA content correlates with long-term survival after pancreatoduodenectomy. Ann Surg 214:648, 1990
43. Bose KK, Allison DC, Hruban RH et al: A comparison of flow cytometric and absorption cytometric DNA values as prognostic indicators for pancreatic carcinoma. Cancer 71:691, 1993
44. Porschen R, Remy U, Bevers G et al: Prognostic significance of DNA ploidy in adenocarcinoma of the pancreas. Cancer 71:3846, 1993
45. Böttger TC, Störkel S, Wellek S et al: Factors influencing survival after resection of pancreatic cancer. A DNA analysis and a histomorphologic study. Cancer 73:63, 1994

46. Parsons L Jr, Palmer CH: How accurate is fine-needle biopsy in malignant neoplasia of the pancreas. Arch Surg 124:681, 1989
47. Edoute Y, Lemberg S, Malberger E: Preoperative and intraoperative fine-needle aspiration cytology of pancreatic lesions. Am J Gastroenterol 86:1015, 1991
48. Blandamura S, Costantin G, Nitti D, Boccato P: Intraoperative cytology of pancreatic masses. A 10-year experience. Acta Cytol 39:23, 1995
49. Yang GC, Slott S, LiVolsi VA, Gupta PK: Rapid assessment of Diff-Quik-stained pancreatic aspirates. A retrospective study of 40 intraoperative fine-needle aspiration consultations, with measurement of nuclear size of look-alike small tissue fragments by image analysis. Acta Cytol 38:37, 1994
50. Shibata D, Almoguera C, Forrester K et al: Detection of c-K-ras mutations in fine-needle aspirates from human pancreatic adenocarcinomas. Cancer Res 50:1279, 1990
51. Silverman JF, Geisinger KR: Ancillary studies in FNA of the liver and pancreas. Diagn Cytopathol, in press
52. Banner BF, Myrent KL, Memoli VA, Gould VE: Neuroendocrine carcinoma of the pancreas diagnosed by aspiration cytology. Acta Cytol 29:442, 1985
53. Hsiu JG, D'Amato NA, Sperling MH et al: Malignant islet-cell tumor of the pancreas diagnosed by fine needle aspiration biopsy. A case report. Acta Cytol 29:576, 1985
54. Bell DA: Cytologic features of islet tumors. Acta Cytol 31:485, 1987
55. Sneige N, Ordonez NG, Veanattukalathil S, Samaan NA: Fine needle aspiration cytology in pancreatic endocrine tumors. Diagn Cytopathol 3:35, 1987
56. Shaw JA, Vance RP, Geisinger KR, Marshall RB: Islet cell neoplasms: a fine needle aspiration cytology study with immunocytochemical correlations. Am J Clin Pathol 94:142, 1990
57. Al-Kaisi N, Weaver MG, Abdul-Karim FW, Siegler E: Fine needle aspiration cytology of neuroendocrine tumors of the pancreas. A cytologic immunocytochemical and electron microscopic study. Acta Cytol 36:655, 1992
58. Labate AM, Zakowski MF: Cytologic features of acinar and islet cell tumors. Acta Cytol 38:858, 1994
59. Tao L-K: Transabdominal Fine-Needle Aspiration Biopsy. Igaku-Shoin, New York, 1990, p. 133
60. Benning TL, Silverman JF, Berns LA, Geisinger KR: Fine needle aspiration of metastatic and hematologic malignancies clinically mimicking pancreatic carcinoma. Acta Cytol 36:471, 1992
61. Carson HJ, Green LK, Casteli MJ et al: Utilization of fine-needle aspiration biopsy in the diagnosis of metastatic tumors to the pancreas. Diagn Cytopathol 12:8, 1995
62. Reyes CV, Wang T: Undifferentiated small cell carcinoma of the pancreas: a report of five cases. Cancer 47:2500, 1981
63. Nguyen-Ho P, Nguyen G-K, Jewell LD: Oncocytic neuroendocrine carcinoma of the pancreas. Report of a case with needle aspiration cytology, immunocytochemistry and electron microscopy. Acta Cytol 38:611, 1994
64. Villanueva RR, Nguyen-Ho P, Nguyen G-K: Needle aspiration cytology of acinar-cell carcinoma of the pancreas: report of a case with diagnostic pitfalls and unusual ultrastructural findings. Diagn Cytopathol 10:362, 1994
65. Dodd LG, Evans DB, Symmans F, Katz RL: Fine-needle aspiration of pancreatic extramedullary plasmacytoma: possible confusion with islet cell tumor. Diagn Cytopathol 10:371, 1994
66. Walts AE, Said JW, Shintaku IP, Lloyd R: Chromogranin as a marker of neuroendocrine cells in cytologic material—an immunohistochemical study. Am J Clin Pathol 84:273, 1985
67. Nguyen G-K, Rayani NA: Hyperplastic and neoplastic endocrine cells of the pancreas in aspiration biopsy. Diagn Cytopathol 2:204, 1986
68. Gala I, Atkinson BF, Nicosia RF, Hermann GA: Fine-needle aspiration cytology of idiopathic pancreatic islet cell adenosis. Diagn Cytopathol 9:453, 1993
69. Pinto MM, Monteiro NL, Tizol DM: Fine needle aspiration of pleomorphic giant cell carcinoma of the pancreas. Case report with ultrastructural observations. Acta Cytol 30:430, 1986
70. Silverman JF, Dabbs DJ, Finley JL, Geisinger KR: A fine needle aspiration biopsy of pleomorphic (giant cell) carcinoma of the pancreas. Cytologic, immunocytochemical, and ultrastructural findings. Am J Clin Pathol 89:714, 1988
71. Collins KA, Geisinger KR, Wakely PE, Jr et al: Extragonadal germ cell tumors: a fine-needle aspiration biopsy study. Diagn Cytopathol 12:223, 1995
72. Rosai J: Carcinoma of pancreas simulating giant cell tumor of bone. Electron-microscopic evidence of its acinar cell origin. Cancer 22:333, 1968
73. Jeffrey I, Crow J, Willis BL: Osteoclast-type giant cell tumor of the pancreas. J Clin Pathol 36:1165, 1983
74. Silverman JF, Finley JL, MacDonald KG, Jr: Fine needle aspiration cytology of osteoclastic giant cell tumor of the pancreas. Diagn Cytopathol 6:336, 1990
75. Walts AE: Osteoclast-type giant cell tumor of the pancreas. Acta Cytol 37:500, 1983

76. Manci EA, Gardner LL, Pollock WJ, Dowling EA: Osteoclastic giant cell tumor of the pancreas. Aspiration cytology, light microscopic, and ultrastructural with review of the literature. Diagn Cytopathol 1:105, 1985
77. Silverman JF, Finley JL, Berns L et al: Significance of giant cells in fine needle aspiration biopsies of benign and malignant lesions of the pancreas. Diagn Cytopathol 5:388, 1989
78. Pettinato G, Maniuel JC, Ravetto C et al: Papillary cystic tumor of the pancreas. A clinicopathologic study of 20 cases with cytologic, immunohistochemical, ultrastructural, and flow cytometric observations, and a review of the literature. Am J Clin Pathol 98:478, 1992
79. Lieber MR, Lack EE, Roberts JR, Jr et al: Solid and papillary epithelial neoplasm of the pancreas. An ultrastructural and immunocytochemical study of six cases. Am J Surg Pathol 11:85, 1987
80. Bondeson L, Bondeson A-G, Genell S et al: Aspiration cytology of a rare solid and papillary epithelial neoplasm of the pancreas: light and electron microscopic study of a case. Acta Cytol 28:605, 1984
81. Foote A, Simpson JS, Stewart RJ et al: Diagnosis of the rare solid and papillary epithelial neoplasm of the pancreas by fine needle aspiration biopsy: light and electron microscopy study of a case. Acta Cytol 30:519, 1986
82. Chen KTK, Workman RD, Efird TA, Cheng AC: Fine needle aspiration cytology diagnosis of papillary tumor of the pancreas. Acta Cytol 30:523, 1986
83. Stachura J, Popiela T, Pietron M et al: Cytology of solid and papillary epithelial neoplasms of the pancreas: a case report. Diagn Cytopathol 4:339–341, 1988
84. Cappellari JO, Geisinger KR, Albertson DA et al: Malignant papillary cystic tumor of the pancreas. Cancer 66:193, 1990
85. Katz LBK, Ehya H: Aspiration cytology of papillary cystic neoplasm of the pancreas. Am J Clin Pathol 94:328, 1990
86. Mendonca ME, Bivar-Weinholtz J, Soares J: Fine needle aspiration cytology of a solid and papillary epithelial neoplasm of the pancreas. Acta Cytol 35:258, 1991
87. Wilson MB, Adams DB, Garen PD, Gansler TS: Aspiration cytologic, ultrastructural, and DNA cytometric findings of solid and papillary tumor of the pancreas. Cancer 69:2235, 1992
88. Greenberg ML, Rennie Y, Grierson JM et al: Solid and papillary epithelial tumor of the pancreas: cytologic case study with ultrastructural and flow cytometric evaluation. Diagn Cytopathol 9:541, 1993
89. Skarda JS, Honick AB, Gibbins CS et al: Papillary-cystic tumor of the pancreas in a young woman: fine-needle aspiration cytology, ultrastructure and DNA analysis. Diagn Cytopathol 10:20, 1994
90. Morohoshi T, Kanda M, Horie A et al: Immunocytochemical markers of uncommon pancreatic tumors: Acinar cell carcinoma, pancreatoblastoma, and solid cystic (papillary-cystic) tumor. Cancer 59:739, 1987
91. Compagno J, Oertel JE: Mucinous cystic neoplasms of the pancreas with overt and latent malignancy (cystadenocarcinoma and cystadenoma). A clinicopathologic study of 41 cases. Am J Clin Pathol 69:573, 1978
92. Jones EC, Suen KC, Grant DR, Chan NH: Fine-needle aspiration cytology of neoplastic cysts of the pancreas. Diagn Cytopathol 3:238, 1987
93. Gupta RK, Scally J, Steward RJ: Mucinous cystadenocarcinoma of the pancreas: diagnosis of fine needle aspiration cytology. Diagn Cytopathol 5:408, 1989
94. Yamada M, Kozuka S, Yamao K et al: Mucin-producing tumor of the pancreas. Cancer 68:159, 1991
95. Rickaert F, Cremer M, Devière J et al: Intraductal mucin-hypersecreting neoplasms of the pancreas. A clinicopathologic study of eight patients. Gastroenterology 101:512, 1991
96. Kawasada Y, Yano T, Yamamoto T et al: Intraductal mucin-producing tumors of the pancreas. Am J Gastroenterol 87:634, 1992
97. Nishihara K, Fukuda T, Tsuneyoshi M et al: Intraductal papillary neoplasm of the pancreas. Cancer 72:689, 1993
98. Santini D, Campione O, Gullo L et al: Intraductal papillary-mucinous neoplasm of the pancreas. A clinicopathologic entity. Arch Pathol Lab Med 119:209, 1995
99. Nagai E, Ueki T, Chijiwa K et al: Intraductal papillary mucinous neoplasms of the pancreas associated with so-called "mucinous ductal ectasia": histochemical and immunohistochemical analysis of 29 cases. Am J Surg Pathol 19:576, 1995
100. Klimstra DS, Heffess CS, Oertel JE, Rosai J: Acinar cell carcinoma of the pancreas. A clinicopathologic study of 28 cases. Am J Surg Pathol 16:815, 1992
101. Klimstra DS, Rosai J, Heffess CS: Mixed acinar-endocrine carcinomas of the pancreas. Am J Surg Pathol 18:765, 1994
102. Ishihara A, Sanda T, Takanari H et al: Elastase-1-secreting acinar cell carcinoma of the pancreas. A cytologic, electron microscopic and histochemical study. Acta Cytol 33:157, 1989
103. Frable WJ, Still WJS, Kay S: Carcinoma of the pancreas, infantile type. Cancer 27:667, 1971

104. Taxy JB: Adenocarcinoma of the pancreas in childhood. Cancer 37:1508, 1976
105. Benjamin E, Wright DH: Adenocarcinoma of the pancreas of childhood: a report of two cases. Histopathology 4:87, 1980
106. Buchino JJ, Castelo FM, Nagaraj HS: Pancreatoblastoma: a histochemical and ultrastructural analysis. Cancer 53:963, 1984
107. Ichijima K, Akaishi K, Toyoda N et al: Carcinoma of the pancreas with endocrine component in childhood: a case report. Am J Clin Pathol 83:95, 1985
108. Rich RH, Weber JL, Shandling B: Adenocarcinoma of the pancreas in a neonate managed by pancreatoduodenectomy. J Pediatr Surg 21:806, 1986
109. Palosaari D, Clayton F, Seaman J: Pancreatoblastoma in an adult. Arch Pathol Lab Med 110:650, 1986
110. Silverman JF, Holbrook CT, Pories WJ et al: Fine needle aspiration cytology of pancreatoblastoma with immunocytochemical and ultrastructural studies. Acta Cytol 34:632, 1990
111. Shorten SD, Hart WR, Petras RE: Microcystic adenomas (serous cystadenomas) of pancreas. A clinicopathologic investigation of eight cases with immunohistochemical and ultrastructural studies. Am J Surg Pathol 10:365, 1986
112. Alpert LC, Truong LD, Bossart MI, Spjut HJ: Adenoma (serous cystadenoma) of the pancreas. A study of 14 cases with immunohistochemical and electron-microscopic correlation. Am J Surg Pathol 12:251, 1988
113. George DH, Murphy F, Michalski R, Ulmer BG: Serous cystadenocarcinoma of the pancreas: a new entity? Am J Surg Pathol 13:61, 1989
114. Hittmair A, Pernthaler H, Totsch M, Schmid KW: Preoperative fine needle aspiration cytology of a microcystic adenoma of the pancreas. Acta Cytol 35:546, 1991
115. Nguyen G-K, Vogelsang PJ: Microcystic adenoma of the pancreas. A report of two cases with fine needle aspiration cytology and differential diagnosis. Acta Cytol 37:908, 1993

5 Kidney

The use of aspiration cytology for the diagnosis of renal neoplasia remains controversial despite a recent increase in the number of publications of fine needle aspiration (FNA) biopsy of renal masses.[1–13] In the United States, FNA of the kidney is used primarily for the identification and treatment of benign cysts and abscesses.[14] Non-neoplastic renal mass lesions include cysts, tuberculosis, abscesses, and infarcts. The combination of modern imaging techniques with percutaneous FNA biopsy has eliminated the need for surgical exploration of some of these lesions. At some centers, however, percutaneous FNA biopsy of renal mass lesions, including neoplasms, has gained increasing acceptance and may contribute considerably to patient care and management.[15] The indications and complications for FNA of the kidney are outlined on page 172.

Whether FNA of non-neoplastic or neoplastic lesions is aspirated, it is important for the cytologist to recognize the benign cellular components of the kidney. FNA of normal renal tubular cells will demonstrate a sheet-like arrangement of the cells, having round, centrally, or eccentrically placed nuclei with a moderate amount of finely granular to pale cytoplasm[15] (Fig. 5-1). Collecting duct tubules and distal tubular cells are generally not as apparent in the smears. These cells resemble proximal tubular cells but are smaller and have darker cytoplasm. The potential exits to confuse hepatocytes with proximal renal tubular cells, but the liver cells generally have more centrally located nuclei. Benign proximal renal tubular cells could also be confused with the granular cells of some renal cell carcinomas. However, the benign cells are arranged in a more uniform cohesive monolayered sheets with lower nuclear to cytoplasmic ratios and no significant atypia. Glomeruli can occasionally be appreciated in the smears; they are generally arranged in tight, lobulated clusters consisting of spindle-shaped endothelial cells with indistinct cytoplasm (Fig. 5-1).

RENAL CYSTS

Single or multiple renal cysts may be congenital or acquired.[10] Cysts complicated by infection or hemorrhage may result in pain and hematuria.[6] Since renal cell carcinoma can also undergo either cystic degeneration or hemorrhage, or both, it is useful to aspirate cystic renal lesions for a more definitive diagnosis.[6,10,15–17]

Aspiration smears of benign renal cysts are rather hypocellular and reveal predominantly scattered macrophages, having finely vacuolated cytoplasm with eccentric round to bean-shaped bland nuclei. Often hemosiderin-laden macrophages are present, indicative of old hemorrhage. Epithelial sheets with reparative changes are possible. These atypical reparative epithelial cells and reactive macrophages may be potential sources of false-positive diagnoses[6,7,15,18,19] (Fig. 5-2). Features favoring non-neoplastic renal cysts include benign cells in small cohesive, polarized groups or individually scattered and the absence of necrotic cells and debris.[15] A smooth cyst lining radiologically also supports a benign diagnosis.

An uncommon cause of a renal cyst is hydatid disease. Frydman et al.[20] reported a patient with pulmonary and renal hydatid disease diagnosed by

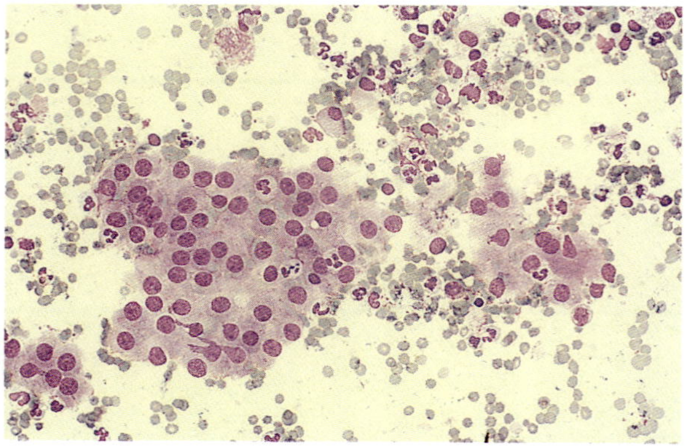

A

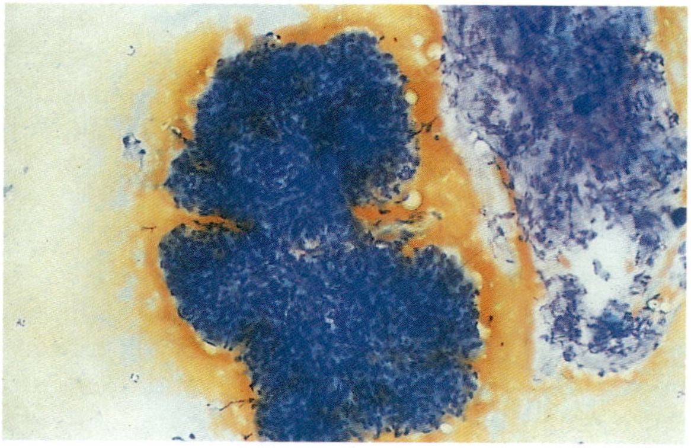

B

Fig. 5-1. (**A**) FNA of acute pyelonephritis, showing sheets of neutrophils and uniform benign-appearing renal tubular cells arranged in a monolayer sheet (Diff-Quik). (**B**) Groups of benign glomeruli cells, having a lobulated appearance (Diff-Quik).

Indications for FNA of Suspected Renal Neoplasms

- Equivocal radiographic findings
- Candidates for renal arterial embolization
- Type malignant neoplasm: nephrectomy for renal cell carcinoma; nephrectomy and ureterectomy for transitional cell carcinoma
- Identify benign and non-neoplastic masses
- Confirm metastatic tumor or lymphoma
- Confirm primary renal cell carcinoma in case with metastatic disease, inoperable or high surgical risk
- Diagnosis of Wilms' tumor in patient receiving preoperative chemotherapy
- Therapeutic evacuation of simple cysts

Complications of FNA of the Kidney

- Hematuria—minor, transient
- Pain and fever
- Very rarely—hemorrhage requiring nephrectomy
- Needle track seeding—rare reports, most using large-bore needles (14- to 20-gauge)

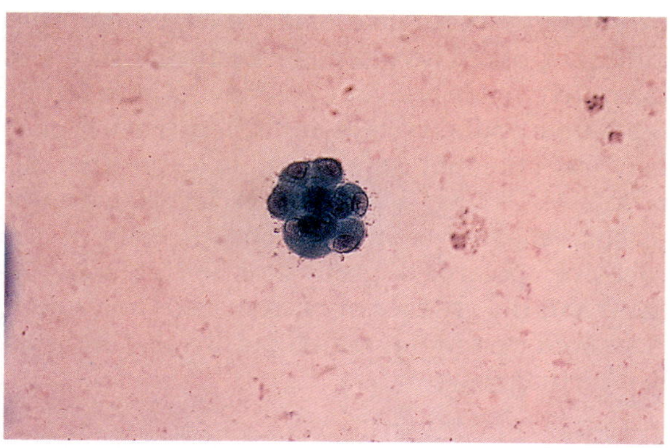

A

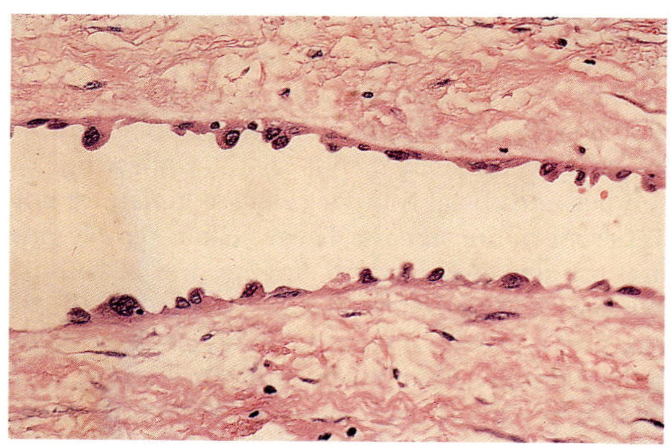

B

Fig. 5-2. (**A**) FNA of cystic lesion of the kidney in which small clusters of mildly atypical epithelial cells showing some slight depth of focus and scalloped external margins are present. Note that the nuclear to cytoplasmic ratios are not appreciably increased (Diff-Quik). (**B**) Histology of the resected multilocular cyst revealing source of the atypical cells seen in the FNA biopsy (H&E).

FNA biopsy. The renal hydatid cyst fluid contained scoleces and hooklets of *Echinococcus*. Another unusual finding in renal aspiration is the presence of Liesegang rings in hemorrhagic renal cysts. Sneige et al.[21] reported two such cases. The periodic ring-like structures varied in size from 8 to 200 μm in diameter and demonstrated regularly striated double walls (Fig. 5-3). Liesegang rings may be easily confused with the putative, giant kidney worm, *Dioctophyma renale*. Although uncertain, Liesegang rings are thought to represent insoluble precipitates in hemorrhagic and cystic lesions.[21] We have personally examined a similar FNA specimen.[22] The nonpolarizable spherical structures, which had double walls with equally spaced radial striation, were present in the direct smears and in the cell block.

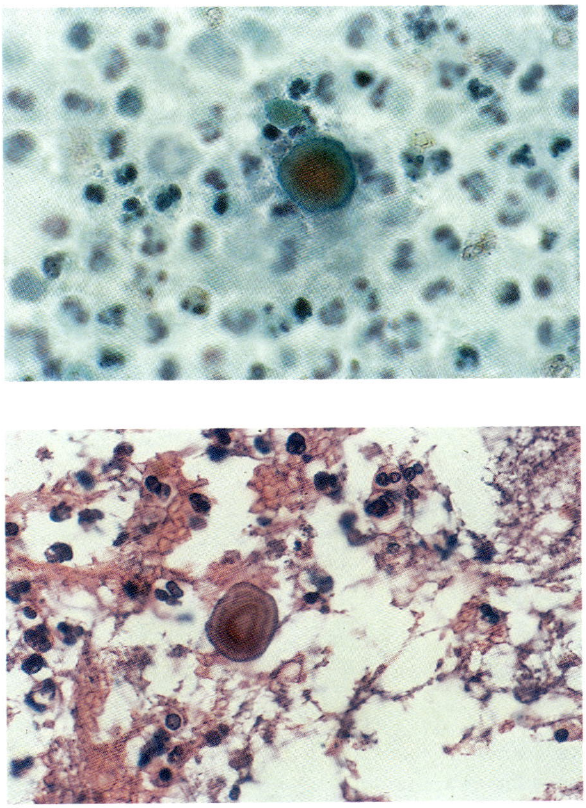

A

B

Fig. 5-3. (**A**) FNA of kidney revealing Liesegang rings, having regularly arranged, striated double walls (Papanicolaou). (**B**) Liesegang ring, having laminated striated walls in FNA of kidney, along with scattered neutrophils in the background (cell block, H&E).

RENAL ABSCESS

The clinical presentation of patients with renal abscesses include chills, fever, severe costovertebral angle pain, tenderness, and pyuria. Most often, these lesions occur in patients with acute pyelonephritis or a systemic inflammatory process.[23] FNA biopsy provides not only diagnostic material, but therapeutic drainage as well.[24] FNA cytologic examination will demonstrate a purulent exudate, consisting of loose sheets of neutrophils, scattered histiocytes, and necrotic debris in the background (Fig. 5-1). The bacteria can often be seen in Diff-Quik–stained smears. Special stains for organisms, including Gram for bacteria, Ziehl-Neelsen for *Mycobacterium*, and Gomori methenamine silver (GMS) for fungi, should be performed along with submission of the aspirated material for aerobic, anaerobic, mycobacterial, and fungal cultures.

OTHER BENIGN RENAL LESIONS

Xanthogranulomatous pyelonephritis is an inflammatory process that poses a real threat for potential false-positive diagnoses of malignancy,[22] via misinterpretation of benign, vacuolated histiocytes for the tumor cells of clear cell carcinoma of the kidney (Fig. 5-4). In general, an aspirate from well-differentiated renal cell carcinoma consists of a more cellular specimen in which the clear cells are arranged in sheets and groups, many of which are transversed by interspersed delicate branching capillaries (Fig. 5-5). Furthermore, although malignant cells in well-differentiated clear cell carcinoma often possess low nuclear to cytoplasmic ratios with abundant clear cytoplasm, some nuclear enlargement and well-developed nucleoli should alert the pathologist to cancer. The vacuolated appearance of clear cell carcinoma is due to excessive cytoplasmic lipid and glycogen.[25–27] The differential diagnosis includes histiocytes from xanthogranulomatous pyelonephritis and fat necrosis, which are often isolated and associated with other inflammatory cells.[28] Histiocytic nuclei tend to be eccentrically positioned within the cell and have a coffee bean (reniform) shape. In difficult cases, immunocytochemical studies for low-molecular-weight cytokeratin and/or epithelial membrane antigen (EMA) will decorate the malignant cells, whereas histocytes are uniformly negative.

In aspirates from renal infarcts, groups of cells showing atypical changes due to regeneration and repair

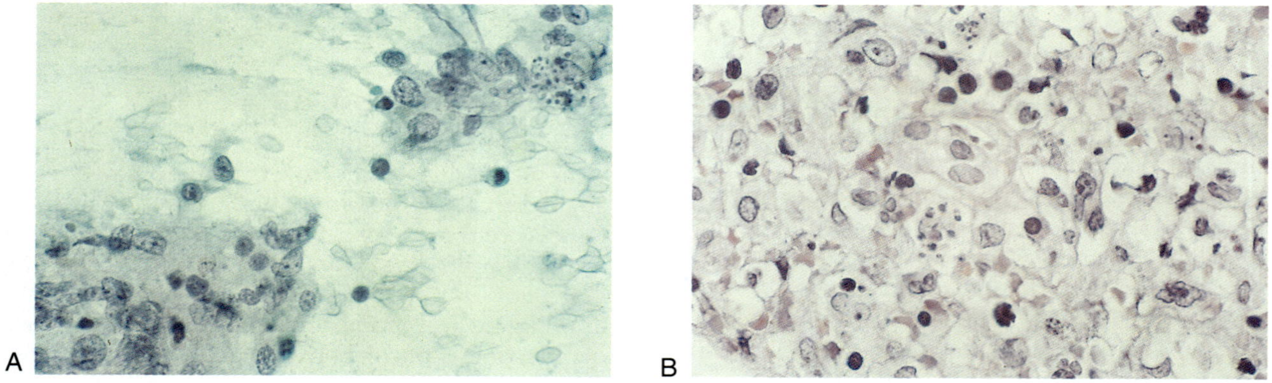

Fig. 5-4. (**A**) FNA of xanthogranulomatous pyelonephritis, consisting of small clusters of histiocytes, having a moderate amount of pale cytoplasm with eccentrically placed oval nuclei (Papanicolaou). (**B**) Cell block of xanthogranulomatous pyelonephritis, consisting of pale histiocytes admixed with a few inflammatory cells. The potential exists for a false-positive diagnosis of clear cell carcinoma of the kidney (H&E).

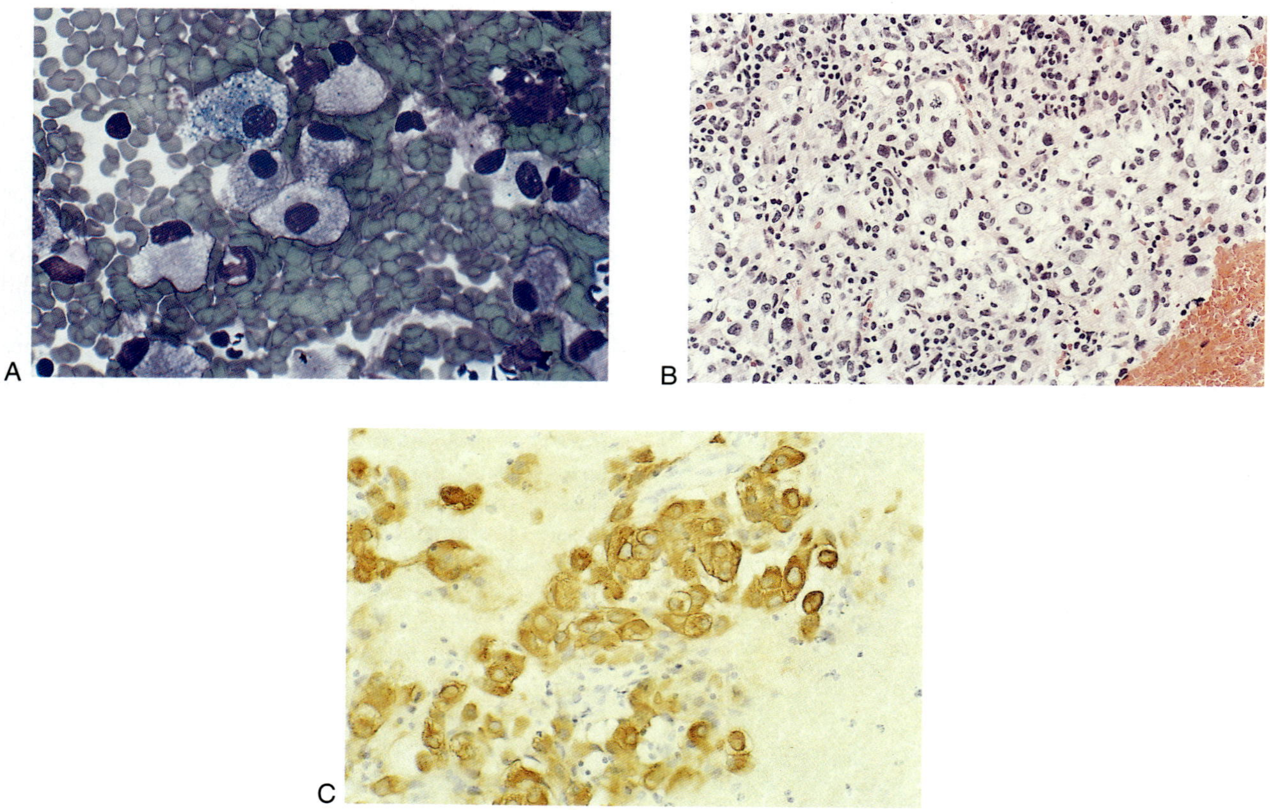

Fig. 5-5. (**A**) FNA of renal cell carcinoma, consisting of scattered clear cells with abundant cytoplasm. Because of the bland appearance of the nuclei, clear cytoplasm, and low nuclear to cytoplasmic ratios, the cells in this aspirate could be confused with histiocytes from xanthogranulomatous pyelonephritis (Diff-Quik). (**B**) Cell block from aspirate of clear cell carcinoma, demonstrating clear cell carcinoma admixed with inflammatory cells. The bland appearance of the cells could lead to confusion with xanthogranulomatous pyelonephritis (H&E). (**C**) Immunoperoxidase stains for cytokeratin decorates many of the clear cells, confirming the diagnosis of clear cell carcinoma (immunoperoxidase).

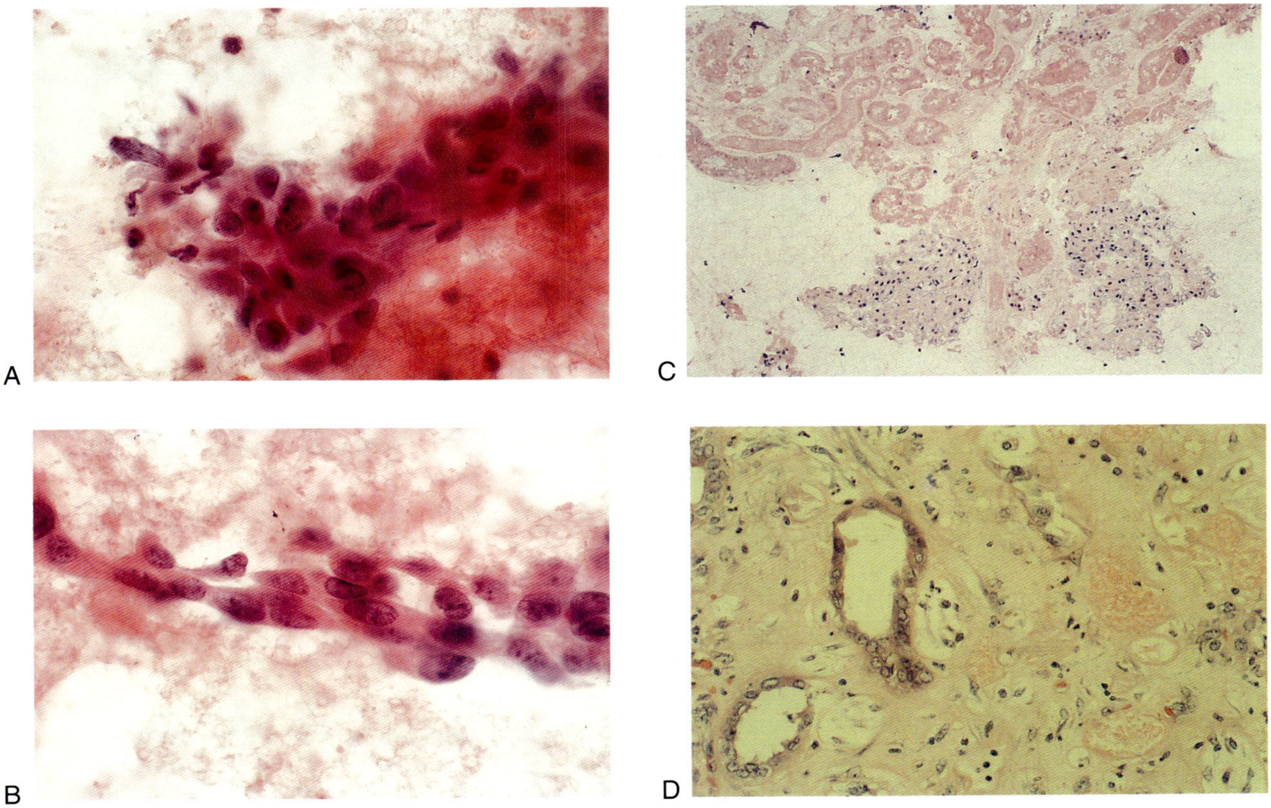

Fig. 5-6. (A) Aspirate of renal infarct, containing renal tubular cells arranged in a repair-like grouping (Papanicolaou). (B) Cluster of renal tubular cells, showing a repair reaction characterized by streaming pattern of the epithelial cells with no loss of polarity and well-defined cell borders. The potential exists for false-positive diagnosis of malignancy due to nuclear enlargement and the presence of nucleoli (Papanicolaou). (C) Cell block of renal infarct, showing necrotic renal tubules occupying the majority of field (H&E). (D) Histology of the resected kidney, showing renal tubules demonstrating a repair reaction along with necrotic renal tubular cells (H&E).

can be misinterpreted as malignant[29] (Fig. 5-6). A conservative diagnostic approach is warranted if the overall cellularity is limited. Correlation with clinical and radiologic findings is also critical to avoid a false-positive diagnosis of malignancy.

RENAL NEOPLASMS

Renal Cell Carcinoma

Renal cell carcinoma is a malignant neoplasm that is generally restricted to adults, with an average age of 55 to 60 years and a male-to-female predominance of 2:1. Renal cell carcinomas rarely occur in the pediatric population. A familial form of renal cell carcinoma has been described, as well as an increased prevalence in von Hippel-Lindau disease, acquired cystic disease, the adult form of polycystic renal disease, and multicystic nephroma. Patients with renal cell carcinoma may present clinically with hematuria, flank pain, abdominal masses, weight loss, anemia, and fever. Occasionally, metastases first call attention to the malignancy. One-third of renal cell carcinomas invade perinephric fat or involve regional lymph nodes, and another one-third of patients have distant metastasis at presentation. The most frequent metastatic sites include lung and bone. The primary treatment of renal cell carcinoma is surgical excision. The overall 5-year survival for renal cell carcinoma is 70 percent, with the prognosis primarily related to stage, presence or absence of distant metastasis, and microscopic grade. Overall, clear cell carcinomas behave less aggressively than do those predominanted by granular cell differentiation. Associations between the DNA ploidy and nuclear grade with prognosis have also been suggested.[29]

Renal cell carcinomas generally yield cellular aspiration smears, often with evidence of necrosis and hemorrhage.[9,10,25,30] In well-differentiated carcinoma, the tumor cells are arranged in small sheets and groups and/or individually (Fig. 5-7). The tumor cells have moderate to abundant pale cytoplasm, may be finely granular or vacuolated, and may be fragile, resulting in frayed cell borders.[31] Occasionally, well-vascularized groups of atypical clear cells are present, as are cells adhering to basement membrane type material (Fig. 5-8). In FNAs of low-grade renal cell carcinoma, the relatively bland nuclear appearance and cytoplasmic vacuoles may lead to a false-negative diagnosis of foam cells from renal cysts or histiocytes from xanthogranulomatous pyelonephritis (Fig. 5-9). FNA of the clear cell type of renal cell carcinoma can also be potentially confused with other clear cell tumors, such as adrenocortical neoplasms. In renal cell carcinoma, abundant glycogen and fat can be revealed by periodic acid-Schiff (PAS) and oil red O stains, respectively (Fig. 5-10). Renal cell carcinomas are positive for cytokeratin, EMA, and/or vimentin, in contrast to adrenal lesions, which are negative for the epithelial antibodies.[32] Although the renal carcinomas typically have cells with low nuclear to cytoplasmic ratios,

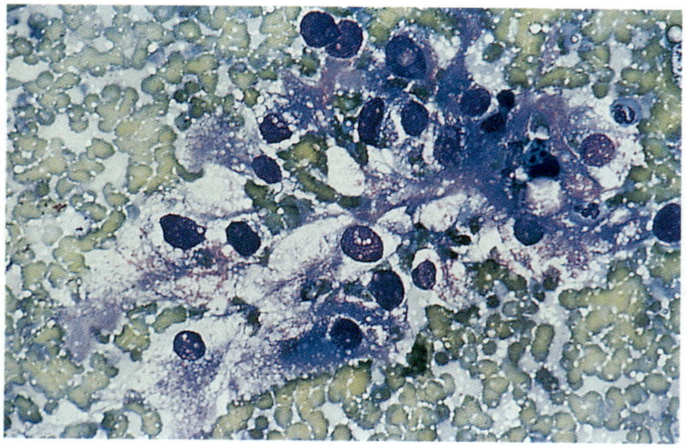

Fig. 5-7. FNA of renal cell carcinoma, consisting of monolayer of clear cells having round to oval nuclei and abundant clear cytoplasm with frayed cytoplasmic membranes. The cells in this tumor would be Fuhrman's nuclear grade 1 (Diff-Quik).

Features of Renal Cell Carcinoma

- Cellular smears with necrosis and hemorrhage
- Single cells, groups, papillary and/or sheets
- Vascularized tumor fragments
- Basement membrane material
- Moderate to abundant clear cytoplasm with frayed cytoplasmic membranes
- Cells with eosinophilic granular cytoplasm
- Nuclear to cytoplasmic ratios low to high
- Occasional intranuclear inclusions, round nuclei, anisonucleosis
- Inconspicuous to prominent nucleoli

Nuclear Grading of Renal Cell Carcinoma (Fuhrman's)

Grade 1—Small, round, uniform nuclei approximately 10 μm with inconspicuous or absent nucleoli

Grade 2—Larger nuclei (15 μm) with irregular outlines and nucleoli

Grade 3—Larger, with prominent nuclear irregularity and large nucleoli

Grade 4—Similar to grade 3, plus bizarre cells, often lobated nuclei, and spindle-shaped cells "sarcomatoid renal cell carcinoma"

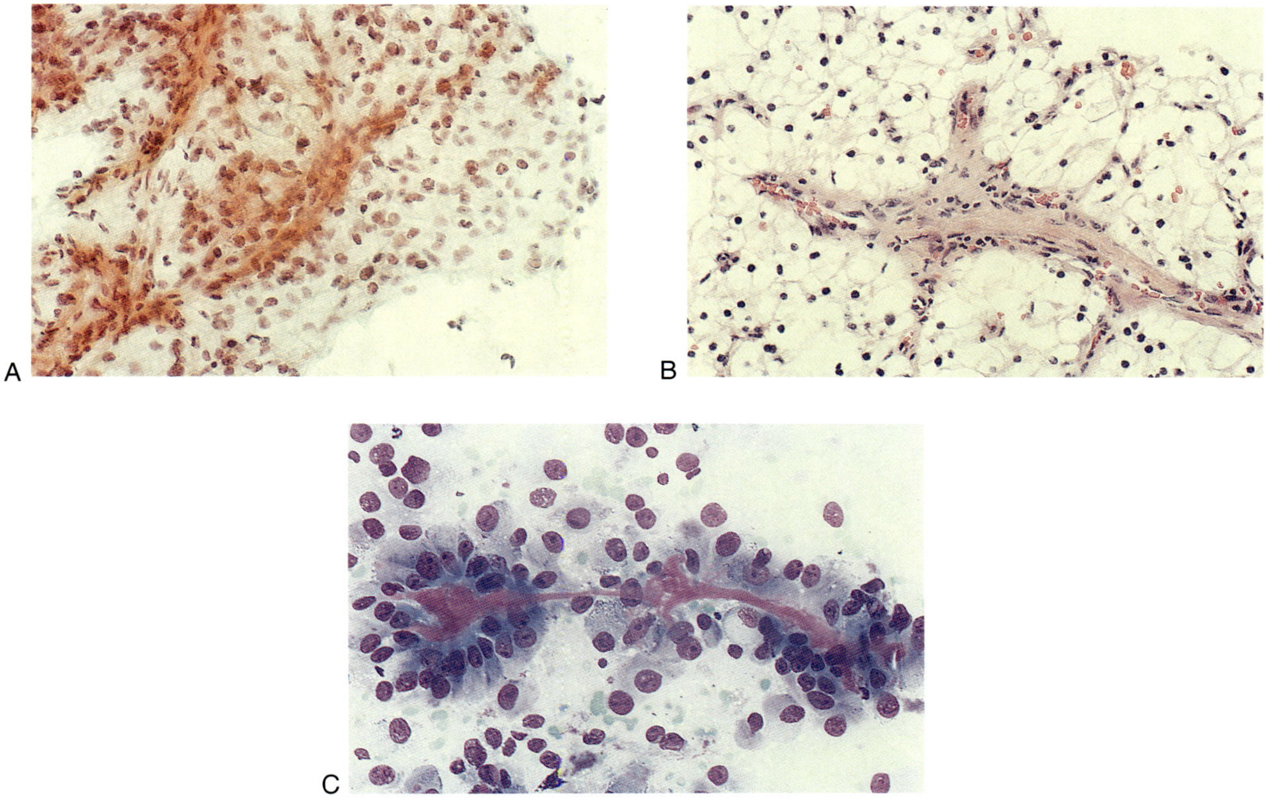

Fig. 5-8. (**A**) FNA of clear cell carcinoma of the kidney, showing sheets of malignant clear cells with interspersed branching capillary channels (Papanicolaou). (**B**) Histologic correlate of the previous cytologic finding, showing low-grade clear cell carcinoma, consisting of cells with small nuclei and abundant clear cytoplasm. Note the interspersed vascular channels (H&E). (**C**) Atypical clear cells from FNA of renal cell carcinoma, aligned along metachromatically staining basement membrane material (Diff-Quik).

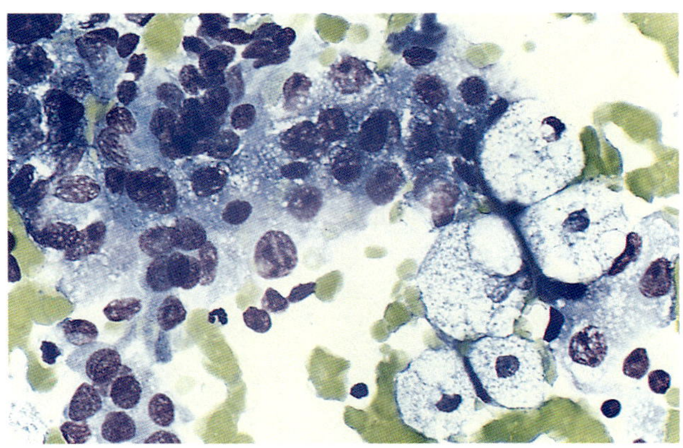

Fig. 5-9. FNA of renal cell carcinoma showing malignant clear cells, having increased nuclear to cytoplasmic ratios, with nuclei possessing small but prominent nucleoli. Note the nearby benign histiocytes, having smaller round to convoluted nuclei and abundant granular pale cytoplasm (Diff-Quik).

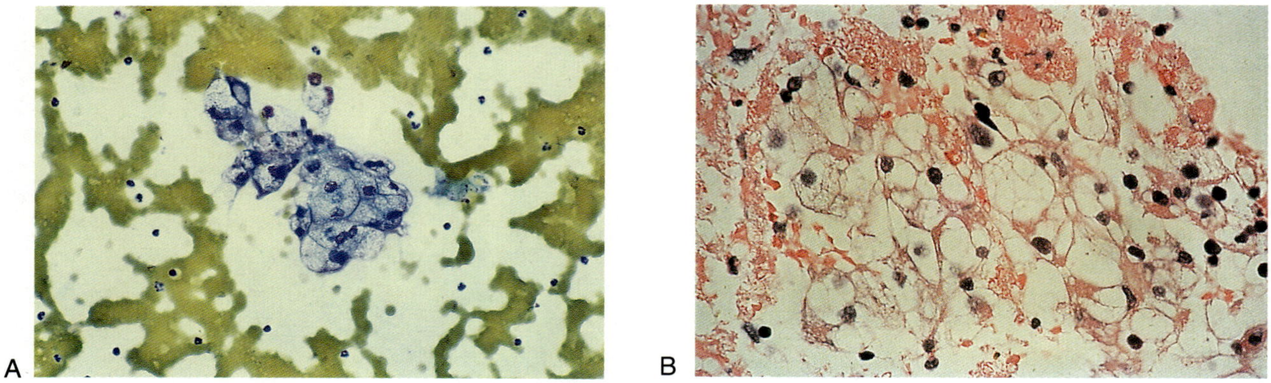

Fig. 5-10. (**A**) FNA of clear cell carcinoma of the kidney, demonstrating cells with moderate to abundant clear cytoplasm and well-defined cell membranes (Diff-Quik). (**B**) Histologic correlate from cell block of FNA of the kidney, demonstrating prominent clearing of the cytoplasm due to abundant fat and glycogen (H&E).

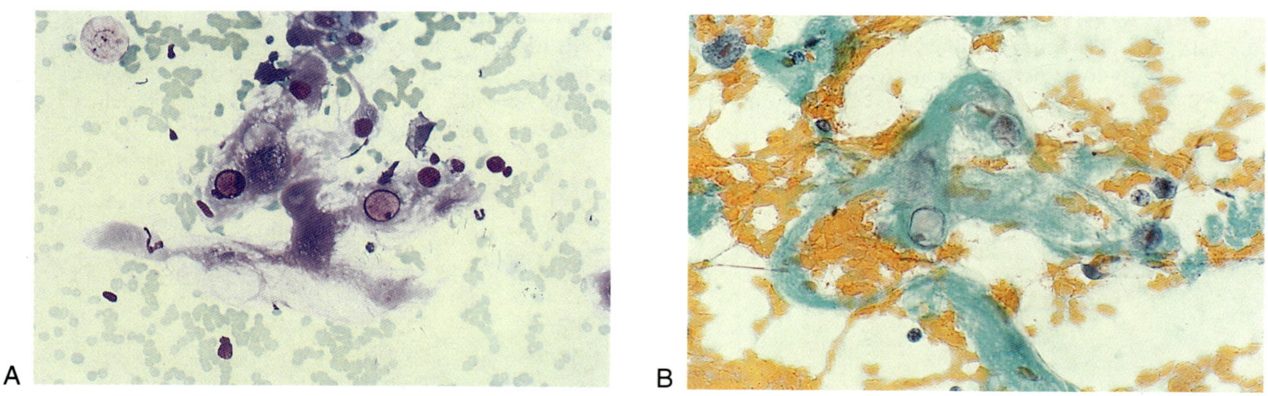

Fig. 5-11. (**A**) Clear cell carcinoma of the kidney showing abundant wispy clear cytoplasm and nuclei possessing intranuclear inclusions (Diff-Quik). (**B**) Prominent intranuclear inclusions and abundant wispy clear cytoplasm in FNA of clear cell carcinoma (Papanicolaou).

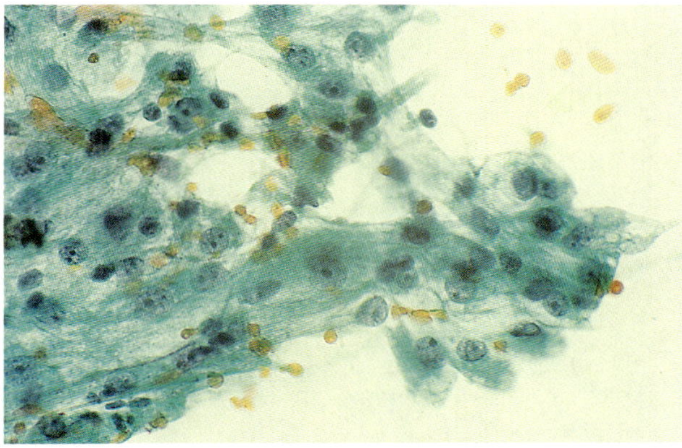

Fig. 5-12. FNA of moderately differentiated clear cell carcinoma of the kidney, showing high nuclear to cytoplasmic ratios and a greater degree of nuclear irregularity and prominence of nucleoli (Papanicolaou).

anisokaryosis can be seen. Ocassionally, intranuclear inclusions may also be present (Fig. 5-11). As histologic and nuclear grade increases, so does the nuclear to cytoplasmic ratio. Nuclear grading can be readily performed on the aspirate, using Fuhrman's nuclear grading system,[30] as outlined on page 176. Grade 1 cells have small, round, uniform nuclei (approximately 10 μm), with inconspicuous or absent nucleoli (Fig. 5-7). Grade 2 cells have larger nuclei (15 μm), with some irregularities in contour with nucleoli (Fig. 5-12). Grade 3 consists of cells with even larger nuclei, more prominent nuclear irregularities, and more prominent nucleoli (Fig. 5-13). Grade 4 cells are similar to grade 3 but also have bizarre, often lobated, nuclei with large chromatin clumps associated with sarcomatoid spindle-shaped cells. A recent study reported nuclear grading of FNA biopsies of renal cell carcinoma and its good concordance with histology and flow cytometric data.[30] The FNA cytologic findings of sarcomatoid renal cell carcinoma have also recently been described.[33] Poorly differentiated renal cell carcinomas tend to show higher nuclear grade, increased nuclear to cytoplasmic ratio, greater degree of cellular discohesion, and bizarre shapes of the malignant cells (Fig. 5-14). Entities considered in the cytologic differential diagnosis of renal cell carcinoma are presented below.

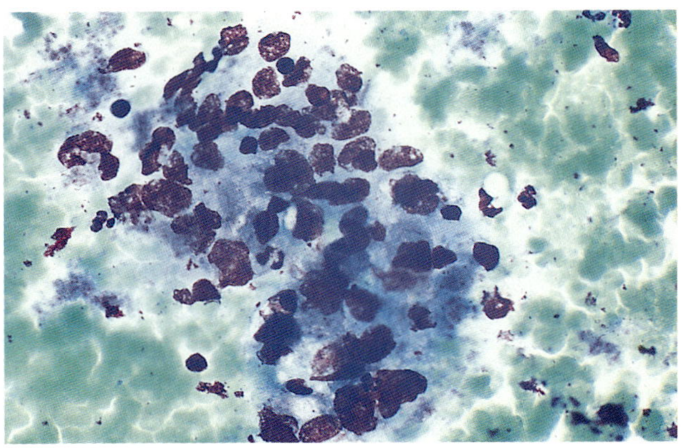

Fig. 5-13. FNA of poorly differentiated clear cell carcinoma of the kidney, showing high nuclear to cytoplasmic ratios with irregular nuclei and small but prominent nucleoli (Diff-Quik).

Differential Diagnosis of Renal Cell Carcinoma

- Adrenal cortical cells vs. clear cells of RCC
- Normal renal tubular cells vs. granular cells of oncocytoma, RCC
- Hepatocytes vs. granular cells of RCC
- Xanthogranulomatous pyelonephritis—histiocytes vs. clear cells of RCC
- Angiomyolipoma—atypical endothelial and/or smooth muscle cells, fat vs. sarcoma, sarcomatoid RCC
- Repair reaction vs. RCC
- Renal oncocytoma vs. granular type of RCC
- Sarcoma, squamous cell cancer versus sarcomatoid RCC
- Transitional cell carcinoma vs. papillary variant of RCC
- Metastatic carcinoma to kidney vs. primary carcinoma

Abbreviation: RCC, renal cell carcinoma.

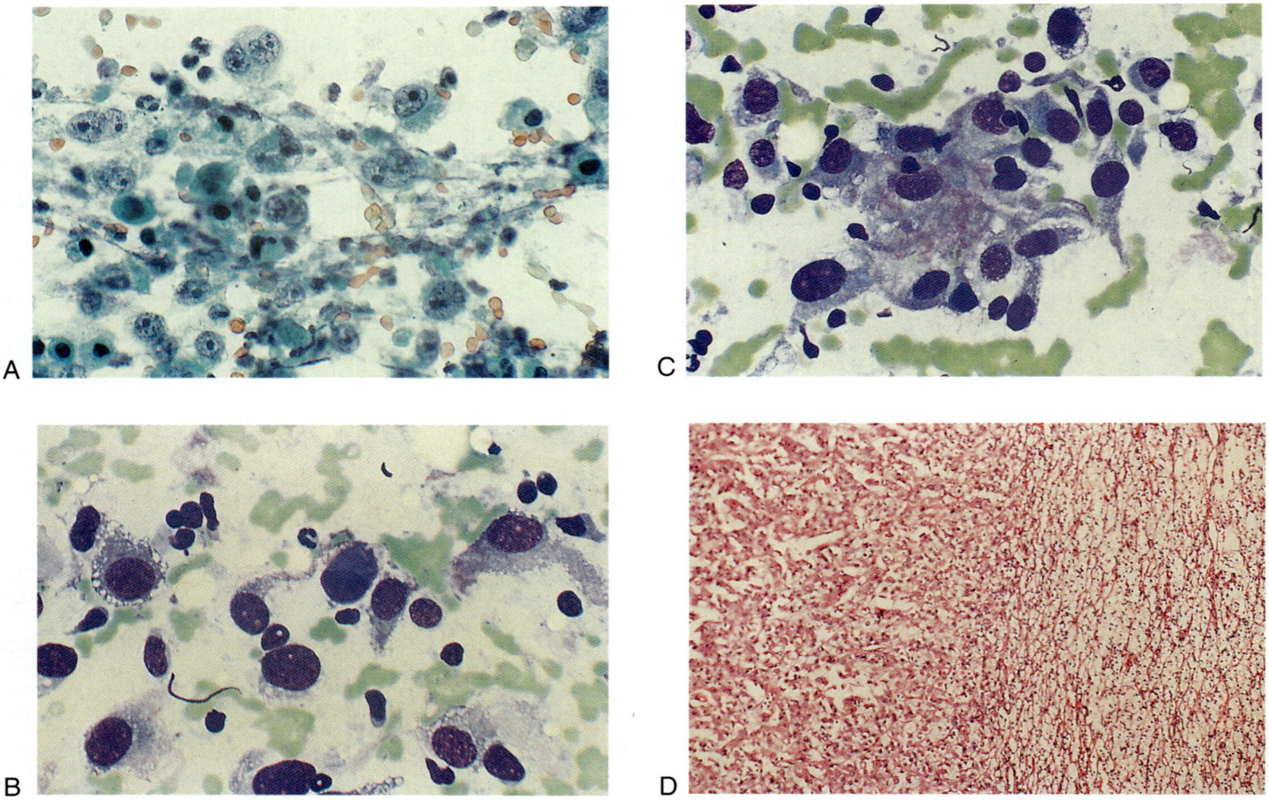

Fig. 5-14. (**A**) FNA of sarcomatoid renal cell carcinoma, showing dissociative pattern of poorly differentiated malignant cells which vary from polygonal to spindle. A few cells show multinucleation (Papanicolaou). (**B**) Scattered malignant cells in FNA of sarcomatoid renal cell carcinoma. Note the polygonal to spindled configurations of the cells and high nuclear to cytoplasmic ratios. A few cells show some clearing of the cytoplasm and vacuolization (Diff-Quik). (**C**) Rare cluster in FNA of sarcomatoid renal cell carcinoma, showing aggregate of cells demonstrating a moderate amount of clear cytoplasm. Note nearby loosely cohesive malignant cells, having a spindled configuration (Diff-Quik). (**D**) Histology of the resected sarcomatoid renal cell carcinoma, demonstrating more conventional clear cell type of renal cell carcinoma in a portion of the field, while the majority of the specimen consists of the high-grade sarcomatoid renal cell carcinoma (H&E).

Renal Oncocytoma

Renal oncocytoma is believed by some to be a very well-differentiated (grade 1) renal cell carcinoma associated with a significantly improved prognosis, as they rarely metastasize.[34] Oncocytic tumors of higher nuclear grade are associated with a greater risk of metastasis and thus need to be classified as renal cell carcinomas of granular (or oncocytic) type. FNAs of renal oncocytoma consist of uniform cells with abundant granular eosinophilic cytoplasm and small bland nuclei with low nuclear to cytoplasmic ratios[35,36] (Fig. 5-15). Generally, a uniform pattern is seen, and necrosis is not evident. However, caution is advised in making a specific diagnosis of renal

Features of Renal Oncocytoma

- Uniform cells with abundant eosinophilic granular cytoplasm
- Small bland nuclei
- Low nuclear to cytoplasmic ratios
- No necrosis
- No prominent nucleoli
- No clear cell change

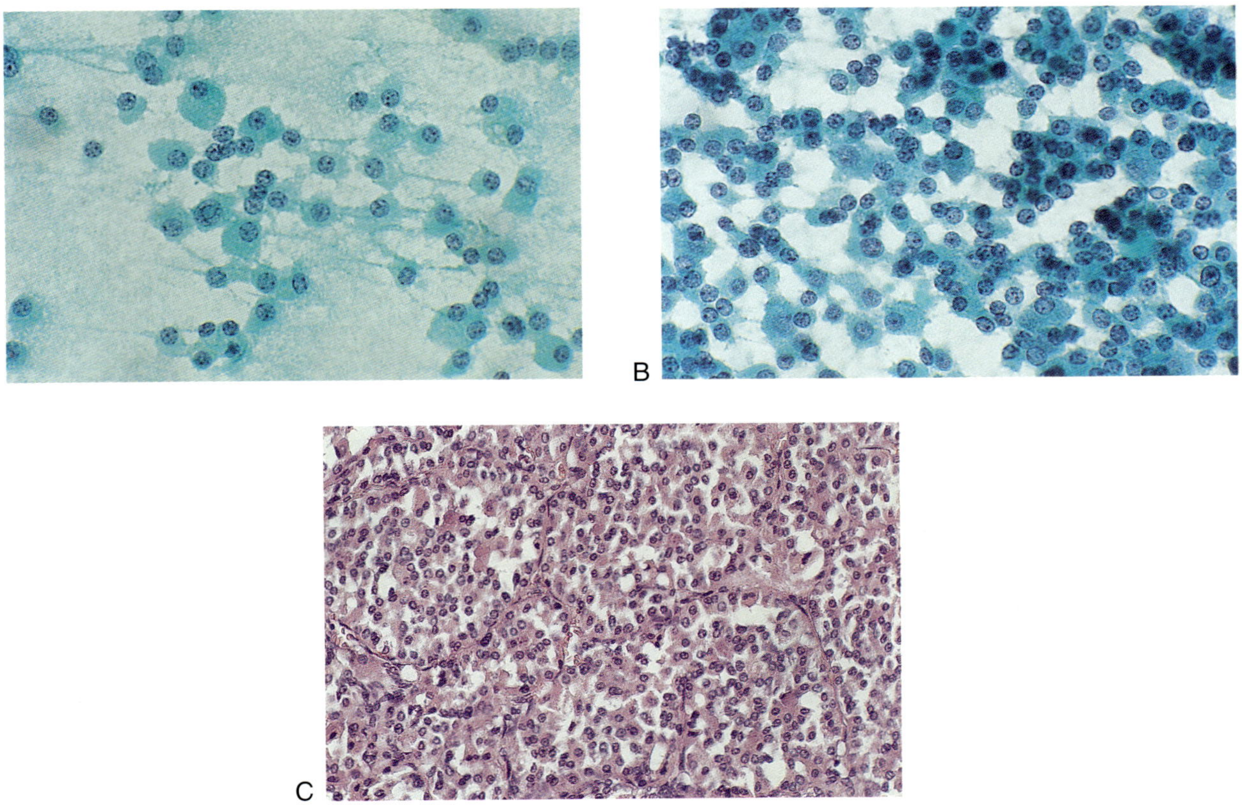

A B C

Fig. 5-15. (**A**) FNA of oncocytoma of the kidney, showing uniform cells with a moderate to abundant amount of granular eosinophilic cytoplasm and small bland nuclei with low nuclear to cytoplasmic ratios (Papanicolaou). (**B**) FNA of oncocytoma demonstrating a cellular smear pattern, consisting of loosely cohesive to individually scattered cells, having a moderate amount of granular cytoplasm. Note the uniform appearance of the nuclei and lack of necrosis in the background (Papanicolaou). (**C**) Resected oncocytoma of kidney, showing uniform population of oncocytic cells with no evidence of nuclear pleomorphism, papillary formation, clear cells, or necrosis (H&E).

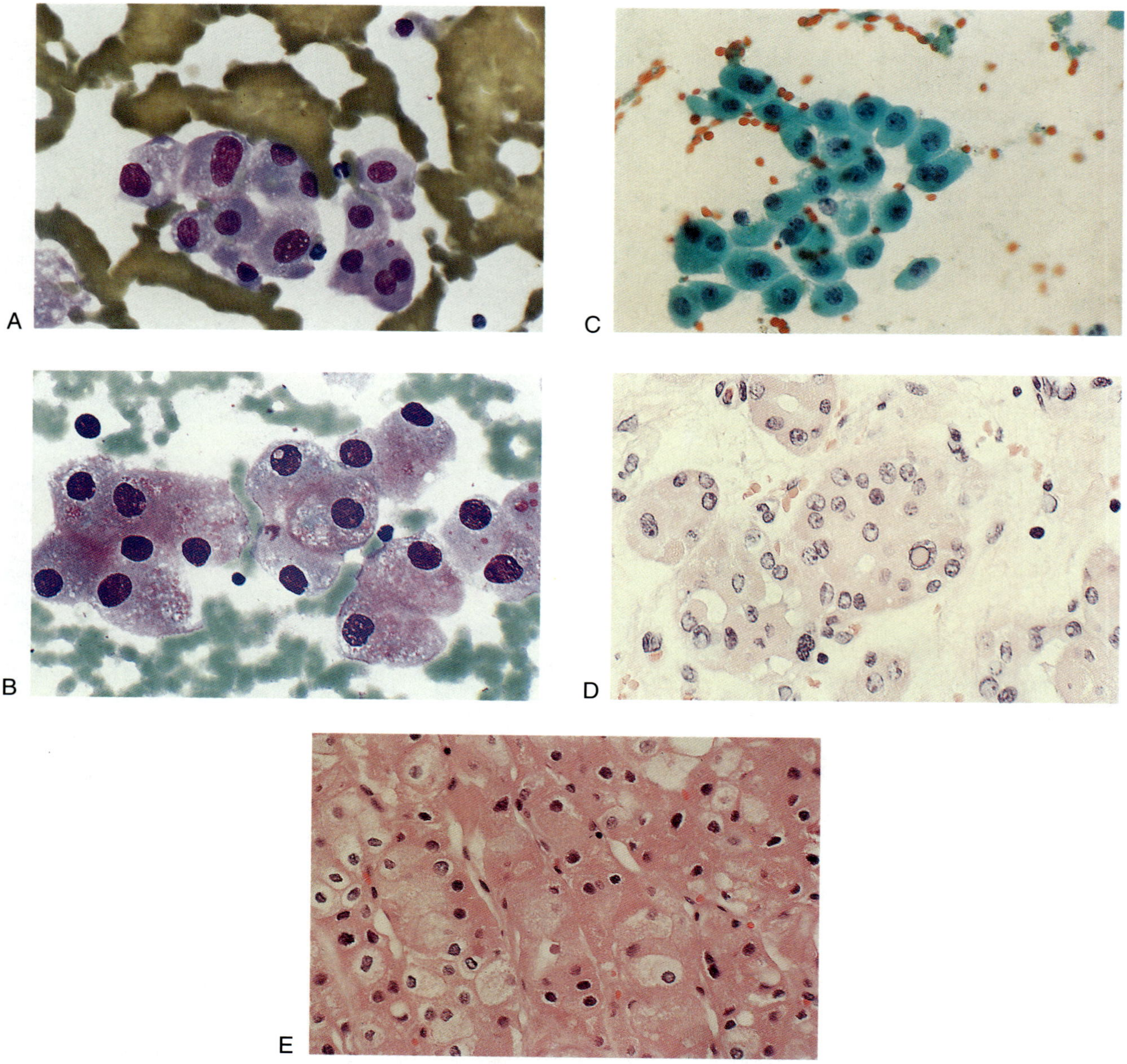

Fig. 5-16. (**A**) FNA of granular cell type of renal cell carcinoma that could potentially be confused with an oncocytoma. Note anisonucleosis, binucleation, and some slight nuclear irregularity (Diff-Quik). (**B**) FNA of granular cell type of renal cell carcinoma that could be confused, cytologically, with an oncocytoma. Only minimal variation in nuclear size is seen, with tumor cells having low nuclear to cytoplasmic ratios. The diagnosis should be deferred until examination of the resected specimen (Diff-Quik). (**C**) FNA of granular cell type of renal cell carcinoma, demonstrating increased nuclear to cytoplasmic ratios, granular cytoplasm, and small but prominent nucleoli (Papanicolaou). (**D**) Histologic examination of resected granular cell type of renal cell carcinoma, showing variation in nuclear size, binucleation, prominent nucleoli, and intranuclear inclusion. Note the abundant eosinophilic granular cytoplasm (H&E). (**E**) Other fields in the resected granular cell type of renal cell carcinoma show more bland appearance of the tumor cells that could potentially be confused with an oncocytoma (H&E).

oncocytoma in FNA biopsy, as some renal cell carcinomas can contain areas histologically indistinguishable from an oncocytoma. We recommended that, in this specific setting, a diagnosis of renal granular cell neoplasm be rendered, with the differential diagnosis of renal oncocytoma versus renal cell carcinoma, granular cell type. Final classification should await examination of the resected specimen. However, the granular type of renal cell carcinoma should be suspected when a greater degree of nuclear pleomorphism occurs and/or associated papillary formation, clear cells, or necrosis is present (Fig. 5-16). Renal oncocytomas also stain for low-molecular-weight cytokeratin but are negative for vimentin. Ultrastructurally, the cells are densely packed with mitochondria.

Papillary Renal Cell Carcinoma

Papillary renal cell carcinoma is a distinct clinicopathologic entity.[37,38] One diagnostic classification requires a minimum of 75 percent papillary architecture in sections to qualify a renal neoplasm as papillary.[39] Histologically, papillary renal cell carcinoma is characterized by fibrovascular stalks lined by neoplastic cells. Necrosis, cyst formation, and numerous macrophages are

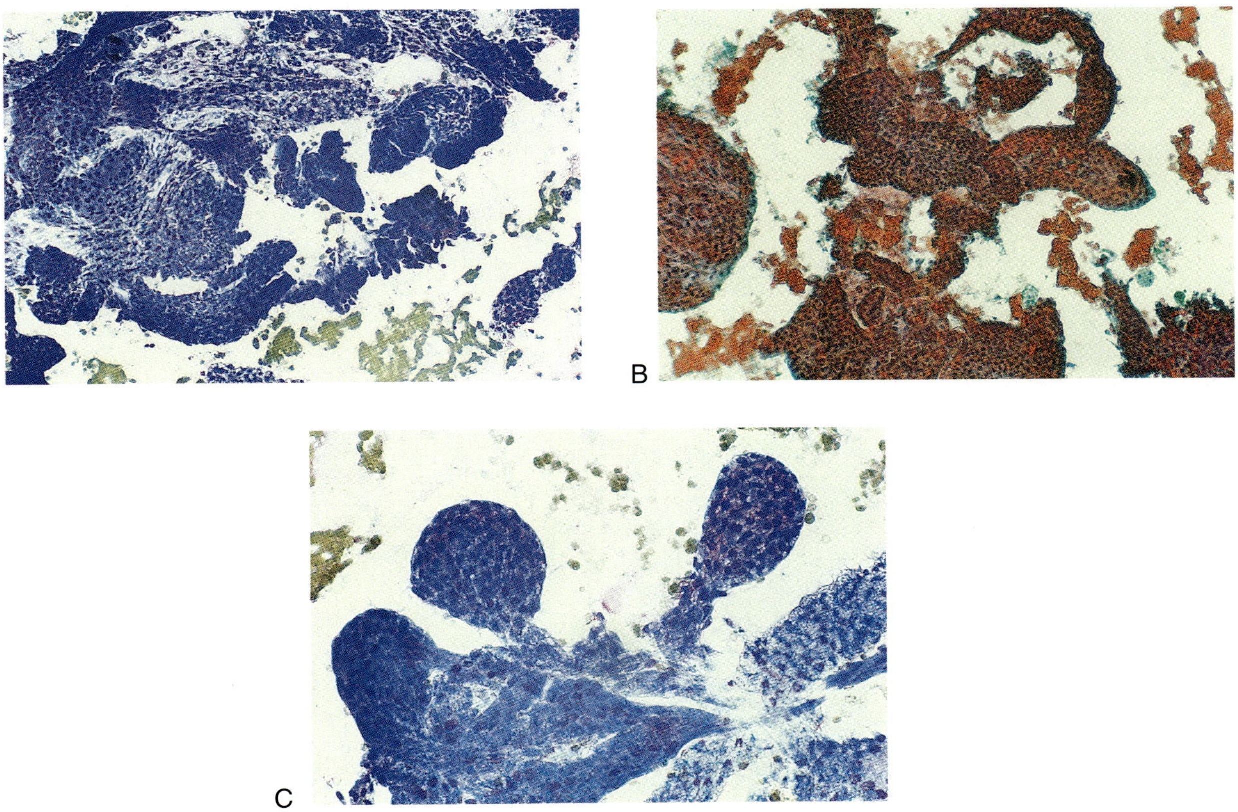

A B C

Fig. 5-17. (**A**) FNA of papillary renal cell carcinoma, demonstrating folded sheets of neoplastic cells, along with other groups of cells arranged in a papillary pattern (Diff-Quik). (**B**) Papanicolaou stain of papillary renal cell carcinoma, demonstrating three-dimensional groups of cells, along with other cells having finger-like papillary projections (Papanicolaou). (**C**) High power of papillary renal cell carcinoma, showing finger-like branching of papillary groups in which there is some depth of focus (Diff-Quik).

often present. Recently, Dekmezian et al.[37] reported FNA cytologic findings of 15 cases. The smears contained numerous papillary clusters with fibrovascular cores (Fig. 5-17), as described below. Relatively few isolated malignant cells were present. The tumors cells had moderate to scant volumes of cytoplasm, with uniform small nuclei with single small nucleoli and occasional nuclear grooves. In some specimens, histiocytes and psammoma bodies were present. Hemosiderin pigment was occasionally noted in some neoplastic cells (Fig. 5-18).

Features of Papillary Renal Cell Carcinoma

- Papillary clusters with fibrovascular cores
- Very few single cells
- Moderate to scant amount of cytoplasm
- Uniform small nuclei with single small nucleoli and occasional nuclear grooves
- Occasional histiocytes and/or psammoma bodies
- Occasional cytoplasmic hemosiderin deposition

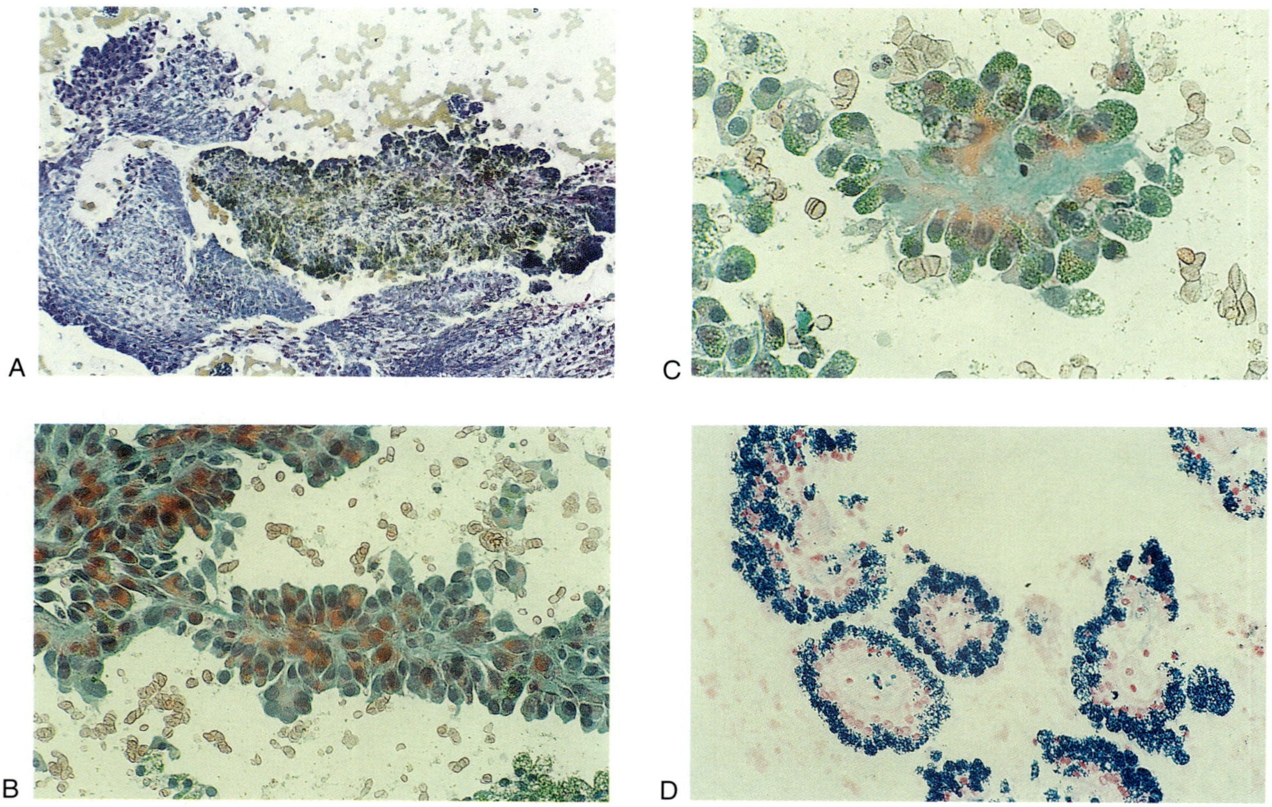

Fig. 5-18. (**A**) Papillary renal cell carcinoma, consisting of vascularized clusters lined by cells having abundant hemosiderin pigment, imparting a refractile yellowish brown appearance to the cytoplasm (Diff-Quik). (**B**) Neoplastic cells aligned along thin vascularized cores. Note intracytoplasmic refractile pigment, corresponding to hemosiderin deposition (Papanicolaou). (**C**) High power of papillary renal cell carcinoma showing neoplastic cells, having abundant golden refractile pigment corresponding to intracytoplasmic iron. Note the alignment of the cells along fibrovascular cores (Papanicolaou). (**D**) Positive staining for hemosiderin in many of the tumor cells (Prussian blue, iron stain).

Transitional Cell Carcinoma of Renal Pelvis and Ureter

Although transitional cell carcinomas of the renal pelvis occur mostly in adults, children are rarely involved. Known associations include analgesic abuse with renal papillary necrosis, administration of Thorotrast, or cyclophosphamide therapy. Most transitional cell carcinomas are moderate to poorly differentiated, with large cells arranged individually and in clusters (Fig. 5-19). Papillary structures are only rarely seen. Nuclei are generally hyperchromatic with prominent nucleoli. Cytoplasm can be dense, with squamous differentiation a not infrequent finding (Fig. 5-20). Occasional cases have a sarcomatoid appearance, with prominent spindling of the malignant cells. A recently described renal neoplasm, collecting duct carcinoma, has features of invasive transitional cell and papillary carcinomas plus an in situ arrangement in the adjacent medullary collecting ducts.[39–47]

Features of Transitional Cell Carcinoma of Renal Pelvis

- Large; pleomorphic cells—single, clusters and/or papillary
- Pleomorphic nuclei with prominent nucleoli
- Irregular shape, dense cytoplasm, eccentric nuclei

Angiomyolipoma

Angiomyolipoma is a rare benign predominantly renal neoplasm that also occurs uncommonly in organs and sites other than the kidney. One-third of patients with renal angiomyolipoma have the characteristic neurologic and cutaneous features of tuberous sclerosis. The neoplasm consists of an admixture of fat, blood vessels, and smooth muscle (Fig. 5-21). Accordingly, FNA will demonstrate vascularized fragments in which adipose tissue and smooth muscle may be seen.[48,49] Potential pitfalls for a misdiagnosis of carcinoma or sarcoma result from misinterpreting the adipocytes as the clear cells of renal cell carcinoma or lipoblasts of liposarcoma, respectively.[50] Occasionally, the smooth muscle component has atypical, even bizarre-appearing, nuclei that may be misinterpreted, resulting in a diagnosis of leiomyosarcoma (Fig. 5-22). A correct diagnosis should be made when a renal mass is aspirated that contains an admixture of large to small blood vessels, mature adipose tissue, and smooth muscle cells (Fig. 5-23). The smooth muscle cells can be highlighted with immunoperoxidase stains for smooth muscle actin (HHF) and HMB-45.[50,51]

Features of Angiomyolipoma

- Fat, blood vessels, and smooth muscle (spindle) cells
- Occasional nuclear atypia of smooth muscle cells
- Smooth muscle cells positive for actin (HHF) and HMB-45

Wilms' Tumor and Other Primary Pediatric Neoplasms

Wilms' tumor (nephroblastoma) is the most frequent and important renal tumor of childhood.[52] Classically, Wilms' tumor is defined as a mixed (triphasic) renal neoplasm composed of metanephric blastema, stromal cells (including skeletal muscle differentiation), and epithelial derivatives (including tubules and glomeruloid structures). Monophasic variants consisting exclusively or predominantly of blastemal or tubular elements are also considered Wilms' tumor. The incidence of Wilms' tumor is approximately 7.6 cases per million children per year. A small proportion of Wilms' tumors are associated with congenital anomalies, such as hemihypertrophy, aniridia, and genitourinary anomalies. Other primary renal tumors in children include cystic partially differentiated nephroblastoma, congenital (conventional) mesoblastic nephroma, atypical mesoblastic nephroma, malignant rhabdoid tumor of kidney, and clear cell sarcoma of kidney.

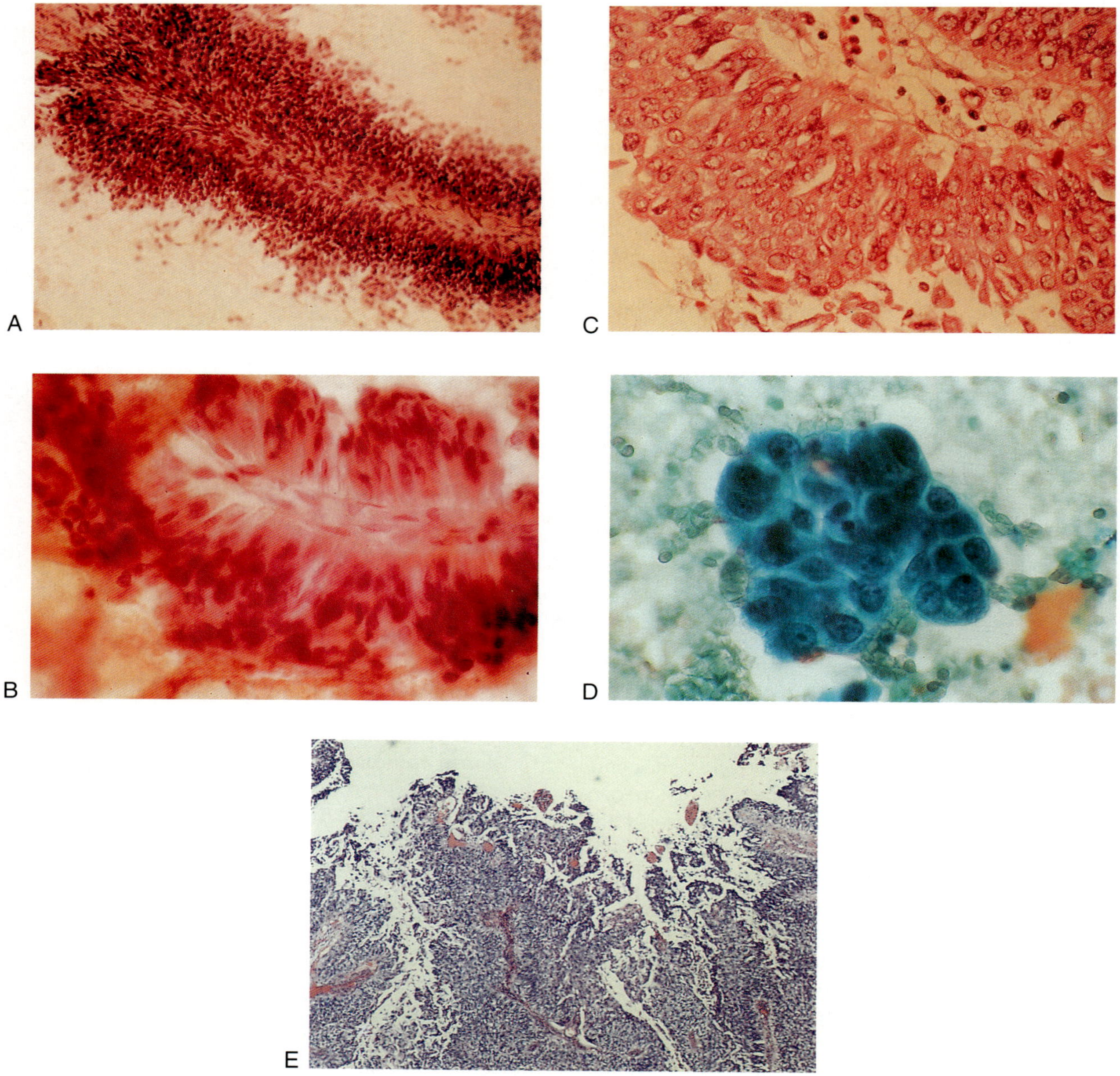

Fig. 5-19. (**A**) FNA of well-differentiated papillary transitional cell carcinoma, demonstrating transitional cells aligned along fibrovascular cores. Note the columnar shape of the cells with nuclear pseudostratification (Papanicolaou). (**B**) High power of transitional cell carcinoma of renal pelvis showing pseudostratification of the nuclei and cells aligned along delicate fibrovascular cores (Papanicolaou). (**C**) Histology of the resected tumor following FNA, confirming the presence of transitional cell carcinoma of the renal pelvis (H&E). (**D**) FNA of poorly differentiated transitional cell carcinoma, demonstrating somewhat three-dimensional group of poorly differentiated malignant cells, having well-defined cell borders and increased nuclear to cytoplasmic ratios (Papanicolaou). (**E**) Histologic examination, showing poorly differentiated transitional cell carcinoma (H&E).

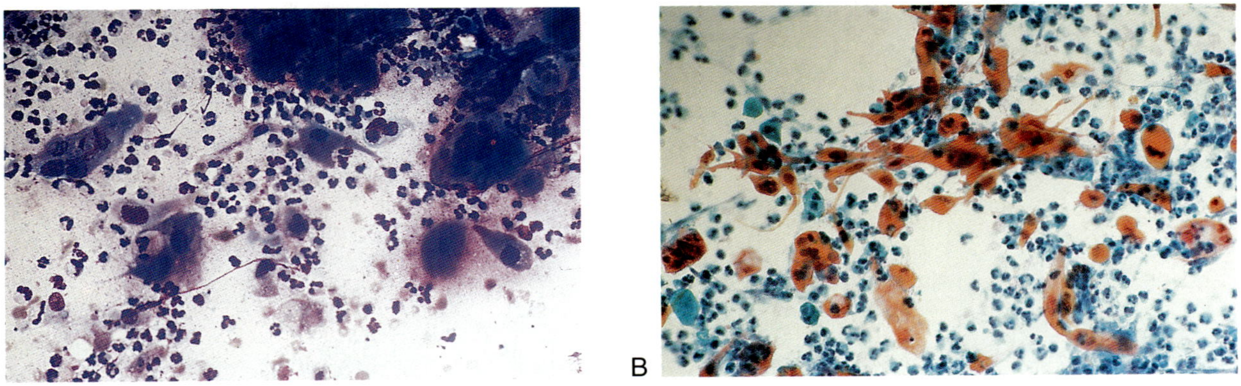

A B

Fig. 5-20. (A) FNA of transitional cell carcinoma of renal pelvis, showing squamous differentiation. Note the bizarre shapes and dense cytoplasm of the malignant cells. A scattering of neutrophils are in the background. (Diff-Quik). (B) Poorly differentiated transitional cell carcinoma of the bladder, demonstrating squamous differentiation with evidence of intracytoplasmic keratinization (Papanicolaou).

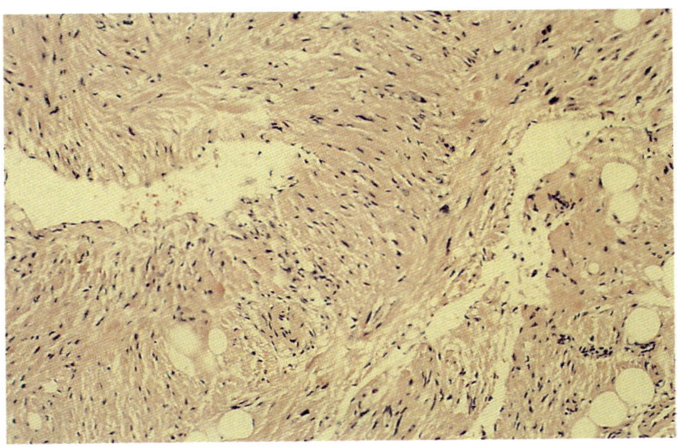

Fig. 5-21. Resected angiomyolipoma of the kidney following FNA biopsy. Note the triphasic cell population, consisting of admixed fat, blood vessels, and smooth muscle cells (H&E).

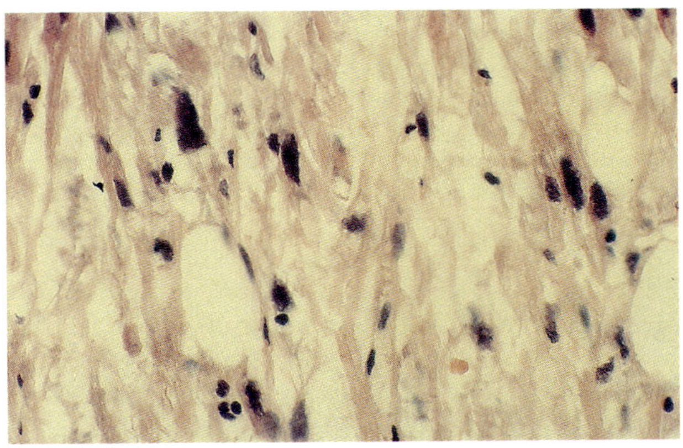

Fig. 5-22. Resected angiomyolipoma, revealing atypical smooth muscle nuclei that could potentially be confused with a sarcoma (H&E).

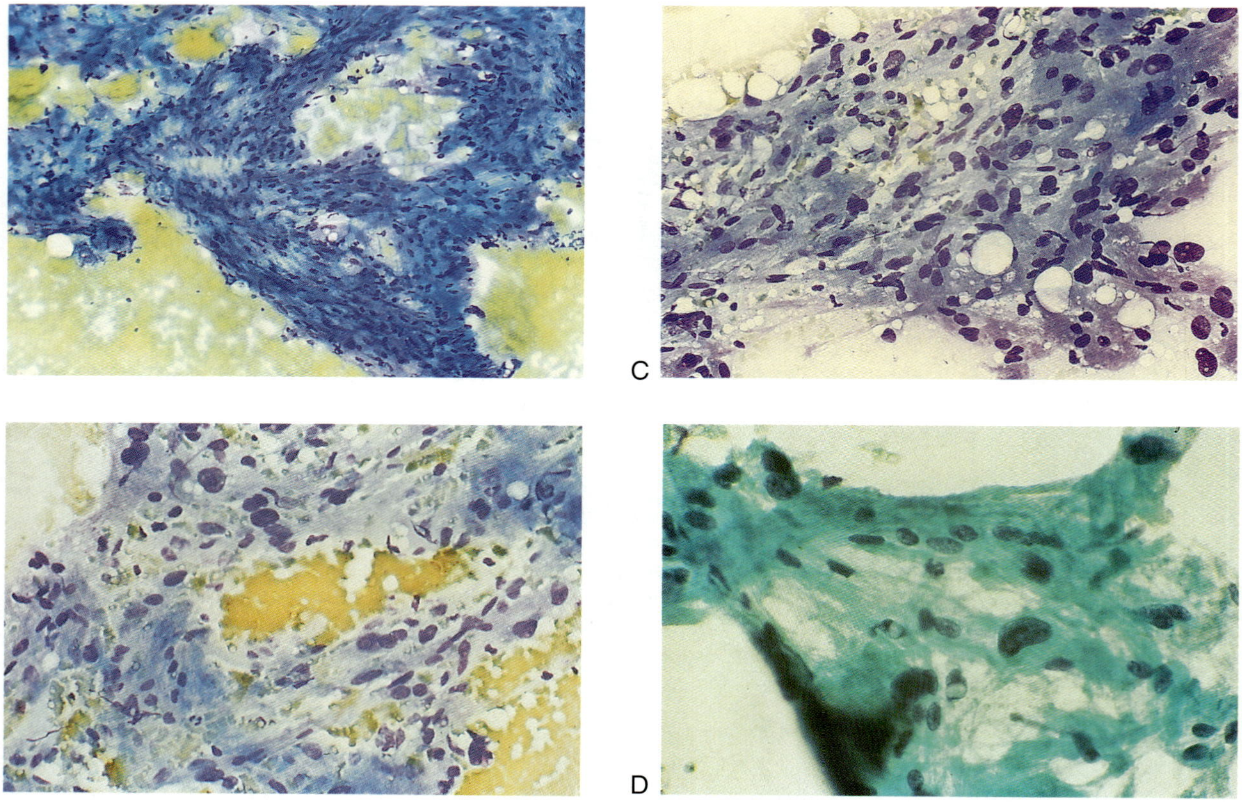

A C B D

Fig. 5-23. (**A**) Low power of FNA of angiomyolipoma, demonstrating branching groups of smooth muscle cells aligned along vascular channels (Diff-Quik). (**B**) Atypical smooth muscle cells in FNA of angiomyolipoma. Note the anisonucleosis and slight nuclear irregularity that potentially could be a source of a misdiagnosis of malignancy (Diff-Quik). (**C**) FNA of angiomyolipoma, consisting of smooth muscle cells, including some showing some slight nuclear atypicality along with admixed fat (Diff-Quik). (**D**) FNA of angiomyolipoma, demonstrating a few scattered cells with nuclear atypicality characterized by nuclear enlargement and irregularity and small nucleoli (Papanicolaou).

FNA cytologic findings of Wilms' tumor almost always include small blastemal cells characterized by solitary round hyperchromatic nuclei surrounded by only a thin rim of cytoplasm.[26,53–55] Within cohesive aggregates, nuclear molding may be evident (Fig. 5-24). The mesenchymal stromal component of Wilms' tumor consists of fragments of fibromyxoid tissue, with interspersed spindle-shaped cells having a generally bland nuclear appearance (Fig. 5-25). The epithelial component consists of larger cells. Their cytoplasm is more abundant with round to polygonal contours. Nuclei may be slightly larger than those in the blastemal elements. This component may be

Features of Wilms' Tumor

- Cellular smears with blastemal, tubules, and mesenchymal cells
- Blastemal cells—small, high nuclear to cytoplasmic ratios, occasional nuclear molding
- Tubules—simple to complex branching
- Mesenchyme—bland spindle cells—background matrix
- Anaplasia—nuclear enlargement (3× smaller blastemal cells), nuclear hyperchromasia, multipolar mitotic figures

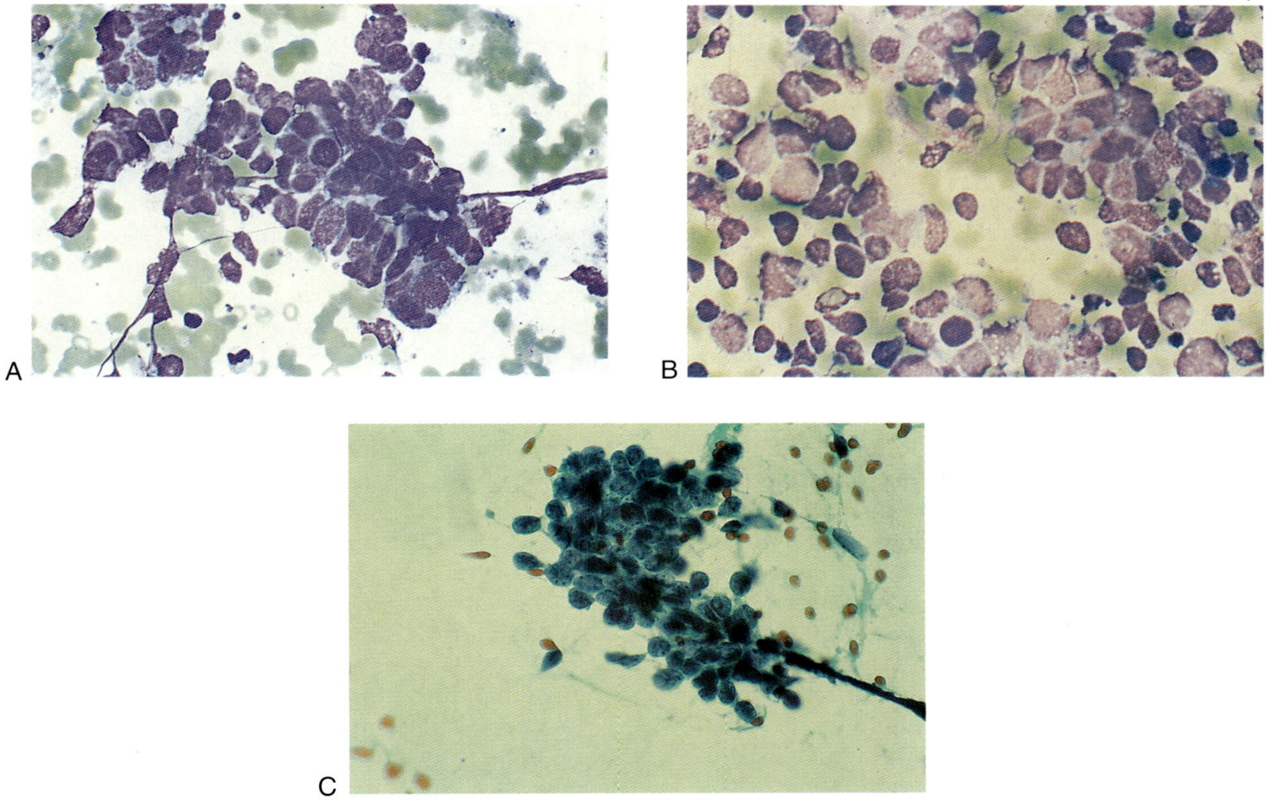

Fig. 5-24. (**A**) FNA of Wilms' tumor, showing blastemal cells arranged in tight clusters in which there is nuclear molding. Some individually scattered tumor cells are present. Note the high nuclear to cytoplasmic ratios of the blastemal cells (Diff-Quik). (**B**) FNA of Wilms' tumor, showing blastemal cells arranged in clusters and individually. Note nuclear molding, which could potentially be confused with an aspirate from neuroblastoma (Diff-Quik). (**C**) FNA of Wilms' tumor, consisting of loose cluster of blastemal cells. Some nuclear molding is noted, while other cells have a more spindled configuration (Papanicolaou).

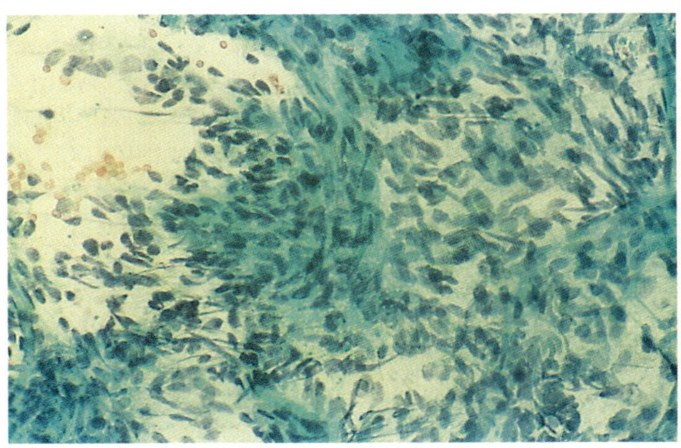

Fig. 5-25. FNA of Wilms' tumor, showing spindle cell component of the tumor. Note the bland appearance of the cells (Papanicolaou).

arranged in simple to complex branching configurations (Fig. 5-26). The differential diagnosis of these epithelial tubules includes Homer Wright rosettes, which may be present in smears of neuroblastoma. However, Homer Wright rosettes are less complex, with the cells tightly arranged around the fibrillary neuropil, rather than a true lumen. In some aspirates of blastemal-type Wilms' tumor, the primitive cells can be arranged around branching blood vessels simulating complex tubules. This pattern can also been seen in other small round cell tumors of childhood. In addition, nuclear molding by this primitive blastemal cells may be confused with similar changes in neuroblastoma. Electron microscopic examination of Wilms' tumor is primarily helpful in excluding other small round cell tumors of childhood, such as neuroblastoma and rhabdomyosarcoma, by demonstrating a lack of neurosecretory granules and muscle fibrils, respectively.[56,57] Rather, ultrastructural features of Wilms' tumor include well-formed intercellular junctions, oligocilia, microvilli, flocculent basement membrane-like material, and phagolysosomes (Fig. 5-27). Immunocytochemical staining is also used to exclude other small round cell tumors (e.g., a negative LCA reaction excludes lymphoma). Wilms' tumor will be positive for vimentin and low-molecular-weight cytokeratin and occasionally, some of the cells show focal NSE positivity. Overall, epithelial differentiation, as confirmed by either immunocytochemistry or electron microscopy, is most helpful, as it substantiates Wilms' tumor and excludes other small round cell tumors.

Anaplasia in Wilms' tumor is defined histologically as the combination of nuclear enlargement equal to at least three times the typical-size nuclei of adjacent cells, nuclear hyperchromasia, and multipolar mitotic figures.[58] Anaplasia in stage II to IV Wilms' tumor is associated with aggressive behavior and decreased survival. Recognition of anaplastic Wilms' tumor is important, as these neoplasms need to be treated more intensely. As expected, FNA cytology of anaplastic Wilms' tumor will demonstrate enlargement of the neoplastic nuclei, so that they are at least three times the diameter of adjacent nuclei of the same cell type, along with atypical mitotic figures and hyperchromasia[58] (Fig. 5-28). Potential pitfalls for overdiagnosing anaplasia include confusing multinucleation with macronucleation, especially when overlapping nuclei are present. A number of artifacts, more often applicable to tissue sections than to cytology smears, may also contribute to misidentifying anaplasia, such as thick sections, poor fixation, or staining of histologic sections, smeared DNA strands, calcification, stain precipitate, and basophilic extracellular mucinous material imitating large nuclei. There is also a potential to underdiagnose anaplasia in an FNA biopsy due to inadequate sampling. However, in the study by Zuppan et al.[57] of anaplastic Wilms' tumors, nearly one-half the tumors had evidence of anaplasia in at least 75 percent of tumor sections; therefore, they concluded that anaplasia will often be detected in a limited biopsy. The presence of anaplasia in Wilms' tumor is associated with a significantly worse prognosis when present in extrarenal tumor sites or predominantly in blastemal-type neoplasms.

One of the current contraindications of pediatric FNA biopsy in the United States is the primary diagnosis of resectable Wilms' tumor, based on the fear of tumor seeding or rupturing the capsule of the kidney, and thereby increasing the stage of Wilms' tumor.[60] However, in the adult FNA literature, tumor seeding along the needle tract of the needle has not been shown to be a complication of the procedure.[61] As noted by Akhtar et al.[53] the role of FNA biopsy in the diagnosis and management of Wilms' tumor is controversial, as most children with a renal mass will undergo nephrectomy, regardless of the

Role of FNA Biopsy in Wilms' Tumor

- Unresectable—diagnosis before chemotherapy
- Metastatic disease
- Bilateral disease
- Document recurrence
- Poor condition of patient
- Large tumor, site uncertain
- Primary diagnosis for resectable lesion—controversial

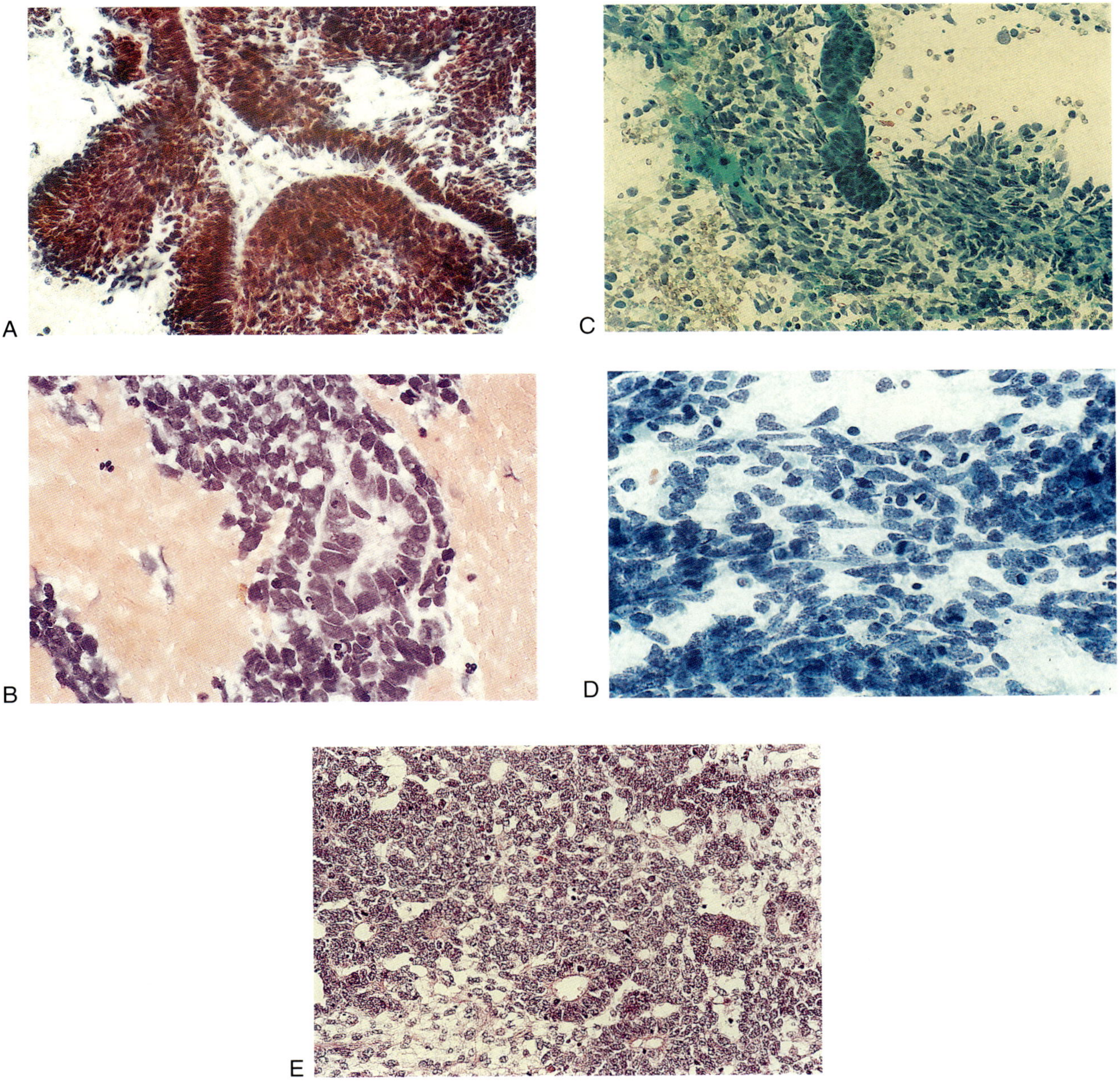

Fig. 5-26. (**A**) FNA of Wilms' tumor, showing complex branching tubules with nearby blastemal cells (Papanicolaou). (**B**) Cell block from FNA of Wilms' tumor, demonstrating blastemal cells and tubules (H&E). (**C**) Aspirate of Wilms' tumor, demonstrating tubules set in a spindle cell background (Papanicolaou). (**D**) FNA of recurrent Wilms' tumor involving renal bed demonstrating blastemal cells, having spindled configurations (Diff-Quik). (**E**) Triphasic histology of Wilms' tumor: tubules, mesenchyme, and blastemal cells (H&E).

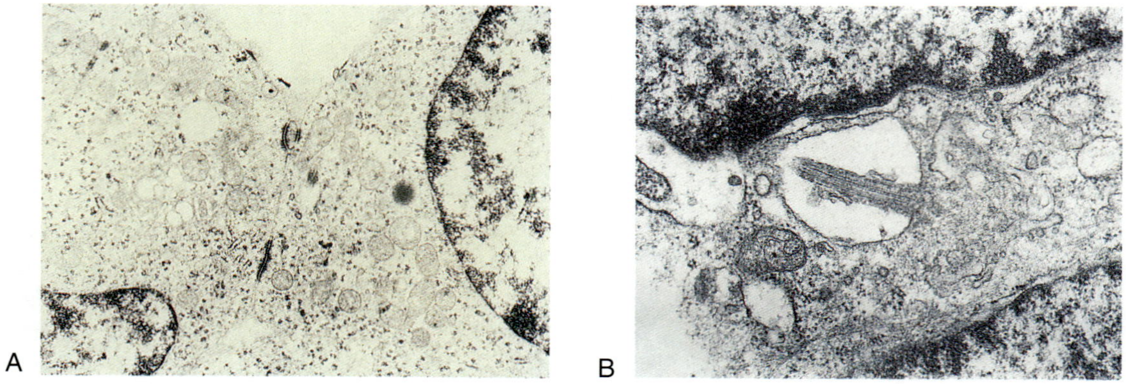

Fig. 5-27. (**A**) Electron microscopy performed on aspirate of Wilms' tumor, demonstrating epithelial junctions and basement membrane-like flocculent material. (**B**) Electron microscopy performed on aspirate of Wilms' tumor demonstrating oligocilia.

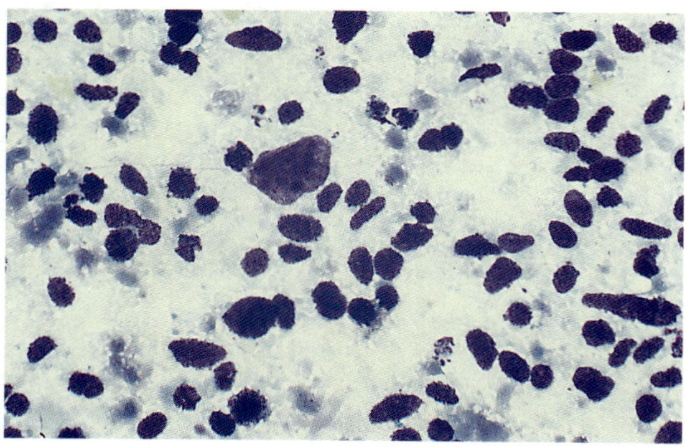

Fig. 5-28. FNA of Wilms' tumor, showing anaplastic features, including marked nuclear enlargement of some of the cells as well as hyperchromasia (Diff-Quik).

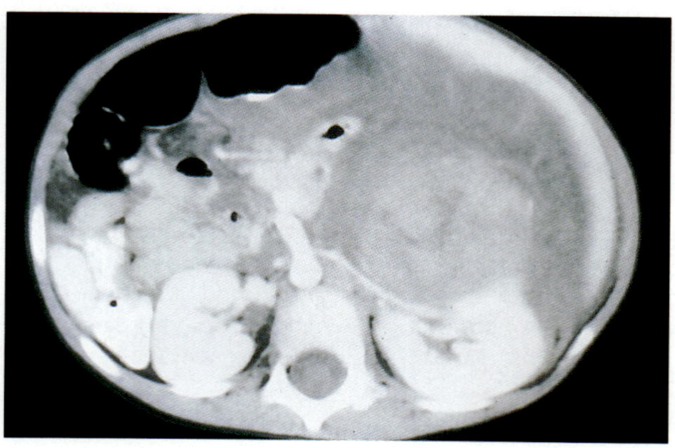

Fig. 5-29. Computed tomography scan of unresectable Wilms' tumor.

nature of the neoplasm. However, FNA biopsy of Wilms' tumor may be quite beneficial in those patients with unresectable Wilms' tumor, stage IV disease, and bilateral (stage V) tumors[51,62,63] (Fig. 5-29), as highlighted on page 190. The International Society of Pediatric Oncology believes that preoperative chemotherapy makes subsequent surgery safer for many patients, as chemotherapy should shrink the tumor mass; furthermore, the neoplasm may become encapsulated by dense bands of fibrous connective tissue.[64–69] Therefore, FNA biopsy may have a valuable role for the pretreatment diagnosis of Wilms' tumor in this setting.

Less Common Renal Tumors in Childhood

Malignant rhabdoid tumors account for approximately 2 percent of the renal tumors submitted to the National Wilms' Tumor Study (NWTS).[70–77] Malignant rhabdoid tumors typically present in children younger than those with Wilms' tumor with a median age of 13 months. Also in contrast to Wilms' tumor, the tumors result in a very aggressive clinical course. Patients with malignant renal rhabdoid tumors also have a high probability of developing a second primary neoplasm in the central nervous system, with cerebellar medulloblastoma the most common tumor. FNA cytology can document primary and metastatic malignant renal rhabdoid tumor.[70] Smears consist of cells arranged individually and in clusters having a single large nucleus with a prominent nucleolus and abundant surrounding cytoplasm.[70,73] Electron microscopic examination of the aspirated material will demonstrate whorled filamentous cytoplasmic inclusions, which may be seen as cytoplasmic "blobs" by light microscopy. Immunohistochemical studies are helpful in the workup of rhabdoid tumors.[70]

Another prognostically unfavorable renal neoplasm sampled by FNA biopsy is clear cell sarcoma. It differs from conventional Wilms' tumor not only by its distinctive histologic appearance but also by a more aggressive clinical course, including a strong tendency to metastasize to bones. FNA cytology will consist of malignant cells arranged singly and in small clusters.[75] The tumor cells are polygonal, stellate, or spindle-shaped with a pale cytoplasm and round to oval, uniform nuclei. With the Diff-Quik stain, mucoid intercellular material staining light pink to deep purple is present in the background.

Recently, case reports describing the FNA findings of mesoblastic nephroma and cystic nephroma have been published.[76–78] The cytomorphologic features of mesoblastic nephroma consisted of spindle cells with uniform bland nuclei arranged both individually and in clusters. The prognosis of this lesion is excellent following nephrectomy. The FNA findings of cystic nephroma in an adult included markedly atypical epithelium cells in papillary clusters that could be misinterpreted as a renal cell carcinoma.[77] Cytologic features suggestive of cystic nephroma included low smear cellularity, a lack of necrosis, and relatively few spindle cells in an aspirate from a cystic renal mass. Drut[79] reported a FNA cytologic finding of an aspirate of a cystic nephroma in 3-year-old child. Benign epithelial cells were arranged in uniform sheets as well as isolated cells.

METASTATIC AND SECONDARY NEOPLASMS

Although secondary involvement of the renal parenchyma is not very frequent, the more common primary sources in adults include carcinomas of lung, breast, and gastrointestinal tract and, in children, adrenal and retroperitoneal neuroblastoma and lymphoma[80] (Fig. 5-30). It is generally not difficult to separate renal cell carcinoma from metastatic carcinoma. Occasionally, immunocytochemistry and/or electron microscopy can be helpful in challenging cases. Cytologic features such as neuropil and Homer Wright rosettes may be seen in neuroblastoma.[56,57,81] Often, ancillary studies such as electron microscopy and immunocytochemistry will be needed to differentiate the blastemal-type Wilms' tumor from a neuroblastoma. Malignant lymphoma rarely involves the kidneys in the initial phase of the disease, although Burkitt's lymphoma can cause a nodular or diffuse infiltrate within the kidney (or usually both) at presentation. We have used FNA biopsy to diagnose Burkitt's lymphoma involving the kidney. The characteristic cytologic picture of disassociated malignant blasts with prominent cytoplasmic basophilic and vacuoles,

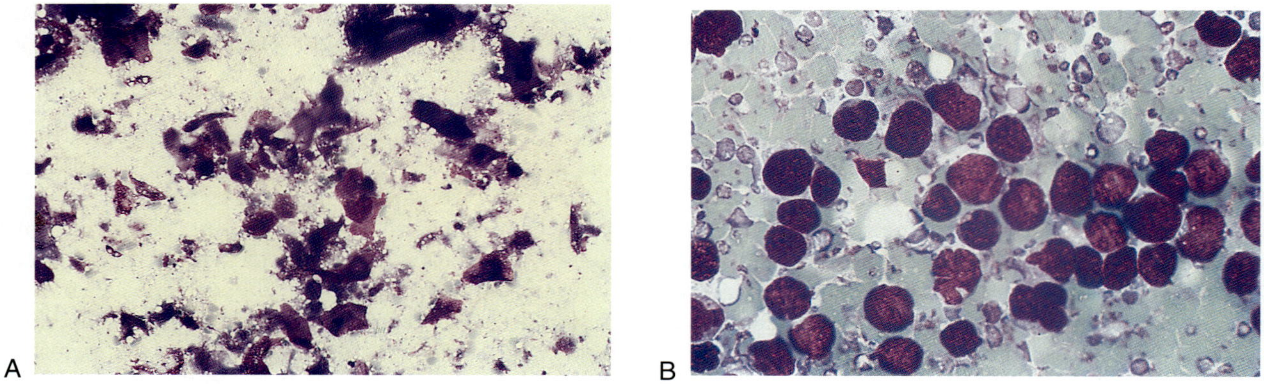

Fig. 5-30. (A) FNA of metastatic squamous cell carcinoma of the lung to the kidney. Note the keratinized cells, having irregular shapes and dense cytoplasm. The potential exists to confuse this cytologic appearance with a transitional cell carcinoma of the renal pelvis, demonstrating squamous differentiation (Diff-Quik). (B) FNA of large cell lymphoma of the kidney, showing dissociative pattern of atypical lymphoid cells. Note the nuclear irregularity and small nucleoli. Numerous lymphoglandular bodies are present in the background (Diff-Quik).

multiple nucleoli, and lymphoglandular bodies was present.[82] Cytogenetic studies performed on the aspirated material demonstrated the characteristic 8:14 translocation. Flow cytometric and/or immunocytochemical analysis of cell surface membranes revealed a B-cell phenotype.

REFERENCES

1. Wehle MJ, Grabstald H: Contraindications to needle aspiration of a solid renal mass: tumor dissemination by renal needle aspiration. J Urol 136:446, 1986
2. Kiser GC, Totonchy M, Barry JM: Needle tract seeding after percutaneous renal adenocarcinoma aspiration. J Urol 136:1292, 1986
3. Schreeb Tvon, Arner O, Skovsted G et al: Renal adenocarcinoma. Is there a risk of spreading tumour cells in diagnostic puncture? Scand J Urol Nephrol 1:270, 1967
4. Auvert J, Abbou CC, Lavarenne V: Needle tract seeding following puncture of renal oncocytoma. p. 597. In Renal Tumors: Proceedings of the First International Symposium on Kidney Tumors. Alan R. Liss, New York, 1982
5. Gibbons RP, Bush WH, Jr, Burnett LL: Needle tract seeding following aspiration of renal cell carcinoma. J Urol 118:865, 1977
6. Koss LG, Woyke S, Olszewski W: Aspiration biopsy. Cytologic Interpretation and Histologic Bases. Igaku-Shoin, New York, 1985
7. Helm CW, Burwood RJ, Harrison NW et al: Aspiration cytology of solid renal tumours. Br J Urol 55:249, 1983
8. Juul N, Torp-Pederson S, Grnvall S: Ultrasonically guided fine needle aspiration biopsy of renal masses. J Urol 133:579, 1985
9. Linsk JA, Franzen S: Aspiration cytology of metastatic hypernephroma. Acta Cytol 28:250, 1984
10. Katz RL: Kidney, adrenal and retroperitoneum. p. 771. In Bibbo M (ed): Comprehensive Cytopathology. WB Saunders, Philadelphia, 1991
11. Orell SR, Langlois SLP, Marshall VR: Fine needle aspiration cytology in the diagnosis of solid renal and adrenal masses. Scand J Urol Nephrol 19:211, 1985
12. Pilotti S, Rilke F, Alasio L et al: The role of fine needle aspiration in the assessment of renal masses. Acta Cytol 32:1, 1988
13. Niceforo J, Coughlin BF: Diagnosis of renal cell carcinoma: value of fine-needle aspiration cytology in patients with metastases or contraindications to nephrectomy. AJR 161:1303, 1993
14. Kline TS: Handbook of Fine Needle Aspiration Biopsy Cytology. 2nd Ed. Churchill Livingstone, New York, 1988
15. Tao L-C: Transabdominal Fine-Needle Aspiration biopsy. Igaku-Shoin, New York, 1990
16. Ambrose SS, Lewis EL, O'Brien DP III et al: Unsuspected renal tumors associated with renal cysts. J Urol 117:704, 1977
17. Deweerd JH: Percutaneous aspiration of selected expanding renal lesions. J Urol 87:303, 1962

18. Plowden KM, Erozan YS, Frost JK: Cellular atypia associated with benign lesions of the kidney as seen as fine needle aspirates. Acta Cytol 28:648, 1984
19. Horwitz CA, Manivel JC, Inampudi S, Kaye K: Diagnostic difficulties in the interpretation of needle aspiration material from large renal cysts. Diagn Cytopathol 11:380, 1994
20. Frydman CP, Raissi S, Watson CW: An unusual pulmonary and renal presentation of echinococcosis. Report of a case. Acta Cytol 33:655, 1989
21. Sneige N, Dekmezian R, Zaatari GS: Liesegang-like rings in fine needle aspirates of renal/perirenal hemorrhagic cysts. Acta Cytol 32:547, 1988
22. Silverman JF: Guides to Clinical Aspiration Biopsy. Infectious and Inflammatory Diseases and Other Nonneoplastic Disorders. Igaku-Shoin, New York, 1991.
23. Baniel J, Manning A, Leiman G: Fine needle cytodiagnosis of renal tuberculosis. J Urol 146:689, 1991
24. Conrad MR, Sanders RC, Mascardo AD: Perinephric abscess aspiration using ultrasound guidance. AJR 128:459, 1977
25. Meisels A: Cytology of carcinoma of the kidney. Acta Cytol 7:239, 1963
26. Silverman JF, Gurley AM, Holbrook CT et al: Pediatric fine-needle aspiration biopsy. Am J Clin Pathol 95:653, 1991
27. Suen KC: Guides to Clinical Aspiration Biopsy: Retroperitoneum and Intestine. Igaku-Shoin, New York, 1988
28. Sugie S, Tanaka T, Nishikawa A et al: Fine-needle aspiration cytology of xanthogranulomatous pyelonephritis. Urology 37:376, 1991
29. Silverman JF, Gurley AM, Harris JP et al: Fine needle aspiration cytology of renal infarcts. Cytomorphologic findings and potential diagnostic pitfalls in two cases. Acta Cytol 35:736, 1991
30. Cajulis RS, Katz RL, Dekmezian R et al: Fine needle aspiration biopsy of renal cell carcinoma. Cytologic parameters and their concordance with histology and flow cytometric data. Acta Cytol 37:367, 1993
31. Unger P, Hague K, Klein G et al: Fine needle aspiration of a renal cell carcinoma with eosinophilic globules. A case report. Acta Cytol 37:201, 1993
32. Kumar D, Kumar S: Adrenal cortical adenoma and adrenal metastasis of renal cell carcinoma: immunohistochemical and DNA ploidy analysis. Mod Pathol 6:36, 1993
33. Auger M, Katz RL, Sella A et al: Fine-needle aspiration cytology of sarcomatoid renal cell carcinoma: a morphologic and immunocytochemical study of 15 cases. Diagn Cytopathol 9:46, 1993
34. Triche TJ, Askin FB: Neuroblastoma and the differential diagnosis of small-, round-, blue-cell tumors. Hum Pathol 14:569, 1983
35. Nguyen G-K, Amy RW, Tsang S: Fine needle aspiration biopsy cytology of renal oncocytoma. Acta Cytol 29:33, 1985
36. Alanen KA, Tyrkko JES, Nurmi MJ: Aspiration biopsy cytology of renal oncocytoma. Acta Cytol 29:859, 1985
37. Dekmezian R, Sneige N, Shabb N: Papillary renal cell carcinoma: fine needle aspiration of 15 cases. Diagn Cytopathol 7:198, 1991
38. Kovacs G: Papillary renal cell carcinoma: a morphologic and cytogenetic study of 11 cases. Am J Pathol 134:27, 1989
39. Fuzesi L, Cober M, Mittermayer C: Collecting duct carcinoma: cytogenetic characterization. Histopathology 21:155, 1992
40. Cromie WJ, Davis CJ, DeTune FA: Atypical carcinoma of kidney. Possibly originating from collecting duct epithelium. Urology 13:315, 1979
41. O'Brien PK, Bedard YC: A papillary adenocarcinoma of the renal pelvis in a young girl. A light- and electron-microscopic study. Am J Clin Pathol 73:427, 1980
42. Hai MA, Diaz-Perez R: Atypical carcinoma of kidney originating from collecting duct epithelium. Urology 19:89, 1982
43. Fleming S, Lewi HJE: Collecting duct carcinoma of the kidney. Histopathology 10:1131, 1986
44. Aizawa S, Kikuchi Y, Suzuki M, Furusato M: Renal cell carcinoma of lower nephron origin. Acta Pathol Jpn 37:567, 1987
45. Kennedy S, Robertson C, Merino M: Collecting duct adenocarcinoma (tumors of the duct of Bellini): a pathologic, immunohistochemical, and ultrastructural study of 6 cases. Lab Invest 58:47A, 1988
46. Bane B, Grignon DJ, Friedman G et al: Collecting duct carcinoma (CDC) of the kidney: a distinct clinicopathologic entity. Lab Invest 60:5A, 1989
47. Layfield LJ: Fine-needle aspiration biopsy of renal collecting duct carcinoma. Diagn Cytopathol 11:74, 1994
48. Glenthoj A, Partoft S: Ultrasound-guided percutaneous aspiration of renal angiomyolipoma. Report of two cases diagnosed by cytology. Acta Cytol 28:265, 1984
49. Nguyen G-K: Aspiration biopsy cytology of renal angiomyolipoma. Acta Cytol 28:261, 1984
50. Wadih GE, Raab SS, Silverman JF: Fine-needle aspiration cytology of renal and retroperitoneal angiomyolipoma: cytologic findings and clinicopathologic pitfalls in diagnosis. Acta Cytol 39:945, 1995
51. Hoon V, Thung SN, Kaneko M, Unger PD: HMB-45 reactivity in renal angiomyolipoma and lymphangioleiomyomatosis. Arch Pathol Lab Med 118:732, 1994
52. D'Angio GJ, Beckwith JB, Breslow N et al: Wilms' tumor (nephroblastoma, renal embryoma). p. 583. In Pizzo PA, Poplack DG (eds): Principles and Practices of Pediatric Oncology. Philadelphia, JB Lippincott, 1989

53. Akhtar M, Ali MA, Sackey K et al: Aspiration cytology of Wilms' tumor: correlation and cytologic and histologic features. Diagn Cytopathol 5:269, 1989
54. Bray GL, Pendergrass TW, Schaller RT, Jr et al: Preoperative chemotherapy in the treatment of Wilms' tumor diagnosed with the aid of fine needle aspiration biopsy. Am J Pediatr Hematol Oncol 8:75, 1986
55. Saarinen UM, Wikstrom S, Koskimies O et al: Percutaneous needle biopsy preceding preoperative chemotherapy in the management of massive renal tumors in children. J Clin Oncol 9:406, 1991
56. Silverman JF, Joshi VV: FNA biopsy of small round cell tumors of childhood: cytomorphologic features and the role of ancillary studies. Diagn Cytopathol 10:245, 1994
57. Silverman JF, Berns LA, Holbrook CT et al: Fine-needle aspiration of primitive neuroectodermal tumors (PNET). Acta Cytol 8:235, 1992
58. Drut R, Pollono D: Anaplastic Wilms' tumor. Initial diagnosis by fine needle aspiration. Acta Cytol 31:774, 1984
59. Zuppan CW, Beckwith JB, Weeks DA et al: The effects of preoperative therapy on the histologic features of Wilms' tumor. An analysis of cases from the Third National Wilms' Tumor Study. Cancer 68:385, 1991
60. National Wilms' Tumor Study #4, Pediatric Oncology Group (POG) #8650, CCSG#461, NCI-INT-0070, p. 33, November 6, 1987
61. Berg JW, Robbins GF: Late look at the safety of aspiration biopsy. Cancer 15:826, 1962
62. Bracken RB, Sutow WW, Jaffe N et al: Preoperative chemotherapy for Wilms' tumor. Urology 19:55, 1982
63. Geisinger KR, Wakely PE, Wofford MM: Unresectable stage IV nephroblastoma: a potential indication for fine-needle aspiration biopsy in children. Diagn Cytopathol 9:197, 1993
64. Hanash KA, Sackey K, Sabbah RS et al: Surgical treatment of bilateral synchronous Wilms' tumors. J Surg Oncol 34:172, 1987
65. Lemerle J, Vonte PA, Tournade MF et al: Effectiveness of preoperative chemotherapy in Wilms' tumor. Results of an International Society of Paediatric Oncology (SIOP) clinical trial. J Clin Oncol 1:604, 1983
66. Lemerle J, Vonte PA, Tournade MF et al: Preoperative versus postoperative radiotherapy, single versus multiple courses of actinomycin D, in the treatment of Wilms' tumor. Preliminary results of a controlled clinical trial conducted by the International Society of Paediatric Oncology (S.I.O.P.). Cancer 38:647, 1976
67. McLorie GA, McKenna PH, Greenberg M et al: Reduction in tumor burden allowing partial nephrectomy following preoperative chemotherapy in biopsy proved Wilms' tumor. J Urol 146:509, 1991
68. Montgomery BT, Kelalis PP, Blute ML et al: Extended followup of bilateral Wilms' tumor: results of the National Wilms' Tumor Study. J Urol 146:514, 1991
69. Sabbah RS, Aur RJA, Hanash K et al: Management of the debilitated child with massive Wilms' tumor. King Faisal Spec Hosp Med J 2:77, 1982
70. Wakely PE, Jr, Giacomantonio M: Fine needle aspiration cytology of metastatic rhabdoid tumor. Acta Cytol 30:533, 1986
71. Weeks DA, Beckwith JB, Mierau GW: Rhabdoid tumor. An entity or a phenotype? Arch Pathol Lab Med 113:113, 1989
72. Perez JS, Perez-Guillermo, Bernal AB et al: Malignant rhabdoid tumor of soft tissues: a cytopathological and immunohistochemical study. Diagn Cytopathol 8:369, 1992
73. Drut R: Malignant rhabdoid tumor of the kidney diagnosed by fine-needle aspiration cytology. Diagn Cytopathol 6:124, 1990
74. Akhtar M, Ali MA, Sackey K et al: Fine-needle aspiration biopsy diagnosis of malignant rhabdoid tumor of the kidney. Diagn Cytopathol 7:36, 1991
75. Akhtar M, Ali MA, Sackey K et al: Fine-needle aspiration biopsy of clear-cell sarcoma of the kidney: light and electron microscopic features. Diagn Cytopathol 5:181, 1989
76. Drut R: Cytologic characteristics of congenital mesoblastic nephroma in fine needle aspiration cytology: a case report. Diagn Cytopathol 8:374, 1992
77. Clark SP, Kung ITM, Tang SK: Fine-needle aspiration of cystic nephroma (multilocular cyst of the kidney). Diagn Cytopathol 8:349, 1992
78. Dey P, Srinivasan R, Nijhawan R et al: Fine needle aspiration cytology of mesoblastic nephroma. A case report. Acta Cytol 36:404, 1992
79. Drut R: Cystic nephroma: cytologic findings in fine-needle aspiration biopsy. Diagn Cytopathol 18:593, 1992
80. Bennington JL, Beckwith JB: Tumors of the kidney, renal pelvis, and ureter. p. 162. In Atlas of Tumor Pathology. 2nd Series, Fascicle 12. Armed Forces Institute of Pathology, Washington, DC, 1990
81. Silverman JF, Dabbs DJ, Ganick D et al: Fine-needle aspiration cytology of neuroblastoma including peripheral neuroectodermal tumor, with immunocytochemical and ultrastructural confirmation. Acta Cytol 32:367, 1988
82. Gurley AM, Silverman JF, Wiley JE et al: The utility of ancillary studies in pediatric FNA cytology. Diagn Cytopathol 8:137, 1992

6 Adrenal Gland

Fine needle aspiration (FNA) biopsy is a rapid diagnostic method for the morphologic evaluation of adrenal masses as small as 1 cm in diameter. In our opinion, FNA biopsy is the procedure of choice for the initial diagnosis of an adrenal mass.[1–3] Although the surgical pathology of adrenal gland has been well described, relatively few publications discuss the FNA cytologic features of adrenal lesions.[1–13]

NORMAL ADRENAL GLAND

FNA cytology of the normal adrenal gland is characterized by small clusters and cords of uniform cells with polygonal contours and central round to oval nuclei.[9,12] Cytoplasm is foamy and clear with cytoplasmic membranes that are often delicate and/or frayed in the cells of the zona fasciculata and glomerulosa, whereas the cells of the zona reticularis have more compact cyanophilic cytoplasm that may contain finely granular, golden brown, lipofuscin pigment in Papanicolaou stains (Figs. 6-1 and 6-2). Because of the location of the right adrenal gland beneath the diaphragm and the overlying liver, it is important not to misinterpret inadvertent sampling of the liver in adrenal aspirates[9,11] (Fig. 6-2).

BENIGN ADRENAL CYSTS

Adrenal cysts are apparently rare lesions, most often discovered in middle-aged adults. The prevalence of adrenal cysts is between the 2 and 6 per 10,000 autopsies.[11] Endothelial cysts are the most common type, followed by benign or neoplastic pseudocysts occurring secondary to hemorrhage.[11] By definition, pseudocysts lack a lining of adrenal epithelial cells. Benign adrenal cysts can measure up to 30 cm in diameter. Frequently FNA of adrenal cysts is therapeutic secondary to decompression of the mass. Grossly, the aspirate may be clear, turbid, or bloody. Microscopically, a few epithelial cells will be scattered among leukocytes and hemosiderin-laden macrophages.[10,11]

INFECTIOUS LESIONS

The infectious etiology of adrenal abscesses tends to vary with age, with bacterial abscesses occurring predominantly in children and fungal and mycobacterial infections in adults[11,15] (Fig. 6-3). Although clinically apparent adrenal tuberculosis is currently uncommon in North America, approximately 50 percent of autopsy patients with late generalized tuberculosis may have involvement of the adrenal glands.[14] More than five decades ago, destructive bilateral adrenal tuberculosis was the classic and major cause of Addison's disease, accounting for 70 percent of cases; today, tuberculosis accounts for only 30 percent of cases of Addison's disease.[12,15] FNA findings of adrenal tuberculosis include histiocytes and lymphocytes scattered in a background of caseous necrotic debris. According to Tao,[12] the epithelioid cells and multinucleated giant cells commonly seen in FNA of tuberculosis involving other organs are not usually present in adrenal tuberculosis.

The adrenal gland may be involved in most patients with disseminated histoplasmosis and paracoccidioidomycosis.[15] As with tuberculosis, fungal infections usually cause bilateral adrenal

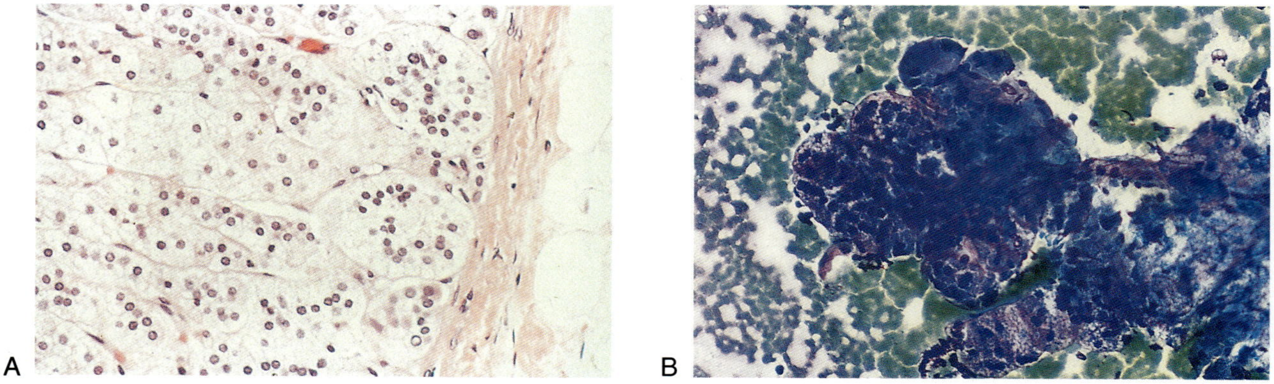

Fig. 6-1. (**A**) Histologic examination of adrenal cortex, showing zona glomerulosa having a lobulated appearance and underlying zona fasciculata, consisting of cords and trabeculae of pale vacuolated cells (H&E). (**B**) FNA of normal adrenal cortex, revealing fragment of zona glomerulosa having a lobulated appearance with cells possessing pale cytoplasm. Some vascularized tissue is also present (Diff-Quik).

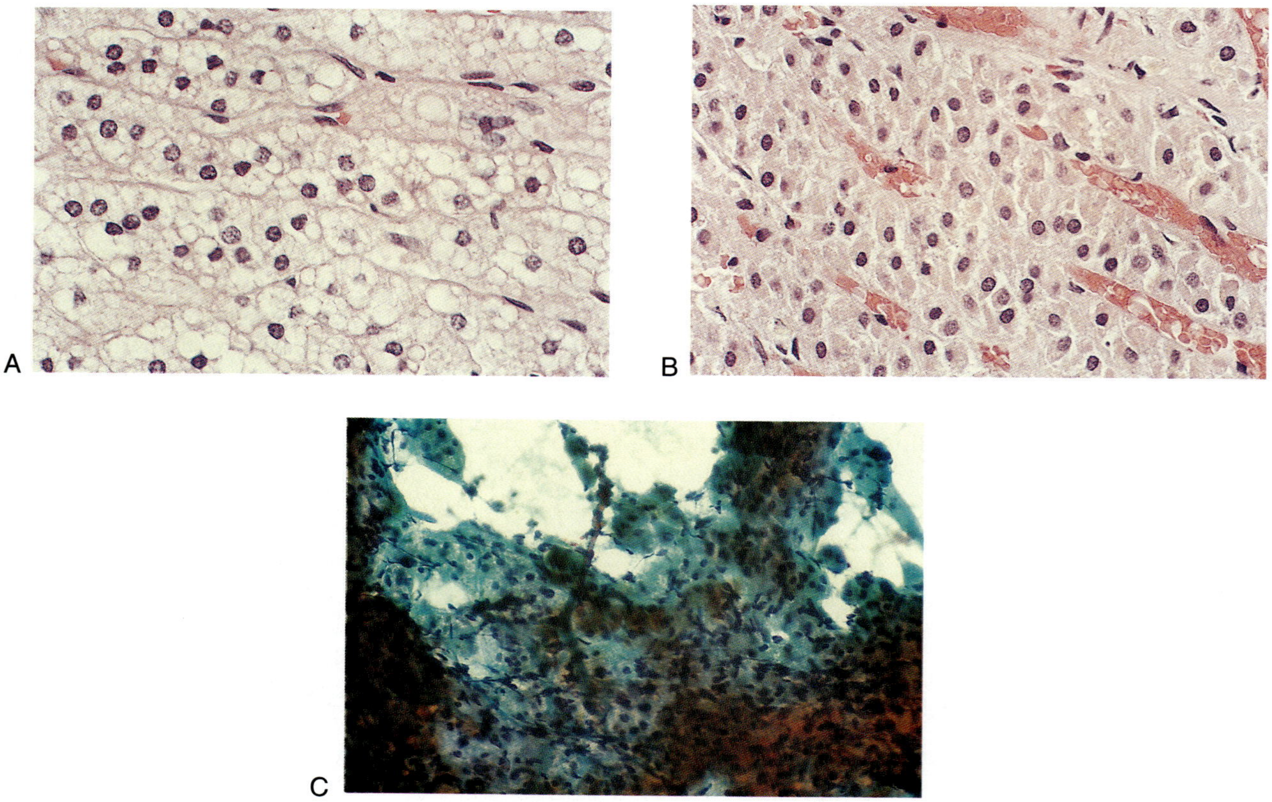

Fig. 6-2. (**A**) Histologic examination of zona fasciculata, consisting of trabeculae of pale cells having vacuolated cytoplasm. The nuclei are round, with delicate chromatin (H&E). (**B**) Histology of zona reticularis, consisting of trabeculae of pale cells having pigmented cytoplasm (H&E). (**C**) FNA of normal adrenal cortex revealing pale cells from zona fasciculata and pigmented cells from zona reticularis (Papanicolaou). (*Figure continues.*)

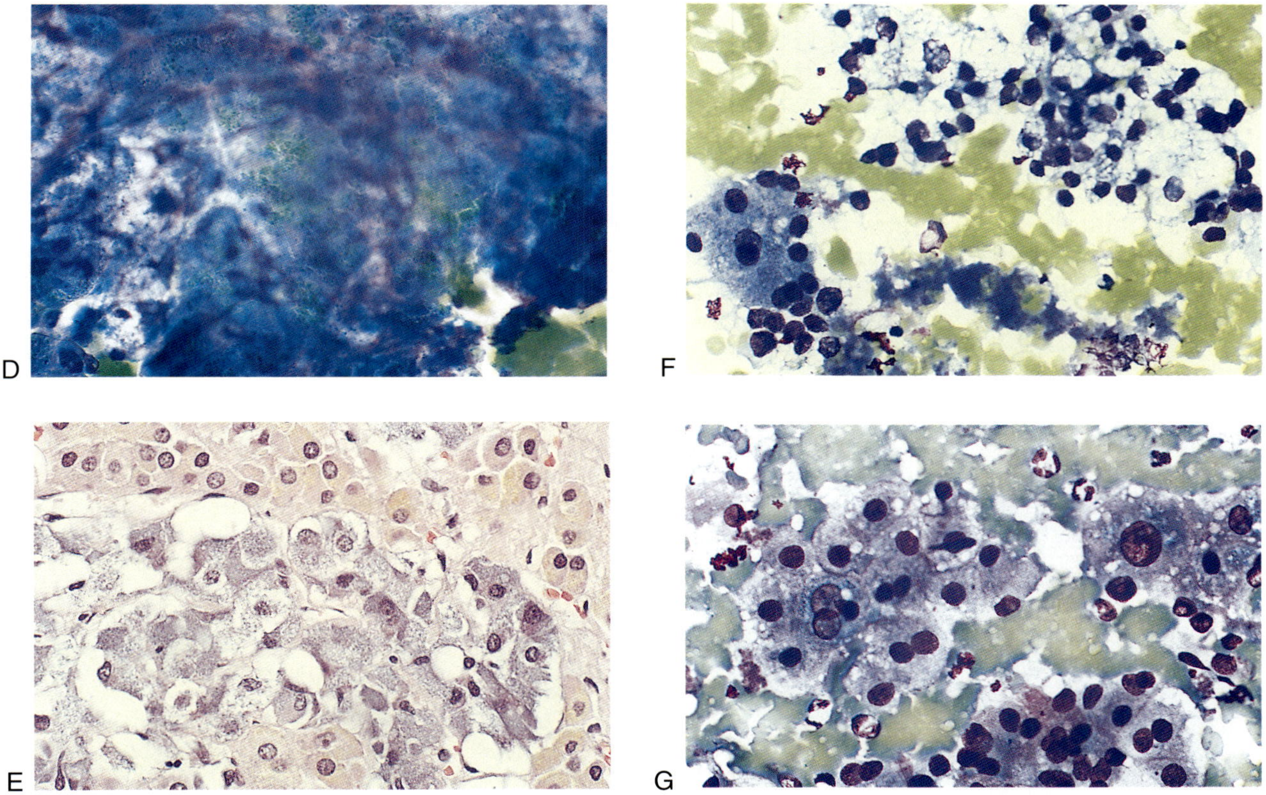

Fig. 6-2 (*Continued*). (**D**) Central portion of the field, showing thick layer of zona reticularis, consisting of cells with golden brown pigment representing lipofuscin (Diff-Quik). (**E**) Histology of medullary region of the benign adrenal gland, consisting of chromocytes having granular cytoplasmic pigment (H&E). (**F**) FNA of right adrenal gland containing vacuolated adrenal tissue and a cluster of benign liver tissue. The liver cells have a cuboidal shape with centrally placed nuclei and surrounding granular cytoplasm (Diff-Quik). (**G**) High-power view of benign liver tissue in aspirate of right adrenal gland. The anisonucleosis and binucleation with the presence of prominent nucleoli could be a potential pitfall for a misdiagnosis of malignancy if the hepatocellular origin of the cells is not appreciated (Diff-Quik).

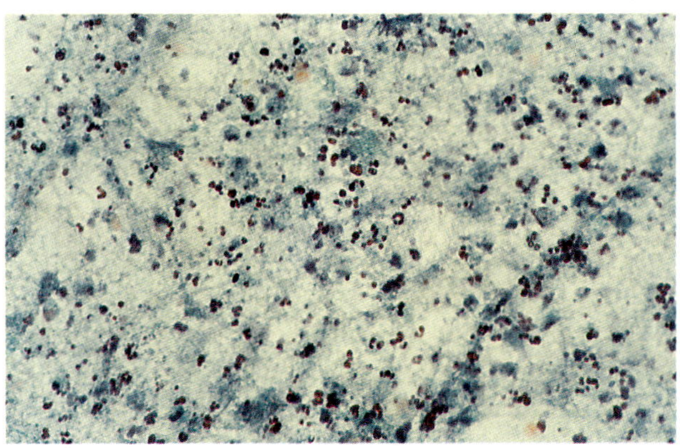

Fig. 6-3. FNA of adrenal abscess consisting of neutrophils with necrotic debris in the background (Papanicolaou).

enlargement. Addison's disease due to fungal infections has been most often associated with *Histoplasma capsulatum*. This complication has been seen in up to 7 percent of disseminated histoplasmosis cases.[15] Our FNA biopsy experience with fungal involvement of the adrenal gland was a case of *Cryptococcus* in which aggregates of histiocytes contained multiple discrete, intracytoplasmic, oval yeast forms surrounded by an area of clearing in the Papanicolaou-stained smears (Fig. 6-4). Extracellular organisms were not easily recognized in the routine smears; the GMS stain, however, revealed not only intracellular yeasts, but even more abundant extracellular organisms, some of which demonstrated classic narrow-based budding. Very few neutrophils and lymphocytes were present, presumably reflecting the host's immunocompromised status. Other disseminated fungal infections that may involve the adrenal gland include blastomycosis and coccidioidomycosis. Reports of adrenal histoplasmosis and cryptococcosis diagnosed by FNA biopsy are rare.[16–19]

ADRENAL MYELOLIPOMA

Adrenal myelolipoma is a rare, benign, nonfunctioning mass composed of a mixture of adipose tissue and bone marrow cells.[5,9,11,13] Infrequently these tumors can cause symptoms related to hemorrhage, necrosis or pressure on nearby structures.[3] There have been occasional reports of giant adrenal myelolipomas measuring up to 34 cm and weighing nearly 6,000 g.[6] Recently, the number of reports detailing CT- and ultrasound-guided FNA biopsies of myelolipoma have escalated.[2–8,20] Diagnostic FNA cytologic features include the presence of

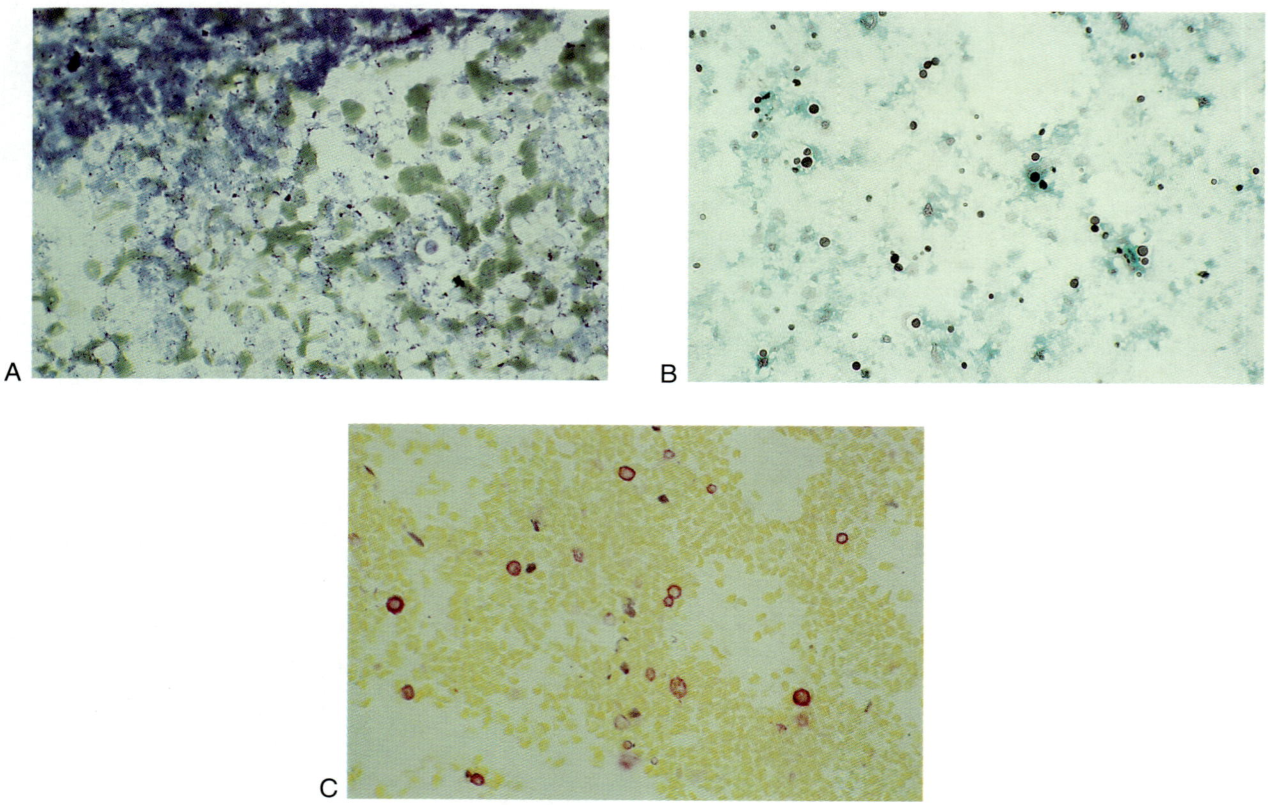

Fig. 6-4. (**A**) FNA of adrenal *Cryptococcus* infection, showing scattered yeast forms with surrounding halos. Note the necrotic debris in the background (Diff-Quik). (**B**) GMS stain performed on aspirate of adrenal *Cryptococcus*, showing scattered yeasts (GMS). (**C**) Mucicarmine stains the capsules of *Cryptococcus* in FNA of an adrenal mass (mucicarmine). (Case courtesy of Dr. Paul Wakely, Medical College of Virginia, Richmond, VA.)

mature and immature myeloid and erythroid cells, lymphocytes, and megakaryocytes (Fig. 6-5). The differential diagnosis of giant myelolipomas include well-differentiated liposarcoma and renal angiomyolipoma.[3,9] Recognition of the diagnostic hematopoietic cells admixed with adipose tissue should clearly differentiate myelolipoma from these other entities.[11]

ADRENOCORTICAL HYPERPLASIA

Hyperplasia of the cortex is almost always bilateral and diffuse or nodular with either micronodules (under 1 cm) and/or macronodules (greater than 1 cm) present.[15] Clinical associations include hypercortisolism, hyperaldosteronism, or virilization or a total lack of symptoms. FNA cytology of adrenocortical hyperplasia resembles that of unremarkable adrenal tissue with groups of vacuolated adrenocortical cells, including many with frayed cytoplasm (Fig. 6-6). Stripped nuclei and frothy lipid material can be seen in the smear background,[21] best appreciated in Diff-Quik preparations. It should be emphasized that these findings are indistinguishable from aspirates of adrenal cortical neoplasms (especially adrenocortical adenoma). Radiologically, bilateral adrenal lesions would favor adrenocortical hyperplasia. Obviously, this stresses the close interplay of clinicians, radiologists, and cytologists.

ADRENOCORTICAL NEOPLASMS

Outside of metastases, features distinguishing adenoma from adrenocortical carcinoma are not absolute.[11] Various classification systems have proposed criteria to separate benign from malignant adrenocortical tumors. Medeiros and Weiss[22] stress histologic findings, whereas others include clinical data as well. Tumor size is a helpful predictor of behavior. Most large lesions are malignant, whereas small neoplasms are usually benign. Lack

A

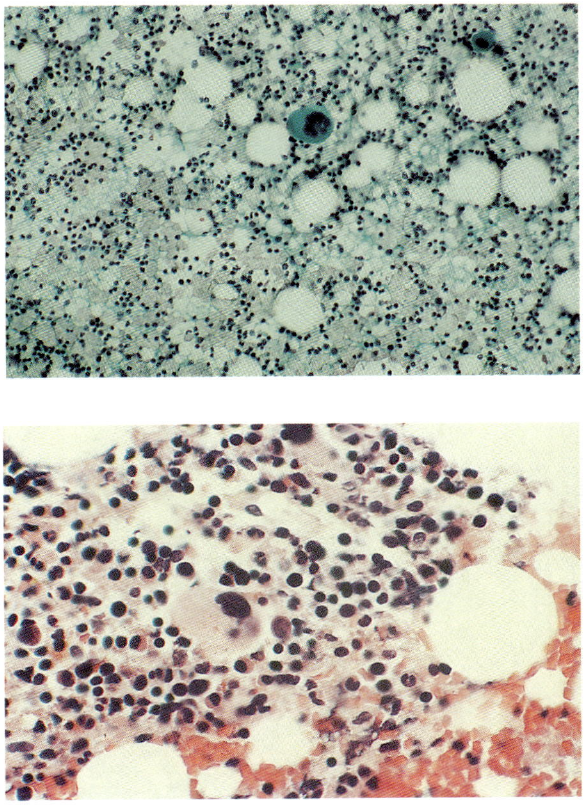

B

Fig. 6-5. (**A**) FNA of adrenal myelolipoma, consisting of hematopoietic cells, including megakaryocytes, mature and immature myeloid cells, and some erythroid precursors, along with interspersed fat (Diff-Quik). (Case courtesy of Dr. Paul Wakely, Medical College of Virginia, Richmond VA). (**B**) Cell block from FNA of adrenal myelolipoma, demonstrating admixed fat and hematopoietic elements (H&E).

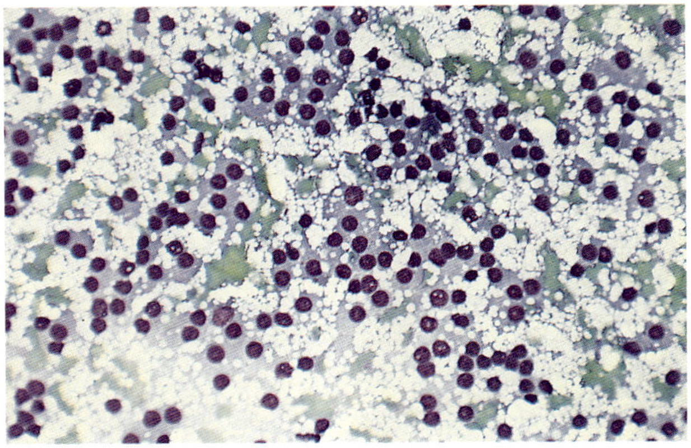

Fig. 6-6. FNA of adrenocortical hyperplasia, demonstrating scattered uniform-appearing adrenocortical cells, including many having frayed vacuolated cytoplasmic borders. Note the frothy lipid material in the background (Diff-Quik). The above findings would be indistinguishable from an aspirate of an adrenocortical adenoma or nodule.

et al.[15] have suggested that any adrenocortical tumor weighing in excess of 100 g should be regarded as a carcinoma. However, size should not be the sole criterion, as large adrenocortical tumors may not metastasize, whereas some small lesions can behave in malignant fashion. According to Medeiros and Weiss,[22] the presence of three or more of the following histologic features predicts malignant behavior: high nuclear grade, mitotic rate greater than 5 per 50 high-power field, atypical mitotic figures, eosinophilic tumor cell cytoplasm, diffuse architecture, necrosis, and invasion of veins, sinusoids, or the capsule. These investigators also believe that the mitotic rate may be the single most important criterion in separating benign from malignant adrenocortical neoplasms.[22] Several of these features can be assessed in the FNA cytology specimens. Obviously, venous, sinusoidal, and capsular invasion cannot be so evaluated.[11] Furthermore, although mitotic figures can be cytologically appreciated, an accurate assessment of the mitotic rate is not possible in FNA specimens. It is our practice that when nuclear atypia is minimal or not present, we generally diagnose the FNA specimen as an adrenocortical neoplasm, favor adenoma or benign cortical nodule. Lesions with a greater degree of nuclear atypia, prominent nucleoli, granular cytoplasm, necrosis, and mitotic activity would favor an adrenocortical carcinoma, especially if the mass is greater than 4 cm in diameter.[9,11] Generally, only a very small number of adrenal cortical carcinomas are under 6 cm in diameter. Yet, regardless of the size of the neoplasm, if diffuse cytologic atypia and pleomorphism are present and are associated with necrosis and mitotic activity, carcinoma should be favored. Clinical findings of virilization, feminization, or a corticoid-secreting adrenal mass causing abdominal pain, fever, or weight loss or failing to become suppressed with high-dose dexamethasone therapy is more consistent with malignancy.[9,11]

Adrenocortical Nodule and Adenoma

Benign adrenocortical nodules, also known as nonhyperfunctioning adenomas, are generally circumscribed, and/or encapsulated small, single lesions found in approximately 2 to 8 percent of the population at autopsy. Most adrenocortical nodules are less than 50 g and 5 cm in maximum dimension.[11] Most benign cortical nodules are not associated with overproduction of adrenocortical steroids.[11] Adrenocortical adenoma is the term reserved for a benign tumor associated with functional syndromes such as primary hyperaldosteronism (Conn's syndrome) or excess cortisol production (Cushing's syndrome).[15] Rarely, virilization may be associated with an adenoma. Histologic examination of an adrenocortical nodule reveals the presence of pale, lipid-laden fasciculata-type cells, as well as compact eosinophilic cells resembling those from the zona reticularis. Nuclei are generally small, round, and uniform.[11]

Aspirates from adrenocortical nodules are cellular, consisting of uniform cells having vacuolated cytoplasm with frayed cytoplasmic borders, along with numerous stripped nuclei (Fig. 6-7), as listed on page 203.[11] The nuclei often cluster and overlap, which at low magnification have a morular pattern (Fig. 6-8). The number of stripped nuclei may be so great that intact cells with pale lipid-laden cytoplasm will be difficult to find. The smear background contains abundant lipid material and cytoplasmic fragments that produce a vacuolated, frothy appearance.[11] Nuclear atypia are unusual, although some anisonucleosis may be present. The nuclei are round to oval and uniform without

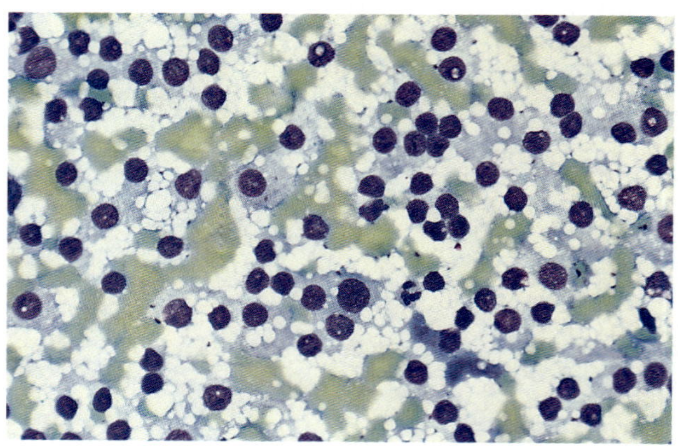

Fig. 6-7. FNA of adrenocortical adenoma, consisting of uniform-appearing adrenocortical cells, having pale to frayed cytoplasm with round to oval nuclei. Note the frothy lipid background (Diff-Quik).

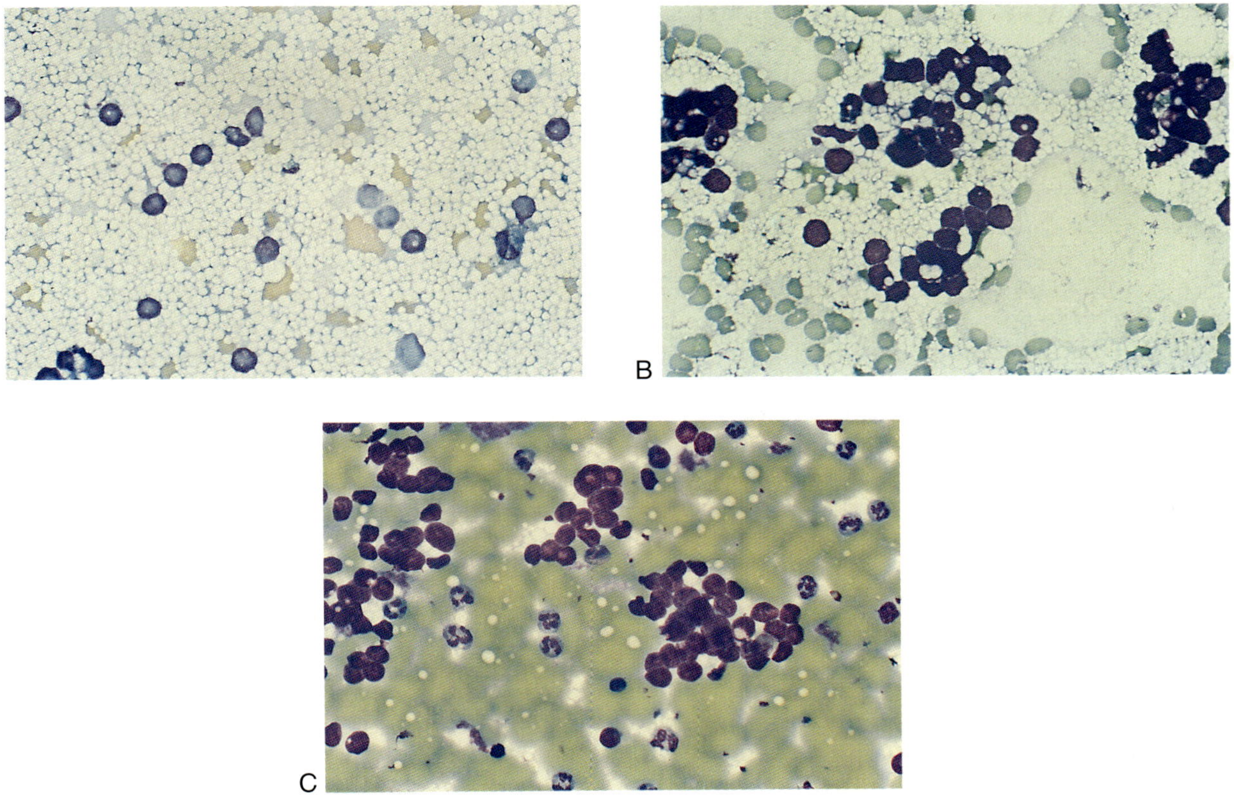

A B C

Fig. 6-8. (**A**) FNA of adrenocortical nodule, consisting of stripped nuclei and abundant frothy lipid material in the background (Diff-Quik). (**B**) Clustering of stripped nuclei in FNA of adrenocortical nodule could potentially be confused with a metastatic small cell carcinoma. Helpful findings include the presence of frothy lipid material in the background and round overlapping nuclei, with no evidence of nuclear molding (Diff-Quik). (**C**) Another example of FNA of adrenocortical nodule that could potentially be confused with a metastatic small cell carcinoma. This case is made more difficult by the lack of frothy lipid material in the bloody background. Helpful diagnostic clues include overlapping nuclei and small nucleoli, rather than nuclear molding and "salt and pepper" chromatin of a small cell carcinoma (Diff-Quik).

molding or prominent nucleoli. Mitotic figures and necrotic debris are absent. FNA cytology of "black adenoma" contains cells (from the zona reticularis) with abundant cytoplasmic lipofuscin granules.[9] Cytologic findings of adrenocortical nodule and adrenocortical hyperplasia are virtually identical.[9] Thus, radiologic findings are more important in making a correct diagnosis. Radiographic evidence of bilateral diffuse or nodular enlargement of the adrenals favors hyperplasia, whereas a dominant unilateral mass predicts an adrenocortical adenoma or benign cortical nodule.[21]

Features of Adrenocortical Adenoma

- Cellular smears with single cells and groups
- Cells with abundant vacuolated cytoplasm and indistinct cell borders
- Uniform, small dark nuclei, anisonucleosis
- Stripped nuclei—potential pitfall for false-positive diagnosis of small cell carcinoma
- Frothy lipid background (best seen in Diff-Quik smears)

Potential diagnostic pitfalls include confusing normal adrenocortical cells with those from a benign cortical nodule.[11] It is therefore essential to review the computed tomography (CT) or ultrasound findings of the adrenal mass and the placement of the needle for correlation with the cytologic features.[23] A well-differentiated adrenocortical carcinoma may also be confused with either an adrenocortical nodule or an adenoma. This is not surprising, as it may be difficult histologically to differentiate these lesions.[22] An adrenocortical nodule can also be confused with a small round cell malignancy due to the fragility of the cells, resulting in stripped overlapping nuclei in the aspirated material.[9,21,23–25] Awareness of this pitfall is especially relevant, as an adrenal nodule is often discovered in the workup of patients with a lung mass. The bare, stripped nuclei derived from an adrenocortical nodule lack the nuclear angularity and true molding of a metastatic small cell carcinoma.[9,11] The presence of small nucleoli will also favor a cortical nodule; metastatic small cell carcinoma will have hyperchromatic "salt and pepper" chromatin and inconspicuous nucleoli. In addition, the characteristic frothy or vacuolated background characterizes the aspirates of the cortical nodule, while metastatic small cell carcinoma has a necrotic background. In FNA of the right adrenal lesions, intact hepatocytes may be confused with adrenocortical cells.[9,11] In general, hepatocytes are larger and more polygonal in shape, with opaque granular cytoplasm, well-defined cell borders, and possibly intracytoplasmic bile pigment. Their nuclei tends to be more centrally positioned and nucleoli more prominent.[9] Large distinct intracytoplasmic fat vacuoles may be present in the hepatocytes, in contrast to the pale, microvesicular cytoplasm of adrenocortical cells. The presence of bile ductular epithelium can also be helpful in suggesting inadvertent sampling of the liver in an adrenal aspirate.

Adrenocortical Carcinoma

Adrenocortical carcinoma is an uncommon malignancy, with only one or two cases per million population, or approximately 0.2 percent of cancers.[11] Adrenocortical carcinomas affect all age groups, with a bimodal distribution: children and young adolescents and older adults in the fifth and seventh decades.[15] The clinical picture of adrenocortical carcinoma is variable but generally includes abdominal pain, a palpable mass, fatigue, weight loss, and intermittent low-grade fever.[11] Because of its retroperitoneal location, signs and symptoms may be delayed until the tumor becomes large. Most adrenocortical carcinomas are quite large with reported weights averaging 705 to 1,210 g; however, neoplasms weighing as little as 39 g have metastasized.[11] The tumor stage has been the most significant prognostic parameter in a number of studies. Others have shown that nuclear grading has some relationship to prognosis. Median survival has been reported as approximately 14 months.[26]

Grossly, adrenocortical carcinoma typically presents as a large mass with foci of necrosis, hemorrhage, and cystic change. Histologically, adrenocortical carcinomas have trabecular, alveolar, or diffuse patterns separated by vascular channels lined with a single endothelial cell layer.[15] Moderate to marked nuclear pleomorphism, brisk mitotic activity, capsular and vascular invasion, necrosis, and hemorrhage may be expected. Generally, tumor cells are negative for low-molecular-weight cytokeratin in formalin-fixed material but may be immunoreactive in frozen tissue.[25] This negative reaction in formalin-fixed material is helpful when the differential diagnosis includes cancers that are generally cytokeratin-positive, such as renal cell carcinoma and hepatoma. The electron microscopic findings supportive of adrenocortical carcinoma include nearly universal shelf-like cristae and tubular cristae or tubulovesicular cristae in approximately 60 percent of cases.[11] Flattened cisternae of rough endoplasmic reticulum are also often present with stacks of parallel lamellae seen in almost all tumors. Flow cytometric DNA ploidy analysis correlates with the histologic diagnosis, but is not a reliable indicator of prognosis.[15,22,26]

Features of Adrenocortical Carcinoma

- Cellular smears
- Variable morphology from vacuolated, lipid-laden cells similar to adrenocortical adenoma to pleomorphic cells
- Poorly differentiated adrenocortical carcinoma: discohesive, pleomorphic cells—cells with eosinophilic cytoplasm, necrosis, spindle cells

Cytomorphologic findings of adrenocortical carcinoma vary, depending in part on the (histologic) grade.[9] In general, aspirates of adrenocortical carcinomas are hypercellular, with a discohesive pattern of single cells, along with occasional groups of malignant cells (Fig. 6-9). Nuclei are pleomorphic, with irregular coarse chromatin and prominent nucleoli. Cytoplasm varies from eosinophilic to vacuolated. Well-differentiated adrenocortical carcinomas have more intracytoplasmic lipid; they could therefore be confused with a cortical adenoma.[27] However, adrenocortical carcinomas have generally larger cells with hyperchromatic nuclei, and mitotic figures may be seen.[9] Poorly differentiated malignancies tend to have a more granular eosinophilic cytoplasm.[9,11] Occasional adrenocortical carcinomas have a sarcomatoid appearance, prompting a differential diagnosis that includes other spindle cell lesions that can involve the adrenal gland, such as pheochromocytomas, sarcomas, and metastases. We previously reported three spindle cell neoplasms encountered in a series of 46 FNA biopsies of the adrenal gland.[28] These neoplasms included a recurrent high-grade adrenocortical carcinoma with a predominant spindle cell pattern, a pheochromocytoma, and a metastatic cutaneous desmoplastic melanoma. Ancillary studies performed on aspirated material were helpful in the differential diagnosis. The adrenocortical adenoma was negative for cytokeratin, neuron-specific enolase (NSE), and muscle-specific actin but showed strong positive cytoplasmic staining for vimentin, as well as ultrastructural evidence of characteristic whorls of

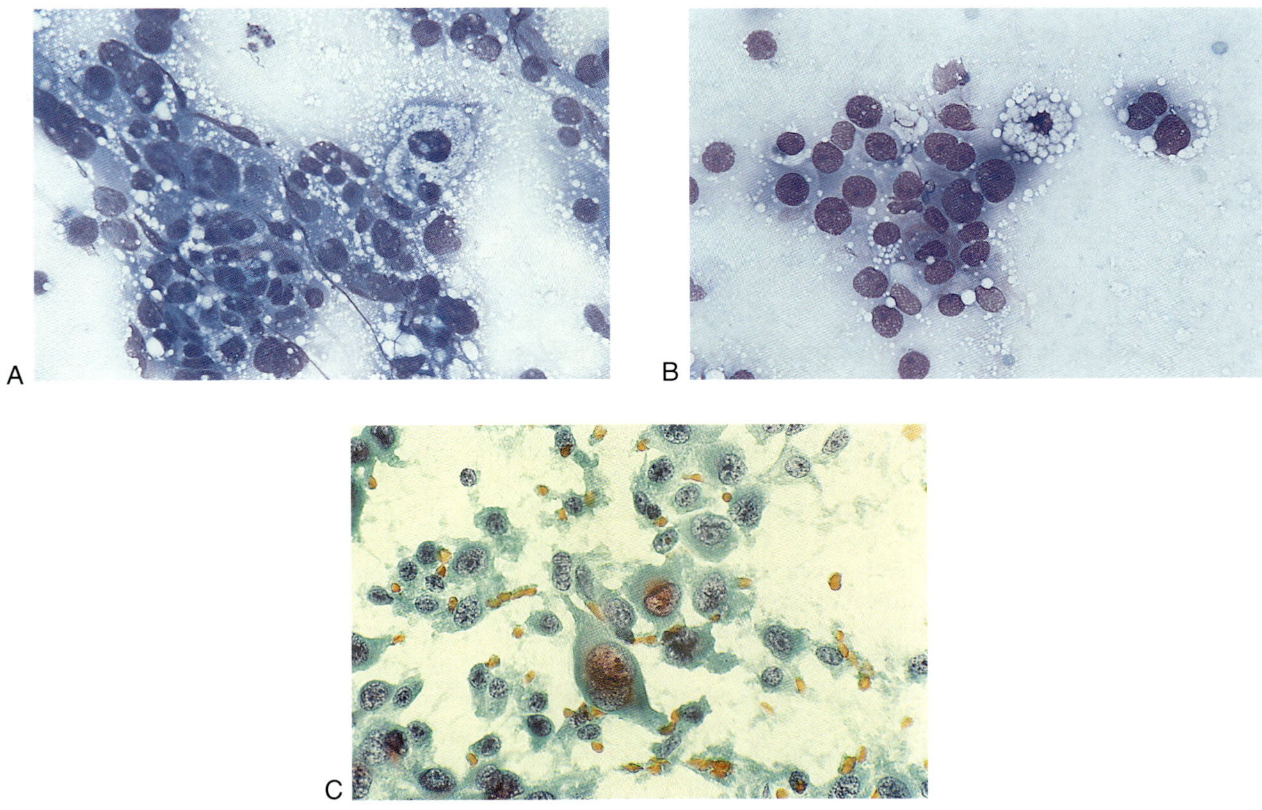

Fig. 6-9. (**A**) FNA of adrenocortical carcinoma, consisting of cells having coarsely vacuolated cytoplasm. The degree of nuclear atypicality is greater than the findings of an FNA of adrenocortical adenoma (Diff-Quik). (**B**) FNA of adrenocortical carcinoma, showing variation in nuclear size and shape, slight nuclear irregularity, increased nuclear to cytoplasmic ratios, and small nucleoli. However, the cells still retain a vacuolated pale cytoplasm (Diff-Quik). (**C**) Poorly differentiated adrenocortical carcinoma, consisting of dissociative smear pattern in which polygonal to spindle-shaped tumor cells are seen (Papanicolaou).

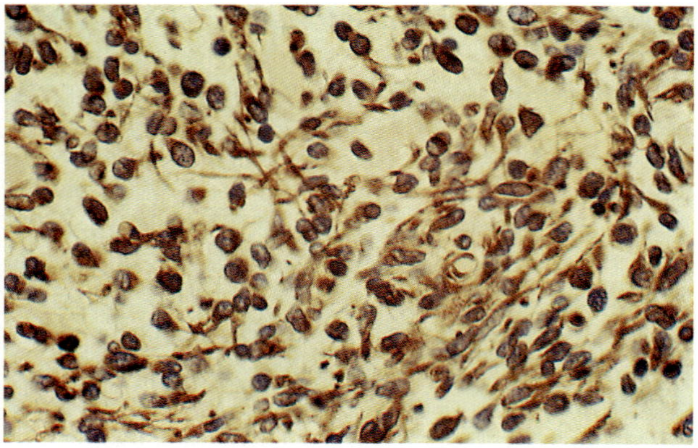

Fig. 6-10. FNA of sarcomatoid-type of adrenocortical carcinoma, having a spindle configuration. The tumor cells were cytokeratin negative but vimentin positive (immunoperoxidase).

smooth endoplasmic reticulin (Fig. 6-10). The pheochromocytoma was positive for NSE and chromogranin. The metastatic desmoplastic malignant melanoma stained positively for S-100 and vimentin; electron microscopic examination revealed rare atypical premelanosomes.[28]

Pheochromocytoma

Pheochromocytomas are neoplasms derived from the chromaffin cells of the medullary portion of the adrenal gland, with an average annual incidence that varies between 1.6 to 2.1 cases per million.[15] Although the finding of clinically unsuspected or undiagnosed cases of pheochromocytoma at surgery or autopsy has been reported in more than one-third of cases, the prevalence of clinically suspected cases has increased due to greater clinical awareness and availability of contemporary imaging techniques and biochemical studies.[11] Most cases are symptomatic due to secretion of excessive amounts of catecholamines, resulting in paroxysmal hypertension, sweating, and palpitation. Increased levels of urinary or plasma vanillylmandelic acid (VMA) and catecholamines occur in up to 90 percent of patients, although as many as 20 percent of tumors may be nonfunctional.[9] The "classic" 10 percent rule indicates that approximately 10 percent of pheochromocytomas are malignant, 10 percent are bilateral, and 10 percent are extra-adrenal.[9,11] Although pheochromocytomas can occur at any age, they are most commonly seen in the fourth and fifth decades.[15] More than 90 percent of cases are sporadic, with the remainder having a familial history as a component of the multiple endocrine neoplasia (MEN) syndrome. The latter present at an earlier age. MEN syndrome type 2A (Sipple) includes pheochromocytoma, medullary carcinoma of the thyroid, and parathyroid hyperplasia. The MEN type 2B syndrome includes pheochromocytoma, medullary thyroid carcinoma, ganglioneuromatosis of the gastrointestinal tract, multiple mucosal neuromas, and connective tissue defects, including a marfanoid habitus.[11]

Sporadic pheochromocytomas are usually solitary, measure from 3 to 5 cm, weigh an average of 100 g, but with a wide weight range.[9,15] Histologic patterns include anastomosing cords (trabecular pattern) and an alveolar or nesting arrangement with an interspersed highly vascularized fibroconnective tissue. When the alveolar or nesting pattern is well developed, a "zellballen" arrangement, resembling that in head and neck paragangliomas can be present. Individual neoplastic cells are generally polygonal in shape, with a moderate amount of pale to granular basophilic cytoplasm.[9,11] Nuclei are eccentrically positioned, with the surrounding cytoplasm having ill-defined or frayed cell borders. Occasional tumor cells have a granular eosinophilic oncocytoid appearance. Other tumor cells can be spindle-shaped, with elongated nuclei showing a coarse chromatin pattern.[9] Some pheochromocytomas include cells resembling ganglion cells with eccentrically placed single or multiple vesicular nuclei with prominent nucleoli, tapering cell processes, and intracytoplasmic basophilic granular material resembling Nissl substance. Occasional cytoplasmic vacuolization may be a potential pitfall in separating pheochromocytomas from an adrenocortical neoplasm or metastatic melanoma. Other findings in pheochromocytomas include nuclear pseudoinclusions and intracytoplasmic hyaline globules.[15] Immunocytochemically, cytoplasm of pheochromocytomas is decorated with chromogranin, NSE, and other neuroendocrine type markers.[15] S-100 staining may highlight the spindled to stellate-shaped sustentacular cells. Ultrastructural examination demonstrates the diagnostic neurosecretory granules. Typically, norepinephrine granules have a prominent eccentric space between the electron-dense cores and surrounding limiting membranes, while epinephrine granules possess narrow, uniform halos around the cores.[15]

Smears of FNA biopsies of pheochromocytomas are generally hypercellular and discohesive, with the cells arranged singly or in loose groups.[9,11] Considerable variability in cell size and shape is immediately evident (Fig. 6-11). Nucleoli may be very prominent. Pleomorphic giant cells with hyperchromatic nuclei often with a degenerative quality and prominent nucleoli can be present. Occasional nuclear pseudoinclusions can be present. Some cases may be composed of more uniform small to intermediate-size cells, nuclei with diffuse, coarsely granular chromatin and lacking prominent nucleoli, and high nuclear to cytoplasmic

Features of Pheochromocytoma

- Hypercellular, single cells or groups
- Variability in cell size and shape: polygonal, pleomorphic to spindle shapes; ill-defined, granular cytoplasm
- Granular cytoplasm, occasional frayed borders
- Can resemble carcinoid
- Neuroendocrine-type nuclei to pleomorphic nuclei with prominent nucleoli
- Occasional pseudoinclusions

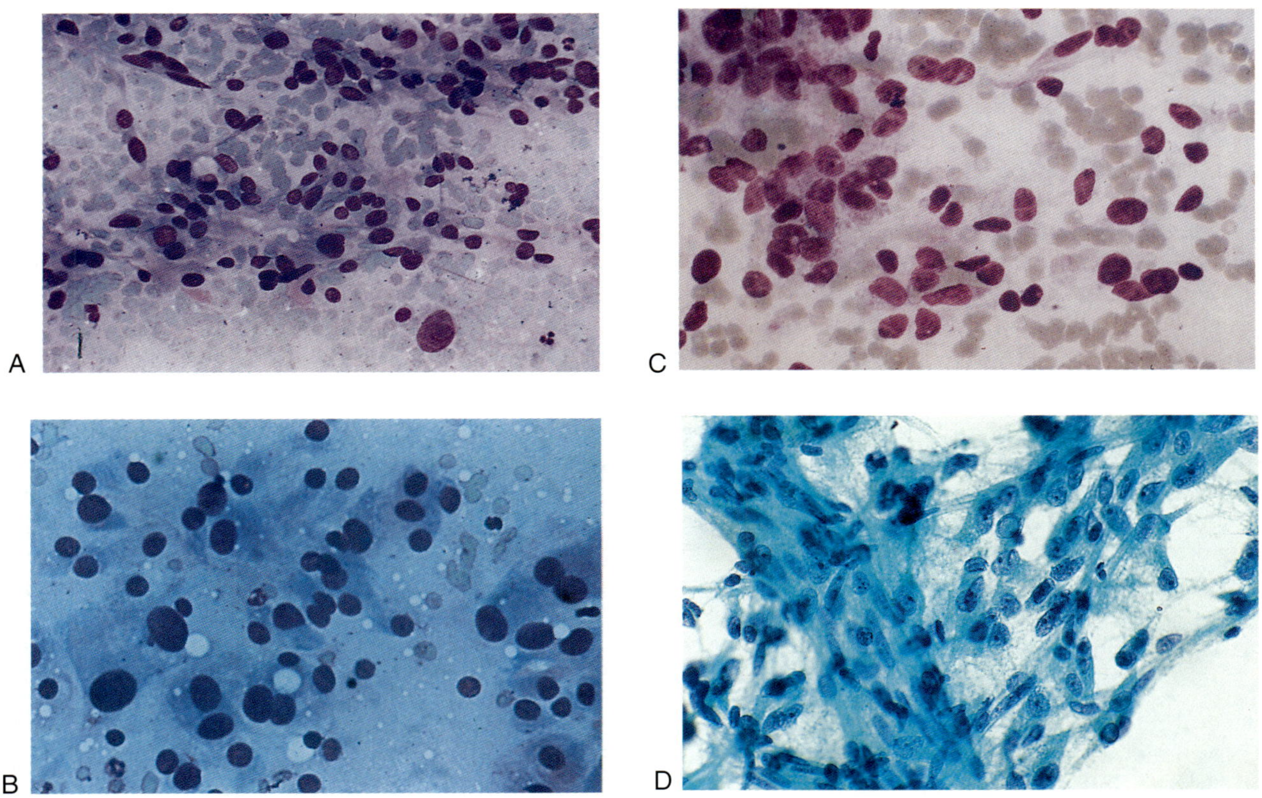

Fig. 6-11. FNA of pheochromocytoma, showing individually scattered cells and loose clusters of polygonal cells in which there is prominent anisonucleosis (Diff-Quik). (**B**) High-power view of pheochromocytoma, showing intermediate to larger cells having abundant granular cytoplasm with frayed borders. Some stripped nuclei are also noted. (Diff-Quik). (**C**) Numerous stripped nuclei and spindle-shaped cells in FNA of pheochromocytoma (Diff-Quik). (**D**) FNA of pheochromocytoma, having a predominantly spindle-cell pattern (Papanicolaou).

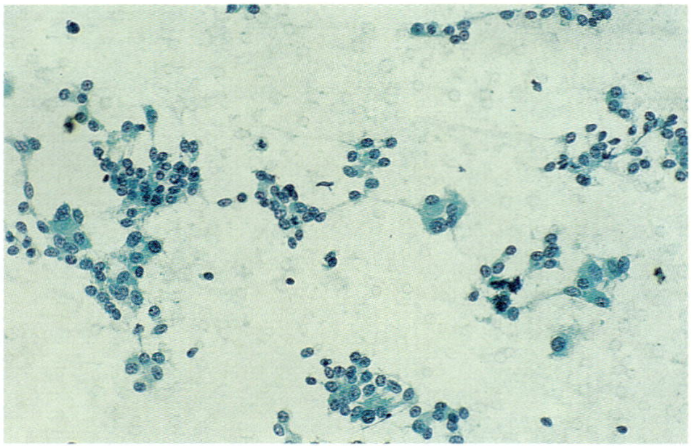

Fig. 6-12. FNA of pheochromocytoma, having a more carcinoidal pattern consisting of loose clusters and individually scattered cuboidal cells with granular cytoplasm and oval eccentric nuclei (Diff-Quik).

ratios, resembling the "prototypical" neuroendocrine cells of a carcinoid tumor (Fig. 6-12).

Potential diagnostic pitfalls include mistaking a pheochromocytoma for sarcomatoid renal cell carcinoma, adrenocortical carcinoma, or a retroperitoneal sarcoma due to the presence of spindled and/or bizarre giant cells. The cytologic feature of sarcomatoid renal call carcinomas are discussed in Chapter 5, while the cytomorphologic findings and ancillary features of retroperitoneum sarcomas are presented in Chapter 7. However, the finding of uniform polygonal-shaped cells with granular cytoplasm, arranged in clusters and singly should suggest a pheochromocytoma. In addition, although pleomorphism may be present, mitoses are not abundant. A rare case of pheochromocytoma may contain cells having intracytoplasmic melanin that potentially could be confused with a metastatic melanoma. Application of ancillary studies including immunocytochemistry and electron microscopic examination should resolve these potentially challenging cases.

Several reports have noted the potential danger of FNA biopsy of pheochromocytoma, due to the possibility of initiating a catecholamine crisis.[1,15,29] Fatal hemorrhagic complications have also been reported.[15] We concur with Cassola et al.[1] that pheochromocytoma can be diagnosed by FNA biopsy, but the radiologist must be familiar with the emergency treatment of a hypertensive crisis.

METASTASES

The adrenal gland is the fourth most common extranodal site of metastatic cancer following lung, liver, and bone.[15] Autopsy studies have documented adrenal metastasis in 9 to 27 percent of cancer patients, with lung and breast carcinoma the most common primary malignancies (Fig. 6-13). When the adrenals are involved, bilateral involvement is frequent. Extraordinary reports of extra-adrenal primary malignancies metastasizing to pheochromocytoma and adrenocortical adenomas exist. With greater application of radiologic imaging techniques for the workup of metastatic cancer, many incidental small adrenal masses have been encountered. In our

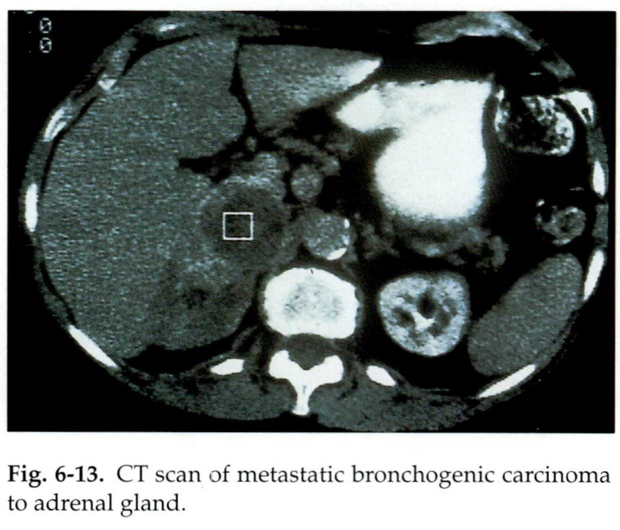

Fig. 6-13. CT scan of metastatic bronchogenic carcinoma to adrenal gland.

initial experience with adrenal FNA biopsies, 23 of 50 aspirates documented metastatic cancer to the adrenal gland.[21] In a similar experience from M.D. Anderson Hospital, 80 cases (42.5 percent) of the total series of 188 cases reported metastatic malignancies, while an almost identical number of primary adrenal lesions (81 cases [43 percent]) were aspirated.[30] These findings indicate the importance of obtaining a morphologic diagnosis of an adrenal mass in patients with a history of an extra-adrenal malignancy. Most metastatic tumors were bronchogenic adenocarcinomas, followed by metastases from the kidney and melanoma (Fig. 6-14). Adrenal involvement by renal cell carcinoma has been reported in approximately 6 percent of cases at nephrectomy and 90 percent of cases at autopsy.[15] Renal cell carcinoma can potentially be confused with some cases of adrenocortical neoplasms. A panel of immunocytochemical studies can help in the differential diagnosis, especially when performed on formalin-fixed cell block material. Adreno cortical tumors fixed in formalin show positive staining with vimentin but negative staining for cytokeratin, epithelial membrane antigen (EMA), and blood group isoantigens. Renal cell carcinomas can show expression of keratin, EMA, vimentin, and/or blood

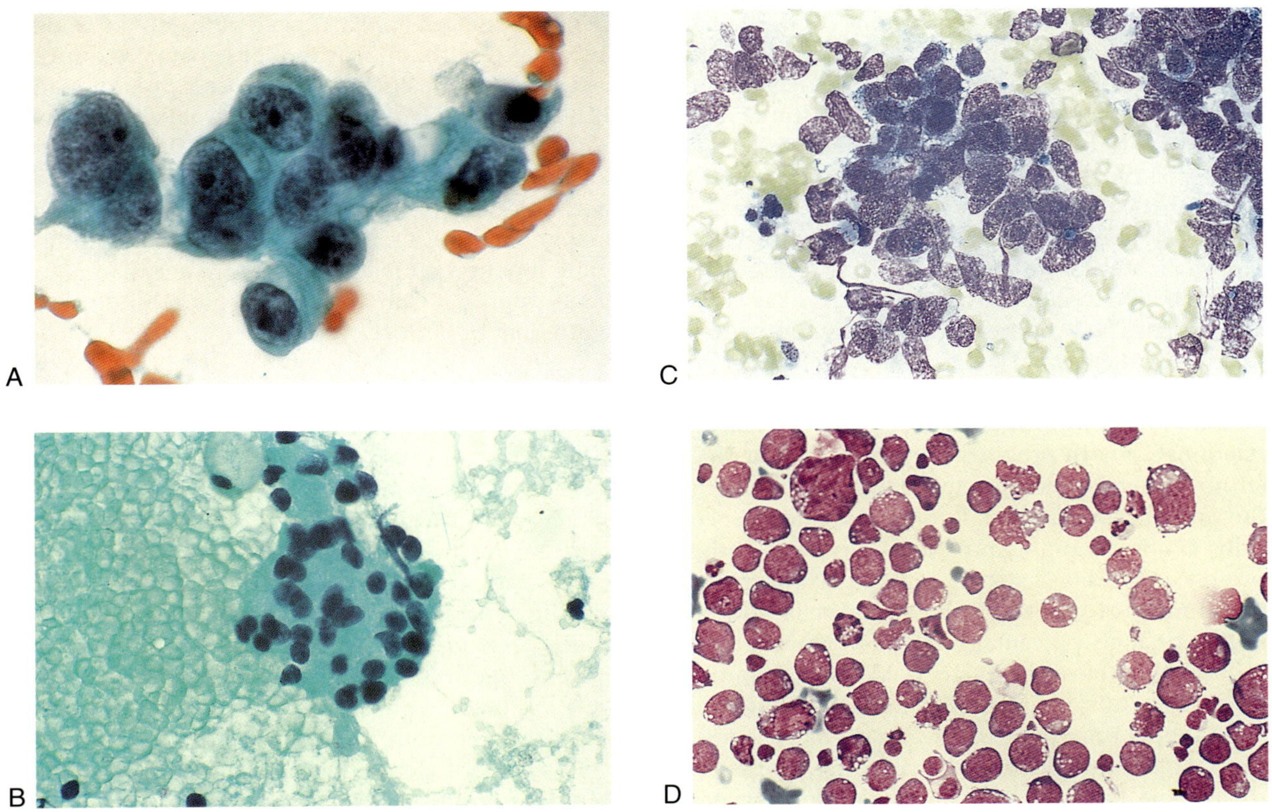

Fig. 6-14. (**A**) Metastatic poorly differentiated adenocarcinoma of the lung to adrenal gland. The tumor cells were cytokeratin and CEA positive. The vacuolated cytoplasm could potentially be confused with intracytoplasmic lipid seen in an aspirate of adrenocortical carcinoma. Positive mucin stain would also exclude a primary adrenocortical carcinoma (Papanicolaou). (**B**) FNA of metastatic prostatic carcinoma to the adrenal gland, revealing microacinar formation. Tumor cells were positive for prostatic specific antigen (Papanicolaou). (**C**) FNA of small cell carcinoma of the lung to the adrenal gland. Note nuclear angularity and molding and lack of frothy lipid material in the background (Diff-Quik). (**D**) FNA of high-grade lymphoma involving the adrenal gland. Note disperse cell pattern, nucleoli, and nuclear irregularity (Diff-Quik).

group isoantigens. In our experience, the cytologic features of metastatic malignancies is straightforward, since the aspirates resembled the primary tumors. Mitchell et al.[31] previously reported that pulmonary adenocarcinoma metastatic to the adrenal gland can mimic normal adrenocortical epithelium in the FNA biopsy specimens. We have not experienced this potential problem.

NEUROBLASTOMA

Neuroblastoma is the third most common malignant neoplasm of childhood and, excluding central nervous system tumors, the most common solid nonlymphoreticular tumor of infancy and childhood.[11] Most are diagnosed before 5 years of age, with approximately 50 percent of patients presenting clinically before 2 years of age.[32–44] Neuroblastoma is the most common malignant tumor of the neonate, with both congenital and fetal cases possible.[33] Neuroblastoma arises from any site containing sympathetic neural tissue; thus, retroperitoneal and adrenal lesions are most commonly observed.[32,33] Approximately one-third of patients present with metastasis to regional lymph nodes, liver, bones, and lungs. The prognosis correlates with the age at diagnosis, the primary site, and stage, rather than histologic grade.[32] However, the relatively recent Shimada classification demonstrates a histologic relationship with prognosis, based on the degree of stromal development, cellular differentiation and the presence of mitotic-karyorrhectic cells.[34] Specifically, a more favorable prognosis correlates with the presence of either a stroma-rich matrix, decrease in the number of mitotic-karyorrhectic cells (low MKI index) or high percentage of differentiating tumor cells. Children less than 1 year old also have a better prognosis. Adrenal neuroblastomas pursue a less favorable course than do extra-adrenal lesions. In contrast to many other pediatric malignancies, neuroblastoma has generally failed to demonstrate an improved clinical course with contemporary chemotherapeutic agents. Consequently, an accurate FNA diagnosis of neuroblastoma is essential.

Grossly, adrenal neuroblastoma generally presents as a single discrete round mass with a smooth external surface or as large multinodular masses.[15] Some tumors may demonstrate extensive necrosis with cystic degeneration and hemorrhage. Histologically, neuroblastic cells are arranged diffusely or in nests separated by thin fibrovascular septae. By our definition, neuroblastoma is composed predominantly (greater than 50 percent) of neuroblasts and neuropils. By contrast, ganglioneuromas are composed exclusively of mature ganglion cells, neurites accompanied by Schwann cells, and fibrous tissue.[45] Ganglioneuroblastoma consists predominantly of ganglioneuromatous elements (greater than 50 percent) and a relatively small neuroblastomatous component.

Features of Neuroblastoma

- Small cells with high nuclear to cytoplasmic ratios
- Fine and coarsely granular chromatin without prominent nucleoli
- Wispy, cytoplasmic, unipolar tags
- Neuropil
- Homer Wright rosettes
- Ganglion cells in ganglioneuroblastoma
- Schwannian-type spindle-shaped cells, collagen, and ganglion cells in ganglioneuroma

FNA cytologic features of neuroblastoma include hypercellular smears with both numerous individually scattered small malignant cells and cohesive aggregates[35] (Fig. 6-15), as shown above. The nuclei of neuroblastic cells vary from oval to slightly irregular, have a finely granular evenly distributed chromatin pattern and inconspicuous nucleoli. Nuclear molding and Homer Wright rosettes are common in cohesive groupings. Some of the neuroblastic cells have wispy, unipolar, cellular processes (Fig. 6-15). A very helpful diagnostic finding is the presence of neuropil seen both in the Diff-Quik and Papanicolaou-stained material (Fig. 6-16). Other features that can be appreciated in the smears include evidence of necrosis, mitotic activity, and rarely calcification. Neoplastic ganglion cells with one to several nucleoli are found in ganglioneuroblastomas. Ganglion cells have a more coarsely granular chromatin, large nucleoli and a moderate volume of cytoplasm. Ganglioneuromas are characterized by ganglion cells mixed with spindle-shaped schwannian-type cells with surrounding collagen (Fig. 6-17). Cells corre-

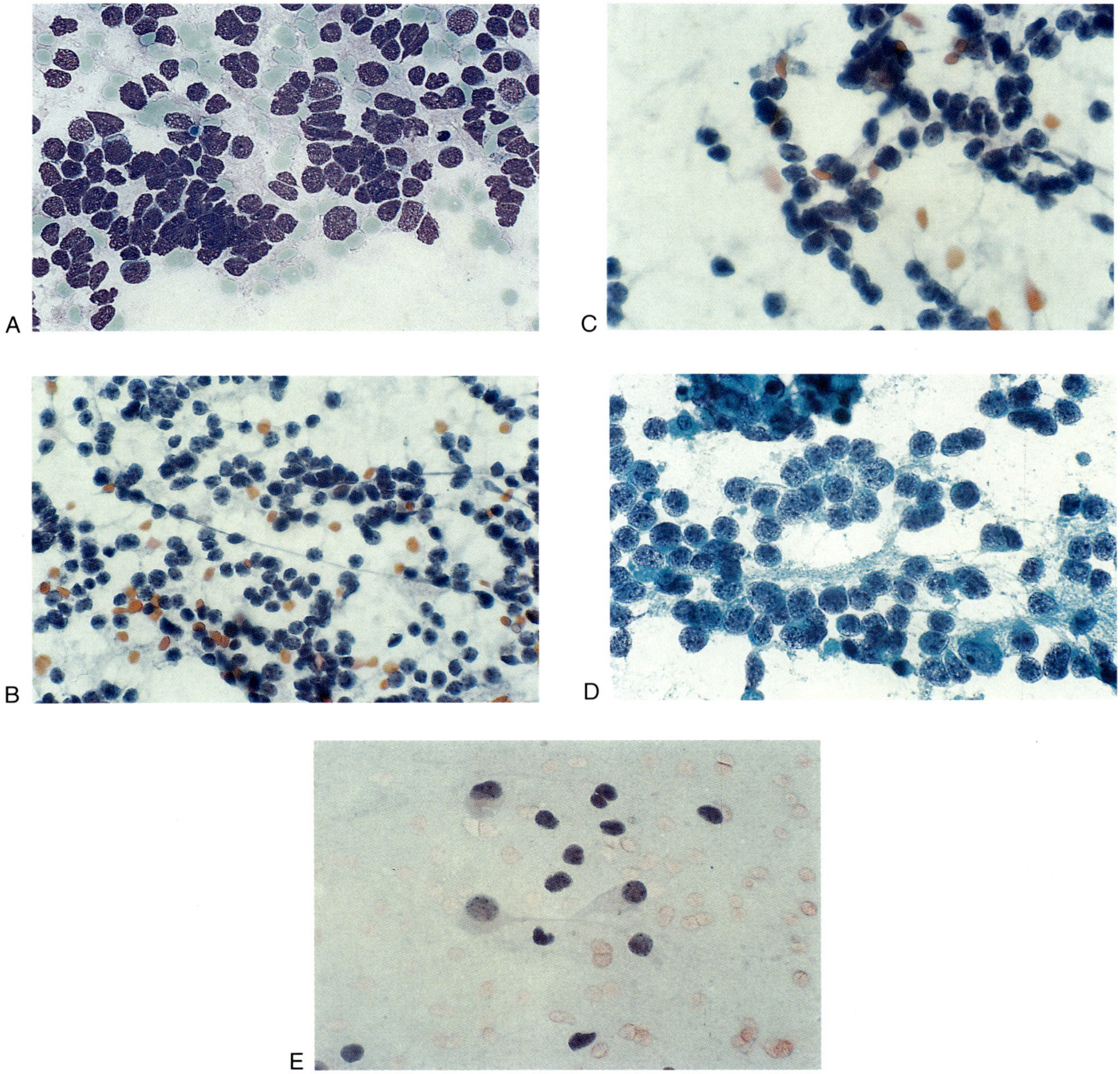

Fig. 6-15. (**A**) FNA of neuroblastoma, consisting of small malignant cells arranged individually and in loose clusters. Note the nuclear molding (Diff-Quik). (**B**) FNA of neuroblastoma, consisting of numerous small malignant cells with high nuclear to cytoplasmic ratios and cells arranged predominately in a dissociative fashion (Papanicolaou). (**C**) FNA of neuroblastoma, showing small malignant cells having protypical "salt and pepper" chromatin with some nuclear angularity (Papanicolaou). (**D**) FNA of neuroblastoma showing disperse neuroblastic cells, including a rare group attempting to form a Homer Wright rosette (Papanicolaou). (**E**) FNA of neuroblastoma, demonstrating some neuroblastic cells with unipolar axonal processes (Papanicolaou).

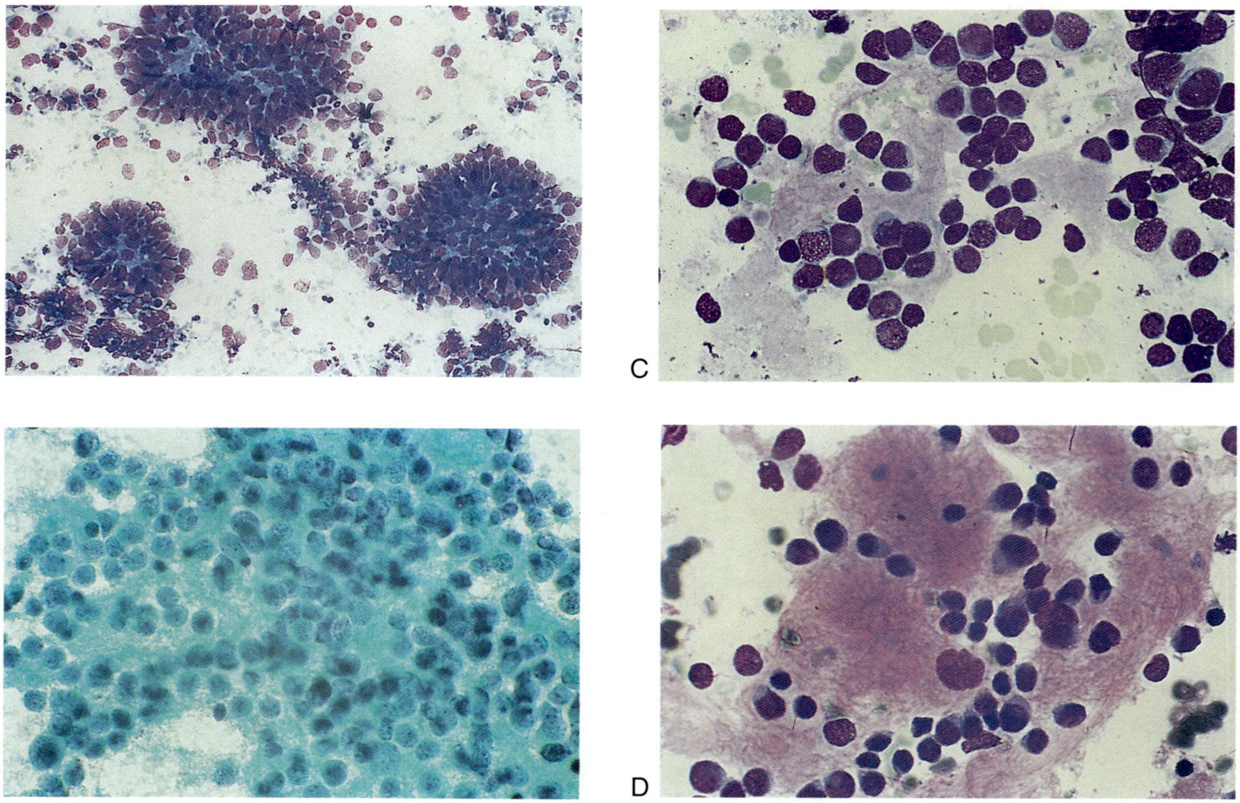

A
B
C
D

Fig. 6-16. (**A**) FNA of neuroblastoma, showing scattered Homer Wright rosettes. Some of the rosettes contain central neuropil (Diff-Quik). (**B**) FNA of neuroblastoma, showing scattered neuroblastic cells interspersed with fibrillary neuropil (Papanicolaou). (**C**) Scattered neuroblastic cells with interspersed fibrillary neuropil (Diff-Quik). (**D**) High power of FNA of neuroblastoma, showing interspersed neuroblastic cells, including some larger differentiated neuroblasts (Diff-Quik).

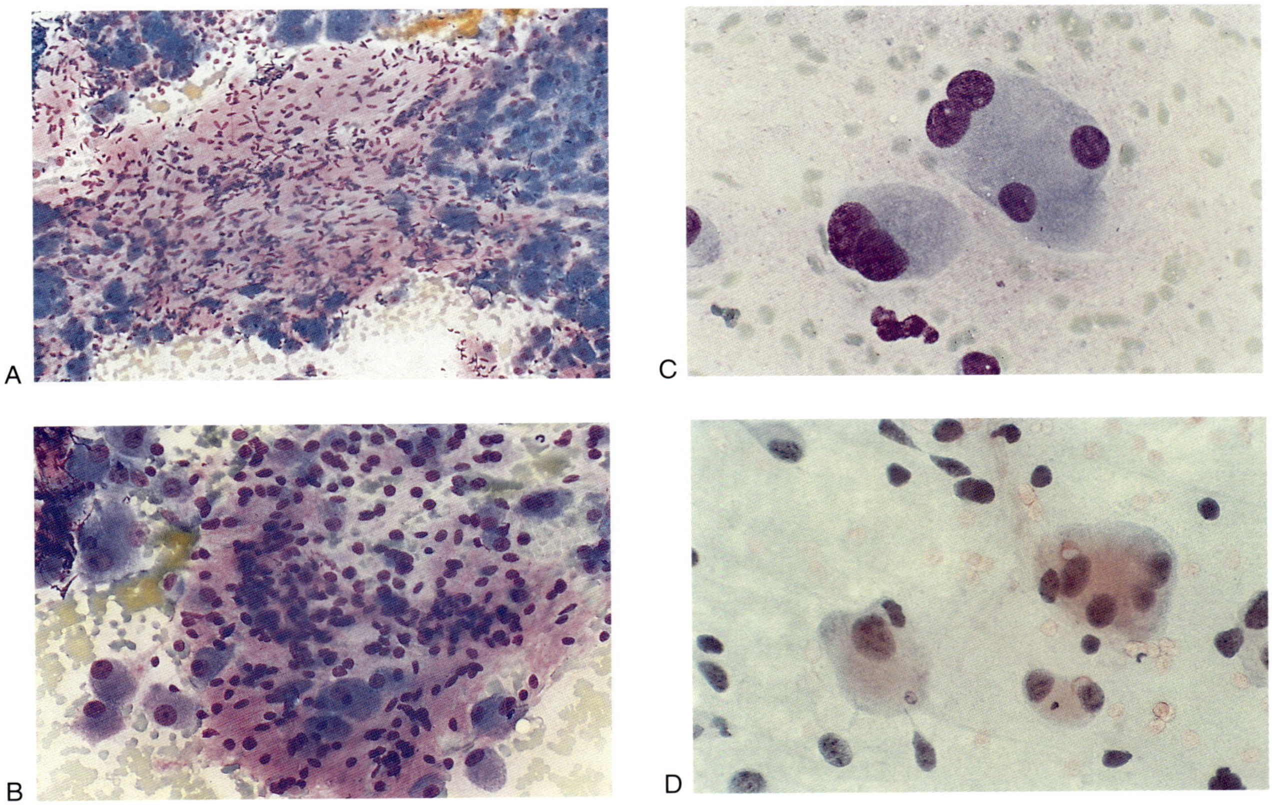

Fig. 6-17. (**A**) FNA of ganglioneuroblastoma, showing spindle-shaped neuronal component with some scattered ganglion cells (Diff-Quik). (**B**) FNA of ganglioneuroblastoma, showing interspersed smaller neuroblastic cells set in the fibrous matrix of the neuronal component (Diff-Quik). (**C**) High-power view of ganglion cells, demonstrating abundant cytoplasm and multinucleation with small nucleoli (Diff-Quik). (**D**) High-power view of scattered multinucleated ganglion cells, with some smaller neuroblastic cells in the background (Papanicolaou).

sponding to differentiating neuroblasts possess greater volumes of cytoplasm.

Ancillary studies are essential in the complete workup of small round tumors[46] (Table 6-1). Characteristic ultrastructural findings of neuroblastoma include cytoplasmic small dense-core neurosecretory type granules and abundant dendritic processes containing microtubules. Immunocytochemical studies supporting the diagnosis of neuroblastoma include positive staining for NSE and/or neurofilament proteins. Potential diagnostic pitfalls include mistaking neuroblastoma for other abdominal pediatric small round cell tumors.[35–50] This includes aspirates of lymphomas such as lymphoblastic and Burkitt's lymphomas, which consist of malignant cells with lymphoid features and containing lymphoglandular bodies in the background of the Diff-Quik–stained smears. Aspirates of lymphoblastic lymphoma, which characteristically presents in cervical or mediastinal lymph nodes, have cells with oval to convoluted nuclei with delicate dispersed nuclear chromatin, inconspicuous nucleoli, and scant surrounding cytoplasm. True cohesion, nuclear molding, and nucleoli should not be present. Burkitt's lymphoma, which frequently present intra-abdominally, yields lymphoid cells of intermediate size, with nuclei having fine to moderately granular chromatin, 2 to 5 nucleoli, and cytoplasm with lipid-laden vacuoles. Scattered histiocytes may also be present in the smears. Wilms' tumor, or nephroblastoma (see Ch. 5), may show the classic triphasic cytologic findings of blastemal type cells, epithelial cells in tubular or glomeruloid groupings, and spindled stromal cells. A monophasic blastemal pattern can be confused with neuroblastoma. Electron microscopic and immunocytochemical studies will separate these two malignancies.[51–54] Intra-abdominal desmoplastic small round cell tumor is a small cell malignancy consisting of small round, oval, spindle-shaped cells. Immunohistochemically, the primitive tumor cells exhibit multidirectional differentiation, staining (e.g., for NSE), and multiple intermediate filaments.[55] The cytologic differential diagnostic features of abdominal small round cell tumor of childhood are presented in Table 6-2.

Table 6-1. Ancillary Studies in the FNA Workup of Small Round Cell Malignancies of Childhood

Cytologic Diagnosis	*Electron Microscopy*	*Immunocytochemistry*	*Cytogenetics*
Neuroblastoma	Relatively frequent DCG and prominent microtubules and cell processes than PNET Cell junctions are primitive	NSE, S-100 Schwann cells, NF	Chromosome 1 deletion, short arm
PNET	Rare DCG Glycogen possible	NSE, vimentin Neurofilaments (NF) β-Microglobulin HBA-71	11;22 translocation
Ewing sarcoma	Glycogen, cell junctions Fine, even chromatin	Vimentin HBA-71	11;22 translocation
Burkitt's lymphoma	No specific EM features; cytoplasmic lipid vacuoles	LCA B-cell phenotype	8;14 translocation
Lymphoblastic lymphoma	No specific EM features	LCA, B- or T-cell phenotype (usually T)	
Rhabdomyosarcoma	Myosin filaments, aberrant sarcomeres Z band Glycogen	Actin, myoglobin, desmin	2;13 translocation in alveolar rhabdomyosarcoma

Abbreviations: DCG, dense-core granules; EM, electron microscopic; LCA, leukocyte common antigen; NSE, neuron-specific enolase.
(Modified from Silverman et al.,[53] with permission.)

Table 6-2. Cytologic Features of Small Round Cell Tumors of Childhood

Cytologic Features	*PNET*	*NB*	*ES*	*BL*	*LL*	*RMS*	*WT*
Nuclear molding	+	+	±	–	–	–	±
Nucleoli	–	–	–	+(2–5)	–	+	–
Ganglion cells	–	±	–	–	–	–	–
Myxoid matrix	–	–	–	–	–	±	–
Cytoplasmic vacuoles	+	–	+ (glycogen)	+ (lipid)	–	–	–
Cytoplasmic tags	±	±	±	–	–	+	–
Horner Wright rosettes	± Rare	+	± Rare	–	–	–	–
Tubules	–	–	–	–	–	–	+
Lymphoglandular bodies	–	–	–	+	+	–	–
Macrophages	±	±	–	+	±	–	–
Neuropil	–	+	–	–	–	–	–
Fibrous matrix	–	–	–	–	–	–	+
Spindle-shaped cells	–	–	–	–	–	+	±
Tadpole or ribbon-shaped cells	–	–	–	–	–	+	±
Dense cytoplasm	–	–	–	–	–	+	±
Karyorrhectic cells	–	+	–	+	±	–	–
Tigroid background	–	–	–	–	–	±	–

Abbreviations: PNET, primitive neuroectodermal tumor; NB, neuroblastoma; ES, Ewing sarcoma; BL, Burkitt's lymphoma; LL, lymphoblastic lymphoma; RMS, rhabdomyosarcoma; WT, Wilms' tumor; +, frequently present or prominent; ±, may be present (rare) or inconspicuous; –, absent.

(Modified from Silverman et al.,[53] with permission.)

REFERENCES

1. Casola G, Nicolet V, vanSonnenberg E et al: Unsuspected pheochromocytoma: risk of blood-pressure alterations during percutaneous adrenal biopsy. Radiology 159:733, 1986
2. deBlois G, DeMay RM: Adrenal myelolipoma diagnosis by computed-tomography-guided fine-needle aspiration. Cancer 55:848, 1985
3. Dunphy CH: Computed tomography-guided fine needle aspiration biopsy of adrenal myelolipoma. Case report and review of the literature. Acta Cytol 35:353, 1991
4. Galli L, Gaboardi F: Adrenal myelolipoma: report of diagnosis by fine needle aspiration. J Urol 136:655, 1986
5. Gould JD, Mitty HA, Pertsemlidis D, Szporn AH: Adrenal myelolipoma: diagnosis by fine-needle aspiration. AJR 148:921, 1987
6. Hruban RH, Bhagavan BS, Epstein JI: Massive retroperitoneal angiomyolipoma. Am J Clin Pathol 92:805, 1989
7. Katsuta K, Nakabayashi H, Kuroda Y, Liu PI: Adrenal myelolipoma: preoperative diagnosis by fine-needle aspiration cytology. Diagn Cytopathol 5:298, 1989
8. Katz RL, Shirkhoda A: Diagnostic approach to incidental adrenal nodules in the cancer patient. Results of a clinical, radiologic, and fine-needle aspiration study. Cancer 55:1995, 1985
9. Katz RL: Kidney, adrenal, and retroperitoneum. p 771. In Bibbo M (ed): Comprehensive Cytopathology. WB Saunders, New York, 1991
10. Katz RL, Patel S, Mackay B, Zornoza J: Fine needle aspiration cytology of the adrenal gland. Acta Cytol 28:269, 1984
11. Silverman JF, Katz R: FNA of the adrenal gland. p.149. In Cytopathology Annual. Williams & Wilkins, Baltimore, 1994
12. Tao L-C: Primary lesions of the adrenals. p. 218. In Tao L-C (ed): Transabdominal Fine-Needle Aspiration Biopsy. Igaku-Shoin, New York, 1990
13. Nguyen G-K: Percutaneous fine-needle aspiration biopsy cytology of the kidney and adrenal. p. 163. In Rosen PP, Fechner RE (eds): Pathology Annual. Vol. 22. Appleton & Lange, E. Norwalk, 1987
14. Scheible W, Coel M, Siemers PT, Siegel H: Percutaneous aspiration of adrenal cysts. AJR 128:1013, 1977
15. Lack EE, Travis WD, Oertel JE: Adrenal cortical nodules, hyperplasia, and hyperfunction. p. 75. In Lack EE (ed): Pathology of the Adrenal Glands. Churchill Livingstone, New York, 1990
16. Anderson CA, Pi WC, Weiss LM: Disseminated histoplasmosis diagnosed by fine needle aspiration biopsy of the adrenal gland. A case report. Acta Cytol 33:337, 1989
17. Valente PT, Calafati SA: Diagnosis of disseminated histoplasmosis by fine needle aspiration of the adrenal gland. Acta Cytol 33:341, 1989

18. Schonfeld AD, Smith DJ, Jackson JA, Hurley DL: Disseminated histoplasmosis with bilateral adrenal enlargement: diagnosis by computed tomography-directed needle biopsy. Tex Med 87:88, 1991
19. Powers CN, Rupp GM, Maygarden SJ, Frable WJ: Fine-needle aspiration cytology of adrenal cryptococcosis: a case report. Diagn Cytopathol 7:88, 1991
20. Pinto MM: Fine needle aspiration of myelolipoma of the adrenal gland. Report of a case with computed tomography. Acta Cytol 29:863, 1985
21. Wadih GE, Nance KV, Silverman JF: Fine-needle aspiration cytology of the adrenal gland: fifty biopsies in 48 patients. Arch Pathol Lab Med 116:841, 1992
22. Medeiros LJ, Weiss LM: New developments in the pathologic diagnosis of adrenal cortical neoplasms. A review. Am J Clin Pathol 97:73, 1992
23. Suen KC: Guides to Clinical Aspiration Biopsy. In Kline T (ed): Retroperitoneum and Intestine. p. 183. Igaku-Shoin, New York, 1987
24. Min K-W, Song J, Boesenberg M, Acebey J: Adrenal cortical nodule mimicking small round cell malignancy on fine needle aspiration. Acta Cytol 32:543, 1988
25. Gaffey MJ, Traweek ST, Mills SE et al: Cytokeratin expression in adrenocortical neoplasia: an immunohistochemical and biochemical study with implications for the differential diagnosis of adrenocortical, hepatocellular, and renal cell carcinoma. Hum Pathol 23:144, 1992
26. Weiss LM, Medeiros LJ, Vickery AL, Jr: Pathologic features of prognostic significance in adrenocortical carcinoma. Am J Surg Pathol 13:202, 1989
27. Chan JKC, Tsang WYW: Endocrine malignancies that may mimic benign lesions. Semin Diagn Pathol 12:45, 1995
28. Nance KV, McLeod DL, Silverman JF: Fine-needle aspiration cytology of spindle cell neoplasms of the adrenal gland. Diagn Cytopathol 8:235, 1992
29. McCorkell SJ, Niles NL: Fine-needle aspiration of catecholamine-producing adrenal masses: a possibly fatal mistake. AJR 145:113, 1985
30. Saboorian H, Katz R, Boyd D, Charnsangavej C: Fine needle aspiration (FNA) cytology of primary and metastatic lesions of adrenal gland. Mod Pathol 5:28A, 1992
31. Mitchell ML, Ryan FP, Jr, Shermer RW: Pulmonary adenocarcinoma metastatic to the adrenal gland mimicking normal adrenal cortical epithelium on fine needle aspiration. Acta Cytol 29:994, 1985
32. Triche TJ, Askin FB: Neuroblastoma and the differential diagnosis of small-, round-, blue-cell tumors. Hum Pathol 14:569, 1983
33. Triche TJ, Askin FB, Kissane JM: Neuroblastoma, Ewing's sarcoma, and the differential diagnosis of small-, round-, blue-cell tumors. p. 145. In Finegold MJ (ed): Pathology of Neoplasia in Children and Adolescents. Vol. 18. WB Saunders, Philadelphia, 1986
34. Shimada H, Chatten J, Newton WA, Jr et al: Histopathologic prognostic factors in neuroblastic tumors: definition of subtypes of ganglioneuroblastoma and age-linked classification of neuroblastomas. J Natl Cancer Inst 73:405, 1984
35. Silverman JF, Dabbs DJ, Ganick DJ et al: Fine needle aspiration cytology of neuroblastoma, including peripheral neuroectodermal tumor, with immunocytochemical and ultrastructural confirmation. Acta Cytol 32:367, 1988
36. Askin FB, Rosai J, Sibley RK, Dehner LP: Malignant small cell tumor of the thoracopulmonary region in childhood. Cancer 43:2438, 1979
37. Linnoila RI, Tsokos M, Triche TJ et al: Evidence for neural origin and PAS positive variants of the malignant cell tumor of thoracopulmonary region ("Askin tumor"). Am J Surg Pathol 10:124, 1986
38. De Chadarevian JP, Vekemans M, Seemayer TA: Reciprocal translocation in small cell sarcomas. N Engl J Med 322:1702, 1984
39. Maletz N, McMorrow LE, Greco A, Wolman S: Ewing's sarcoma: pathology tissue culture and cytogenetics. Cancer 58:252, 1986
40. Whang-Peng J, Triche TJ, Knutsen T et al: Chromosome translocation in peripheral neuroepithelioma. N Engl J Med 311:584, 1984
41. Tsokos M, Linnoila RI, Chandra RS, Triche TJ: Neuron-specific enolase in the diagnosis of neuroblastoma and other small, round cell tumors of childhood. Hum Pathol 15:575, 1984
42. Look AT, Hayes FA, Nitschke R et al: Cellular DNA content as a predictor of response to chemotherapy in infants with uresectable neuroblastoma. N Engl J Med 15:575, 1984
43. Shimada H, Seeger R, Joshi V et al: Histopathology and N-myc gene in advanced neuroblastomas: a report from the Children's Cancer Study Group (CGSG). Pediatr Pathol 8:671, 1988
44. Ganick DJ, Silverman JH, Holbrook CT et al: Clinical utility of the fine needle aspiration in the diagnosis and management of neuroblastoma. Med Pediatr Oncol 16:101, 1988
45. Joshi VV, Silverman JF, Altshuler G et al: Pathology of neuroblastic tumors: a report from the Pediatric Oncology Group. I. Systematization of primary pathologic features. Hum Pathol 24:493, 1993

46. Gurley AM, Silverman JF, Wiley JE et al: The utility of ancillary studies in pediatric FNA cytology. Diagn Cytopathol 8:137, 1992
47. Palombini L, Vertrani A, Vecchione R et al: The cytology of ganglioneuroma on fine needle aspiration smear. Acta Cytol 26:259, 1982
48. Romansky SG, Crocker DW, Shaw KNF: Ultrastructural studies on neuroblastoma: evaluation of cytodifferentiation. Cancer 42:2392, 1978
49. Taylor SR, Nunez C: Fine needle aspiration biopsy in a pediatric population: report of 64 consecutive cases. Cancer 54:1449, 1984
50. Geisinger KR, Silverman JF, Wakely PE: Pediatric Cytopathology. ASCP Press, Chicago, 1995
51. Strausbauch PH, Neill JSA, Benning TL et al: Applications of electron microscopy in fine-needle aspiration cytology. p. 173. In Smith WA (ed): Cytopathology Annual. Williams & Wilkins, Baltimore
52. Silverman JF, Joshi VV: Fine-needle aspiration biopsy of small round cell tumors of childhood: cytomorphologic features and the role of ancillary studies. Diagn Cytopathol 10:245, 1994
53. Silverman JF, Berns LA, Holbrook CT et al: Fine-needle aspiration of primitive neuroectodermal tumors (PNET). A report of three cases. Acta Cytol 36:541, 1992
54. Joshi VV, Silverman JF: Pathology of neuroblastic tumors. Semin Diagn Pathol 11:107, 1994
55. Akhtar M, Ali MA, Sabbah R et al: Small round cell tumor with divergent differentiation: cytologic, histologic, and ultrastructural findings. Diagn Cytopathol 11:159, 1994

7 Retroperitoneum and Pelvis

The retroperitoneum may be loosely defined as the space from the posterior parietal peritoneum back to the deep abdominal wall and from the twelfth thoracic vertebrae to the distal borders of the iliac crest and the sacrum. Organs and tissues normally present in this space include the kidneys, ureters, adrenal glands, pancreas, duodenum, ascending and descending portions of the colon, lymph nodes, and large vascular and neural structures. The clinical aspiration cytomorphology of many of these structures is discussed in Chapters 2, 4, and 5. Some neoplasms that characteristically occur in the retroperitoneum, such as extragonadal germ cell tumors, are also discussed in Chapter 2. Soft tissue sarcomas are one of the major tumor types to occur in the retroperitoneum and pelvis. Much of this chapter is dedicated to the discussion of the aspiration cytology of this group of malignancies.

SOFT TISSUE SARCOMAS

Although soft tissue tumors may arise almost anywhere in the body, a considerable proportion of them arise in the retroperitoneum and pelvis.[1] In our experience, a large proportion of all primary cytologic diagnoses of soft tissue sarcomas result from aspirates of this body site.

Soft tissue tumors may be defined as neoplasms and related proliferations that arise in, and/or that morphologically recapitulate, the nonepithelial, nonlymphoreticular extraskeletal tissues of the body. Both benign and malignant soft tissue tumors have traditionally been classified on a histogenetic basis. Certainly, this does have clinical significance, but from practical diagnostic and prognostic viewpoints, both clinical staging and histologic grading of sarcomas are more important than determining, with relatively few exceptions, the exact cell type comprising the soft tissue mass. Obviously, the separation of benign from malignant soft tissue tumors is absolutely crucial. Compared to other body sites, such as the extremities, a very large proportion of all soft tissue tumors in the retroperitoneum are sarcomatous. In addition, many retroperitoneal sarcomas carry a poor prognosis, as they may attain huge dimensions before their diagnosis. Many surgeons and other physicians believe that formal, open tissue biopsy is the only appropriate procedure for the initial diagnosis of a sarcoma. Fine needle aspiration (FNA) biopsy is a rapid and technically simple procedure that may permit a definitive diagnosis of primary soft tissue neoplasms.[2–11] The major goal of distinguishing benign from malignant proliferations may be accomplished with a relatively high degree of accuracy by aspiration biopsy. Cytomorphologically, the most important feature in this decision process is the overall cellularity of the smears. Sarcomas generally yield moderately to highly cellular specimens (Figs. 7-1 and 7-2), whereas benign mesenchymal lesions usually are more sparsely cellular. Other key cytologic ingredients of sarcomas include a relatively uniform hyperchromasia among the aspirated cells, and the presence of moderate to marked pleomorphism (Fig. 7-2). These features need to be inte-

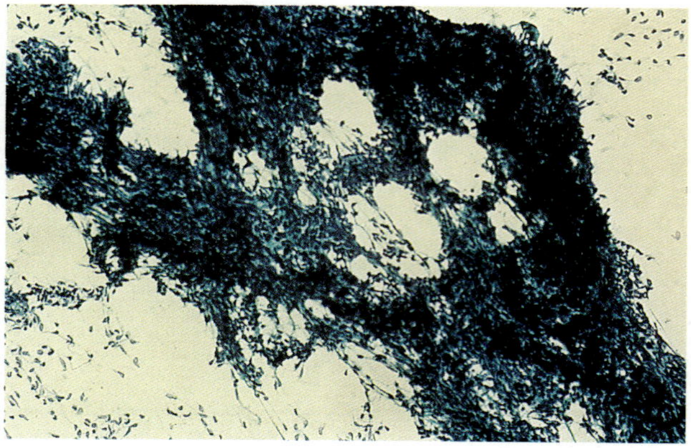

Fig. 7-1. Aspirates of soft tissue sarcomas yield specimens that were at least moderately and generally very highly cellular. Literally thousands of malignant cells are present in this field of an aspirated synovial sarcoma. Many are present in thick three-dimensional clumps, but others are present in loosely arrayed monolayers and as individually dispersed elongated cells (Papanicolaou).

Fine Needle Aspiration Biopsy Classification of Soft Tissue Sarcomas: Predominant Light Microscopic Appearance

- Myxoid sarcomas
- Spindle cell sarcomas
- Pleomorphic sarcomas
- Polygonal cell sarcomas
- Round cell sarcomas

grated with the relevant clinical data including the patient's age, the presumed site of origin of the tumor, and its estimated size.

From an FNA biopsy perspective, most soft tissue sarcomas can be divided into five general categories on the basis of predominant patterns seen at low magnification,[7,10,11] as shown in the classification above. The myxoid sarcoma category requires the recognition of a well-developed extracellular matrix, whereas the other four are based on the predominant appearance of the tumor cells.[12,13] The round cell sarcomas predominate in children and adolescents, whereas the other four occur mostly in adults.[14–19] Compared to tissue sections, the ability to classify accurately a specific sarcoma as to histogenetic type is limited.[20] However, in most settings, the precise histologic type of sarcoma is not absolutely important. These five categories largely correlate with both histologic grade and with prognosis, and thus provide an adequate basis upon which therapeutic decisions can be rendered.

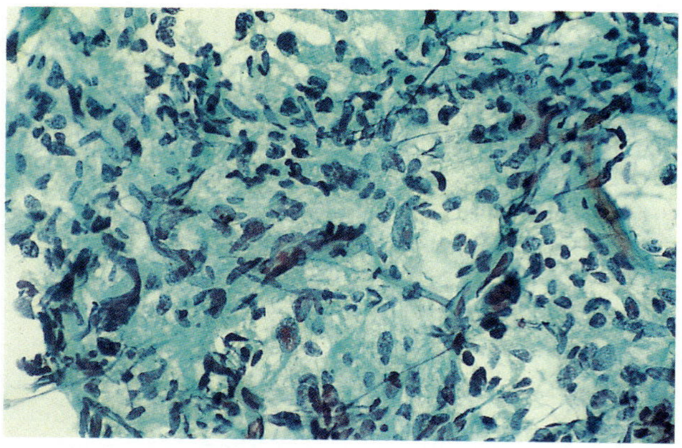

Fig. 7-2. Aspirate of a malignant fibrous histiocytoma, manifesting a moderate level of pleomorphism and uniform hyperchromasia among the neoplastic cells, as well as a high level of smear cellularity. Macronucleoli are present in some of the nuclei (Papanicolaou).

Although almost all soft tissue sarcomas occur with varying frequencies in the retroperitoneum and pelvis, a few of them arise almost exclusively in more superficial sites (e.g., the extremities). Therefore, for completeness's sake, this chapter will discuss the aspiration cytomorphology of all major forms of sarcomas.

MYXOID SARCOMAS

The myxoid soft tissue sarcomas include myxoid malignant fibrous histiocytoma, myxoid liposarcoma, and myxoid chondrosarcoma. Myxoid sarcomas are characterized by abundant extracellular matrix material.[11–13] This matrix stains metachromatically with the Romanowsky stains and thus appears as a rather homogeneous reddish-purple substance (Figs. 7-3 and 7-4). Usually, it is more apparent with the Diff-Quik stain than with the Papanicolaou stain; in the latter, it usually appears as a pale green material (Fig. 7-3). Neoplastic cells, both singly and in small aggregates, appear embed-

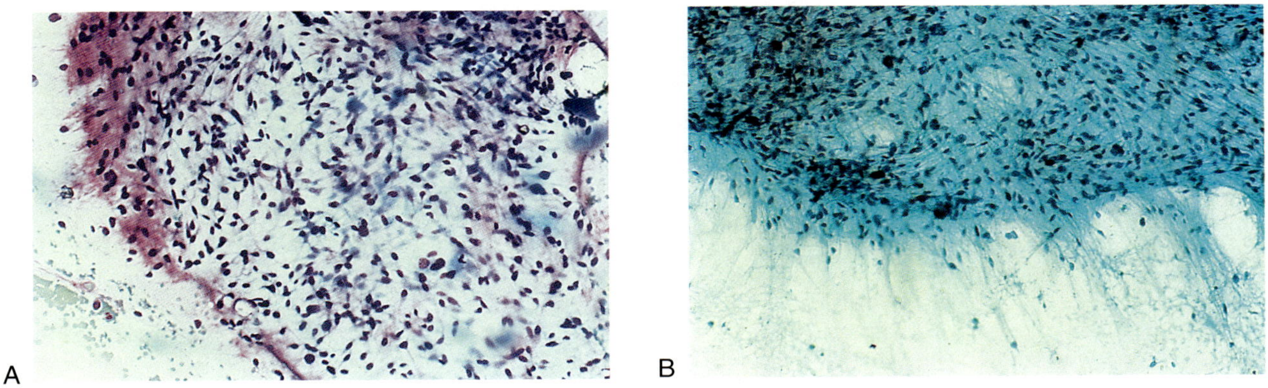

Fig. 7-3. Classic presentation of a myxoid sarcoma in an aspiration biopsy. (**A**) The abundant extracellular matrix stains metachromatically and with varying intensity. Cellularity is marked as numerous small, uniform neoplastic cells criss-cross through the matrix in an apparent haphazard manner. Note the mild level of pleomorphism in this myxoid malignant fibrous histiocytoma. (**B**) Same neoplasm is seen with the Papanicolaou stain. Numerous small tumor cells are embedded within the matrix, which appears bluish green and has a distinct fibrillar quality. With alcohol fixation, the malignant cells may appear even more uniform in size (**A**, Diff-Quik; **B**, Papanicolaou).

ded within this matrix material, as well as scattered nearby in the smear background (Figs. 7-5 to 7-7).

Recently, Wakely et al.[13] evaluated a large series of aspirates of benign and malignant myxoid soft tissue tumors. These investigators used a stringent definition, that is, chondromyxoid matrix had to occupy most of the area in three or more ×10 microscopic fields. In both malignant and benign categories, the matrix may appear translucent and have rather ill-defined contours, conferring an amorphous appearance to the material. In addition, many of the malignant neoplasms had a matrix with a different quality, namely, it appeared denser with well-defined, sharp edges, as listed on page 222. This feature was best appreciated in the extraskeletal chondrosarcomas (Fig. 7-8) and was less evident with the liposarcomas and fibrous histiocytomas. Such opaque stromal matrix was uncommonly found in the smears of benign neoplasms. In Diff-Quik–stained material, Gordon and colleagues[21] have described the matrix of chondrosarcomas as having a different appearance from that of other myxoid neoplasms. Specifically, they stated that the matrix will appear reddish pink in chondrosarcoma while appearing blue to bluish red in the other lesions. This has not been our experience.[13] Although pleomorphism may be apparent among the neoplastic cells in any of the myxoid sarcomas, it is usually relatively minimal in liposarcoma and chondrosarcoma, whereas it may be moderate in fibrous histiocytoma.

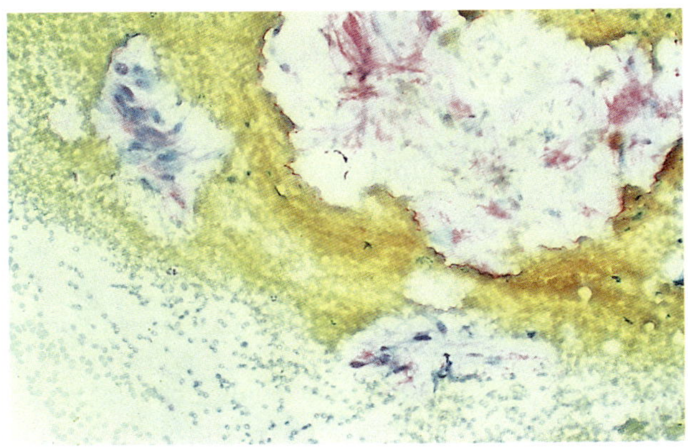

Fig. 7-4. Aspirate of a myxoid malignant fibrous histiocytoma (different from the one depicted in Fig. 7-3). Cellularity is very low in this field, and pleomorphism is not evident. The small groups of malignant cells are dispersed within the myxoid matrix. As is evident, large masses of myxoid material may be seen without any associated tumor cells. This demonstrates why adequate sampling with the needle is so important in achieving the correct malignant diagnosis (Diff-Quik).

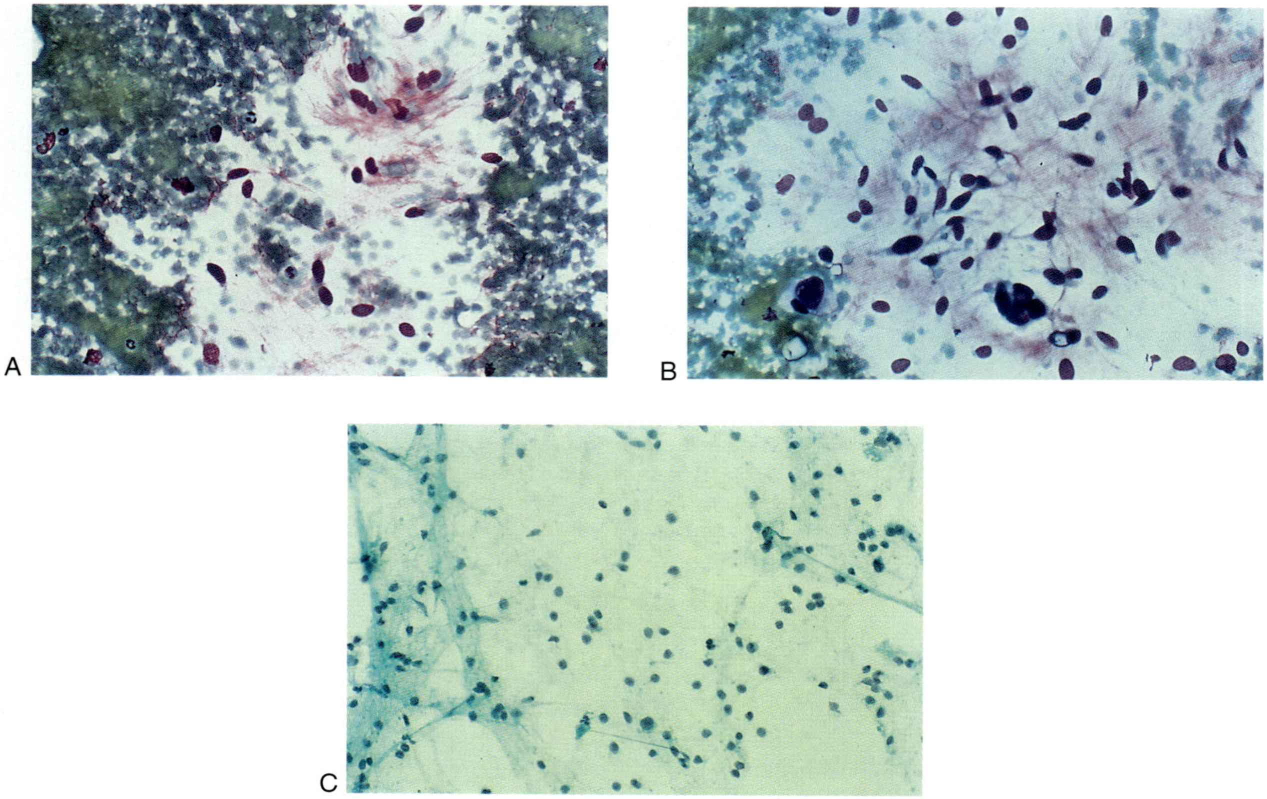

Fig. 7-5. In most myxoid sarcomas, the majority of the neoplastic cells are small, each possessing a single ovoid darkly stained nucleus and a rather high nuclear to cytoplasmic ratio. The neoplastic cells have cytoplasmic tails which appear to blend with the surrounding matrix material. (**A**) There is little evidence of pleomorphism. (**B**) Pleomorphism attains a moderate level. (**C**) From a myxoid liposarcoma. The matrix is poorly stained in this example. Although the cellularity is fairly high, the small malignant cells are strikingly uniform with very high nuclear to cytoplasmic ratios and stellate or spindled configurations (**A & B**, Diff-Quik; **C**, Papanicolaou).

Ninety-five percent of the sarcomas were at least moderately cellular (Fig. 7-3), but this feature was also present in 18 percent of benign cases.[13] More than one-half of sarcomas showed at least a moderate level of nuclear pleomorphism. This trait was seen in only 9 percent of the benign aspirates. Only a few obviously malignant cells will be present in the aspirates of most myxoid sarcomas (Fig. 7-7). Generally, the neoplastic cells are small and are either spindled or stellate in contour (Figs. 7-5 and 7-6). Nuclei are usually small to moderate in size, hyperchromatic, and single. However, multinucleated tumor cells may also be present (Figs. 7-5B and

Cytomorphologic Features of Myxoid Sarcomas

- Abundant extracellular matrix
- Moderate cellularity
- Mild to moderate pleomorphism
- Tumor cells
  - Generally small
  - Stellate or spindled
  - Small hyperchromatic nucleus
  - High nuclear to cytoplasmic ratio

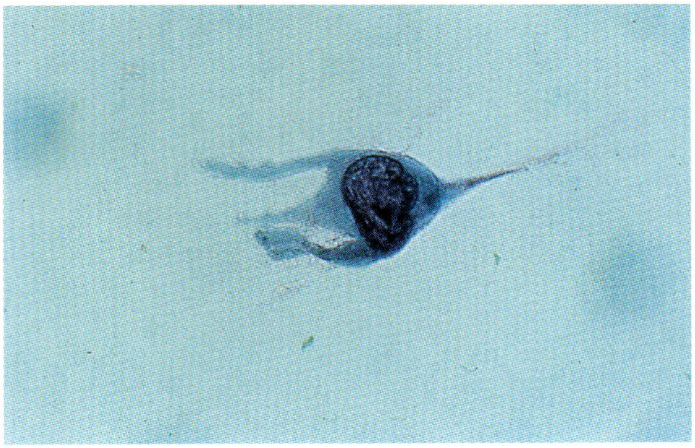

Fig. 7-6. When the malignant cells lie free of the matrix, their stellate contours may be better appreciated. Although this nucleus is hyperchromatic and irregular in shape, nucleoli are inconspicuous (Papanicolaou).

7-7). The nuclear to cytoplasmic ratios are generally high. The tumor cells may lie free within the smears, as well as embedded within the matrix. The long cytoplasmic processes may appear to connect adjacent neoplastic cells and to blend into the adjacent matrix material. The stellate contours of tumor cells may be better appreciated when they lie free of the matrix (Fig. 7-6).

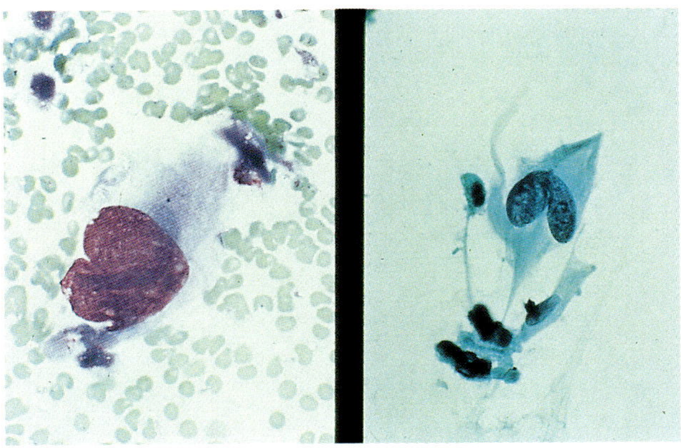

Fig. 7-7. Scattered obviously malignant cells fully consistent with sarcoma are often present in low numbers in smears from myxoid sarcomas. In the setting of abundant myxoid material, one should definitely search smears quite carefully for such elements. In binucleated sarcoma cells, one characteristically sees an acute angle formation between the two malignant nuclei (**A**, Diff-Quik; **B**, Papanicolaou).

Once a myxoid tumor is identified as malignant, it is not always possible to subtype it histologically. However, some aspirates do provide histogenetic clues. If one appreciates a prominent capillary network and identifiable lipoblasts (Figs. 7-9 and 7-10), these features indicate a myxoid liposarcoma. Lipoblasts may be defined as cells having sharply delineated lipid vacuoles in their cytoplasm which displace and distort the nuclei (Figs. 7-9 and 7-10). More specifically, these vacuoles may appear to create a scalloped nuclear contour. Cytoplasmic vacuoles may be seen in other myxoid sarcomas as well, but they are generally small and do not decisively affect nuclear position or shape (Fig. 7-11). In myxoid chondrosarcomas, the neoplastic cells may be present within lacunar spaces, although this is an uncommon phenomenon. Malignant chondrocytes are often oval to round and may be binucleated (Fig. 7-8). Niemann et al.[22] have described eccentric positioning of the neoplastic nuclei as diagnostically distinctive for the myxoid chondrosarcomas. Ancillary diagnostic studies may be helpful in this distinction. Chondrosarcoma, and possibly liposarcoma, will be positive for S-100 protein by immunocytochemistry, whereas malignant fibrous histiocytoma will be nonreactive. Ultrastructural examination of aspirated material has little to offer except for one specific situation: the cytoplasm of myxoid chondrosarcoma cells characteristically possesses parallel arrangements of small tubular structures within distended cisternae of rough endoplasmic reticulum.[21]

Within the retroperitoneum and pelvis, the differential diagnosis of myxoid sarcomas includes entities such as mucinous adenocarcinomas and chordomas. Some adenocarcinomas, especially those arising in the large intestine, may be characterized by abundant extracellular mucin. In smears, this mucin may be indistinguishable from sarcomatous stroma. However, the carcinoma cells may form tight three-dimensional aggregates and glandular structures, confirming their epithelial nature, versus the poorly developed intercellular cohesion in aspirates of the sarcomas (Fig. 7-12). Furthermore, carcinomatous elements have more distinct inter-

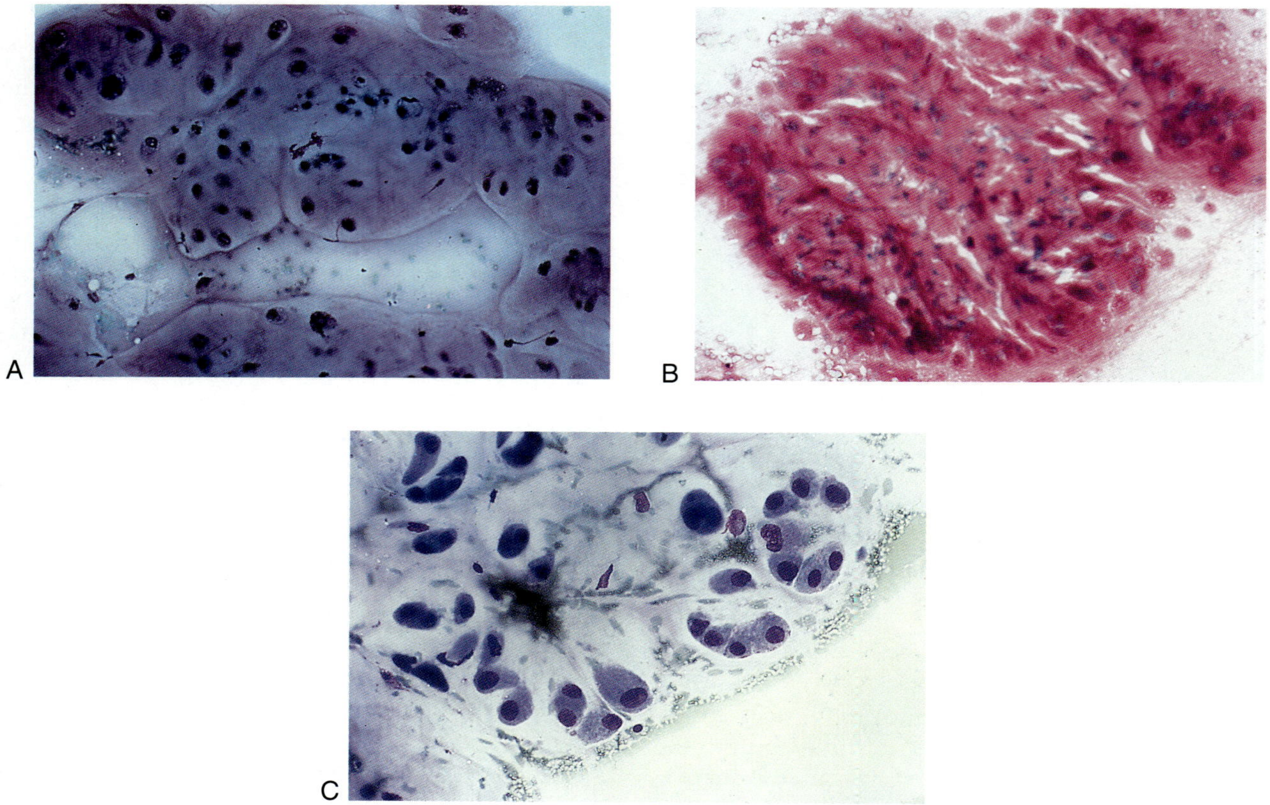

Fig. 7-8. (**A–C**) The matrix material in aspirates of extraskeletal myxoid chondrosarcomas may appear more opaque and have more sharply defined edges than are seen in the other myxoid soft tissue tumors. Some of these malignant cells appear to be present within distinct lacunae. (**C**) The matrix material stains less intensely, allowing the individual cells to be better appreciated. These malignant chondrocytes have more abundant cytoplasm and more sharply defined cell borders than are expected in most myxoid sarcomas (Diff-Quik). (Courtesy of Dr. Paul Wakely, Jr., Richmond, VA.)

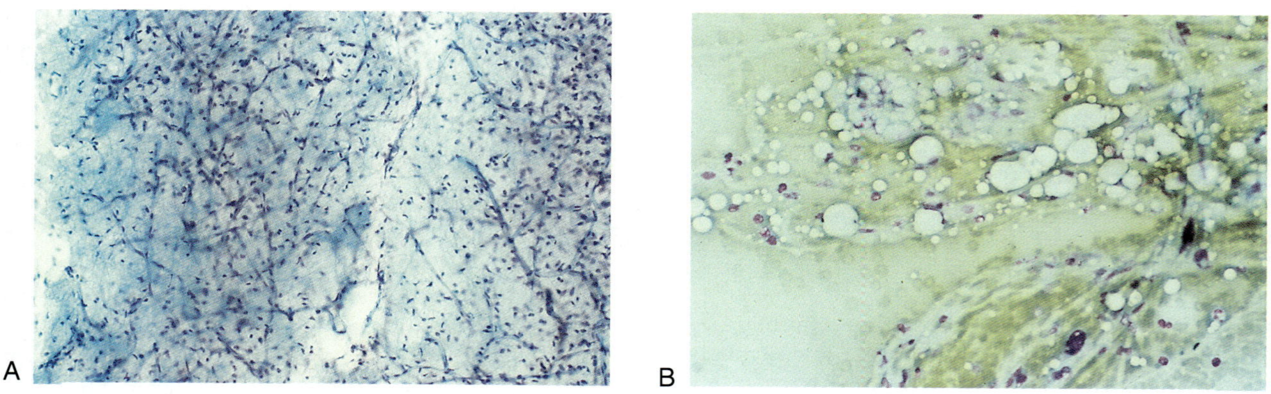

Fig. 7-9. These are aspirates from two different retroperitoneal myxoid liposarcomas. (**A**) The classic (chicken-wire) vasculature of delicate interlacing capillaries is readily apparent. Also present are numerous small spindled and stellate tumor cells set within the myxoid matrix material. (**B**) The outlines of the capillary vascular network can be appreciated as can numerous lipoblasts (**A**, Papanicolaou; **B**, Diff-Quik).

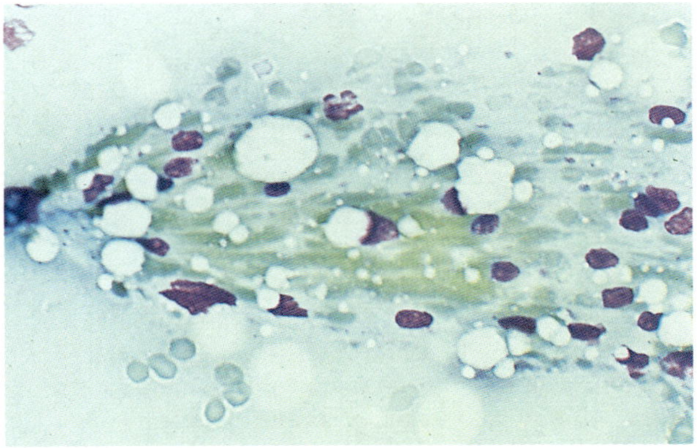

Fig. 7-10. In this myxoid liposarcoma, one or more sharply delineated lipid vacuoles occupy the cytoplasm of the tumor cells and displace and distort the malignant nuclei. Relative to vacuoles seen in the cytoplasm of other soft tissue sarcomas, lipid vacuoles in liposarcoma tend to be larger (Diff-Quik).

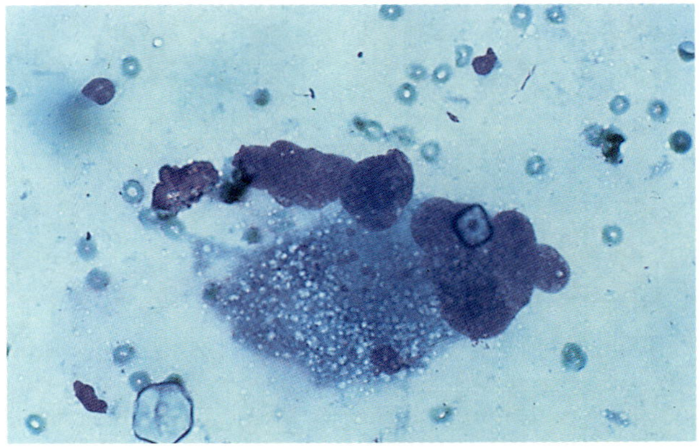

Fig. 7-11. Cytoplasmic vacuoles may be seen in other myxoid sarcomas, but they are generally smaller and do not decisively affect the nuclear position or shape. This tumor cell is from a myxoid malignant fibrohistiocytoma. Numerous minute vacuoles are present that do not appreciably alter nuclear or cellular contours (Diff-Quik).

A 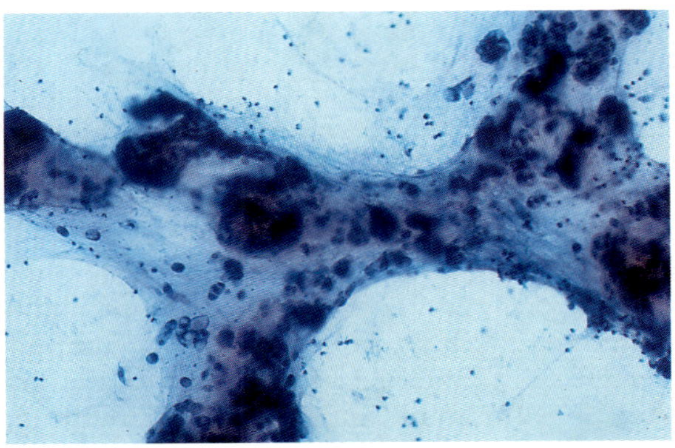

B 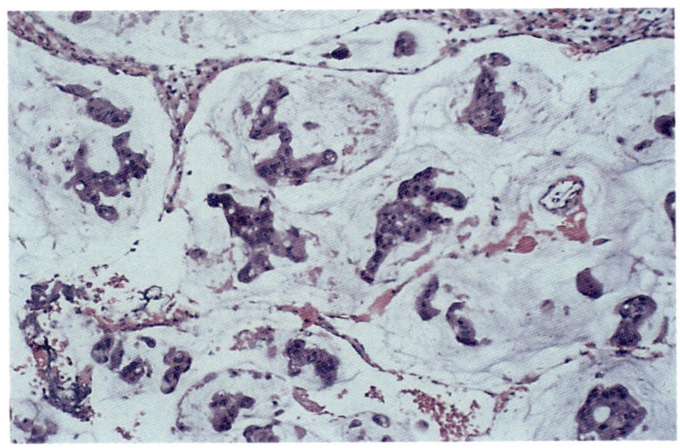

Fig. 7-12. (**A** & **B**) Aspirate from a colloid adenocarcinoma of the cecum. Three-dimensional spherical glandular structures are embedded within the extracellular mucinous material. Smaller cellular aggregates and individually dispersed malignant cells are also present. Many of the latter have abundant cytoplasmic mucin and a peripherally oriented nucleus. (**B**) In the cell block rudimentary glandular lumens can be seen. The cells have polygonal contours and there is a suggestion of a palisading of nuclei (**A**, Papanicolaou; **B**, H&E).

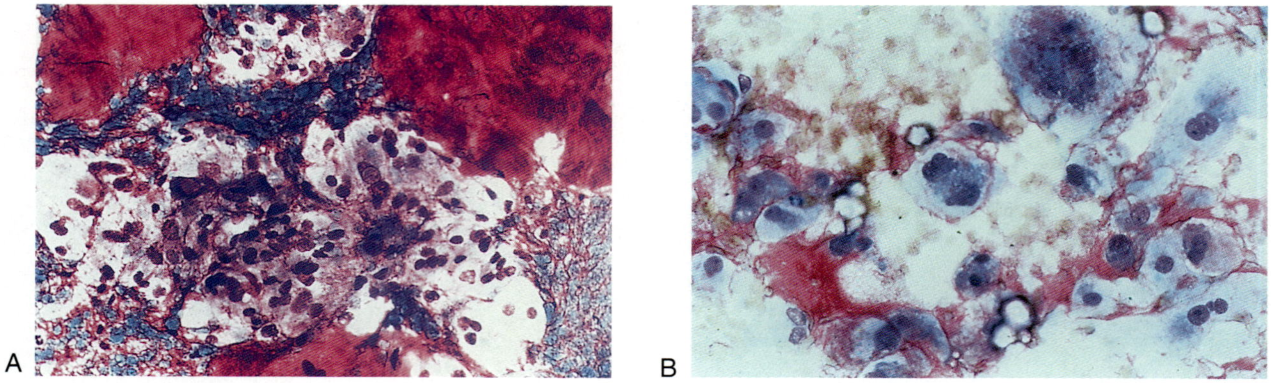

Fig. 7-13. **(A & B)** Aspirate of a primary chordoma of the sacrum. The abundant intensely stained extracellular matrix material could easily cause confusion with a myxoid soft tissue neoplasm. In contrast to the typical appearance of myxoid sarcomas, a distinct separation of matrix material and neoplastic elements is evident. In addition, these neoplastic cells are larger than one usually sees in sarcomas, have more prominent cytoplasm, and hence lower nuclear to cytoplasmic ratios (Diff-Quik). (Courtesy of Dr. Paul Wakely, Jr., Richmond, VA.)

cellular borders and polygonal cell outlines. Nuclei may appear elongated and have prominent nucleoli. Finally, the admixture of neutrophils and debris (dirty necrosis) may be closely associated with both the carcinoma cells and the mucin. This necrotic material is unexpected in the myxoid sarcomas. Ultrastructurally, carcinoma cells will show good evidence of epithelial differentiation, such as well-formed intercellular junctions, prominent microvilli, and cytoplasmic mucin granules. By immunocytochemistry, the carcinoma cells are positive for cytokeratin and carcinoembryonic antigens. Both markers are negative in most sarcomas. Chordomas may be very difficult or impossible to separate from sarcomas purely on cytomorphologic grounds.[23,24] In our relatively limited experience, there may be at least one morphologic clue, however. In contrast to the typical appearance of myxoid sarcomas, which have an intimate admixture of matrix and tumor cells, in chordoma there may be a more distinct separation of matrix and neoplastic elements (Fig. 7-13). Furthermore, the tumor cells from chordomas may be larger, have more prominent cytoplasm, and hence lower nuclear to cytoplasmic ratios. Chordomas can be positive for cytokeratin. Overall, radiographic, cytologic, and immunocytochemical data will need be integrated for the final diagnosis. Many of the benign myxoid lesions that arise in the extremities, such as ganglion cysts, intramuscular myxomas, and nodular fasciitis, do not pose problems in the retroperitoneum.[25–27] It must be remembered that rarely melanoma will metastasize with a myxoid phenotype.[28,29]

SPINDLE CELL SARCOMAS

The spindle cell sarcomas form a category that we believe is the most difficult of the five as it is fraught with the greatest potential for both false-positive and false-negative diagnoses.[11,20,30,31] This group includes malignant fibrous histiocytoma, fibrosarcoma, leiomyosarcoma, and synovial sarcoma.[32] It may be very difficult to distinguish benign spindle cell proliferations from low-grade spindle cell sarcomas in fine needle aspiration biopsies. For example, the distinction between a well-differentiated fibrosarcoma and the fibromatoses may be impossible.[30]

False-negative diagnoses are notorious and may result from several factors. One of the more common factors is the high level of collagenization of some neoplasms, resulting in rather hypocellular smears. Another factor is that the degrees of hyperchromasia and nuclear atypicality may be minimal

Cytomorphologic Features of Spindle Cell Sarcomas

- Moderate to high cellularity
- Individual cells and fragments with indistinct margins
- Variable pleomorphism, often slight
- Tumor cells
  - Elongated contours
  - Single oval to elongated nuclei
  - Uniform hyperchromasia
  - Variable chromatin patterns
  - Cytoplasm: scant, nonspecific
  - High nuclear to cytoplasmic ratio

or unrecognizable. False-positive diagnoses may result from biopsies of highly cellular benign neoplasms with or without cytologic atypia.

The cellularity of aspirates of the spindle cell sarcomas vary from specimen to specimen, as shown above. The aspiration smears will contain numerous isolated malignant cells and/or fragments of tumor (Figs. 7-14 to 7-16). The latter typically have very indistinct edges as the peripherally oriented tumor cells appear to be "falling off of or escaping from" its surface. This is in contrast to the sharp borders that one expects to find in fragments of aspirated carcinomas. Almost by definition, the

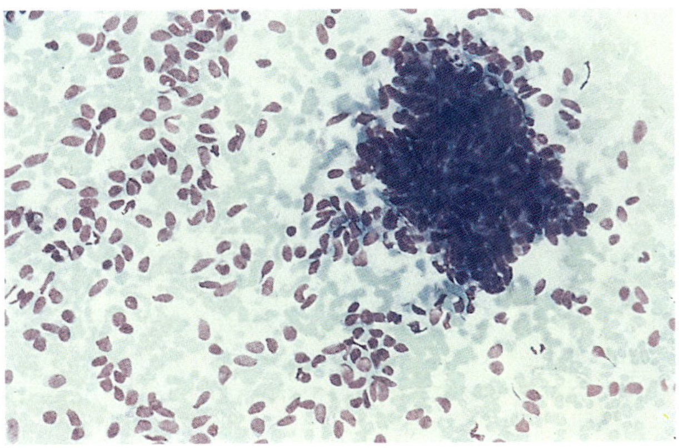

Fig. 7-14. This is the characteristic appearance of an aspirated spindle cell sarcoma, which, in this case, is a fibrosarcoma. Cellularity is obviously high. In addition to many isolated spindle-shaped tumor cells, there is a fragment of neoplasm. Its margins are indistinct, as the peripherally oriented cells appear to be exfoliating from its surface. Individual neoplastic cells have plump ovoid contours, high nuclear to cytoplasmic ratios, and striking uniformity (Diff-Quik).

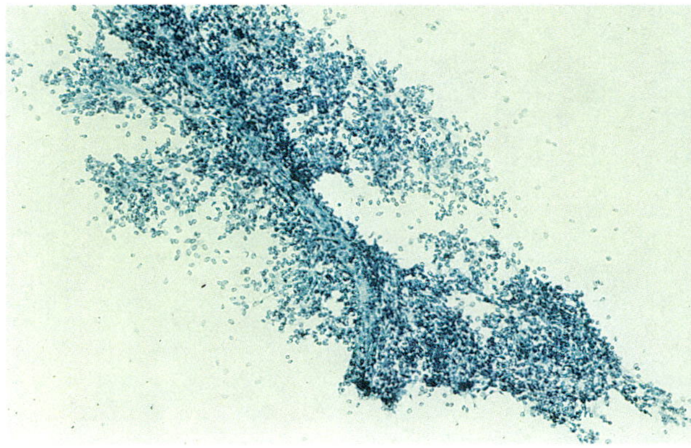

Fig. 7-15. Aspirates of some spindle cell sarcomas yield predominantly large fragments of tumor. This malignant hemangiopericytoma presented as fragments composed literally of hundreds of neoplastic cells. Note the fuzzy or indistinct edges possessed by the aggregate which are in sharp distinction to the sharply defined edges that one expects to see in aspirates of carcinomas (Papanicolaou).

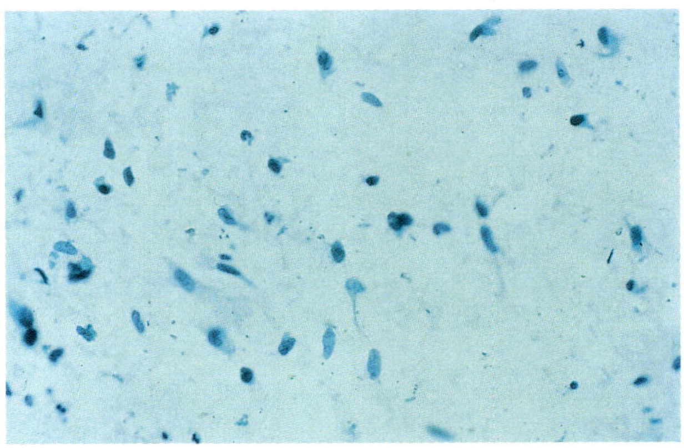

Fig. 7-16. This pelvic leiomyosarcoma presented in this smear almost exclusively as isolated neoplastic cells. Their spindled contours are obvious. With specimens such as this one, there is no way of determining an architectural arrangement of the cells. Hyperchromasia is apparent (Papanicolaou).

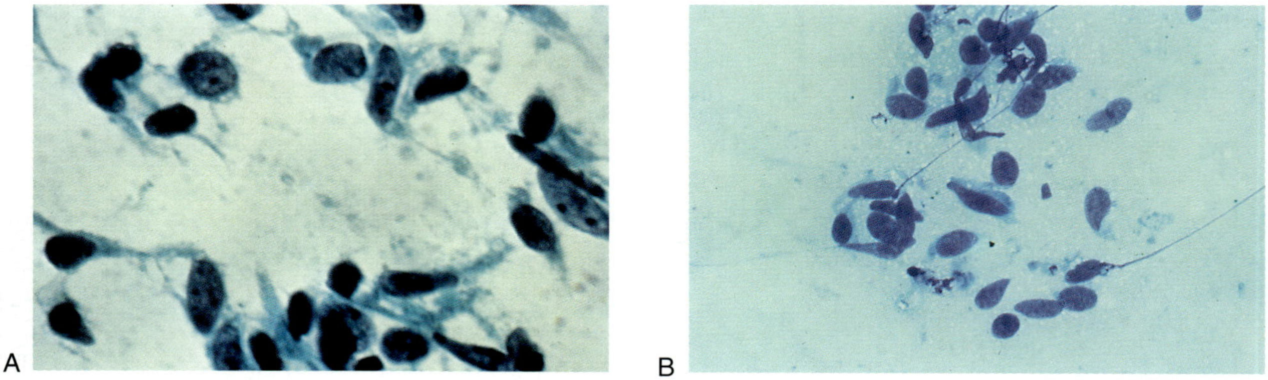

Fig. 7-17. Uniform hyperchromasia characterize the tumor cells in this aspirated synovial sarcoma. Most of the nuclei are elongated but others are more ovoid or almost round in shape. Their chromatin is uniformly finely granular, evenly distributed and darkly stained. Minute nucleoli are suggested. Many of the cells manifest tapering tails of cytoplasm that appear to drift into the smear background. Specimen B is the Diff-Quik–stained mate from specimen A. Hyperchromasia is more difficult to evaluate with this type of preparation. Again, the elongated nuclear contours and the very high nuclear to cytoplasmic ratios are obvious (**A**, Papanicolaou; **B**, Diff-Quik).

neoplastic cells and their nuclei are ovoid to elongated in shape. Most cells possess a single nucleus with rather darkly stained chromatin which may be finely to coarsely granular or even structureless in appearance (Figs. 7-17 and 7-18). Both even and irregular distributions of chromatin granules may be present. Nuclei occupy a substantial fraction of the total cell volume. Multinucleated cells are distinctly uncommon, except in malignant fibrous histiocytoma. The scanty cytoplasm almost never provides specific histogenetic clues in aspirates of spindle cell sarcomas (Figs. 7-19 to 7-21).

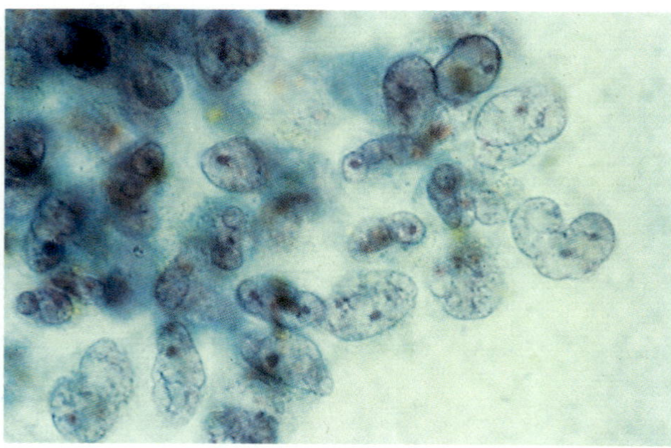

Fig. 7-18. In some tumors, small nucleoli may be apparent in many of the nuclei, as seen in this malignant hemangiopericytoma. The chromatin is very finely granular and pale staining. Although their nuclear membranes are delicate, many of them are quite irregular in shape (Papanicolaou).

Once a spindle cell neoplasm is identified as sarcomatous, an attempt to classify it as to cell type may be undertaken. Nuclear contours may be of some value. In leiomyosarcomas, the nuclei characteristically have blunted or "squared-off" ends; the latter may be associated with perinuclear vacuoles[33–35] (Fig. 7-19). Malignant peripheral nerve sheath nuclei may have a serpentine appearance, pointed tips and a very high length-to-width ratio[36] (Fig. 7-21). The nuclei of malignant fibroblasts (fibrosarcoma and fibrous histiocytoma) will appear somewhere between these two extremes. The nuclei of aspirated synovial sarcomas (Fig. 7-17) tend to have a very plump ovoid contours (low length-to-width ratios). Cytoplasm is usually scanty in volume, cyanophilic, and nondescript. In individual malignant cells, the cytoplasm typically blends imperceptibly into the background of the smear or adjacent matrix material.

In both cell block preparations and to a lesser extent in direct smears, architectural patterns may be per-

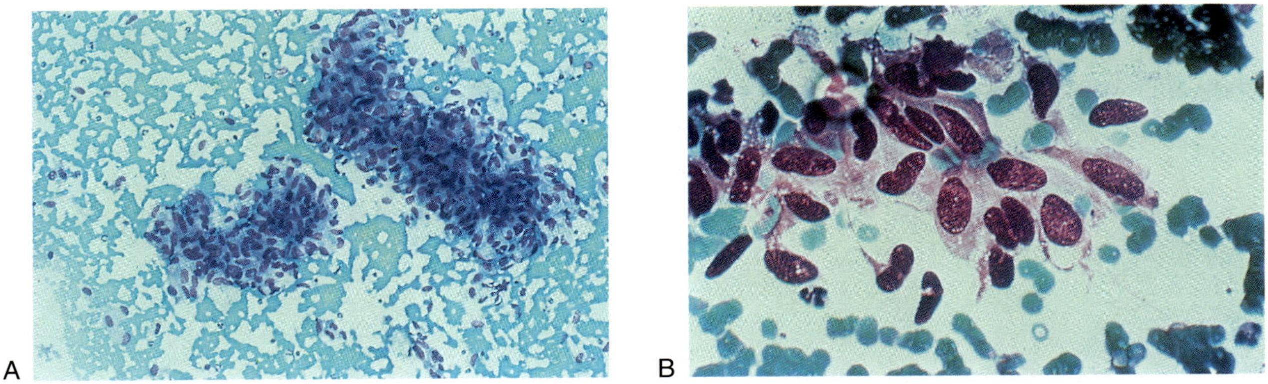

Fig. 7-19. (**A & B**) Aspirated leiomyosarcoma in which large aggregates of malignant cells were present. (**A**) Only a small minority of all of the tumor cells were present as isolated elements. Many of the nuclei have blunt, squared-off contours, but others appear pointed. These tumor cells possess a greater volume of cytoplasm than is often seen in spindle cell sarcomas. Based solely on cytomorphology, it would be quite difficult to designate this specifically as a leiomyosarcoma (Diff-Quik).

ceived (Fig. 7-22). However, it is our opinion that this is much more difficult to appreciate in cytologic preparations than in histologic tissue sections. For example, as pointed out by Dahl et al.[37] in aspirates of leiomyosarcomas, although some of the nuclei may appear in rows, or tandem position, most cells were arranged haphazardly. Similarly, a rhythmic palisading of elongated nuclei into parallel rows separated by cytoplasm and matrix is suggestive of neoplasms of peripheral nerve sheath origin. In our experience, however, this pattern of arrangement is unusual in smears (Fig. 7-21). If foci of cells are discovered with an epithelioid differentiation in aspirates of a spindle cell sarcoma, this biphasic pattern might be quite suggestive of synovial sarcoma.[37,38] In angiosarcomas, plump to elongated malignant

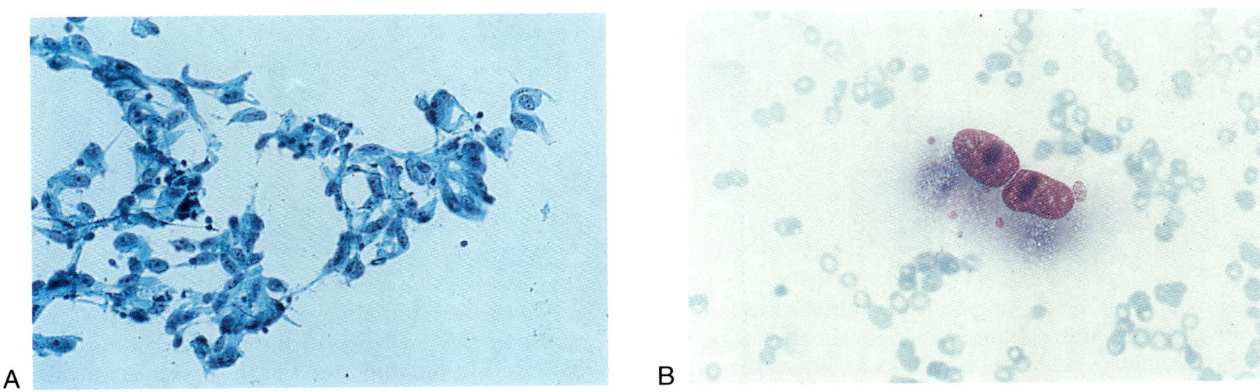

Fig. 7-20. (**A & B**) Plump ovoid nuclei characterize the malignant cells of this aspirated angiosarcoma. The chromatin is uniformly hyperchromatic and small nucleoli are apparent in most cells. Cytoplasm is wispy and cyanophilic; cell borders are indistinct. Within this aggregate there is a suggestion of capillary lumen formation. However, one could not make a definitive diagnosis of angiosarcoma on the basis of this material. (**B**) A binucleated tumor cell is suggested, although this may represent two adjoining neighboring cells. The nuclei are homogeneous with plump ovoid contours, open chromatin, and distinct nucleoli. Cytoplasm is abundant and cell borders are indistinct. Minute vacuoles are present within the cytoplasm. There are no distinguishing histogenetic features (**A**, Papanicolaou; **B**, Diff-Quik).

A

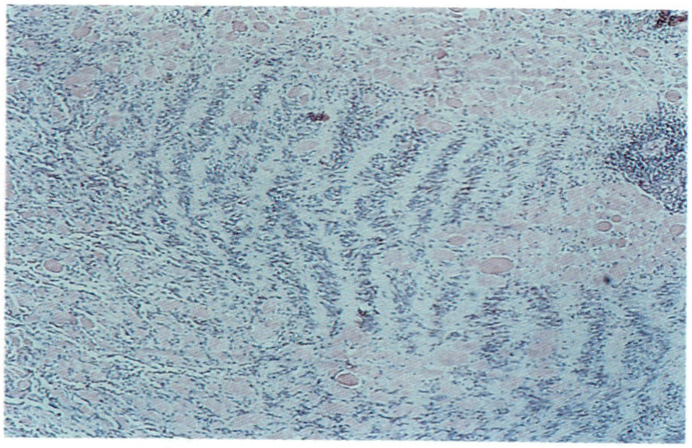

B

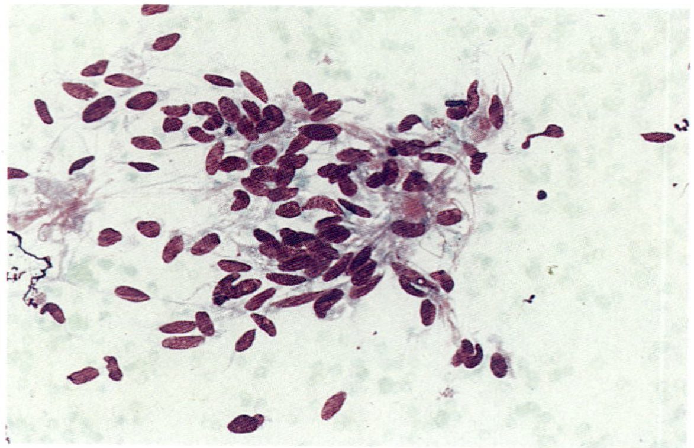

Fig. 7-21. (**A**) Histologically, nuclear palisading is exceptionally well developed in this malignant peripheral nerve sheath tumor, even where it invades skeletal muscle. (**B**) Best example of nuclear palisading, which could be identified in the preoperative aspiration smears. Nuclei are somewhat lined up with one another. The nuclei appear thin with somewhat wavy contours. Cytoplasm forms delicate bipolar tails which tend to merge with those of neighboring cells. However, it would be very difficult to designate this neoplasm specifically as a malignant peripheral nerve sheath tumor based solely on this cytomorphology. Actually, this aspiration biopsy was diagnosed as a spindle cell sarcoma without any further classification cytologically (**A**, H&E; **B**, Diff-Quik).

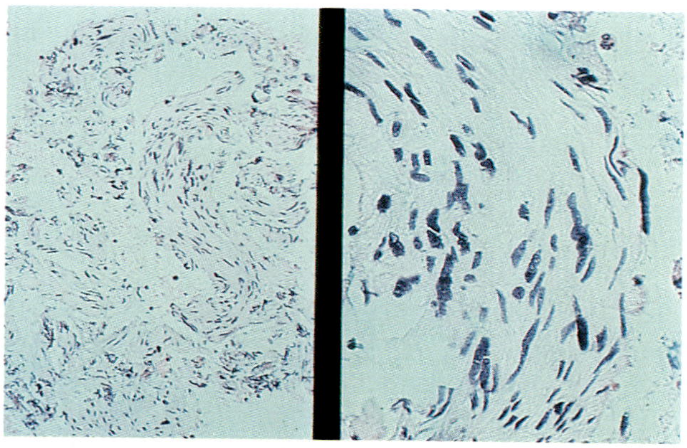

Fig. 7-22. Cell block preparation from an abdominal aspirate of a young adult man. The direct smears are almost totally acellular. The cell block, however, is quite helpful in that it contains fragments of spindle-shaped cells consistent with a low-grade spindle cell sarcoma. The cytologic diagnosis offered the possibility this represented a leiomyosarcoma and this was proved to be correct in the subsequent resection specimen. Cell blocks can also be valuable for immunocytochemistry (H&E).

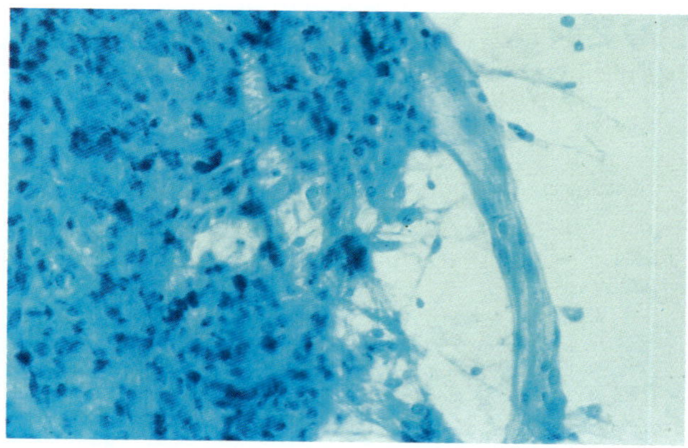

Fig. 7-23. This retroperitoneal angiosarcoma provides an example of the challenge of histogenic classification of spindle cells sarcomas. Much of the specimen consisted of sheets of plump, ovoid to elongated cells. On one side of this field, however, is a structure resembling a blood vessel which is lined by cells resembling those seen in the more solid areas (Papanicolaou).

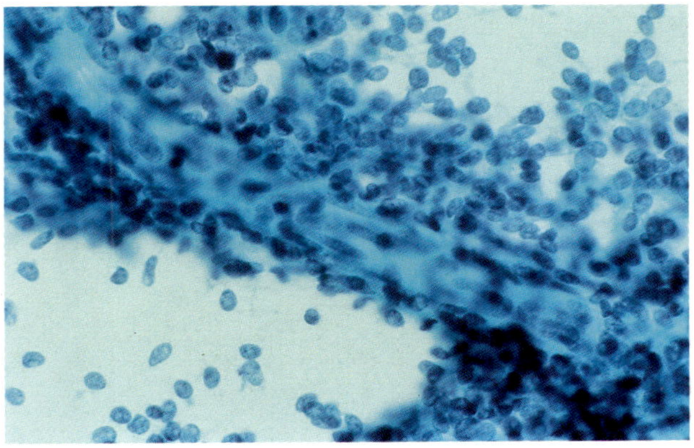

Fig. 7-24. Numerous small uniform plump malignant cells are intimately involved with a small blood vessel. This picture is quite similar to that seen in the angiosarcoma in Fig. 7-23. However, this is a hemangiopericytoma. In cytologic smears, the difference between the two can be very subtle (Papanicolaou).

cells may line structures fully resembling blood vessels, as well as having tumor cells in much more solid areas[39,40] (Fig. 7-23). However, a similar appearance may be seen in aspirates of malignant hemangiopericytomas[41] (Fig. 7-24). In aspiration smears, this difference can be very subtle. In addition, it is important to recall that hemangiopericytomatous arrangements may be found in neoplasms other than hemangiopericytomas. Based purely on cytomorphologic features, we do not believe that Kaposi sarcoma can be specifically identified.[20,42] Immunocytochemistry may be helpful in preoperatively distinguishing among the different spindle cell sarcomas. In general, the neoplastic cells in all of these tumors will be decorated by vimentin. As can be seen in Table 7-1, other reactions may be more helpful in suggesting a specific diagnosis.

Table 7-1. Immunocytochemical Reactions of Spindle Cell Sarcomas

Tumor	*Distinctive Antigen*
Synovial sarcoma	Cytokeratin
Leiyomyosarcoma	Desmin, muscle-specific actin
Fibrosarcoma	—
Nerve sheath sarcoma	S-100 protein
Angiosarcoma	Factor VIII-related antigen, CD34
Kaposi sarcoma	CD34
Hemangiopericytoma	±Muscle-specific actin

PLEOMORPHIC SARCOMAS

In contrast to both myxoid and spindle cell sarcomas, the pleomorphic sarcomas are almost always easily recognized as malignant and usually sarcomatous in nature in aspiration smears, as listed below. The most characteristic feature is the presence of very large to enormous tumor giant cells that possess one or more hyperchromatic nuclei[5,8] (Figs. 7-25 and 7-26). Furthermore, direct smears are usually very cellular with little tendency for the malignant cells to aggregate. Their marked pleomorphism is generally apparent in the scanning lens view (Fig. 7-27). In a single field, one may see an admixture of small round cells, polygonal elements, very elongated cells, and large multinucleated tumor giant cells (Fig. 7-27). The latter may be monstrous and bizarre in appearance. They will be characterized by one or more large, contorted and hyperchromatic nuclei that may or may not possess huge nucleoli. In many pleomorphic sarcomas, the nuclei in the giant cells characteristically form an acute angle with each other (Fig. 7-28). Cytoplasm is also quite variable among the cells. Sometimes the large nuclei appear

Cytomorphologic Features of Pleomorphic Sarcomas

- High cellularity with little cohesion
- Marked pleomorphism of size and shape
- Tumor cells
  - Small to gigantic spectrum
  - Bizarre forms frequent
  - One or more large hyperchromatic irregular nuclei
  - Prominent nuclei
  - Cytoplasm variable in volume and quality
  - Indistinct cell borders

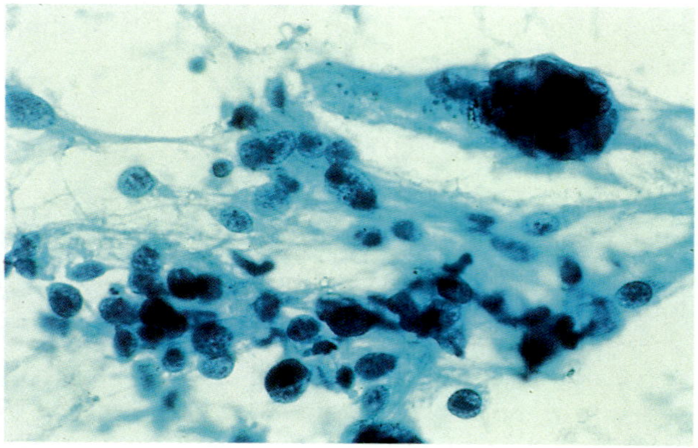

Fig. 7-25. Pleomorphic malignant fibrous histiocytoma. Both high cellularity and pleomorphism are readily apparent. Tumor cells vary from round and polygonal to elongated to multinucleated tumor giant cells. Hyperchromasia is obvious, as is a wide range in nuclear to cytoplasmic ratios (Papanicolaou).

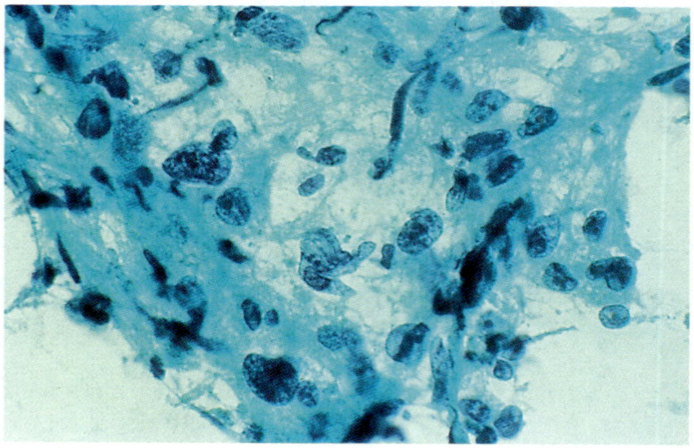

Fig. 7-26. Many of the nuclei in this sarcoma are highly contorted resulting in bizarre configurations. The chromatin is coarsely granular, irregular distributed, and very darkly stained. Cell borders are indistinct and appear to blend with that of adjoining cells and possibly stroma (Papanicolaou).

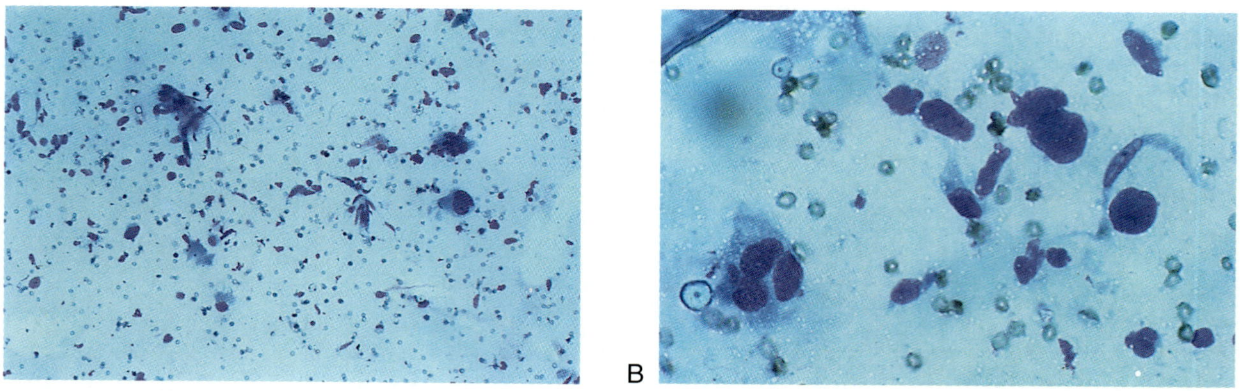

A B

Fig. 7-27. (**A & B**) The pleomorphism in this malignant fibrous histiocytoma is obvious in the scanning lens view. There is a mixture of small round cells, polygonal neoplastic elements, elongated tumor cells, and large multinucleated malignant giant cells. This admixture of cellular appearances is generally quite predictive of a malignant fibrous histiocytoma (Diff-Quik).

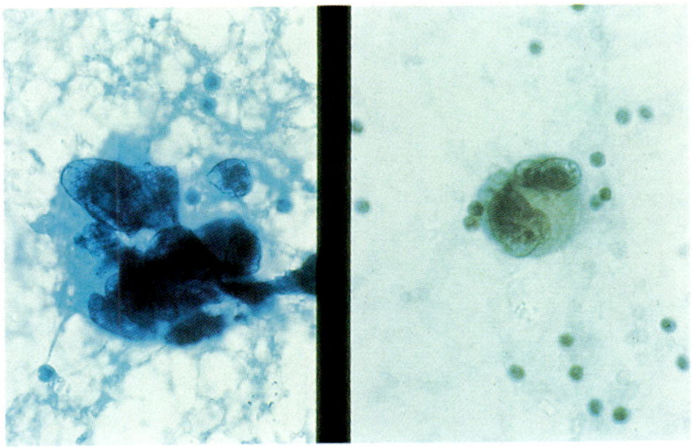

Fig. 7-28. Tumor giant cells often with multiple nuclei, are expected in pleomorphic sarcomas. The binucleated cell (right), obtained from a liposarcoma, manifests an acute angle between the two nuclei. The other cell, from a malignant fibrous histiocytoma, has a grotesque-appearing nucleus with dense, dark chromatin (Papanicolaou).

to be stripped of all their cytoplasm, whereas other cells may have voluminous, generally cyanophilic cytoplasm with frayed or irregular edges. Some of the smaller round or spindle cells may have very high nuclear to cytoplasmic ratios. Cellular borders are indistinct as they tend to blend imperceptibly into the smear background (Fig. 7-26). Necrotic debris may also be seen in the smears.

This category includes the pleomorphic variants of liposarcoma, malignant fibrous histiocytoma, and the very rare rhabdomyosarcoma. As a generalization, very finely vacuolated or granular cytoplasm is typical of the cells of malignant fibrous histiocytoma, whereas larger and more sharply defined vacuoles are characteristic of liposarcoma.[43,44] Theoretically, cytoplasmic cross-striations may be present in rhabdomyosarcomas, but this is an exceedingly rare event in fine needle aspiration smears. Based purely on cytomorphology, one may suggest which of these three forms of pleomorphic sarcoma is present, but it may be very difficult or impossible to be totally certain. One should take comfort knowing that this distinction preoperatively is probably not important; rather, the identification of a mass lesion as a pleomorphic sarcoma of any type provides the needed diagnostic information to initiate the appropriate therapeutic activity.

The differential diagnosis of these retroperitoneal and pelvic soft tissue sarcomas includes a number of other neoplasms, many of which are also malignant. This is in contrast to fine needle aspiration biopsy of pleomorphic sarcomas on the extremities, where other non-neoplastic, reactive proliferations, such as proliferative fasciitis and pleomorphic lipoma, need to be considered. Two primary renal neoplasms are in this list, namely, sarcomatoid renal cell carcinoma and angiomyolipoma. In the latter, the smooth muscle component can be quite bizarre in appearance. Radiographic and immunocytochemical data can provide valuable assistance in this decision process. The smooth muscle component may be actin and HMB-45 positive. Neoplasms of both the cortex and the medulla of the adrenal gland can also simulate pleomorphic sarcomas in aspiration biopsies. With adrenal cortical carcinomas, loosely cohesive and absolutely bizarre appearing cells are present in the aspirate. This problem may be compounded by the fact that these carcinomas are often negative for cytokeratin. Pheochromocytomas may yield highly cellular smears composed of discohesive pleomorphic cells. By immunocytochemistry, neural markers will be positive; if an aliquot of the aspirate is placed in glutaraldehyde, then electron microscopy will demonstrate numerous cytoplasmic dense-core neurosecretory granules. We have seen several examples of giant cell carcinoma of the pancreas which cytologically closely mirror pleomorphic sarcomas. These neoplastic cells typically coexpress keratin and vimentin. Furthermore, knowing the precise needle placement site is crucial. In aspirates of both renal and pancreatic masses, smears need to be carefully scrutinized for evidence of epithelial differentiation. Uncommon highly pleomorphic malignant lymphomas also need to be considered in the differential diagnosis. These may present in smears with numerous individually dispersed bizarre cells; this is especially true of Ki1-positive anaplastic lymphoma (Fig. 7-29). To an extent, lymphoglandular bodies will be present in the background. Tumor

A 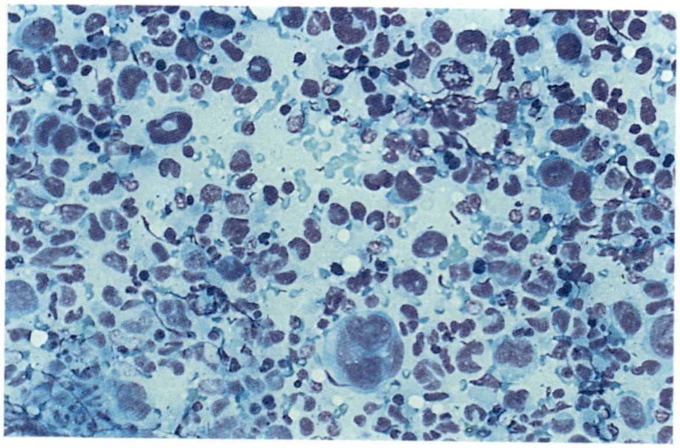 B 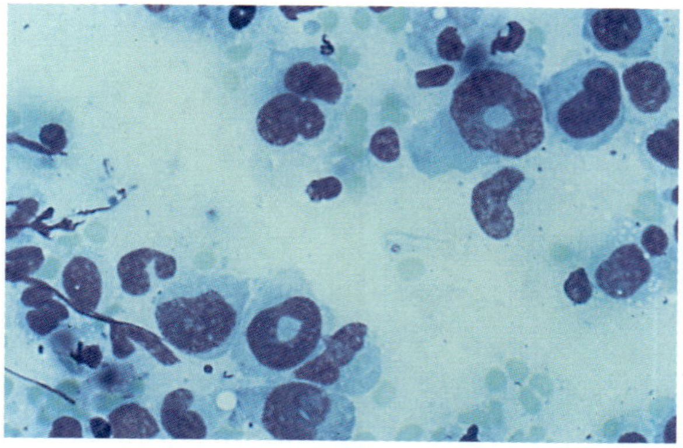

Fig. 7-29. (**A & B**) Highly pleomorphic malignant lymphomas may yield highly cellular specimens with individually dispersed bizarre tumor cells that may closely simulate pleomorphic soft tissue sarcomas. In this example, lymphoglandular bodies are not well developed. In all likelihood, immunocytochemistry would be required to provide the correct diagnosis (Diff-Quik).

cells will be positive for leukocyte common antigen (LCA) and CD_{30}. Finally, metastatic melanoma is the great masquerader and may certainly simulate pleomorphic sarcomas in aspiration biopsies (Fig. 7-30). Microscopic clues include nuclear pseudoinclusions and cytoplasmic pigment. If necessary, ancillary diagnostic procedures can confirm the diagnosis of melanoma.

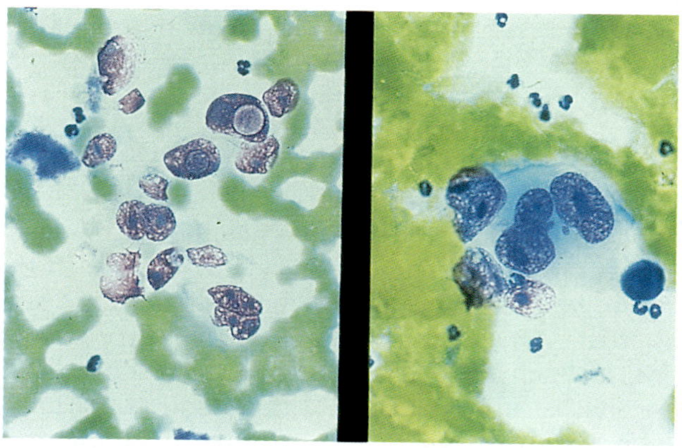

Fig. 7-30. Metastatic melanomas may yield cellular smears composed of discohesive pleomorphic tumor cells. Superficially, at least, these could closely resemble pleomorphic soft tissue sarcomas. Microscopic clues include cytoplasmic pigment and nuclear pseudoinclusions. Of course, the past medical history will be generally helpful, as may immunocytochemistry (Diff-Quik).

POLYGONAL CELLS (EPITHELIAL-LIKE) SARCOMAS

Of the five general categories of soft tissue sarcomas, the least frequent is the polygonal or epithelial-like group. This category includes epithelioid sarcoma, clear cell sarcoma, alveolar soft part sarcoma, and epithelioid variants of other tumors, such as malignant peripheral nerve sheath neoplasms. Perhaps the rare malignant granular cell tumor also best fits in this category.[45] The majority of these originate as soft tissue masses in the extremities and actually are quite uncommon as primary tumors in the retroperitoneum and pelvis. Conversely, they may metastasize to these sites.

Aspiration biopsies yield moderately to highly cellular smears in which there is little in the way of intercellular cohesion[45–49] (see box on p. 237). That is, one will see numerous individually dispersed neoplastic cells and small loosely cohesive aggregates (Figs. 7-31 to 7-34). Generally, distinctive architectural patterns of arrangement will not be recognized. The individual tumor cells have polygonal to round contours, distinct cellular borders, and moderate volumes of cytoplasm. The latter will vary in quality from dense and

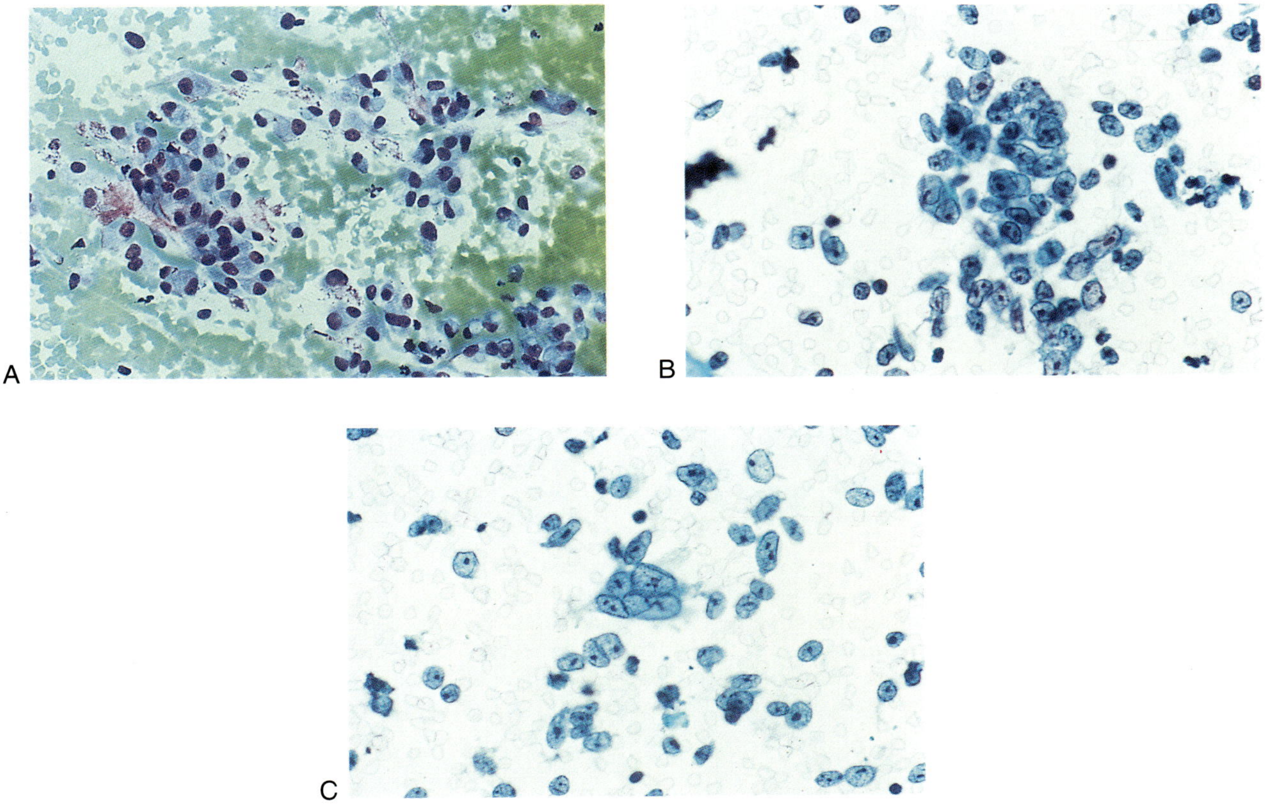

Fig. 7-31. (**A–C**) This is an aspirate of a clear cell sarcoma (from a soft tissue mass in the leg of a middle-aged woman). Cellularity is fairly high. Only blood is present in the smear background; that is, there is only scanty matrix material. The cells are large and uniform with polygonal contours and distinct cellular edges. Each possesses a solitary generally eccentrically positioned nucleus. Nucleoli are prominent and are much more easily recognized in Papanicolaou-stained preparations. Tumor cells present in loose clusters and individually. Where cohesion is well maintained, cell borders are quite distinct, as seen in C (**A**, Diff-Quik; **B & C**, Papanicolaou).

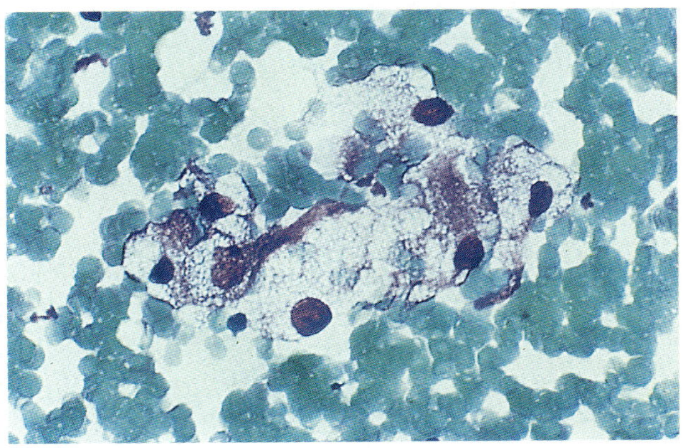

Fig. 7-32. Lung metastasis containing cells from a metastatic clear cell sarcoma. Cytoplasm is abundant and optically clear and has a somewhat reticulated appearance. Cell borders are well defined compared to most soft tissue sarcomas. Each cell possesses a single eccentric round nucleus with a well-defined nucleolus (Diff-Quik).

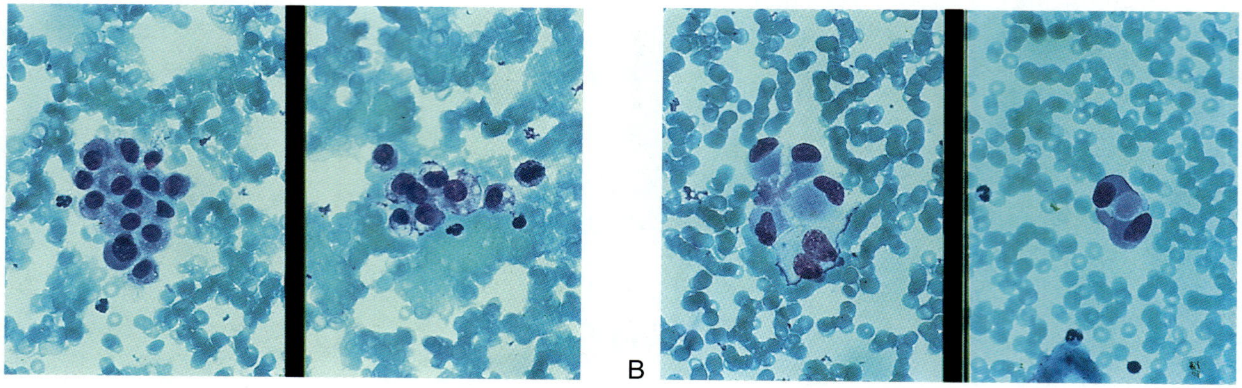

A B

Fig. 7-33. (**A & B**) Primary aspirate of an epithelioid sarcoma (of the forearm). Most of the malignant cells are present in small monolayer-like aggregates. Cytoplasm is moderate in volume, dense, and sharply defined. Most of these round nuclei are peripherally located in the cells. Nucleoli are apparent. Their resemblance to an epithelial neoplasm is obvious. Intercellular cohesion is less well maintained in Fig. B than it is in Fig. A. Cell borders are quite distinct, and contours are polygonal. The nuclear to cytoplasmic ratios are relatively low. The sharply defined cytoplasmic blobs probably correspond to whorled bundles of vimentin filaments (Diff-Quik).

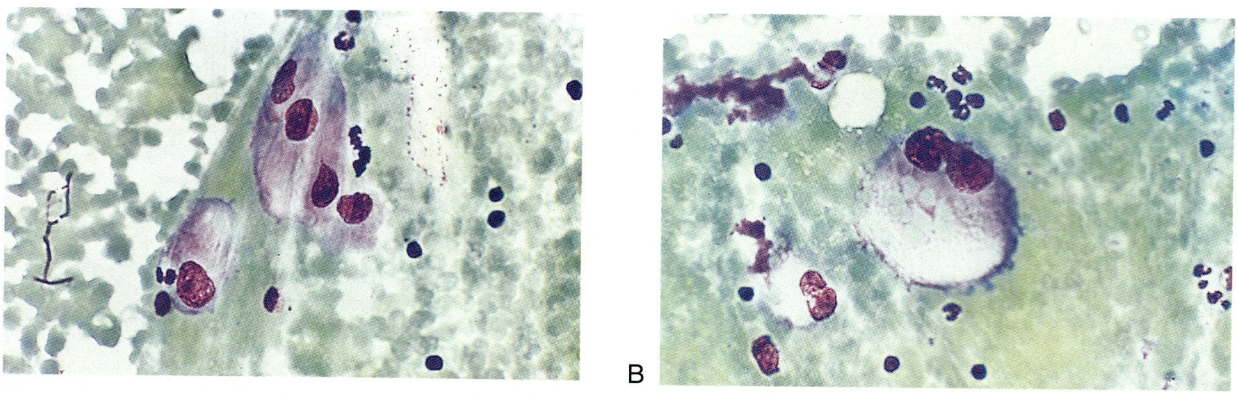

A B

Fig. 7-34. (**A & B**) Alveolar soft part sarcoma presented almost exclusively in the smears as individually dispersed large neoplastic cells. They are characterized by voluminous somewhat foamy cytoplasm and by one or, less often, two very eccentrically positioned nuclei. The latter are round, have immature chromatin and a prominent nucleolus. Based purely on cytomorphologic features, on a cell-for-cell basis, it would be difficult to distinguish this tumor from some forms of adenocarcinomas (Diff-Quik).

Cytomorphologic Features of Polygonal Cell Epithelial-like Sarcomas

- Moderate to high cellularity with little cohesion
- Tumor cells
  - Polygonal to round
  - Distinct borders
  - Cyanophilic to clear cytoplasm
  - Solitary round nucleus
  - Vesicular chromatin
  - Prominent nucleolus

cyanophilic to optically clear. It will be obviously and coarsely granular in granular cell tumors.[45] Most of the cells have a single round nucleus, in which the chromatin is generally pale stained, but it may appear hyperchromatic. One or more prominent nucleoli are characteristic. A minority of the tumor cells have more elongated or spindled contours.

ROUND CELL SARCOMAS

The soft tissue sarcomas of childhood are generally characterized as the round cell type.[14–19] The two most common considerations are rhabdomyosarcoma and extraskeletal Ewing's sarcoma. As the name implies, most neoplastic cells and the nuclei appear round. By far, rhabdomyosarcoma is the most frequent in children and adolescents. The most common clinicopathologic variant is embryonal rhabdomyosarcoma, which is also the type most likely to arise in the retroperitoneum and pelvis, while the alveolar variant usually arises in an extremity. Ancillary diagnostic procedures, for example, electron microscopy and immunocytochemistry, are essential to make the specific correct FNA diagnosis.[50–52]

Aspiration biopsies of embryonal rhabdomyosarcoma generally yield moderately to very highly cellular smears with numerous individually dispersed malignant cells,[52–54] as detailed below. Pleomorphism of cellular size and shape is apparent (Figs. 7-35 to 7-37). Many of the tumor cells may appear undifferentiated, possessing a single round nucleus and a very high nuclear to cytoplasmic ratio. Other cells are larger and have greater volumes of cytoplasm. This may include elongated cells with eccentrically positioned nuclei, resulting in the characteristic (but unfortunately uncommon) strap cell. Multinucleation is not infrequent at all. In fact, binucleated and multinucleated tumor cells are a helpful clue to separating embryonal rhabdomyosarcoma from other forms of small cell tumors of childhood (Fig. 7-36). On the basis of the volume of cytoplasm, Ahktar and colleagues[54] have subdivided rhabodymyoblasts cytologically into early, intermediate, and late forms. Although this is arbitrary and subjective, it does call attention to the fact that malignant rhabdoymyoblasts may show a spectrum of pleomorphism uncommon in other round cell sarcomas (Figs. 7-35 to 7-37). Most nuclei are round, but they may be more pleomorphic than are seen in other small cell malignant tumors of childhood. Chromatin is typically finely granular. Nucleoli may be well developed (Fig. 7-36). Very infrequently, the smear background may have a frothy or tigroid appearance, as is characteristic of aspirates of germinomas (Fig. 7-37). By immunocytochemistry, at least some of the malignant cells will demonstrate myogenous differentiation, characterized by positive staining for muscle specific actin and/or desmin. The more primitive elements will be positive only for vimentin. By electron microscopy, some of the neoplastic cells will demonstrate at least rudimentary muscle differentiation with myosin filaments and/or Z-band material.

Cytomorphologic Features of Embryonal Rhabdomyosarcoma

- Cellular smears with varying degrees of intercellular cohesion
- Small to intermediate-sized cells with round polygonal or spindled contours
- Nuclei: round to oval; hyperchromatic; nucleoli may be well developed; often eccentric; may be two or more per cell
- Cytoplasm variable in amount

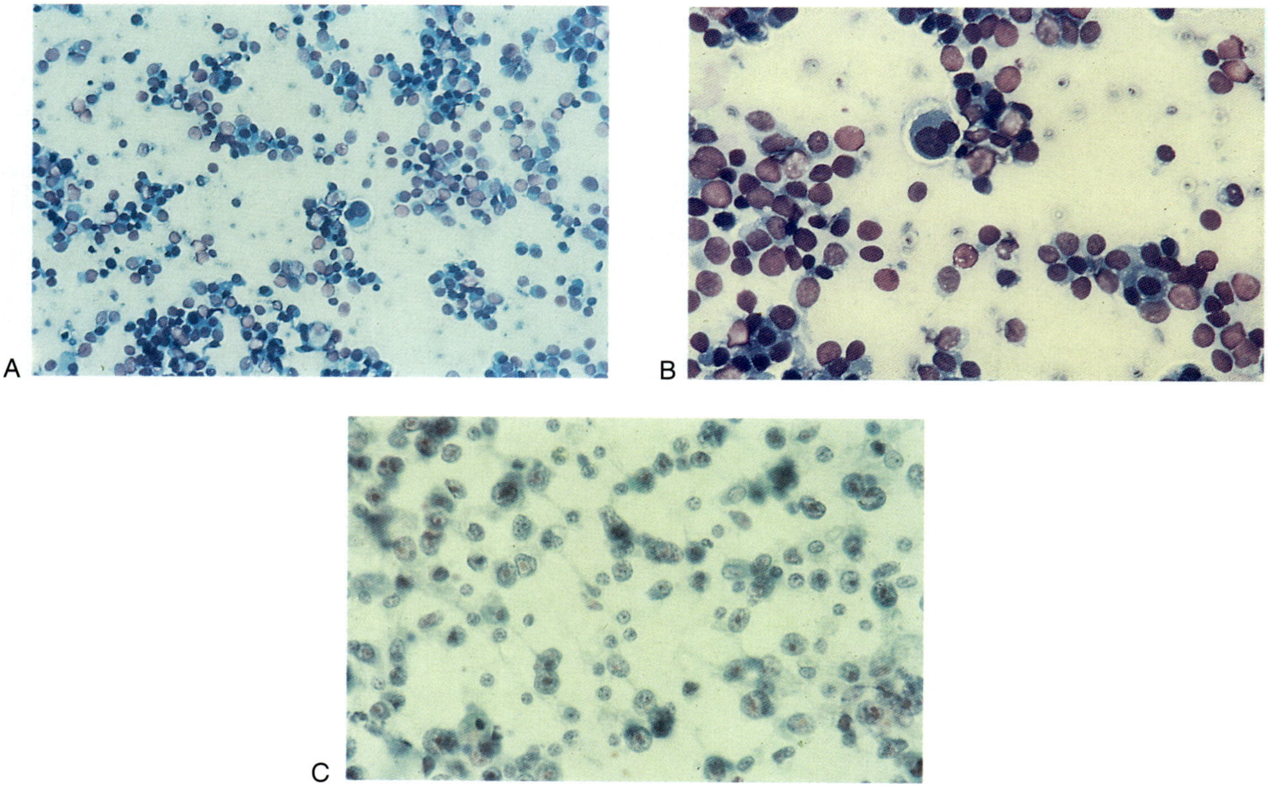

Fig. 7-35. Embryonal rhabdomyosarcoma yields cellular smears with a discohesive pattern of individually dispersed tumor cells, as well as small loosely cohesive aggregates. In this example, most of the tumor cells are uniform with a solitary round nucleus and exceedingly high nuclear to cytoplasmic ratios. A few of the tumor cells possess greater volumes of cytoplasm, and at least one is binucleated. Fig. B is a higher magnification of the center portion of Fig. A. In general, nucleoli are more easily discernible in Papanicolaou-stained preparations. (**C**) Almost all of the tumor cells possess at least one large and at times irregularly contoured nucleolus. Hyperchromasia is evident (**A & B**, Diff-Quik; **C**, Papanicolaou).

Ewing's sarcoma also yields highly cellular smears predominated by dispersed small malignant cells,[15,52,55] as listed on page 240. In contrast to rhabdomyosarcoma, pleomorphism is almost nonexistent. In fact, the striking uniformity among the tumor cells is characteristic (Figs. 7-38 and 7-39). Each cell possesses a single round nucleus and a very high nuclear to cytoplasmic ratio. The chromatin is very finely stipled and evenly dispersed throughout the nucleus. Nucleoli are absent or at least inconspicuous. With the Diff-Quik stain, only a thin rim of nondescript basophilic cytoplasm will surround many of the nuclei. However, in some tumor cells, cytoplasmic volume appears slightly increased and contains sharply delineated small clear vacuoles (Figs. 7-38 and 7-40). The latter reflect the presence of prominent cytoplasmic accumulations of glycogen. Ahktar et al. have described small blunt cytoplasmic tags as characteristic of Ewing's sarcoma.[15] Unfortunately, with alcohol fixation, the cytoplasmic vacuolization is lost. With the Papanicolaou stain, cytoplasm is cyanophilic and nondescript (Fig. 7-39). Although intercellular cohesion is generally not well preserved in aspiration smears, both small and large aggregates are occasionally seen. Within these cohesive clusters, pseudorosette formation may be recognized (Fig. 7-40). Many of these features are shared by the primitive neuroectodermal tumor, a related small round cell neoplasm of childhood.

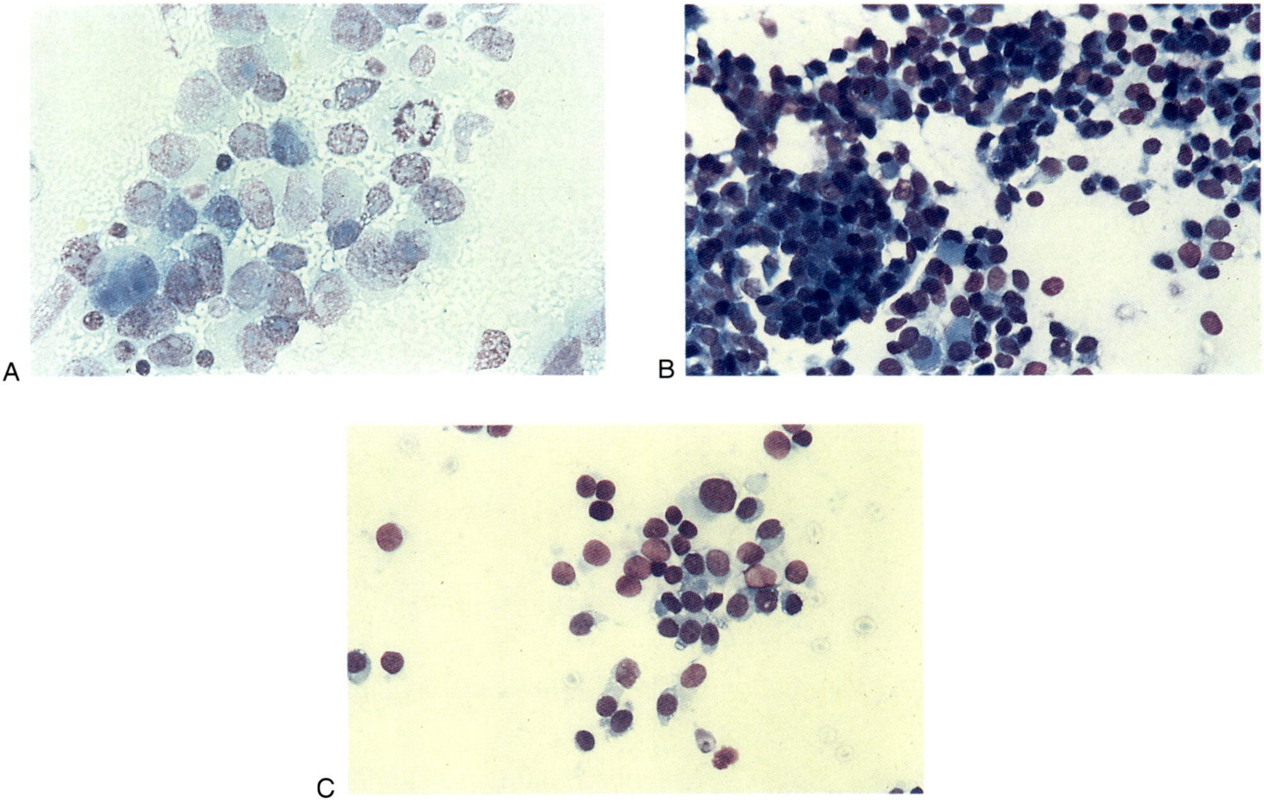

Fig. 7-36. (**A**) Most of the malignant cells reveal no distinctive features of differentiation. However, nucleoli are well developed; this would not be expected in Ewing's sarcoma. Also unexpected for Ewing's sarcoma is the binucleated tumor cell. The nuclear to cytoplasmic ratios are high. (**B & C**) From a different patient. The nuclei in the cells are round and generally have relatively little cytoplasm. In many of the cells in which cytoplasm is present to a greater extent, the nuclei appear somewhat larger and are eccentrically positioned. (**B**) A few of the tumor cells have elongated tails of cytoplasm (comet cells). Overall, pleomorphism is present at a low level (Diff-Quik).

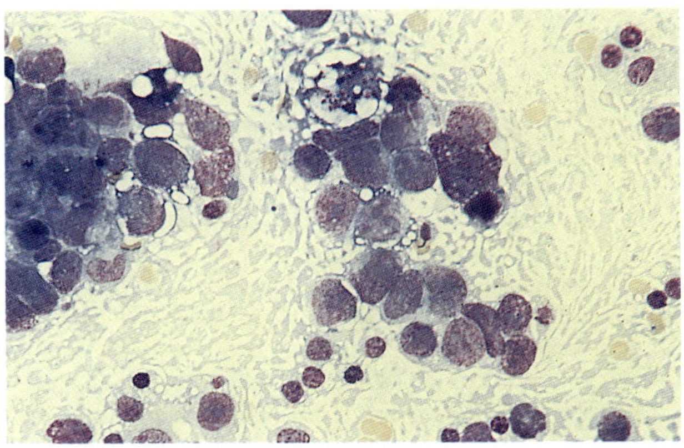

Fig. 7-37. Rarely, the smear background in an aspirated rhabdomyosarcoma closely resembles the tigroid appearance characteristic of germinomas (Diff-Quik).

> **Cytomorphologic Features of Ewing's Sarcoma**
>
> - High cellularity with numerous uniform single mononucleated cells
> - Nuclei: round; fine even chromatin; inconspicuous nucleoli
> - Cytoplasm: scanty; glycogen vacuoles; blebs
> - Pseudorosettes may be present

Ewing's sarcoma may be considered the prototypic example of these malignant small cell tumors of childhood.[52] Ancillary diagnostic procedures are absolutely required for a specific diagnosis. Electron microscopy helps exclude other small cell tumors of childhood and to support the contention of Ewing's sarcoma. Ultrastructural features include large pools of glycogen within the cytoplasm, primitive intercellular junctions, and uniformly distributed euchromatin. Recently, a monoclonal antibody directed against the glycoprotein product of the MIC2 gene has been suggested as a rather specific marker for

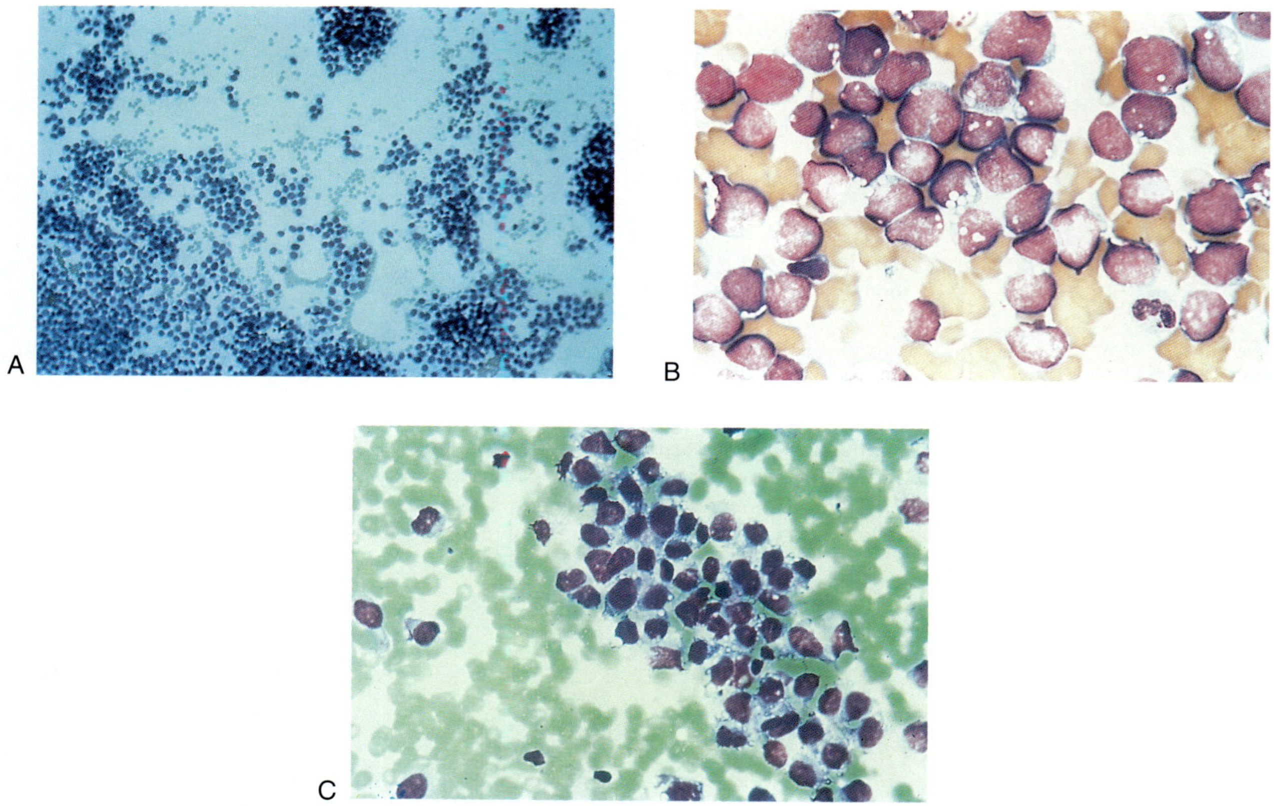

Fig. 7-38. Ewing's sarcoma generally yields highly cellular smears composed of dispersed small homogeneous neoplastic cells. (**A**) Most of the tumor cells are present in monolayer sheets; there is no evidence of an organoid arrangement within them. (**B**) The tumor cells are exceedingly uniform. Each possesses a single round nucleus and a very high nuclear to cytoplasmic ratio. The chromatin is finely reticulated and nucleoli are distinctly inconspicuous. Some of the tumor cells have sharply punched-out cytoplasmic vacuoles containing glycogen. (**B**) Note the absence of intercellular cohesion and of lymphoglandular bodies. (**C**) Intercellular cohesion, however, may be evident as seen in the loosely arrayed sheet. Some of the individually dispersed tumor cells have blunt cytoplasmic tags. Cytoplasmic glycogen vacuoles are evident in some of the neoplastic cells (Diff-Quik).

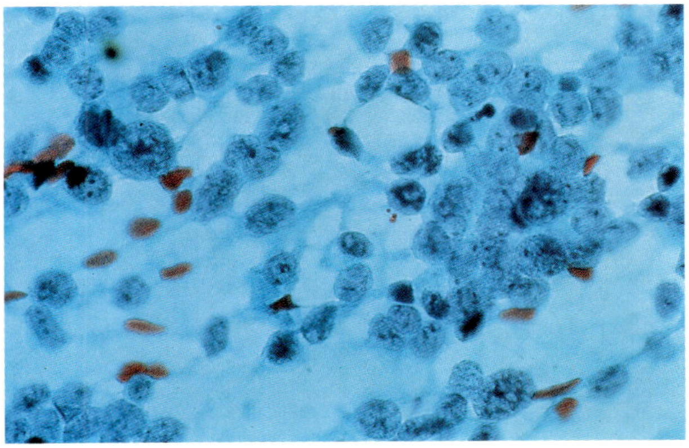

Fig. 7-39. Some of the distinctive features seen in Fig. 7-38 are lost with alcohol fixation. In particular, cytoplasmic vacuolization is not recognized. With the Papanicolaou stain, the chromatin is finely granular and evenly distributed throughout the nucleus. Nucleoli are poorly visible. There is no evidence of specific differentiation (Papanicolaou).

Ewing's sarcoma. Cytogenetic data may be even more specific. If an aliquot of the aspirate is submitted for karyotypic analysis, one may demonstrate the consistent 11;22 translocation.

The cytologic differential diagnosis of both rhabdomyosarcoma and Ewing's sarcoma includes malignant lymphomas, mesenchymal chondrosarcoma, neuroblastoma, nephroblastoma, and intra-abdominal desmoplastic small cell tumor. Both neuroblastoma and nephroblastoma have been described in earlier chapters. Based on a combination of cytomorphology, clinical and radiographic data, immunocytochemistry, and other ancillary diagnostic procedures, the distinction should be straightforward. We have never seen an example of an intra-abdominal mesenchymal chondrosarcoma in a fine needle biopsy.

BURKITT'S LYMPHOMA

Malignant lymphomas comprise a major component of the differential diagnosis of round cell sarcomas in aspiration biopsies. In an intra-abdominal location, the major type would be small noncleaved cell or Burkitt's lymphoma. The tumor occurs predominately in patients in the first two decades of life and in our experience is seen most frequently in

A

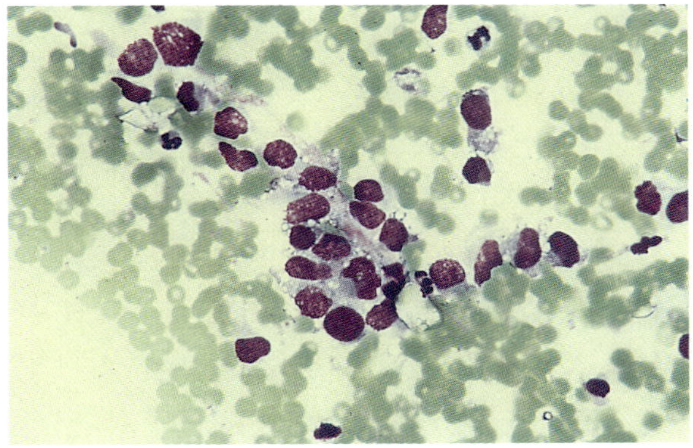

B

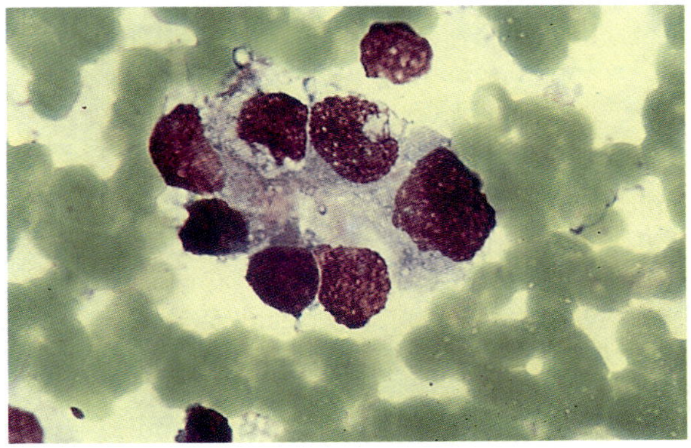

Fig. 7-40. (**A & B**) Rarely, in aspirates of Ewing's sarcoma, pseudorosettes are evident. Perhaps this reflects the close relationship that this sarcoma has with primitive neuroectodermal tumors (Diff-Quik).

Cytomorphologic Features of Burkitt's Lymphoma

- High cellularity with noncohesive uniform blasts
- Blasts two to three times size of mature lymphocyte
- Nuclei: round; moderately clumped chromatin; two or more nucleoli
- Cytoplasm: moderate volume; basophilic; lipid vacuoles
- Tingible body macrophages
- Lymphoglandular bodies

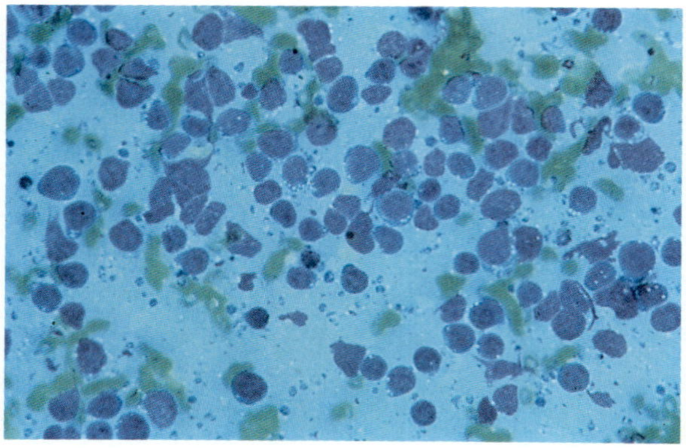

A

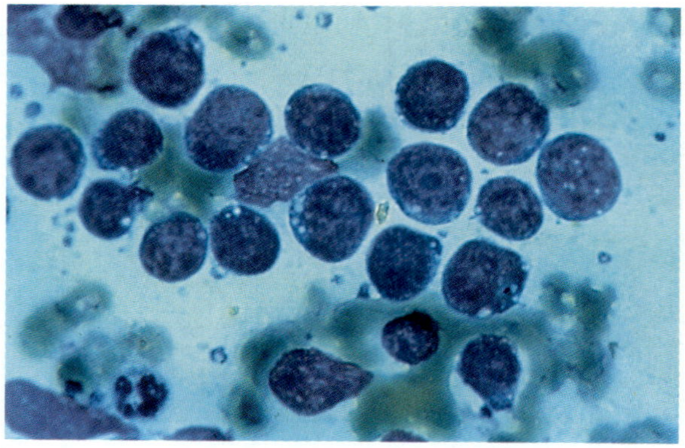

B

Fig. 7-41. (**A & B**) Aspiration smears of Burkitt's lymphoma almost always yield exceedingly cellular smears composed of uniform appearing, individually dispersed neoplastic cells. Nuclei are often perfectly round with finely to moderately clumped chromatin and multiple prominent nucleoli. Cytoplasm varies from scanty to somewhat moderate in volume and is deeply basophilic. Tumor cells may have cytoplasmic vacuoles that are lipid positive. Lymphoglandular bodies are present in the smear background (Diff-Quik).

children less than 11 years of age. In the United States, Burkitt's lymphoma is predominantly an extranodal infradiaphragmatic tumor. Most frequently, it involves the distal ileum, appendix, large intestine, ovaries, and kidneys. Histologically, uniform lymphoid elements with round nuclei and multiple nucleoli infiltrate tissues. Burkitt's lymphomas have very high proliferative rates and, accordingly, mitotic figures are numerous. The classic starry-sky pattern may be seen in which benign histiocytes are interspersed among the more dark-stained neoplastic cells. Burkitt's lymphoma is a high-grade lymphoma that shows a variable clinical response to chemotherapy.

Aspiration biopsies produce a characteristic picture (see box on p. 241). These smears are highly cellular and composed of monotonous noncohesive tumor cells[52] (Figs. 7-41 and 7-42). Uniform blasts are two to three times the diameter of a small mature-appearing lymphocyte. Most often, their solitary nuclei are round, have moderately clumped chromatin, and two or more evident nucleoli. A thin to moderate rim of cytoplasm surrounds the centrally positioned nucleus. With the Diff-Quik stain, it has a deeply basophilic appearance and may contain lipid-positive vacuoles (Fig. 7-41). This intense basophilia and vacuolization is not evident in Papanicolaou-stained material (Fig. 7-42). As expected from the histopathology, tingible-body macrophages and mitotic figures are frequent. The smear background contains numerous lymphoglandular bodies.

Although the cytomorphology is characteristic, the diagnosis can be confirmed through the use of ancillary procedures. As with most other lymphomas, the tumor cells are positive for leukocyte common antigen. Burkitt's lymphoma is a B-cell neoplasm that can be confirmed immunologically. By immunocytochemistry or flow cytometry, tumor

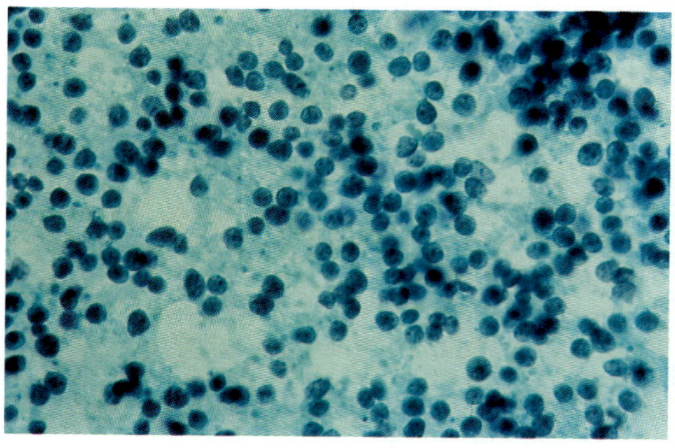

Fig. 7-42. With the Papanicolaou stain, some of these morphologic clues are lost. In particular, cytoplasmic basophilia and vacuolization are not apparent. On the other hand, the tumor cells appear more uniform and nucleoli may be even more prominent (Papanicolaou).

cells will be positive for CD_{10}, CD_{19}, CD_{20}. The majority of Burkitt's lymphomas possess an 8;14 translocation that can be demonstrated cytogenetically in aspirated material.

INTRA-ABDOMINAL DESMOPLASTIC SMALL CELL TUMOR

Although quite rare, another neoplasm that needs to considered in the differential diagnosis of round cell sarcomas is the recently described entity, the intra-abdominal desmoplastic small cell tumor.[56,57] This primitive malignancy typically affects adolescent males as a primary enlargement of the peritoneum and/or omentum. Visceral involvement is not evident. Multidirectional differentiation is witnessed by immunostaining for multiple intermediate filaments and neuroendocrine markers. Although their histogenesis is uncertain, it has been proposed that these tumors represent primitive malignancies of mesothelial origin. Histologically, one sees variably sized aggregates of small round malignant cells separated by cellular desmoplastic reactive stroma.

Aspiration biopsies are variably cellular, presumably reflecting the prominent stroma within the neoplasm.[58–60] Small homogeneous-appearing malignant cells are present in both aggregates and as isolated cells (Fig. 7-43). On a cell-for-cell basis, they resemble very closely those seen in Ewing sarcoma and thus reveal no specific differentiation. Each cell possesses a solitary round nucleus and a high nuclear to cytoplasmic ratio. Generally, nucleoli are inconspicuous. Large fragments of variably cellular collagenized stroma may be a helpful diagnostic clue.

GYNECOLOGIC TUMORS

One of the more frequent indications for aspiration biopsies of the retroperitoneum and especially the pelvis is the investigation of patients with known or suspected gynecologic malignancies. In Europe, FNA biopsies are used for the primary diagnosis of cancer, in particular ovarian carcinoma.[61] In the United States, aspiration biopsy has been used only infrequently to diagnosis primary neoplasms.[62–69] Most of the experience in the United States, including ours, has been used to diagnose recurrent or metastatic tumors in women who have been previously treated for gynecologic malignancies. A number of publications have focused on the value of FNA biopsies in this particular setting, attesting to its value. Overall, this is a very accurate technique to investigate both superficial and deep-seated pelvic mass lesions. Reported values for diagnostic specificity have ranged form 95 percent to 100 percent.[67–69] These and other studies attest to the very low rate of false-positive aspiration diagnoses. The single such diagnosis in the series by Belinson et

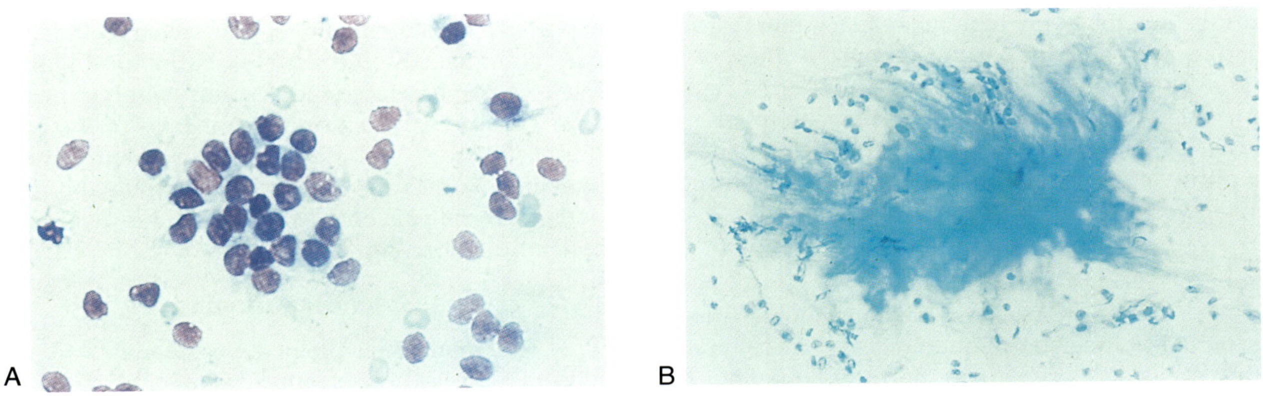

Fig. 7-43. (**A & B**) Aspirates of the intra-abdominal desmoplastic small cell tumor may not be highly cellular due to the prominent fibroblastic and collagenous stroma. The malignant cells are small and homogeneous, each possessing a single round nucleus, finely reticulated chromatin, inconspicuous nucleoli, and very high nuclear to cytoplasmic ratios. The aspirates may also contain fragments of variably cellular collagenized mesenchymal tissue (**A**, Diff-Quik; **B**, Papanicolaou).

al.[64] resulted from a misinterpretation of reactive histiocytes in a superficial lymph node as metastatic adenocarcinoma in a woman with extramammary Paget's disease. In one of the largest published series, Moriarity et al.[67] had three false-positive aspiration diagnoses. One occurred in a cul-de-sac specimen in a woman with well-differentiated endometrial adenocarcinoma; a colovaginal fistula related to a diverticulum was the aspiration target. A second false-positive diagnosis resulted in a misdirected aspirate that sampled a primary cervical carcinoma in a woman with a vaginal nodule. The latter subsequently was determined to be adenosis. The final patient had a carcinosarcoma; an aspirate of a leg mass was considered a metastasis but was proved histologically to be a benign lipoma.[67] Diagnostic sensitivities have not fared as well, with reported values varying from 65 percent to 91 percent, as false-negative diagnoses are a major limitation of this procedure.[67–69] In many instances, the false-negative results occur in women with prior radiation therapy, which results in prominent fibrosis.[62,69] As elsewhere in the body, dense, sclerotic connective tissue may prohibit the aspiration of malignant cells. False-negative diagnoses may also result from misdirected aspirates and from sampling of large necrotic tumors. With this in mind, Moriarity and colleagues[67] provided an important concept; namely, applying adequacy criteria to these specimens. Specifically, their criteria required finding at least two groups of cells appropriate for the biopsy target on at least two different slides. This results in a relatively high inadequacy rate but does significantly and importantly improve the predictive value of negative test results. Reported figures for this have ranged from 57 percent to 91 percent.[67–69] Values reported for the predictive value of a positive test result have ranged from 98 percent to 100 percent.[67–69]

Cervical Carcinoma

One of the more common gynecologic tumors sampled is metastatic squamous cell carcinoma of the cervix. As elsewhere in the body, squamous cell carcinomas may show a range of histologic and cytologic differentiation. In well-differentiated squamous cell carcinomas, a large proportion of the neoplastic cells have abundant, dense cytoplasm, often showing evidence of keratinization. In fact, although we generally use the Diff-Quik stain for immediate interpretations of aspirates, a rapid Papanicolaou technique is quite useful in demonstrating cytoplasmic keratin. Large, sharply angulated nuclei with pyknotic-like chromatin patterns are typical. Well-differentiated squamous carcinomas may be associated with extensive keratinization so that aspirate smears may contain mostly anucleated squames. In less differentiation squamous carcinomas, the nuclear to cytoplasmic ratios are increased, keratinization is reduced, and both coarse chromatin patterns and nucleoli are expected. As stated above, the smear cellularity may be rather low secondary to radiation-induced sclerosis. We have also seen examples of highly cellular smears in which only a minority of the cells present were neoplastic; most represented neutrophils in a necrotic background, presumably secondary to necrosis of the tumor, at times related to recent radiation (Fig. 7-44).

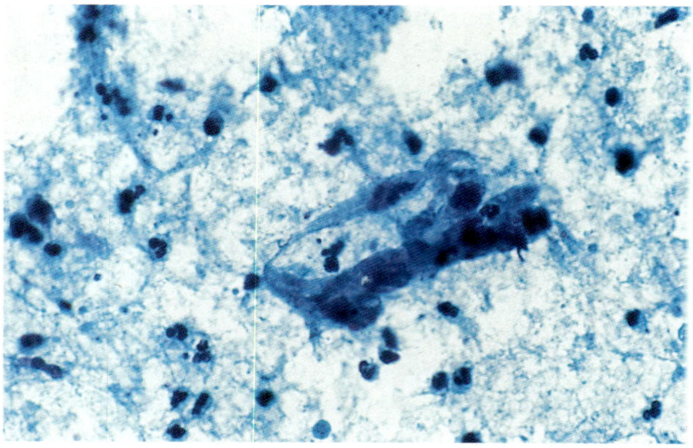

Fig. 7-44. Metastatic squamous cell carcinoma presented as a partially cystic retroperitoneal mass following radiation for primary cervical tumor. Rare malignant spindled squamous cells are present in a sea of neutrophils (Papanicolaou).

Adenocarcinoma

Several distinct histologic types of adenocarcinomas may arise in the female genital tract. To an extent, the aspiration cytomorphology is also distinct and correlates well with the histologic picture. In general, the cellularity of these aspirates is moderate to high. In serous adenocarcinomas, most of the neoplastic cells are present in cohesive aggre-

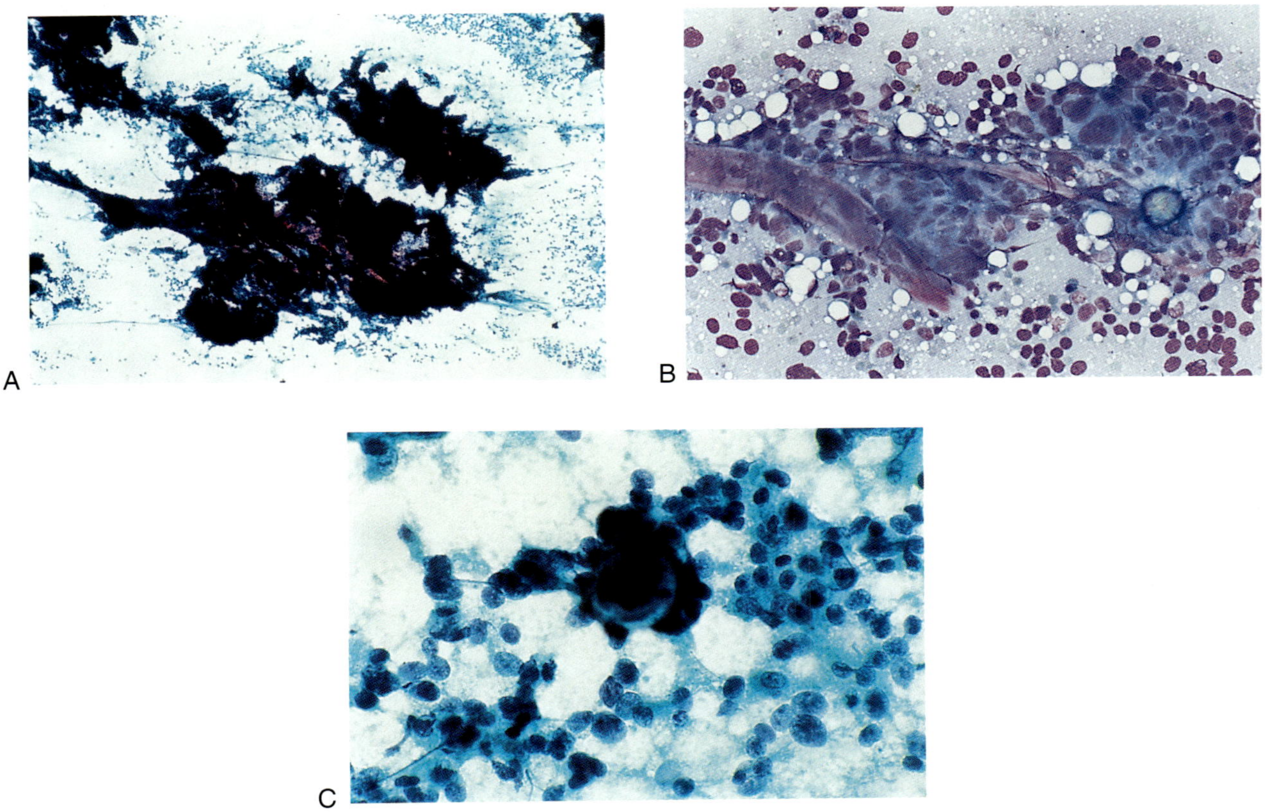

Fig. 7-45. FNA biopsy of the omentum with metastatic well-differentiated papillary serous cystadenocarcinoma from the ovary. (**A**) High cellularity is evident. In addition, a complex papillary configuration is recognizable in this scanning lens view. Although many of the tumor cells are present in large three-dimensional cohesive aggregates, many others are individually dispersed throughout the smear. (**B**) The connective tissue core of a papillae is evident. It is encased by moderately pleomorphic adenocarcinoma cells. (**C**) Tumor cells are relatively uniform with round or slightly ovoid nuclei and finely granular darkly stained chromatin. A minority of the nuclei have obvious nucleoli. At the center of the field, a very darkly stained three-dimensional structure is evident. It probably represents a fragment of a papillae (**A & C**, Papanicolaou; **B**, Diff-Quik).

gates.[70] Although many of these may be two-dimensional sheets, characteristic branched papillary structures may also be present (Figs. 7-45 to 7-48). The cytologic features of individual cells are best seen at the periphery of the cellular aggregates. These cells are characterized by solitary, large nuclei with hyperchromatic chromatin and well-developed nucleoli. The nuclear to cytoplasmic ratios are usually quite high, as the cytoplasm tends to be scanty with little evidence of secretory activity (vacuoles are unusual). Glandular or acinar structures may also be evident. Psammoma bodies may appear free within the smear as well intimately associated with the cellular aggregates (Figs. 7-46 and 7-47). In aspirates of metastatic borderline serous tumors, the cytologic features of malignancy may not be as well developed (Fig. 7-47). In particular, nucleoli may be inconspicuous and nuclear size may be reduced. Cytomorphologically, it is impossible to distinguish papillary serous adenocarcinomas arising in the ovary or endometrium from histologically similar-appearing carcinomas arising as a primary tumor of the peritoneum.[71] As stated by Ramzy and Delaney,[70] it may be very difficult or impossible to distinguish a serous adenocarcinoma from an endometrioid adenocarcinoma. In this latter neoplasm, malignant cells also tend to present in cohesive aggregates, but well-defined papillary

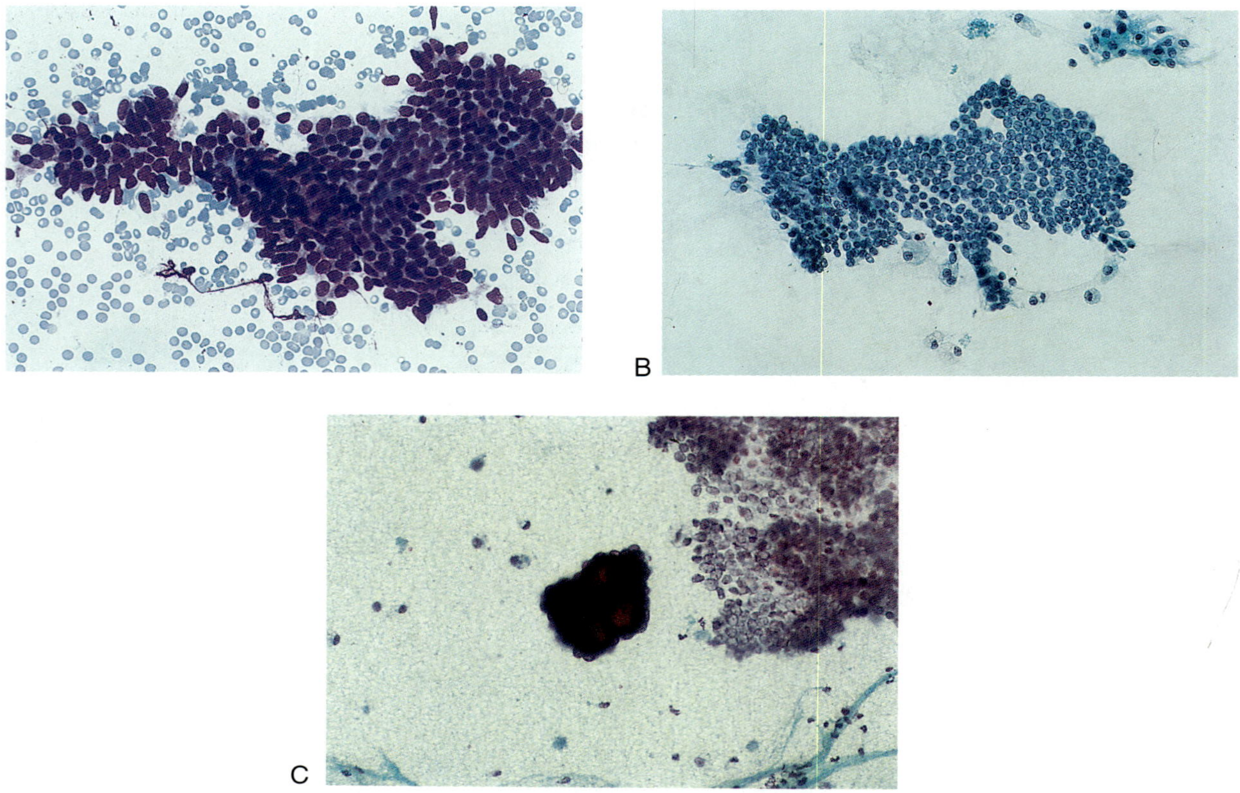

A B C

Fig. 7-46. (**A–C**) Aspirates of well-differentiated papillary serous cystadenocarcinoma typically include flat sheets of uniform relatively small neoplastic cells. To an extent, they resemble sheets of benign mesothelial cells seen in pelvic wash specimens. A honeycomb arrangement of the cells is often present, suggesting that polarity is not markedly altered. (**C**) There appears to be a fusion of several psammoma bodies (**A**, Diff-Quik; **B & C**, Papanicolaou).

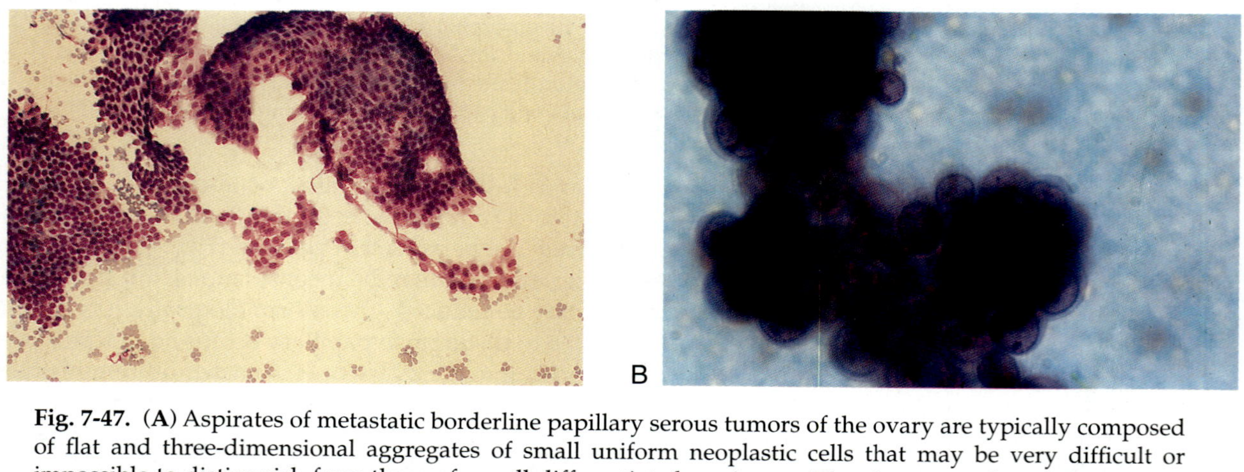

A B

Fig. 7-47. (**A**) Aspirates of metastatic borderline papillary serous tumors of the ovary are typically composed of flat and three-dimensional aggregates of small uniform neoplastic cells that may be very difficult or impossible to distinguish from those of a well-differentiated carcinoma. They have round or slightly ovoid nuclei, high nuclear to cytoplasmic ratios, and small nucleoli. (**B**) The chromatin is finely granular and evenly distributed throughout the nuclei, which may contain inconspicuous nucleoli. Psammoma bodies consist of concentric layers of calcified material which are often surrounded by a solitary layer of tumor cells (**A**, Diff-Quik; **B**, Papanicolaou).

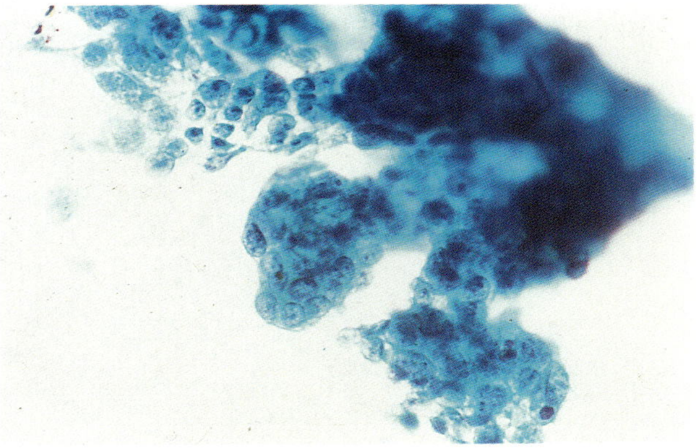

Fig. 7-48. In high-grade (poorly differentiated) papillary serous cystadenocarcinoma, there may be a reduction in the degree of intercellular cohesion compared to the more differentiated counterparts. Usually, however, some cohesion persists with the formation of papillary structures. Compared to the better differentiated tumors, nuclear pleomorphism and the presence of larger and multiple nucleoli are characteristic (Papanicolaou).

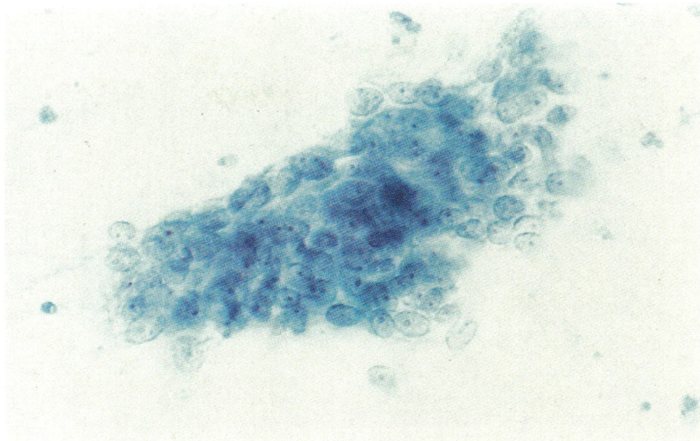

Fig. 7-49. FNA biopsy of a metastatic endometrioid carcinoma from the ovary. This aggregate is densely packed by relatively uniform-appearing malignant cells. Each tumor cell has a solitary hyperchromatic nucleus and a distinct nucleolus. The nuclear to cytoplasmic ratios are quite high. This would be basically impossible to distinguish from a serous adenocarcinoma (Papanicolaou).

structures are less frequent (Fig. 7-49). Individual neoplastic cells will closely resemble those seen in the serous neoplasms, although chromatin may appear coarser. In addition, psammoma bodies will be far less common.

Mucinous adenocarcinomas generally yield a larger proportion of individually dispersed malignant cells and smaller aggregates.[70] As these malignancies are characterized by abundant intracellular mucin, the cytoplasm appears more abundant than in either serous or endometrioid tumors. Distinct vacuoles may be seen or the cytoplasm may have a granular appearance to it (Fig. 7-50). Consequently, the nuclear to cytoplasmic ratios are lower. In addition, nucleoli tend to be less well developed.

In our limited experience, the aspiration cytomorphology of clear cell adenocarcinoma is distinctive, especially with the Diff-Quik stain.[71a] The neoplastic cells may be intimately associated with and attached to the external surfaces of fragments of a basement membrane-like stroma (Fig. 7-51). This material may be deeply eosinophilic with the above-mentioned stain. With the Papanicolaou stain, it has a pale green appearance. Similar to the mucinous carcinomas, the nuclear to cytoplasmic ratios are relatively low. This is the consequence of abundant clear cytoplasm. Other dis-

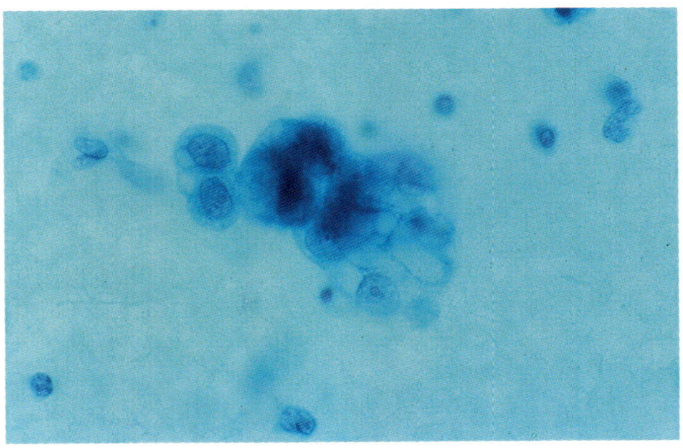

Fig. 7-50. Mucinous adenocarcinoma includes tumor cells with abundant cytoplasmic mucin. This typically distends the cytoplasm with homogeneous-appearing and optically clear regions. Nucleoli may or may not be well developed (Papanicolaou).

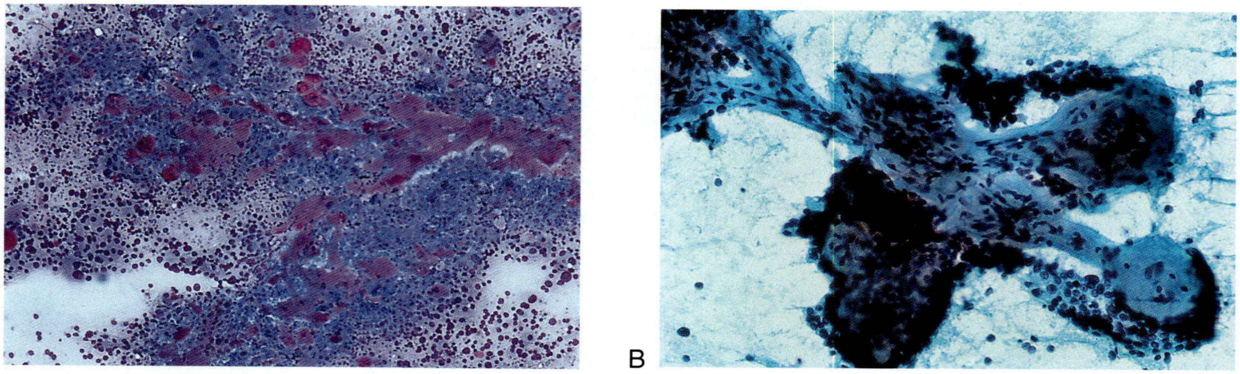

Fig. 7-51. Perhaps the most characteristic aspect of aspirates of clear cell adenocarcinoma of the female genital tract is the connective tissue framework. This scaffolding consists of branching arrays of dense, homogeneous matrix material. (**A**) With the Diff-Quik, it readily stands out as a magenta or pink material. (**B**) It has a glassy cyanophilic texture with the Papanicolaou stain (**A**, Diff-Quik; **B**, Papanicolaou).

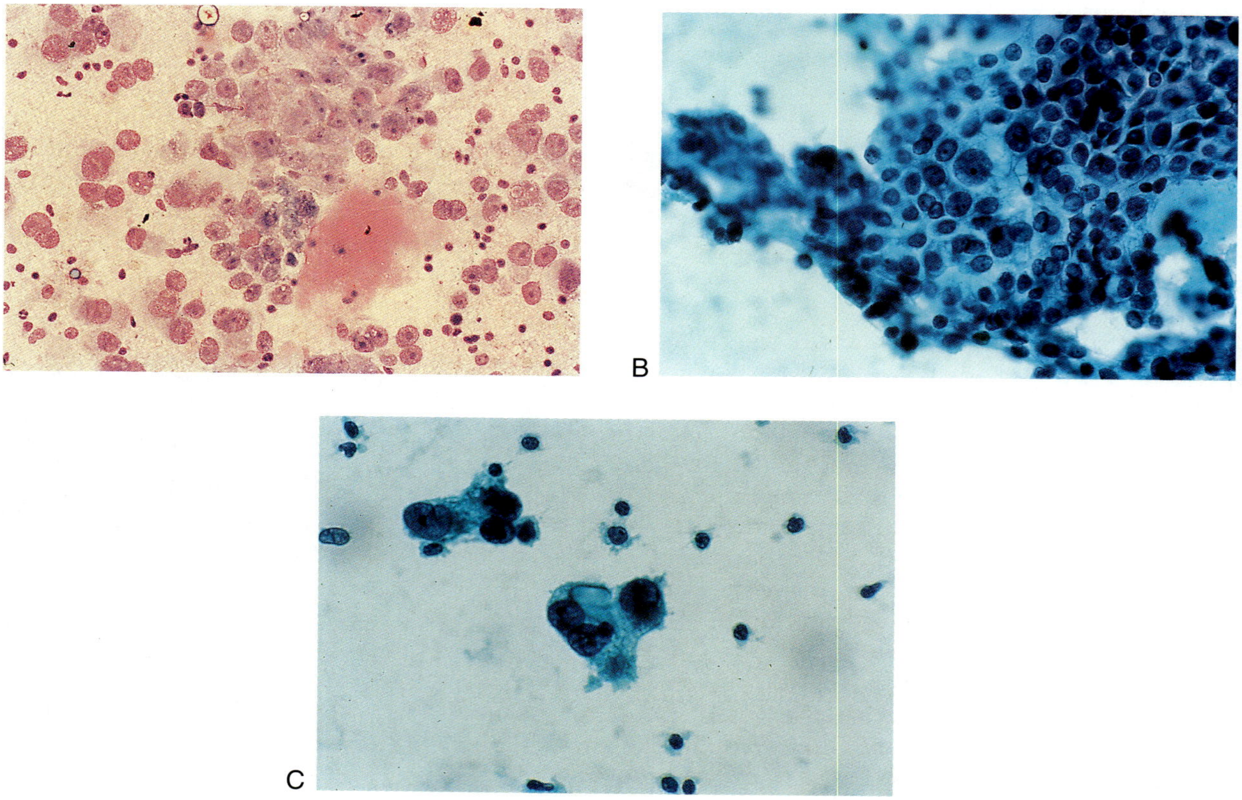

Fig. 7-52. (**A–C**) Very prominent nucleoli are another characteristic feature of the clear cell adenocarcinomas. Cytoplasm is poorly stained, representing the abundant glycogen that may be present (**A**, Diff-Quik; **B & C**, Papanicolaou).

tinctive features include very prominent nucleoli, distinct cellular borders, and a hobnail arrangement. In the latter situation, the neoplastic cells have a bulbous appearance in that the nucleus is oriented away from the basal portion of the cell (Fig. 7-52).

Nonepithelial Ovarian Tumors

Ovarian germ cell tumors may also be sampled by aspiration biopsy. The cytomorphology of these neoplasms is indistinguishable from that of extragonadal germ cell neoplasms, described in Chapter 2.[72,73]

As with most other workers, our experience with aspiration cytology of sex cord-stromal tumors of the ovary is limited.[74,75] Our exposure has been restricted essentially to samples of primary and metastatic granulosa cell tumors. Aspirates are at least moderately cellular and consist of a monotonous population of individually dispersed and cohesive neoplastic cells. The cells are characterized by a solitary ovoid nucleus with a distinct delicate membrane, inconspicuous nucleoli, and a very finely granular chromatin pattern (Fig. 7-53). The latter may appear very homogeneous, somewhat reminiscent of the nuclei characteristic of papillary adenocarcinoma of the thyroid. Cytoplasm is scanty and nondescript. Pleomorphism is minimal. Although nuclear grooves may be well developed, we agree with Ramzy and colleagues that they may not be prominent.[74,75] The cellular aggregates may be quite

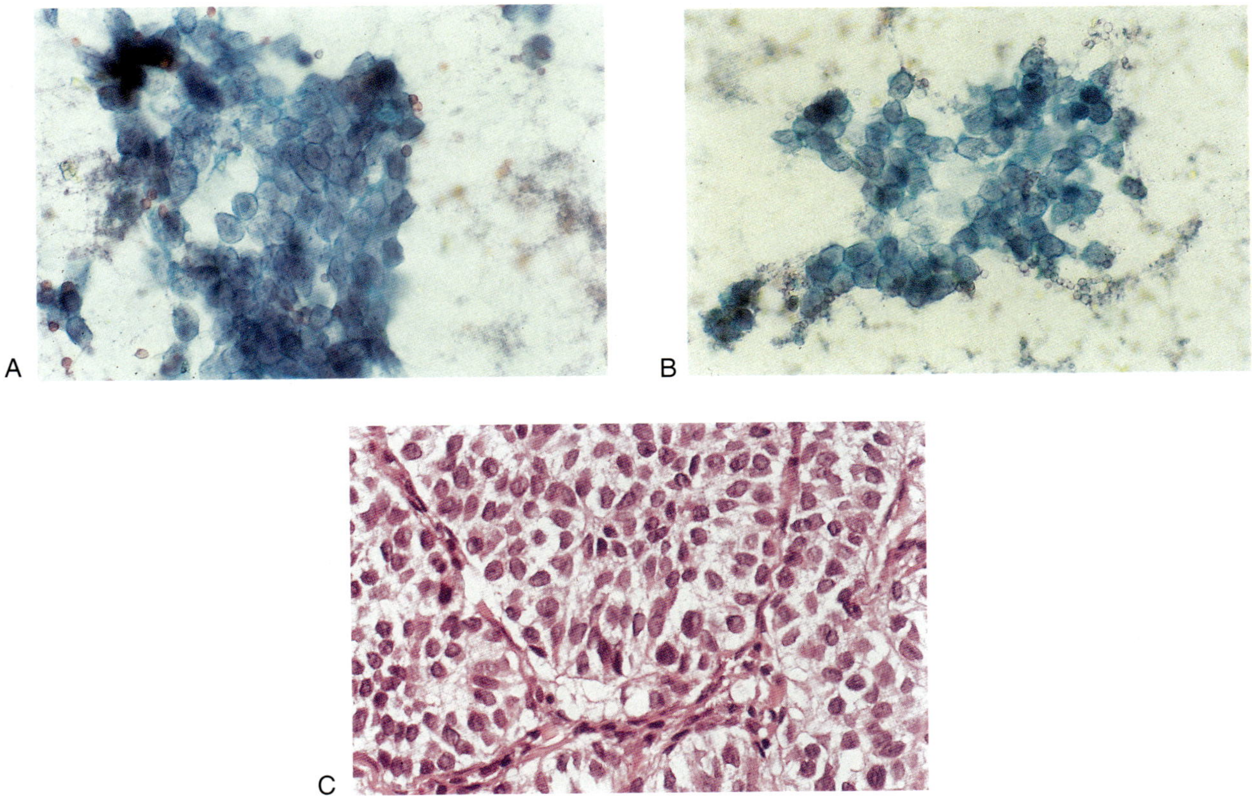

Fig. 7-53. (**A–C**) The neoplastic cells in aspirates of granulosa cell tumors are strikingly uniform from field to field and from specimen to specimen. They have ovoid contours, delicate but prominent nuclear membranes, very finely granular (powdery) chromatin, generally no obvious nucleoli, and infrequently longitudinal grooves. The nuclear to cytoplasmic ratios are higher than expected in the surface epithelial carcinomas of the ovary (**A & B**, Papanicolaou; **C**, H&E).

large and typically have sharply defined margins. Infrequently, Call-Exner bodies are evident as small gland-like structures surrounding dense structureless material. These bodies may be more evident in cell block preparations. We have seen amorphous basement membrane-like material throughout the aspirate of a granulosa cell tumor. Ramzy et al.[74] described the aspiration cytomorphology of a Sertoli cell tumor. The tumor cells had uniform ovoid nuclei, irregularly distributed chromatin, and well-developed nucleoli. In the smears, the neoplastic cells were arranged both within acini and trabeculae. Cytoplasm was moderate in volume.

Sarcomas

Sarcomatous neoplasms may arise within the organs of the female genital tract. Histologically, they form a spectrum of diagnostically challenging neoplasms, predominantly in the endometrium and myometrium. Experience with aspiration biopsies of these gynecologic sarcomas has been almost completely restricted to the identification of metastatic or recurrent tumors. Cytomorphologically, many of these mirror the appearances seen in primary soft tissue sarcomas. Recently, we reviewed our FNA biopsy experiences with gynecologic sarcomas, including malignant mixed müllerian tumors.[76] Aspirates of low-grade stromal sarcomas were moderately cellular and consisted of bland, polygonal, and spindle-shaped cells arranged singly and in variably sized clusters[76,77] (Fig. 7-54). The chromatin was finely granular and evenly distributed throughout the ovoid to elongated nuclei. Nuclear to cytoplas-

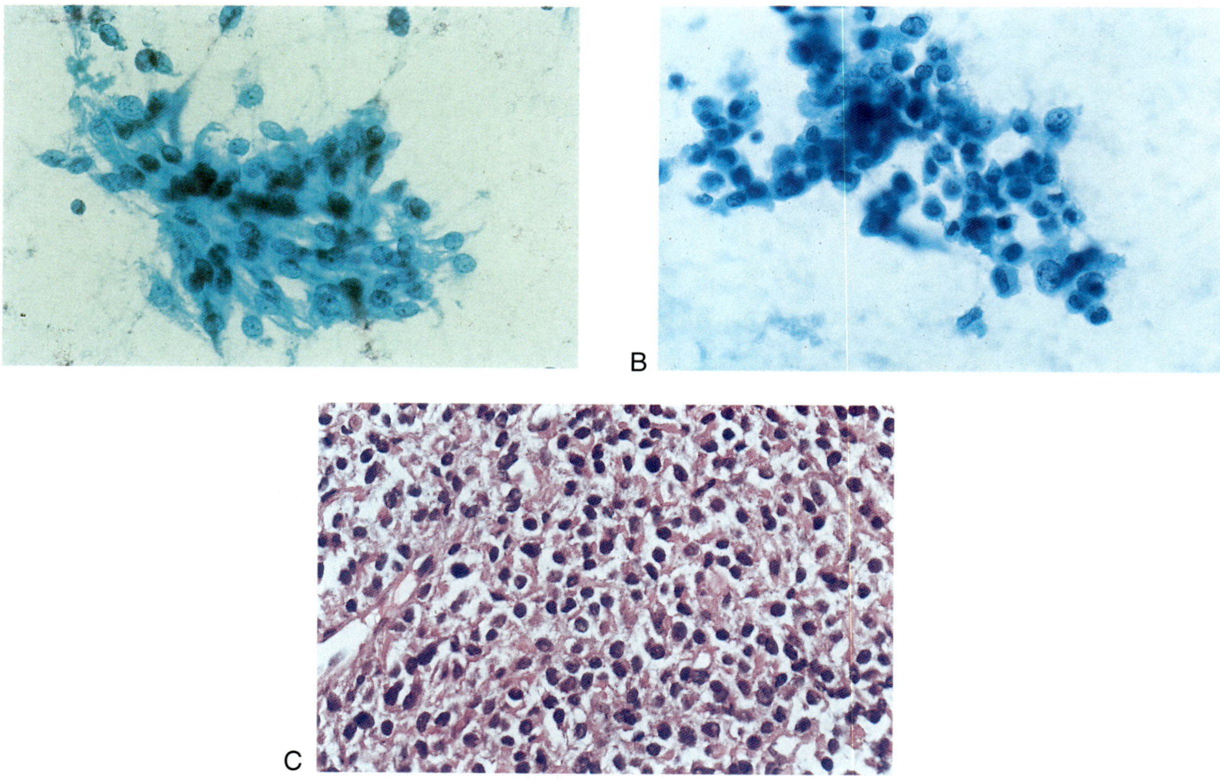

A B C

Fig. 7-54. FNA biopsies of two low-grade endometrial stromal sarcomas both yield loosely cohesive aggregates of rather uniform-appearing neoplastic cells. (**A**) The cells and nuclei tend toward elongated or spindled contours. (**B**) The neoplastic cells are predominantly round to polygonal. In both, the chromatin is finely granular and evenly distributed throughout the nuclei. Small nucleoli are visible in a few of the nuclei. Cytoplasm is cyanophilic and delicate (**A & B**, Papanicolaou). (**C**) Cell block from the same aspirate of the tumor in Fig. B. The uniformity of the polygonal cellular contours and the round nuclear shapes is readily apparent (H&E).

mic ratios were generally high. Cytoplasm was filmy in quality. Each cell possessed a single nucleus with a small nucleolus. The aggregates of tumor cells appeared loosely cohesive. One-half of the aspirates of the malignant mixed müllerian tumors yielded an admixture of epithelial and stromal elements (Fig. 7-55). At times, there was a suggestion of a blending of the two cell types. The carcinomatous component resembled most closely that of a poorly differentiated serous adenocarcinoma. Although the mesenchymal elements were generally nonspecific and spindled in appearance, malignant cartilage was recognized in one sample. Mourad and colleagues recently reported the aspiration cytomorphologic features of 16 metastatic and recurrent malignant mixed müllerian tumors.[78] Most of the specimens were procured from abdominal sites. Carcinomatous elements were present in all 16 specimens but the sarcomatous component was noted in only 5 (31 percent). For the most part, the carcinoma was considered to be very high grade, with definite glandular differentiation in only a small minority (19 percent). The malignant mesenchymal component in the smears had a homologous spindle cell sarcomatous appearance without any evidence of heterologous differentiation. Necrotic debris often filled the smear backgrounds. The authors pointed out that the cell block preparation was especially good for recognition of the biphasic nature of these neoplasms.[78] Other sarcomas may present distinctive features (Fig. 7-56).

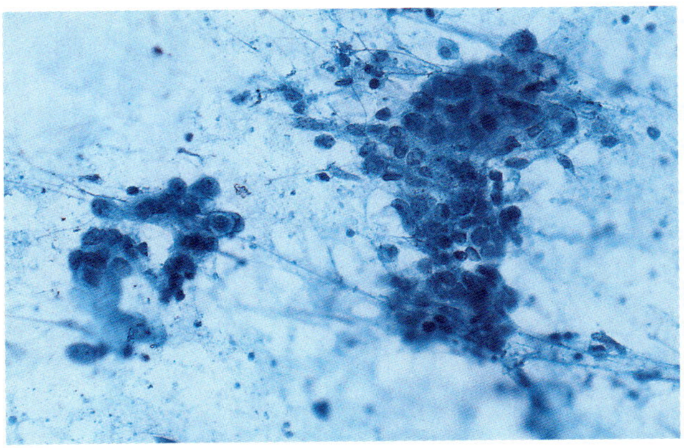

Fig. 7-55. In aspirates of carcinosarcomas, it is not always possible to determine whether the obviously malignant cells are sarcomatous or carcinomatous. In this field, the largest aggregate is probably epithelial in nature. Intercellular cohesion is well preserved and cell borders are well defined. The smallest cellular collection here may be sarcomatous, in that there is a syncytial arrangement of cells that have indistinct edges (Papanicolaou).

A

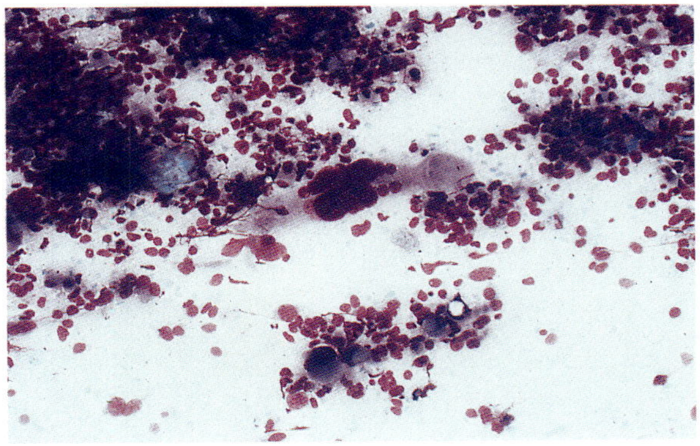

B

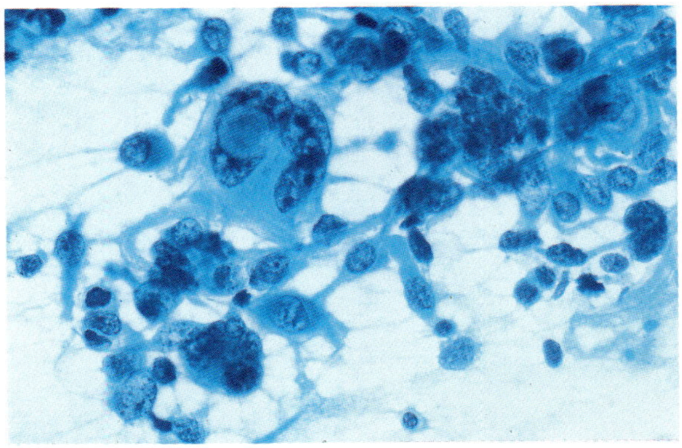

Fig. 7-56. (**A & B**) Aspiration biopsy of a metastatic pleomorphic rhabdomyosarcoma primary in the uterine corpus. (**A**) The marked cellular pleomorphism is apparent as most of the malignant elements are small, possess a solitary round nucleus, and very high nuclear to cytoplasmic ratios; that is, they appear generally undifferentiated. In addition, there are several large tumor giant cells. This includes an enormous multinucleated tumor giant cell with abundant dense cytoplasm arrayed in a broad tail (Diff-Quik). (**B**) The marked pleomorphism of cellular sizes and shapes is apparent. Chromatin is coarsely granular and very darkly stained. The largest multinucleated tumor giant cell includes a huge nuclear pseudoinclusion (Papanicolaou).

Cysts

Whereas aspiration biopsy of ovarian masses suspicious for malignancy is thought to be contraindicated by many individuals in the United States, there is a growing popularity and utilization of aspiration cytology to evaluate the cells comprising ovarian cysts, which are clinically benign. Most often, these aspirates are performed during laparotomy or laparoscopy. Frequently such patients are being evaluated for infertility or pelvic pain. These non-neoplastic cystic lesions include follicular cysts, luteal cysts, surface inclusion cysts, and cystic endometriosis.

The aspiration cytomorphology of follicular cysts of the ovary has recently been detailed in a number of publications, in part due to the potentially alarming appearance of these specimens.[79–84] The cellularity of these aspirates is highly variable. In some cases these specimens consist of clear or blood-tinged fluid containing only a few histiocytic elements and blood; some workers have suggested that such specimens should be considered insufficient or inadequate for diagnosis.[82,84] Nunez and Diaz[82] claimed that some of these sparsely cellular specimens may be derived from atretic follicular cysts with a thin lining. If such specimens are excluded from consideration, then the cellularity of aspirates of ovarian follicular cysts are moderate to marked (Fig. 7-57). The predominant cell type seen in these smears is the granulosa cell, which is the component lining the

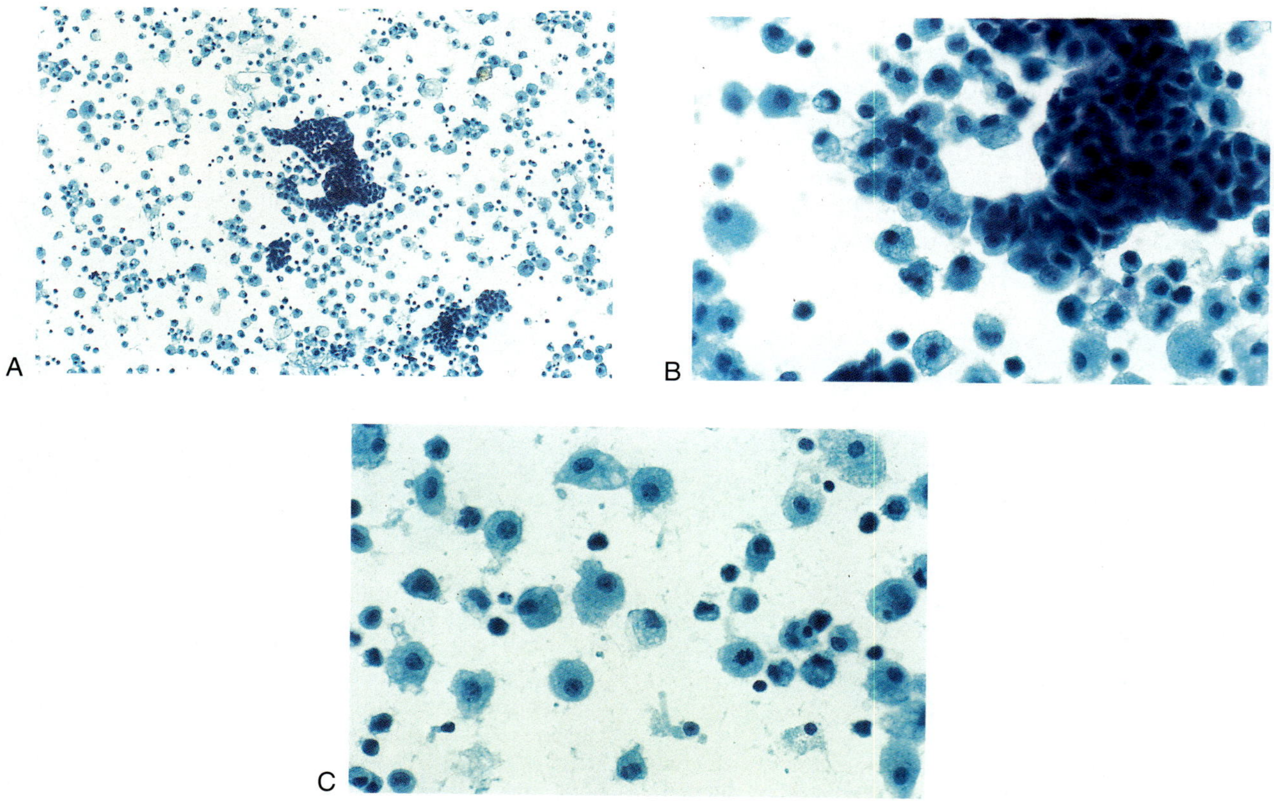

Fig. 7-57. (**A**) The cellularity of aspirated ovarian follicular cysts is generally quite high and may include both numerous isolated cells and variably sized and contoured cellular clumps. (**B**) Most of the aspirated cells obtained from follicular cysts of the ovary are granulosa cells with varying states of luteinization. They are characterized by solitary round nuclei, finely granular chromatin, and foamy to granular cytoplasm. Where the cells abut against one another, cell borders are well defined. (**C**) A solitary mitotic figure is evident (**A–C**, Papanicolaou).

cysts. These elements may be present both as solitary cells and in variably sized and shaped aggregates. The aggregates may be round with smooth edges, irregular, or even papillary in appearance. Overall, these aggregates appear loosely cohesive. The granulosa cells are characterized by a solitary (rarely binucleated) nucleus and a relatively high nuclear to cytoplasmic ratio (Fig. 7-57). The nucleus will have a smooth delicate membrane, evenly distributed chromatin, and small to inconspicuous nucleoli. Well-developed longitudinal nuclear grooves are unusual. Characteristically, the nuclei are somewhat eccentrically positioned within the cytoplasm. The latter is eosinophilic and varies from granular to foamy. Granulosa cells with the latter appearance often have more abundant cytoplasm, representing luteinized forms. Chromatin appearances vary from finely to coarsely granular. Mitotic figures are usually present; Nunez and Diaz[82] calculated an average of 7 mitotic figures per 10 high power fields in these specimens. If one is unaware of the cytomorphologic picture of these specimens, then they may be mistaken for cancer.[80,81] DNA ploidy analysis may be of some assistance in this regard.[85]

Corpus luteum cysts may also yield cellular smears containing both isolated and aggregated granulosa cells[84] (Fig. 7-58). The latter may have abundant luteinized cytoplasm. Although nucleoli may be more prominent than in follicular cysts, mitotic activity will be reduced.

Endometriotic cysts yield hemosiderin-laden histiocytes, endometrial glandular and stromal cells, and both fresh and old blood (Fig. 7-59). One must find endometrial elements in order to make the specific diagnosis of endometriosis. Endometrial glandular cells will be present in flat or three-dimensional small aggregates; the cells have a solitary dark-stained nucleus and a high nuclear to cytoplasmic ratio. Nucleoli may be evident. The stromal component will consist of loosely cohesive arrays of cells with greater volumes of foamy or granular cytoplasm and small vesicular oval to bean-shaped nuclei. Their cytoplasm generally does not contain blood pigment.

MULTICYSTIC MESOTHELIOMA

This rare benign lesion occurs most often in young adults with a marked preponderance for women.[86,87] Histologically, these masses are dominated by cystic spaces lined by uniform, bland mesothelial cells. Some examples are more solid and have foci closely resembling adenomatoid tumors. Although they may be multifocal or recur following

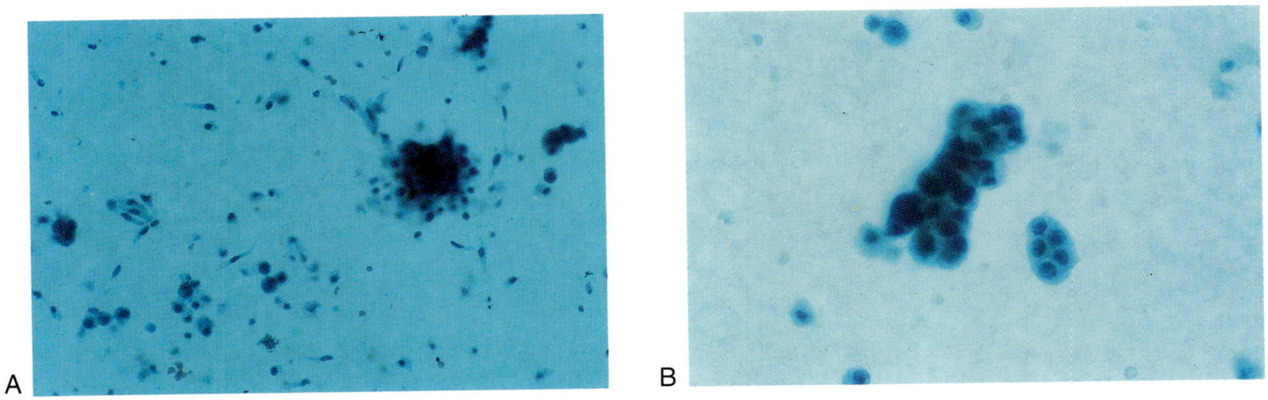

Fig. 7-58. (**A**) Moderately cellular smear, including clustered and dispersed cells. The former possess an epithelial appearance with abundant indistinct cytoplasm. Some of the isolated cells resemble those in the aggregates, but others are elongated or spindled, perhaps representing thecal elements. (**B**) These granulosa cells are characterized by benign round nuclei and well-defined borders. Nuclear grooves are not evident (**A & B**, Papanicolaou).

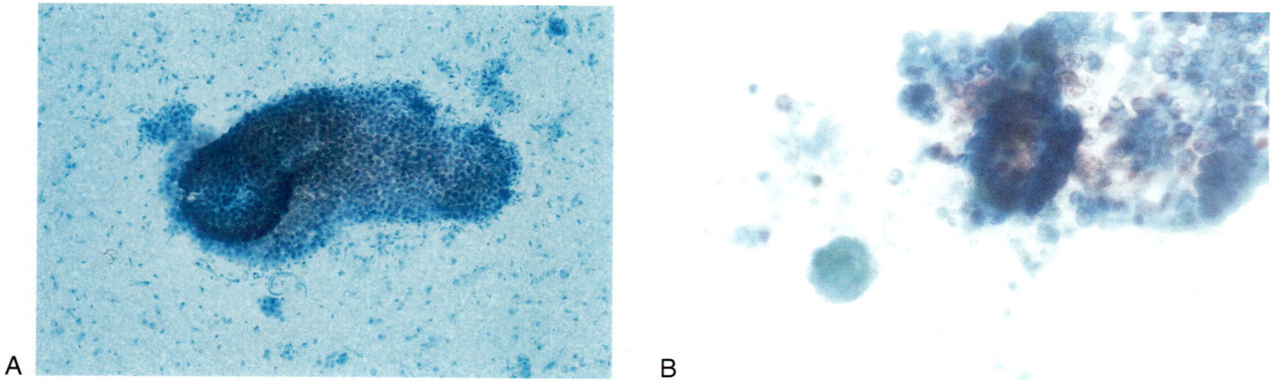

Fig. 7-59. (A) Endometrial-like glandular cells are arrayed in a large sheet with a suggestion of gland formation. It is surrounded by dispersed stromal and histiocytic elements. (B) Pale chromatin and nucleoli characterize the nuclei of the glandular cells. The isolated stromal cell may contain hemosiderin in its cytoplasm (A & B, Papanicolaou).

surgical resection, they are benign. Some authors do not even consider them as a true neoplasm.[87]

Reports of aspiration biopsies are uncommon.[88,89] They describe homogeneous, benign mesothelial cells in sheets and lining delicate connective tissue septae. There is no evidence of atypia, necrosis, or mitotic activity. We have seen a single example of a FNA biopsy of a multicystic mesothelioma. The smears were moderately cellular with many obvious mesothelial cells (Fig. 7-60). Nuclei were round with fine chromatin and small nucleoli. These were misinterpreted as contaminants. Other cells with similar nuclei were embedded in a dense eosinophilic matrix. A minority had large cytoplasmic vacuoles, which when focally numerous resembled an adenomatoid tumor.[90]

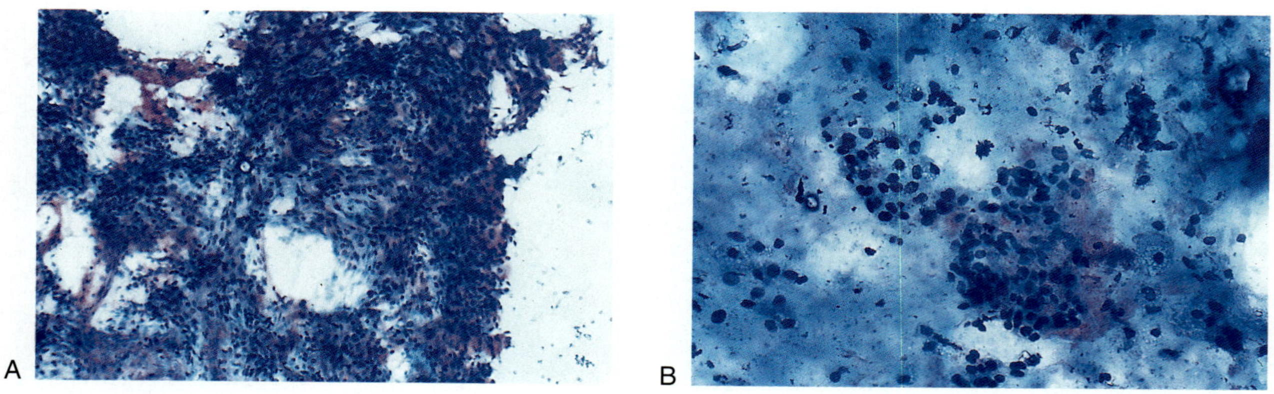

Fig. 7-60. (A & B) FNA biopsy of a multicystic mesothelioma yielded fragments of collagen-containing connective tissue with numerous mesothelial-like cellular elements. These fragments were apparently derived from the septae of this subsequently resected multilocular cystic mass. The constituent cells had uniformly round nuclei and polygonal to more ovoid cellular contours. In addition, the smear background contained a dense proteinaceous material in which similar-appearing cells were suspended. Where the cells adjoin one another, articulations that are characteristic of mesothelial cells can be appreciated. The cytoplasm of a minority of cells had prominent vacuoles (A & B, Diff-Quik).

MISCELLANEOUS LESIONS

Compared to the workup of patients with gynecological malignancies, FNA biopsy is used less often to evaluate potential metastatic lesions in male patients with known genitourinary cancers.[91,92] However, prostatic, testicular, and penile neoplasms most frequently metastasize to intra-abdominal sites. Radiographically directed FNA biopsies can play an important role in the investigation of such masses.

Rarely, primary tumors arise in the retroperitoneum which histologically are indistinguishable from surface epithelial neoplasms of the ovary. Recently, Motoyama et al.[93] reported the aspiration cytomorphology of two examples of retroperitoneal mucinous cystic tumors. Grossly, the aspirated material had a mucoid quality. Neoplastic epithelial cells were rare to absent in the aspiration smears. The major cellular components were neutrophils and macrophages. One of the two cases contained many acellular hyalin globules. Following resection, one was diagnosed as a mucinous neoplasm with focal cyst adenocarcinoma, the other as a borderline mucinous neoplasm.[93]

Recently, Young et al.[94] described the cytologic features of a retrocystic hamartoma (tailgut cyst) in a pediatric patient. The smears contained abundant squamous epithelial cells, neutrophils, and lymphocytes.

REFERENCES

1. Mack TM: Sarcomas and other malignancies of soft tissue, retroperitoneum, peritoneum, pleura, heart, mediastinum, and spleen. Cancer 75:211, 1995
2. Akerman M, Iduall I, Rydhaom A: Cytodiagnosis of soft tissue tumors and tumor-like conditions by means of fine needle aspiration biopsy. Arch Orthop Traumat Surg 96:61, 1980
3. Rydholm A, Akerman M, Iduall I, Persson BM: Aspiration cytology of soft tissue tumours. A prospective study of its influence on choice of surgical procedure. Intern Orthop 6:209, 1982
4. Layfield LJ, Anders KH, Glasgow BJ, Mirra JM: Fine-needle aspiration of primary soft-tissue lesions. Arch Pathol Lab Med 110:420, 1986
5. Miralles TG, Gosalbez F, Menédez P, et al: Fine needle aspiration cytology of soft-tissue lesions. Acta Cytol 30:671, 1986
6. Kindblom L-G, Walaas L, Widéhn S: Ultrastructural studies in the preoperative cytologic diagnosis of soft tissue tumors. Semin Diagn Pathol 3:317, 1986
7. Suen KC: Retroperitoneum and intestine. p. 31. In Kline TS (ed): Guides to Clinical Aspiration Biopsy. Igaku-Shoin, New York, 1987
8. Oland J, Rosen A, Reif R et al: Cytodiagnosis of soft tissue tumors. J Surg Oncol 37:168, 1988
9. Nguyen G-K: What is the value of fine-needle aspiration biopsy in the cytodiagnosis of soft tissue tumors? Diagn Cytopathol 4:352, 1988
10. Gonzalez-Campora R, Munoz-Arias G, Otal-Salaverri C et al: Fine needle aspiration cytology of primary soft tissue tumors. Morphologic analysis of the most frequent types. Acta Cytol 36:905, 1992
11. Cohen MB, Layfield LJ: Fine needle aspiration biopsy of soft tissue tumors. p. 101. In Schmidt W (ed): Cytopathology Annual. ASCP Press, Chicago, 1994
12. Gonzalez-Campora R, Otal-Salaverri C, Hevia-Vázquez A et al: Fine needle aspiration in myxoid tumors of the soft tissues. Acta Cytol 34:179, 1990
13. Wakely PE, Jr, Geisinger KR, Cappellari JO et al: Fine needle aspiration cytology of soft tissue: chondromyxoid lesions. Diagn Cytopathol 12:10, 1995
14. Geisinger KR, Hajdu SI, Helson L: Exfoliative cytology of nonlymphoreticular neoplasms in children. Acta Cytol 28:16, 1984
15. Akhtar M, Ali A, Sabbah R et al: Fine needle aspiration biopsy diagnosis of round cell malignant tumors of childhood cancer. Cancer 55:1805, 1985
16. Valkov I, Bojikin B: Fine-needle aspiration biopsy of abdominal and retroperitoneal tumors in infants and children. Diagn Cytopathol 3:129, 1987
17. Wakely PE, Jr, Kardos TF, Frable WJ: Application of fine needle aspiration biopsy to pediatrics. Hum Pathol 19:1383, 1988
18. Gorczyca W, Bedner E, Juszkiewicz P et al: Aspiration cytology in the diagnosis of malignant tumors in children. Am J Pediatr Hematol Oncol 14:129, 1992
19. McGahey BE, Moriarty AT, Nelson WA et al: Fine needle aspiration biopsy of small round blue cell tumors of childhood. Cancer 69:1067, 1992
20. Hajdu SI, Ehya H, Frable WJ et al: The value and limitations of aspiration cytology in the diagnosis or primary tumors. Acta Cytol 33:741, 1989
21. Gordon MD, Vetto J, Meshul CK, Schmidt WA: FNA of extraskeletal myxoid chondrosarcoma: cytomorphologic EM and X-ray microanalysis features. Diagn Cytopathol 10:352, 1994
22. Neimann TH, Bottles K, Cohen MB: Extraskeletal myxoid chondrosarcoma: fine-needle aspiration biopsy findings. Diagn Cytopathol 11:363, 1994.
23. O'Dowd G, Schumann GB: Aspiration cytology and cytochemistry of coccygeal chordoma. A case report and review of the literature. Acta Cytol 27:178, 1983

24. Kontozoglou T, Quizilbash AH, Sianos J, Steed R: Chordomas: Cytologic and immunocytochemical study of four cases. Diagn Cytopathol 2:55, 1986
25. Oertel YC, Beckner ME, Engler WF: Cytologic diagnosis and intrastructure of fine-needle aspirates of ganglion cysts. Arch Pathol Lab Med 110:938, 1986
26. Caraway NP, Staerkel GA, Fanning CV et al: Diagnosing intramuscular myxoma by fine-needle aspiration: a multidisciplinary approach. Diagn Cytopathol 11:255, 1994
27. Dahl I, Akerman M: Nodular fascitis. A correlative cytologic and histologic study of 13 cases. Acta Cytol 25:215, 1981
28. Rocamora A, Carrillo R, Vives R, Solera JC: Fine needle aspiration biopsy of myxoid metastasis of malignant melanoma. Acta Cytol 32:94, 1988
29. Lindholm K, de la Torre M: Fine needle aspiration biopsy cytology of myxoid metastatic malignant melanoma. Acta Cytol 32:719, 1988
30. Raab SS, Silverman JF, McLeod DL et al: Fine needle aspiration biopsy of fibromatoses. Acta Cytol 37:323, 1993
31. Powers CN, Berardo MD, Frable WJ: Fine-needle aspiration biopsy: pitfalls in the diagnosis of spindle-cell lesions. Diagn Cytopathol 10:232, 1994
32. Shmookler BM. Retroperitoneal synovial sarcoma. A report of four cases. Am J Clin Pathol 77:686, 1982
33. Dahl I, Hagmar B, Angervall L: Leiyomyosarcoma of the soft tissue: A correlative cytological and histological study of 11 cases. Acta Pathol Microbiol Scand Sect A 89:285, 1981
34. Tao L-C, Davidson DD: Aspiration biopsy cytology of smooth muscle tumors: a cytologic approach to the differentiation between leiomyosarcoma and leiomyoma. Acta Cytol 37:300, 1993
35. Smith MB, Silverman JF, Raab SS et al: Fine-needle aspiration cytology of hepatic leiomyosarcoma. Diagn Cytopathol 11:321, 1994
36. Vendraminelli R, Cavazzana AO, Poletti A et al: Fine-needle aspiration cytology of malignant nerve sheath tumors. Diagn Cytopathol 8:559, 1992
37. Aisner SC, Seidman JC, Burke KC, Young JWR: Aspiration cytology of biphasic and monophasic synovial sarcoma. A report of two cases. Acta Cytol 37:413, 1993
38. Sapi Z, Bodó M, Megyesi J, Rahóty P: Fine needle aspiration cytology of biphasic synovial sarcoma of soft tissue. Report of a case with ultrastructural, immunohistologic and cytophotometric studies. Acta Cytol 34:69, 1990
39. Abele JS, Miller T: Cytology of well-differentiated and poorly differentiated hemangiosarcoma in fine needle aspirates. Acta Cytol 26:341, 1982
40. Silverman JF, Lannin DR, Larkin EW et al: Fine-needle aspiration cytology of postirradiation sarcomas, including angiosarcoma, with immunocytochemical confirmation. Diagn Cytopathol 5:275, 1989
41. Geisinger KR, Silverman JF, Cappellari JO, Dabbs DJ: Fine needle aspiration cytology of malignant hemangiopericytomas with ultrastructural and flow cytometric analysis. Arch Pathol Lab Med 114:705, 1990
42. Hales M, Bottles K, Miller T et al: Diagnosis of Kaposi's sarcoma by fine-needle aspiration biopsy Am J Clin Pathol 88:20, 1987
43. Walaas L, Kindblom L-G: Lipomatous tumors: a correlative cytologic and histologic study of 27 tumors examined by fine needle aspiration cytology: Hum Pathol 16:6, 1985
44. Akerman M, Rydholm A: Aspiration cytology of lipomatous tumors: a ten-year experience at an orthopedic oncology center. Diagn Cytopathol 3:295, 1987
45. Geisinger KR, Kawamoto EH, Marshall RB et al: Aspiration and exfoliative cytology, including ultrastructure, of a malignant granular-cell tumor. Acta Cytol 29:593, 1985.
46. Pohar-Marinsek Z, Zidar A: Epithelioid sarcoma in FNAB smears. Diagn Cytopathol 11:367, 1994.
47. Shabb N, Sneige N, Fanning CV, Dekmezian R: Fine-needle aspiration cytology of alveolar soft part sarcoma. Diagn Cytopathol 7:293, 1991
48. Caraway NP, Fanning CV, Wojcik EM et al: Cytology of malignant melanoma of soft parts: fine needle aspirates and exfoliative specimens. Diagn Cytopathol 9:632, 1993
49. Almeida MM, Nunes AM, Frable WJ: Malignant melanoma of soft tissue. A report of three cases with diagnosis by fine needle aspiration cytology. Acta Cytol 38:241, 1994
50. Gurley AM, Silverman JF, Lassaletta MM et al: The utility of ancillary studies in pediatric FNA cytology. Diagn Cytopathol 8:137, 1992
51. Akhtar M, Bedrossian CW, Ali MA et al: Fine-needle aspiration biopsy of pediatric neoplasms: correlation between electron microscopy and immunocytochemistry in diagnosis and classification. Diagn Cytopathol 8:258, 1992
52. Geisinger KR, Silverman JF, Wakely PW: Pediatric Cytopathology. ASCP Press, Chicago, 1994, p. 250
53. Seidal T, Walaas L, Kindblom L-G, Angervall L: Cytology of embryonal rhabdomyosarcoma: a cytologic, light microscopic, electron microscopic, immunohistochemical study of seven cases. Diagn Cytopathol 4:292, 1988
54. Ahktar M, Ali MA, Bakry M et al: Fine-needle aspiration biopsy diagnosis of rhabdomyosarcoma: cytologic, histologic, and ultrastructural correlations. Diagn Cytopathol 8:465, 1992

55. Brehant LE, Anderson LH, Taylor DA: Extraskeletal Ewing's sarcoma. Diagnosis of a case by fine needle aspiration cytology. Acta Cytol 30:683, 1986
56. Gerald WL, Miller HK, Battifora H et al: Intra-abdominal desmoplastic small round cell tumor. Report of 19 cases of a distinctive type of high-grade polyphenotypic malignancy affecting young individuals. Am J Surg Pathol 15:499, 1991
57. Layfield LJ, Lenarsky C: Desmoplastic small cell tumors of the peritoneum coexpressing mesenchymal and epithelial markers. Am J Clin Pathol 96:536, 1991
58. Caraway NP, Fanning TV, Bainbridge TC et al: Fine needle aspiration of intraabdominal desmoplastic small cell tumors. Acta Cytol 36:599, 1992
59. Setrakian S, Gupta PK, Heald J et al: Intraabdominal desmoplastic small round cell tumor. Report of a case diagnosed by fine needle aspiration cytology. Acta Cytol 36:373, 1992
60. Thompson EN III, Teot LA, Geisinger KR: Exfoliative and aspiration cytomorphology of intraabdominal desmoplastic small round cell tumor: the first reported case. Acta Cytol 36:599, 1992
61. Kjellgren O, Angstrom T, Bergman F et al: Fine-needle aspiration biopsy in diagnosis, and classification of ovarian carcinoma. Cancer 28:967, 1971
62. Sevin B-U, Greening SE, Nadji M et al: Fine needle aspiration cytology in gynecologic oncology. I. Clinical aspects. Acta Cytol 23:277, 1979
63. Bonfiglio TA, Macintosh PK, Patten SF et al: Fine needle aspiration cytopathology of retroperitoneal lymph nodes in the evaluation of metastatic disease. Acta Cytol 23:126, 1979
64. Belinson JL, Lynn JM, Papillo JL et al: Fine-needle aspiration cytology in the management of gynecologic cancer. Am J Obstet Gynecol 139:148, 1981
65. Flint A, Terhart K, Murad TM, Taylor PT: Confirmation of metastases by fine needle aspiration biopsy in patients with gynecologic malignancies. Gynecol Oncol 14:382, 1982
66. Fortier KJ, Clarke-Pearson DL, Creasman WT, Johnston WW: Fine-needle aspiration in gynecology: Evaluation of extrapelvic lesions in patients with gynecologic malignancy. Obstet Gynecol 65:67, 1985
67. Moriarty AT, Glant MD, Stehman FB: The role of fine needle aspiration cytology in the management of gynecologic malignancies. Acta Cytol 30:59, 1986
68. Nash JD, Burke TW, Woodward JE et al: Diagnosis of recurrent gynecologic malignancy with fine-needle aspiration cytology. Obstet Gynecol 71:333, 1988
69. Layfield LJ, Heaps JM, Berek JS: Fine-needle aspiration cytology accuracy with palpable gynecologic neoplasms. Gynecol Oncol 40:70, 1991
70. Ramzy I, Delaney M: Fine needle aspiration of ovarian masses. I. Correlative cytologic and histologic study of celomic epithelial neoplasms. Acta Cytol 23:97, 1979
71. Chen KTK: Psammocarcinoma of the peritoneum. Diagn Cytopathol 10:224, 1994
71a. Matthews LJ, Cappellari JO, Geisinger KR: The cytopathology of Müllerian clear cell carcinoma. Mod Pathol, in press
72. Akhtar M, Ali MA, Hug M et al: Fine-needle aspiration biopsy of seminoma and dysgerminoma: cytologic, histologic, and electron microscopic correlations. Diagn Cytopathol 6:99, 1990
73. Collins KA, Geisinger KR, Wakely PE, Jr et al: Extragonadal germ cell tumors: A fine-needle aspiration biopsy study. Diagn Cytopathol 12:223, 1995
74. Ramzy I, Delaney M, Rose P: Fine needle aspiration of ovarian masses. II. Correlative cytologic and histologic study of non-neoplastic cysts and noncelomic epithelial neoplasms. Acta Cytol 23:185, 1979
75. Ehya H, Lang WR: Cytology of granulosa cell tumor of the ovary. Am J Clin Pathol 85:402, 1986
76. Finley JL, Silverman JF, Cappellari JO, Geisinger KR: Fine needle aspiration cytology of gynecologic sarcomas and mixed tumors. Cytopathol Annu, in press
77. Nguyen G-K, Berendt RC: Aspiration biopsy cytology of metastatic endometrial stromal sarcoma and extragenital mixed mesodermal tumor. Diagn Cytopathol 2:256, 1986
78. Mourad WA, Sneige N, Katz RL et al: Fine-needle aspiration cytology of recurrent and metastatic mixed mesodermal tumors. Diagn Cytopathol 11:328, 1994
79. Selvaggi SM: Cytology of nonneoplastic cysts of the ovary. Diagn Cytopathol 6:77, 1990
80. Stanley MW, Horowitz CA, Frable WJ: Cellular follicular cyst of the ovary: fluid cytology mimicking malignancy. Diagn Cytopathol 7:48, 1991
81. Selvaggi SM: Fine-needle aspiration cytology of ovarian follicle cysts with cellular atypia from reproductive-age patients. Diagn Cytopathol 7:189, 1991
82. Nunez C, Diaz JI: Ovarian follicular cysts: a potential source of false positive diagnoses in ovarian cytology. Diagn Cytopathol 8:532, 1992
83. Greenebaum E, Mayer JR, Stangel JJ et al: Aspiration cytology of ovarian cysts in in vitro fertilization patients. Acta Cytol 36:11, 1992
84. Wojcik EM, Selvaggi SM: Fine-needle aspiration cytology of cystic ovarian lesions. Diagn Cytopathol 11:9, 1994
85. Greenbaum E, Yee HT, Liu J: DNA ploidy of ovarian and adnexal cyst fluid. A useful adjunct to cytology. Acta Cytol 38:201, 1994

86. Weiss SW, Tavassoli FA: Multicystic mesothelioma: an analysis of pathologic findings and biologic behavior in 37 cases. Am J Surg Pathol 12:737, 1988
87. Ross MJ, Welch WR, Scully RE: Multiocular peritoneal inclusion cysts (so-called cystic mesotheliomas). Cancer 64:1136, 1989
88. Baddoura FK, Varma VA: Cytologic findings in multicystic peritoneal mesothelioma. Acta Cytol 34:524, 1990
89. Devaney K, Kragel PJ, Devaney EJ: Fine-needle aspiration cytology of multicystic mesothelioma. Diagn Cytopathol 8:68, 1992
90. Pérez-Guillermo M, Thor A, Löwhagen T: Paratesticular adenomatoid tumors. The cytologic presentation in fine-needle aspiration biopsies. Acta Cytol 33:6, 1989
91. Oliver RTD, Kellett HMJ, Curling OM et al: The value of fine-needle aspiration cytology in the management of metastatic germ cell tumors. Br J Urol 57:200, 1985
92. Scappini P, Piscioli F, Pusiol T et al: Penile cancer. Aspiration biopsy cytology for staging. Cancer 58:1526, 1986
93. Motoyama T, Chida T, Fujiwara T et al: Mucinous cystic tumor of the retroperitoneum. A report of two cases. Acta Cytol 38:261, 1994
94. Young NA, Neeson T, Bernal D et al: Retrorectal cystic hamartoma diagnosed by fine-needle aspiration biopsy. Diagn Cytopathol 6:359, 1990

8 Spleen

Fine-needle aspiration (FNA) of the spleen has been used mainly in Europe to diagnose systemic medical diseases such as sarcoidosis,[1–3] amyloidosis,[4] myeloid metaplasia,[5,6] metabolic storage disorders,[5–7] and familial hemophagocytic lymphohistiocytosis.[8] Localized splenic lesions diagnosed by FNA biopsy include hemangioma[9,10] and infectious processes, including abscess,[11] kala-azar,[12] malaria, and tuberculosis[13] (Fig. 8-1), as listed below. Approximately 70 percent of patients with splenic abscesses have a concurrent infection.[14–16] Common predisposing causes include endocarditis; urinary tract, abdominal, or pulmonary infections; and intravenous drug use.[15] Aerobic bacteria account for two-thirds of the organisms causing splenic abscesses, followed by anaerobic bacteria.[15] Fungal splenic abscesses are usually seen only in immunocompromised hosts. There have been a few reports of splenic abscesses in patients with acquired immunodeficiency syndrome (AIDS).[16] Splenic abscesses in these patients can be due to histoplasmosis, atypical *Mycobacterium tuberculosis*, *Propionibacterium*, *Pneumocystis carinii*, *Toxoplasma gondii*, *Coccidioidomycosis*, and *Aspergillus*.[16] Other causes of splenic mass lesions in patients with AIDS include bacillary angiomatosis, Kaposi sarcoma, and lymphoma.[16]

Splenic Non-neoplastic Processes

- Myeloid metaplasia
- Granulomatous conditions (e.g., sarcoid, infectious, lymphomas)
- Abscesses
- Metabolic storage diseases (e.g., Gaucher's, Niemann-Pick)
- Langerhans cell histiocytosis
- Amyloidosis
- Epidermoid cyst

NEOPLASTIC DISORDERS

Except for Kline's brief textbook discussion of FNA biopsy of the spleen, there has been virtually no cytologic discussion of this topic in FNA textbooks by North American authors.[17] In their textbook, Linsk and Franzen[5] illustrated a solitary case of metastatic bladder carcinoma involving the spleen. Only rare examples of FNA biopsy of the spleen for the diagnosis of metastatic carcinoma exist.[18,19] For neoplastic disorders, FNA biopsy is used mostly for the diagnosis of malignant lymphomas.[5,20–27] For example, Johnson et al.[22] reported 10 splenic FNA biopsies for the diagnosis of lymphoma, including immunophenotyping by flow cytometry of the aspirated material. Jansson et al.[21] reported 97 FNA biopsies of the spleen in which lymphoma was suspected clinically; lymphoma was confirmed in 25 percent, although no patient had focal splenic lesions.[21] FNA of epidermoid cysts of the spleen has also been reported.[27,28]

We reported a series of eleven FNA biopsies of the spleen for the evaluation of neoplastic disorders.[29] Six patients had documented lymphoma or leukemia (Fig. 8-2). Splenic FNA suggested the initial diagnosis of Hodgkin's disease in a seventh (Fig. 8-3). Each of the remaining four patients had a known carcinoma and a focal splenic lesion. FNA examination docu-

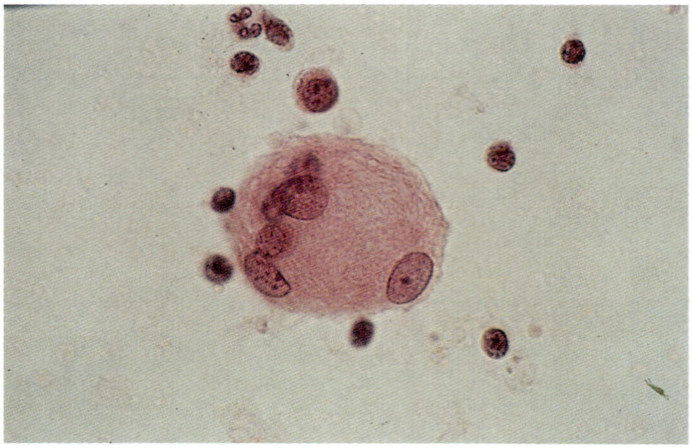

Fig. 8-1. FNA of Gaucher's disease involving the spleen. Note multinucleated cell with voluminous pale striated cytoplasm (H&E). (Case courtesy of Dr. Paul Wakely, Medical College of Virginia, Richmond, VA.)

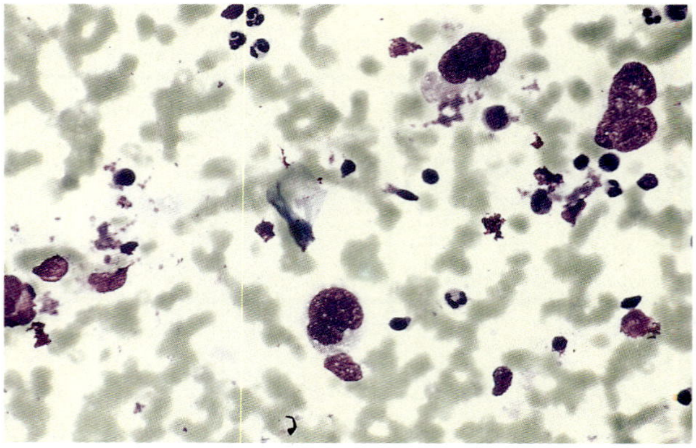

Fig. 8-3. FNA of mass in spleen suggestive for Hodgkin's disease characterized by Reed-Sternberg variants having enlarged, lobated nuclei along with a mixed inflammatory cell background (Diff-Quik).

mented metastasis to the spleen in three. The fourth patient demonstrated necrotic cells which most likely represented metastatic testicular choriocarcinoma. Metastatic disease involving the spleen is considered uncommon, usually occurring late in the course of disseminated carcinoma.[30] At autopsy, carcinoma in the spleen has been reported in 1.6 to 30 percent of patients.[30,31] Lung and breast carcinomas and melanoma have the highest incidence of such metastases[30] (Fig. 8-4). The spleen may also be involved by direct extension of pancreatic carcinoma and retroperitoneal neoplasms. Wolf and Neiman[30] speculated that the incidence of splenic metastasis may be lower than that considered in the literature due to the greater utilization of chemotherapeutic and radiotherapeutic agents. We believe, however, that due to the increased utilization of computed tomography (CT) and ultrasound evaluation in patients with malignancy, a greater number of patients with either splenomegaly and/or focal lesions involving the spleen will be encountered. It is also our experience that some of these patients have an infectious etiology for their focal splenic lesion, most likely occurring secondary to their immunocompromised state because of their immunosuppression.

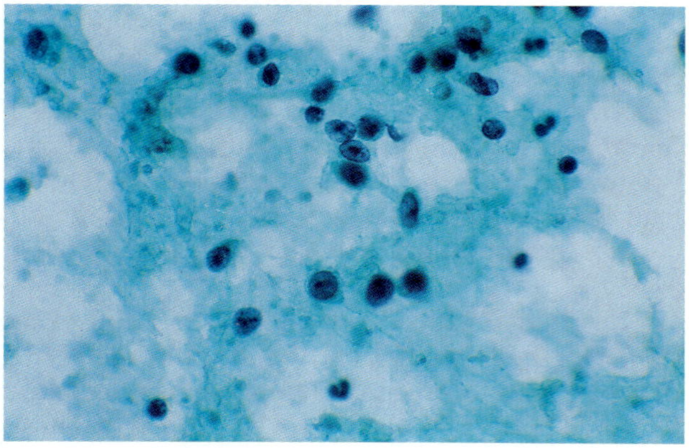

Fig. 8-2. FNA of large cell lymphoma involving spleen characterized by scattered cells demonstrating nuclear irregularity (Papanicolaou).

POTENTIAL COMPLICATIONS

Another potential reason for the apparent underutilization of splenic FNA is the fear of a hemorrhagic complication.[5,17,32] In Soderstrom's[6] extensive experience, however, no hemorrhagic complications occurred when using needles with calibers no larger than 22 gauge. It has been recommended that FNA biopsy of the spleen should not be performed in patients with uncorrected bleeding diathesis, infectious mononucleosis, and certain underlying conditions such as polycythemia vera and megakaryocytic myelosis with thrombocytosis. One of our 11 patients had uncontrolled hemorrhage, necessitating splenectomy the day after the FNA biopsy. No complications occurred in the

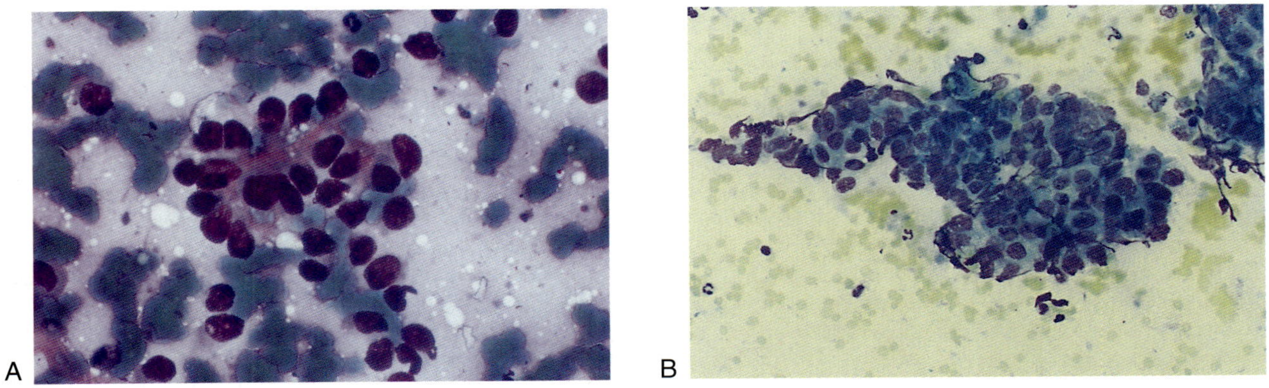

Fig. 8-4. (**A**) FNA of bronchogenic carcinoma metastatic to the spleen. The tumor cells are arranged in an acinar pattern (Diff-Quik). (**B**) Metastatic serous papillary carcinoma of the ovary to the spleen. Note the convoluted papillary configuration of the group of neoplastic cells (Diff-Quik).

remaining 10 patients. Consequently, we believe that splenic FNA biopsy for the evaluation of diffuse and focal lesions in patients with malignancy is safe and should have more widespread usage.

REFERENCES

1. Selroos O: Sarcoidosis of the spleen. Acta Med Scand 200:337, 1976
2. Selroos O, Koivunen E: Usefulness of fine-needle aspiration biopsy of spleen in diagnosis of sarcoidosis. Chest 83:193, 1983
3. Taavitsainen M, Koivuniemi A, Helminen J et al: Aspiration biopsy of the spleen in patients with sarcoidosis. Acta Radiol 28:723, 1987
4. Pasternack A: Fine-needle aspiration biopsy of spleen in diagnosis of generalized amyloidosis. BMJ 2:20, 1974
5. Linsk JA, Franzen S: Diseases of the Lymph Node and Spleen. p. 354. In: Linsk JA, Franzen S: Clinical Aspiration Cytology. 2nd Ed. JB Lippincott, Philadelphia, 1989
6. Soderstrom N: Anatomy and histology of the spleen. p. 224. In Zajicek J (ed): Aspiration Biopsy Cytology. Monographs in Clinical Cytology. Vol. 7. Part 2: Cytology of Infradiaphragmatic Organs. S Karger, Basel, 1979
7. Domanski H, Dejmek A, Ljung R: Gaucher's disease in an infant diagnosed by fine needle aspiration of the liver and spleen. A case report. Acta Cytol 36:410, 1992
8. Henter J-I, Elinder G: Familial hemophagocytic lymphohistiocytosis. Clinical review based on the findings in seven children. Acta Paediatr Scand 80:269, 1991
9. Barbazza R, De Martini A, Mognol M et al: Fine needle aspiration biopsy of a splenic hemangioma. A case report with review of the literature. Haematologica 75:278, 1990
10. Edoute Y, Ben-Haim B, Ben-Arie Y et al: Difficult preoperative diagnosis of a patient with sclerosing splenic hemangioma. Am J Gastroenterol 84:817, 1989
11. Schwerk WB, Maroske D, Roth S, Arnold R: Ultrasound-guided fine-needle puncture in the diagnosis and therapy of liver and spleen abscesses. Dtsch Med Wochenschr 111:847, 1986
12. Haque I, Haque MZ, Krishnani N et al: Fine needle aspiration cytology of the spleen in visceral leishmaniasis. Acta Cytol 37:73, 1993
13. Sharan K, Sinha RKP: Fine needle aspiration biopsy of spleen in kala-azar. Indian Pediatr 27:1287, 1990
14. Allan JD: Splenic abscess: pathophysiology, diagnosis and management. p. 23. In Reminton JS, Swartz MN (eds): Current Clinical Topics in Infectious Disease. Vol. 14. Blackwell Scientific, Boston, 1994
15. Nallathambi MN, Ivatury RR, Lankin DH et al: Pyogenic splenic abscess in intravenous drug addiction. Am Surg 53:342, 1987
16. Scully RE, Mark EJ, McNeely WF, McNeely BU: Case records of the Massachusetts General Hospital. N Engl J Med 332:249, 1995
17. Kline TS: Handbook of Fine Needle Aspiration Biopsy Cytology. 2nd Ed. Churchill Livingstone, New York, 1988, p. 459
18. Cristallini EG, Peciarolo A, Bolis GB, Valenti L: Fine needle aspiration biopsy diagnosis of a splenic metastasis from a papillary serous ovarian adenocarcinoma. Acta Cytol 35:560, 1991

19. Nosanchuk JS, Tyler WS, Terepka RH: Fine-needle aspiration of spleen: diagnosis of a solitary ovarian metastasis. Diagn Cytopathol 4:159, 1988
20. Block M, Jacobson LO: Splenic puncture. JAMA 142:641, 1950
21. Jansson S-E, Bondestam S, Heinonen E et al: Value of liver and spleen aspiration biopsy in malignant diseases when these organs show no signs of involvement in sonography. Acta Med Scand 213:279, 1983
22. Johnson A, Akerman M, Cavallin-Stahl E: Flow cytometric detection of B-clonal excess in fine needle aspirates for enhanced diagnostic accuracy in non-Hodgkin's lymphoma in adults. Histopathology 11:581, 1987
23. Suzuki T, Shibuya H, Yoshimatsu S, Suzuki S: Ultrasonically guided staging splenic tissue core biopsy in patients with non-Hodgkin's lymphoma. Cancer 60:879, 1987
24. Zajicek J (ed): Aspiration Biopsy Cytology. Monographs in Clinical Cytology. Vol. 7. Part 2: Cytology of Infradiaphragmatic Organs. S Karger, Basel 1979
25. Moriarty AT, Schwenk GR, Jr, Chua G: Splenic fine needle aspiration biopsy in the diagnosis of lymphoreticular diseases. A report of four cases. Acta Cytol 37:191, 1993
26. Zeppa P, Vetrani A, Luciano L et al: Fine needle aspiration biopsy of the spleen. A useful procedure in the diagnosis of splenomegaly. Acta Cytol 38:299, 1994
27. Tuccari G, Giuffre G, Muscara M: Epidermoid cyst of the spleen: diagnosis suggested by fine-needle aspiration biopsy. Diagn Cytopathol 8:517, 1992
28. Nerlich A, Permanetter W: Fine needle aspiration cytodiagnosis of epidermoid cysts of the spleen. Report of two cases. Acta Cytol 35:567, 1991
29. Silverman JF, Geisinger KR, Raab SS, Stanley MW: Fine needle aspiration biopsy of the spleen in the evaluation of neoplastic disorders. Acta Cytol 37:158, 1993
30. Wolf BC, Neiman RS: Disorders of the Spleen. Major Problems in Pathology Vol. 20. WB Saunders, Philadelphia, 1989, p. 199
31. Klein B, Stein M, Kuten A et al: Splenomegaly and solitary spleen metastasis in solid tumors. Cancer 60:100, 1987
32. Orell SR, Sterrett GF, Walters MN-I, Whitaker D: Manual and Atlas of Fine Needle Aspiration Cytology. Churchill Livingstone, Edinburgh, 1986, p. 149

9 Interventional Radiology of Deep Organs

THORAX

The major advances in cross-sectional imaging techniques, especially computed tomography (CT), which have evolved over the past decade have both expanded the spectrum of lesions amenable to fine needle aspiration (FNA) biopsy and permitted sampling of very small and very deep-seated lesions.[1] Mass lesions in the hilum of the lung, mediastinum, and pulmonary parenchyma can be sampled safely. Coincident with the advances in cross-sectional imaging, the expanding ability of cytopathologists to interpret relatively small cytologic samples has allowed the use of even thinner needles, increasing the safety of the procedure. In addition, we strongly advocate close cooperation between the cytopathologist and the imaging specialist performing the procedure, thus limiting the number of samples, significantly decreasing the complication rate[2] and increasing accuracy via triage for ancillary procedures.

The key indications for performing FNA biopsy in the thorax are detailed on page 264. The presence of a bleeding diathesis, lack of a safe route, and severe obstructive airways disease are the primary contraindications to performing FNA biopsy in the thorax. The major complications of FNA of the thorax are pneumothorax and hemorrhage.[3]

Most chest lesions amenable to FNA biopsy are identified as a suspicious opacity with chest radiographs or CT scans. CT scans for staging of a known malignancy or surveillance of a patient with a cancer that has a propensity to metatasize to the chest account for another group of patients with lesions amenable to FNA biopsy. Regardless of the method of discovery, we prefer to obtain a CT scan prior to the FNA biopsy, both to aid in choosing the best guidance modality and to choose the safest path to the mass. If a lesion is within the pulmonary parenchyma and is of sufficient size to be seen on a chest radiograph, the modality used to perform the biopsy will generally be fluoroscopy. A fluoroscopically controlled biopsy allows real-time monitoring of the needle, both as it initially enters the lesion and during sampling. Thus, the biopsy can be performed rapidly and accurately. With the aid of a movable tabletop and multidirectional image intensifiers, the positioning of the needle in the lesion, rather than superficial or deep to the lesion, can be guaranteed. One other advantage of fluoroscopic guidance is the immediate identification of a pneumothorax. Fluoroscopy is also helpful in placing a chest tube in the event of a symptomatic pneumothorax (Fig. 9-1).

If a pulmonary or mediastinal mass is in contact with the chest wall, it will usually be amenable to a percutaneous biopsy via ultrasound (US).[4] We prefer

Major Indications for Performing FNA Biopsy in the Thorax

1. To confirm that a suspicious opacity seen on an imaging study is malignant
2. To obtain a culture of a presumed inflammatory lesion
3. To aid in staging cancers that have extended into the mediastinum
4. To confirm the benign nature of a stable opacity

US guidance in this situation because, as with fluoroscopy, it allows real-time guidance during placement of the needle and during specimen acquisition. An added benefit of US guidance is that air is a perfect reflector: that is, an acoustic wave striking an air interface will be completely reflected back to the transducer and will not penetrate the target. Therefore, any air-containing lung between a mass and the transducer will prevent the target from being seen. Conversely, if the target can be seen with US, a safe path exists for the biopsy needle without intervening lung, thus reducing the risk of pneumothorax. For this reason, pneumothoraces from US-guided biopsies are rare. In recent years, small-footprint, high-frequency linear transducers have been developed with attached biopsy guides, which allow superb visualization of the biopsy needle, further decreasing the risk for pneumothorax. However, sonography cannot be used for any lesion completely surrounded by an air-containing lung. In the mediastinum, both real-time and color Doppler sonography aid in differentiating mass lesions from adjacent vascular structures (Fig. 9-2).

CT guidance in the chest is reserved for pulmonary or mediastinal mass lesions that are difficult to visualize with fluoroscopy or that are in close proximity to major vascular structures and that are not in contact with the chest wall. CT is also used for mediastinal masses that are in dangerous locations due to close proximity to major blood vessels or the heart. CT allows the choice of a safe path to the target, as well as confirming that the needle tip is within the lesion of interest. This is especially important for a nonmalignant lesion. The decision to manage a patient conservatively with a benign FNA biopsy diagnosis carries more weight when there is imaging confirmation of accurate needle location within the lesion during sampling. There are drawbacks, however, to CT-guided lung biopsies. The procedure is considerably more time-consuming and costly than either fluoroscopy or US guidance, as the patient must be scanned, the needle placed, and the patient rescanned, to ensure that the trajectory is accurate. While current generation scanners are considerably faster than former units, there is still the need to rescan the patient after each needle manipulation as there is no real-time capability with CT. Each maneuver increases the risk of complication. For this reason, we frequently use a coaxial needle system, so that multiple samples can be obtained while traversing the pleura only once (Fig. 9-3).

When the coaxial technique is used with CT guidance, the larger needle is left in place and the patient allowed to breathe quietly until the first set of slides is reviewed. The greatest risk of pneumothorax occurs with crossing the pleural space, rather than the size or the presence of the needle.

A new technique, which has been described for mediastinal masses close to the pleural surface, is the instillation of saline into the pleural space. Thus, the pleural is deflected to the edge of the lesion, allowing an extrapleural approach. This has proved especially helpful in lesions involving the azygoesophageal recess[5] (Fig. 9-4).

Whichever technique is used for needle placement, once the target is entered, the needle is moved to and fro within the mass during suspended respiration. It is helpful to make this motion several times before applying suction with a syringe. The motion is repeated after suction has been applied. The suction is then released and the needle removed from the chest. This combination of techniques improves the likelihood of obtaining an adequate sample. The pathologist examines selected air-dried Diff-Quik–stained slides for immediate review. Repeat sampling is made only if the pathologist reports insufficient diagnostic material or if special techniques, such as electron microscopy, are required.

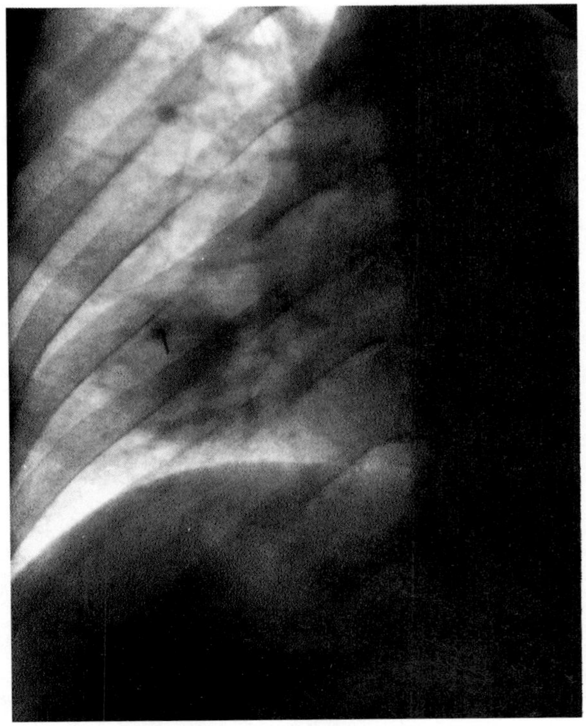

B

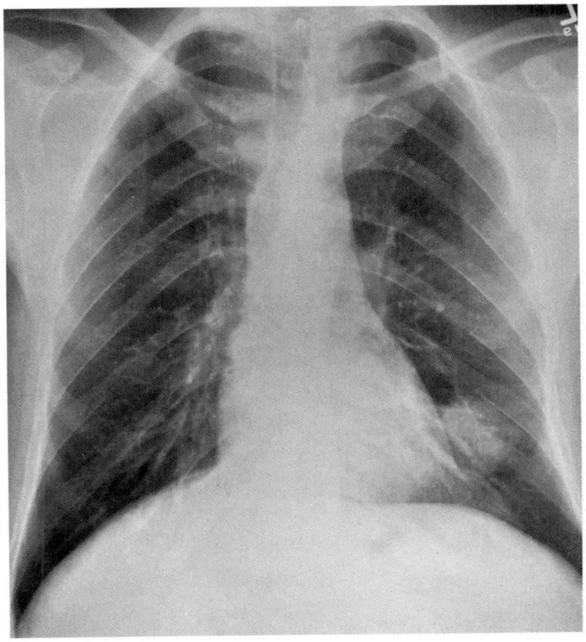

A

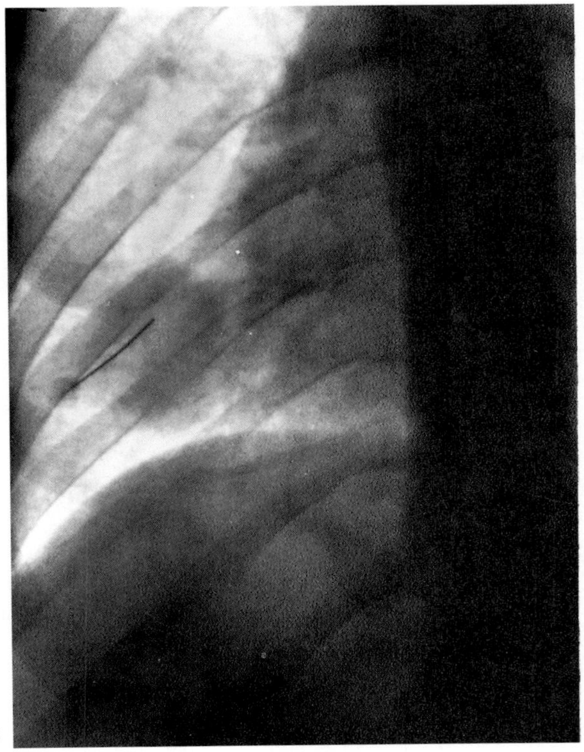

C

Fig. 9-1. (**A**) This chest radiograph demonstrates a mass lesion in the left lower lobe. Since the lesion is large and readily seen with fluoroscopy, this modality was chosen for the biopsy. CT scan (not shown) confirmed that the lesion was in the parenchyma of the lung and posterior in location. (**B**) Fluoroscopic image, demonstrating the needle tip directed toward the lesion. (**C**) Additional images were obtained while rotating the x-ray tube, confirming that the needle tip remained within the lesion. Once the tip was confirmed within the target, the specimen was obtained.

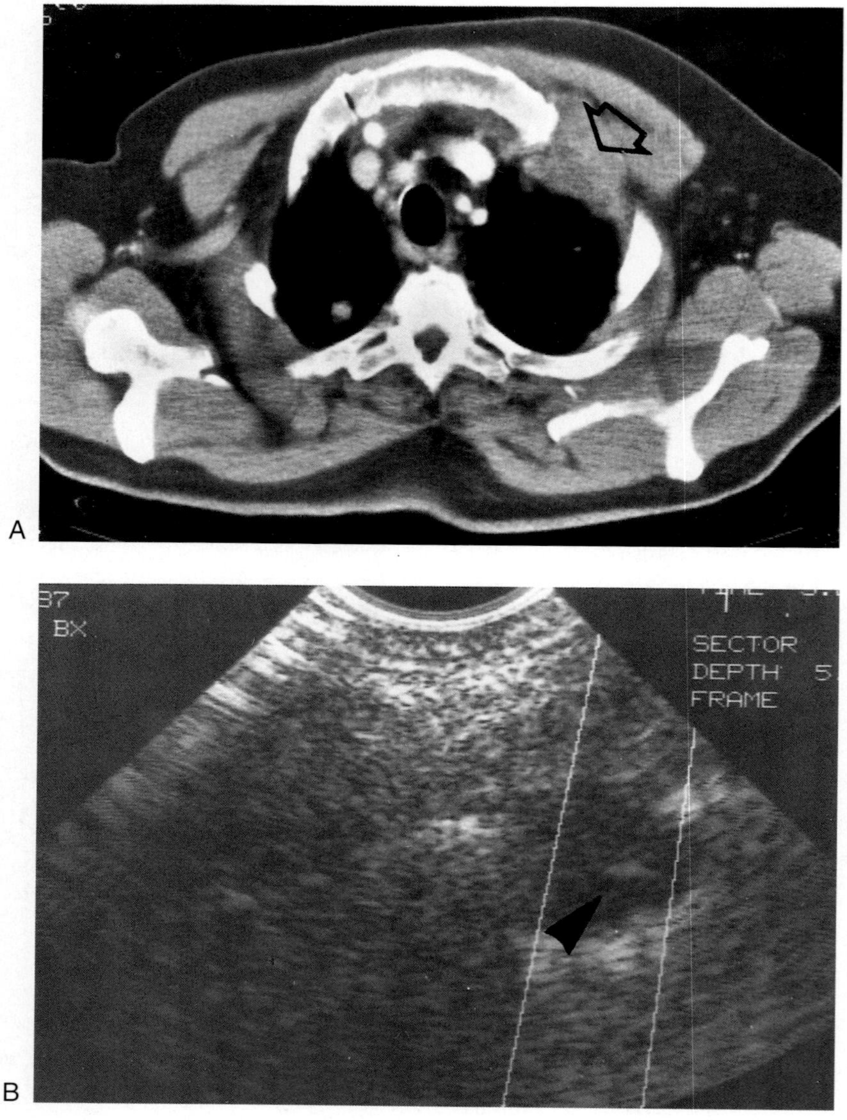

Fig. 9-2. (**A**) CT examination of the chest demonstrating a mass lesions in the left upper lobe contiguous with the chest wall (*arrow*). Since this lesion is in contact with the chest wall, it was amenable to a percutaneous biopsy under ultrasound guidance. (**B**) Ultrasound image demonstrating the needle tip (*solid arrow*) within the hypoechoic mass lesion. The fact that the mass is in contact with the chest wall provided an acoustic window without intervening air-containing lung. The biopsy was performed successfully without complications using this technique.

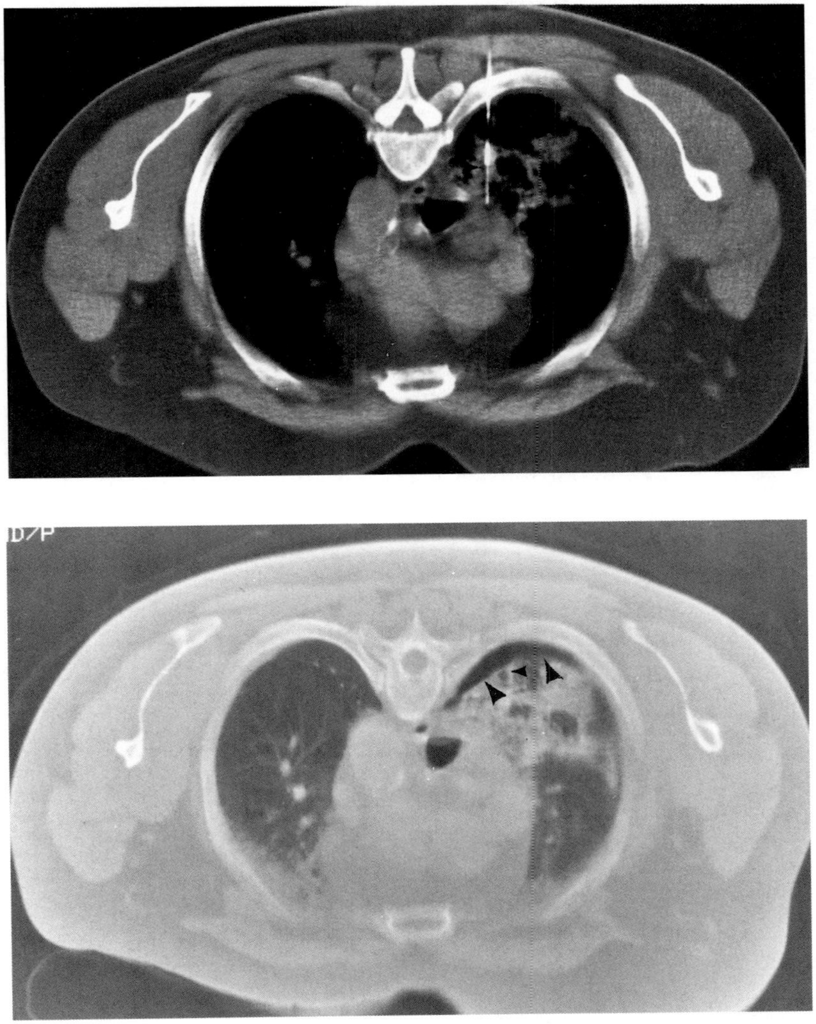

A

B

Fig. 9-3. (**A & B**) Computed tomography (CT) was chosen to sample a lesion in the right lower lobe in this patient with air space disease. The lesion was ill-defined on a chest radiograph, making fluoroscopic guidance difficult; the lesion did not contact the chest wall, making ultrasound guidance impossible. (**A**) CT demonstrating the needle (*arrow*) within the air space opacity. (**B**) A post-biopsy CT image was made with lung windows, demonstrating a small pneumothorax (*solid arrows*) that did not require therapy. (Note left and right reversed, as the prone approach was used, to provide the shortest path to the pathology.)

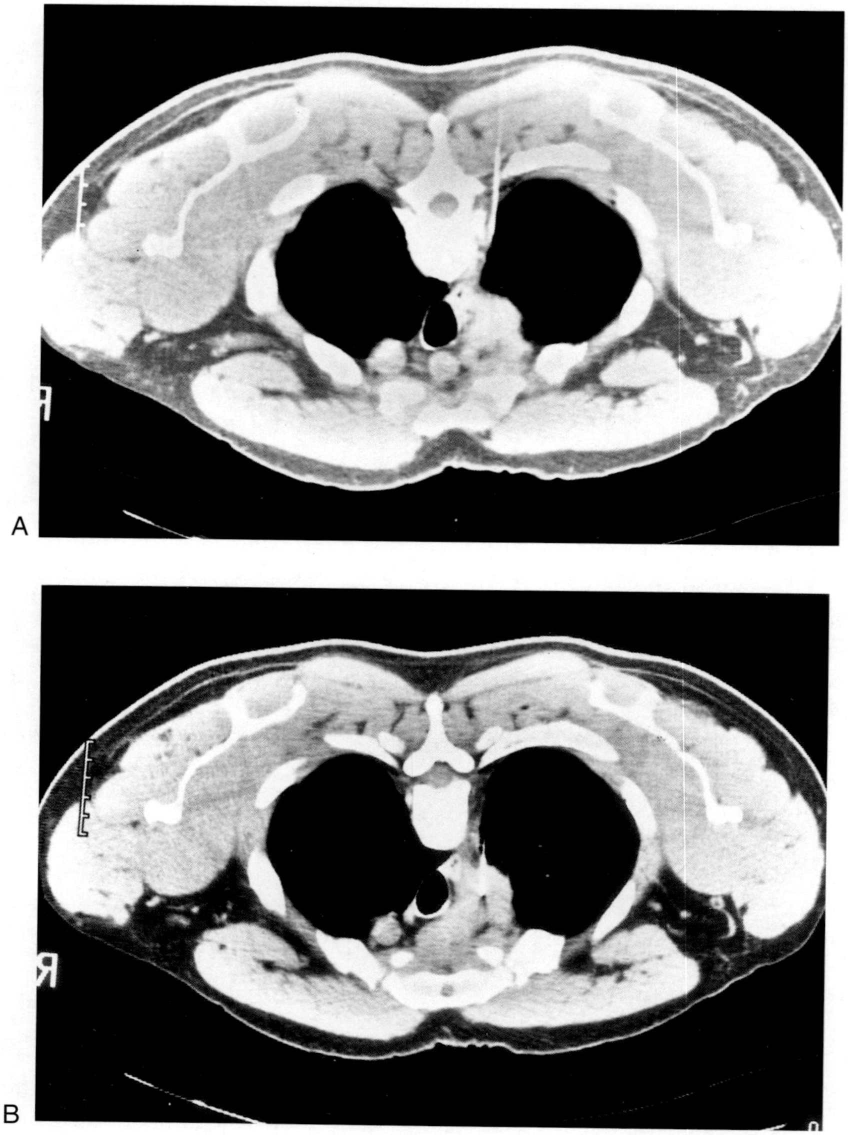

Fig. 9-4. (**A**) A posterior approach was used to sample a mediastinal mass in this patient, using the technique of saline infusion to provide an extrapleural path. Note the needle tip adjacent to the spine within the pleural space following the installation of saline to widen the pleural space. (**B**) The needle was advanced to the target. (*Figure continues.*)

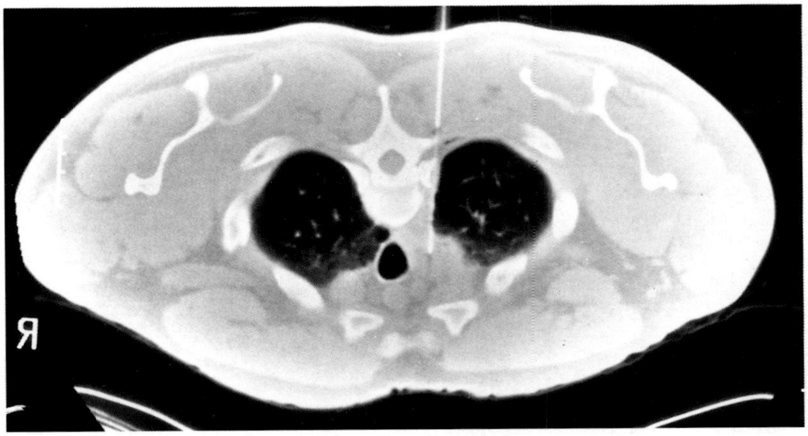

Fig. 9-4 (*Continued*). **(C)** With soft tissue windows, the needle can be seen within the mediastinal adenopathy. The biopsy was performed without a complicating pneumothorax.

The above-described imaging and biopsy techniques make FNA biopsy of the thorax a safe and reliable technique, with sensitivities for the diagnosis of malignancy exceeding 90 percent in most series.[6] The complication rate is variably reported, with most series varying between 10 percent and 35 percent for pneumothorax.[6]

ABDOMEN AND RETROPERITONEUM

As with chest lesions, percutaneous biopsy of abdominal lesions begins with the identification of an abnormality on a cross-sectional imaging technique. Most lesions are initially identified on either CT or US studies, or less commonly on magnetic resonance imaging (MRI) examinations. Often the lesion identified on MRI studies are abnormalities seen on spine examinations in which either metastatic deposits or contiguous soft tissue abnormalities are seen adjacent to the anatomic area of interest. Once an abnormality is identified, a decision is made as to the safest approach and as to the appropriate technique. Although most biopsies are performed to confirm suspected malignancy, the technique has expanded in recent years to characterize a number of benign entities.[7] The suspected pathology—benign versus malignant—will influence both the choice of the biopsy needle and the handling of the material obtained. If an inflammatory lesion is suspected, material will be obtained for culture. If a hemangioma of the liver is suspected, a core specimen may be obtained, which includes the lesion as well as a small amount of normal liver. This technique has proved safe and effective, provided the hemangioma is completely surrounded by liver parenchyma, in contrast to a subcapsular mass. The former will allow for tampanade with bleeding, whereas the later situation may cause a significant subcapsular hemorrhage.[8]

For most suspected malignancies, our preference is to use 20- or 22-gauge needles. These small-caliber needles are safe, permitting both deep and superficial biopsies, with few complications. The thin-walled needles must be handled carefully during the initial placement, as the needle can be quite easily bent. This phenomenon is especially important with US guidance, as any bending will cause the needle to leave the narrow US beam, resulting in a loss of conspicuity of the needle tip. The thin-walled, small-caliber needles are especially important for deep lesions. Occasionally, bowel, including stomach and small bowel, must be traversed to enter the lesion. Every attempt, however, is made to avoid traversing the colon, as this increases the risk of bacterial infection. If lymphoma is suspected from the imaging characteristics, we will occasionally use a larger-bore biopsy needle. The larger tis-

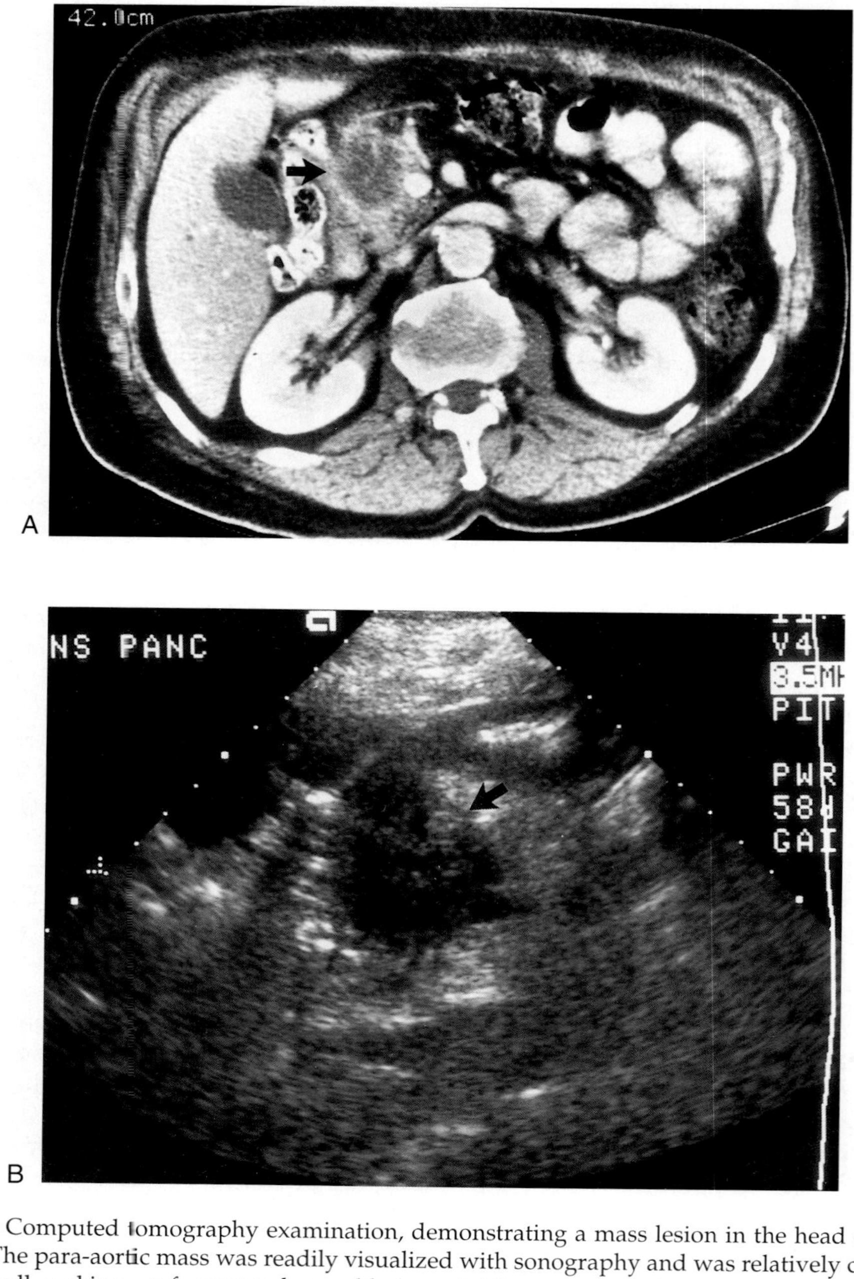

Fig. 9-5. (**A**) Computed tomography examination, demonstrating a mass lesion in the head of the pancreas (*arrow*). (**B**) The para-aortic mass was readily visualized with sonography and was relatively close to anterior abdominal wall, making a safe approach possible (*arrow*). *(Figure continues.)*

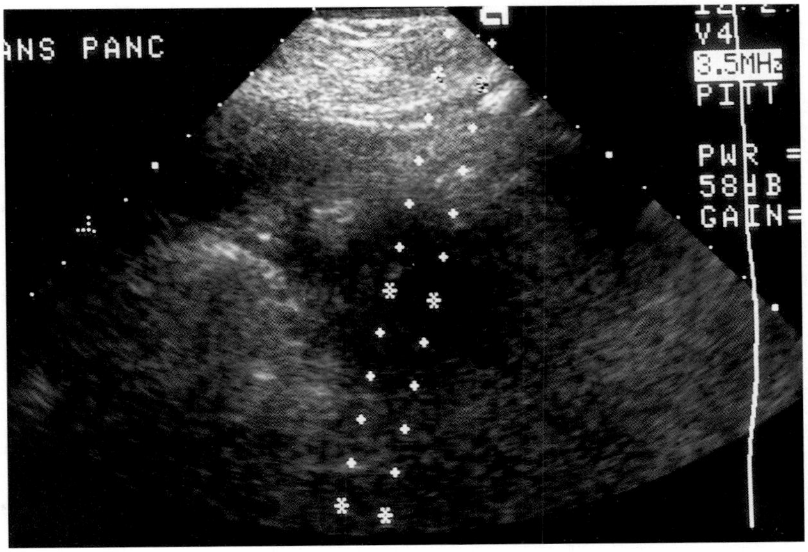

C

Fig. 9-5 (*Continued*). (**C**) Ultrasound image; the dotted lines represent the computer-generated path taken by the needle when placed in the biopsy guide. The actual needle placement and sample acquisition were made with real-time monitoring. The para-aortic biopsy was performed quickly without complication.

sue samples are more likely to result in a positive diagnosis with lymphoma, particularly in Hodgkin's disease. Core specimens are obtained when histology is needed, as with diffuse medical lesions of the liver and with soft tissue components of primary bone tumors, as well as when a neurogenic tumor is suspected. Our experience with retroperitoneal neurogenic tumors in adults has been disappointing with FNA techniques. When this diagnosis is suspected, and the original specimen taken with a thin-walled needle is nondiagnostic, a core specimen obtained with a biopsy gun will be made. The decision as to which guidance system will be used for the biopsy is also made on the basis of the initial imaging study. A safe route for the biopsy avoids the colon and major vessels. Our preference is to use US guidance, when feasible, in a significant percentage of our cases. This technique is relatively inexpensive and fast, and provides real-time monitoring of the needle path and tissue sampling (Fig. 9-5). As with chest lesions, a guidance attachment is used in most biopsies, allowing for accurate determination of the needle path before the needle is actually placed, thereby avoiding major vessels and organs. When the lesion is in a location inaccessible to US because of gas in the bowel or body habitus, CT is used. CT has the advantage of confirming the precise location of the needle relative to the target before the acquisition of material and is unaffected by patient size or the presence of bowel gas (Fig. 9-6). The disadvantages of CT are increased expense, a more time-consuming procedure, and lack of real-time monitoring. Rarely, fluoroscopy will be used for deep abdominal lesions, primarily for targets related to the spine. However, in recent years CT has replaced this modality for biopsies of this type.

SPECIFIC ANATOMIC SITES

Liver

The most frequently biopsied organ in the abdomen is the liver, with metastases from colorec-

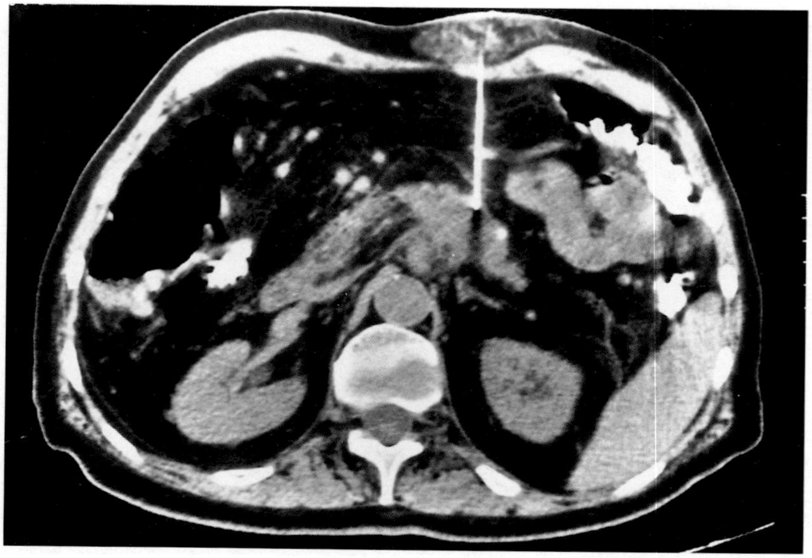

Fig. 9-6. A mass in the body of the pancreas was biopsied with computed tomography guidance in this patient, as the transverse colon obscured the lesion with sonography. The needle tip can be seen in the anterior surface of the focal area of enlargement in the tail of the pancreas. The aspiration biopsy material proved positive for pancreatic carcinoma.

tal and breast carcinomas the most frequent diagnoses. The liver is ideally suited for US guidance, as the acoustic characteristics of the liver are such that even small lesions are readily seen and sampled. Occasionally, tumors within the liver are poorly visualized with US, necessitating the use of CT for the biopsy. The converse is also true. We prefer US guidance for several reasons. The vasculature of the liver is readily seen, especially with the aid of color Doppler, and a biopsy path can be established to avoid large hepatic or portal vein branches. Lesions close to the diaphragm can be approached from an inferior to superior approach; with real-time monitoring of needle placement and sampling position, the lung can be avoided. US provides for a considerably shorter procedure than CT, which is especially helpful when larger-bore needles for core samples are required or when lesions are close to the lung.

When a mass is poorly visualized with US, CT is the modality of choice. Either coaxial needle techniques or multiple tandem needles can be used. With the coaxial technique, a larger needle serves as a guide through which a smaller-caliber needle is placed into the mass. Samples can then be obtained with the smaller-caliber needle and, if necessary, with the larger needle prior to removal. With the tandem technique, the first needle serves as a reference. The patient is rescanned and, if the reference needle is in satisfactory position, a second needle is placed parallel to the reference needle and the sample obtained.[9] Whichever technique is used, slides are made for immediate review by the pathologist, with additional material obtained as needed.

Adrenal Gland

The adrenal gland is a common site for metastatic disease and benign processes such as adenomas or cortical nodules, often identified during surveillance of known malignancies—especially in patients with breast and lung carcinoma. The adrenal gland poses special problems for FNA biopsy, both because of its deep position in the retroperitoneum and because of its proximity to the lung, particularly on the left side.

While adrenal masses can usually be identified with US, a safe approach for obtaining tissue is often difficult (Fig. 9-7). A posterior approach avoids major organs, bowel, and vessels; however, the large paraspinal muscles are often an impediment to visualizing the lesion with US. This approach can be used in thin patients. We have sampled right-sided lesions through the liver, but this requires a rather long needle course through both the anterior and posterior surfaces of the liver and requires a very cooperative patient (Fig. 9-8).

For these reasons, most adrenal masses are sampled with CT guidance rather than US, which provides the advantages of a safer posterior approach and

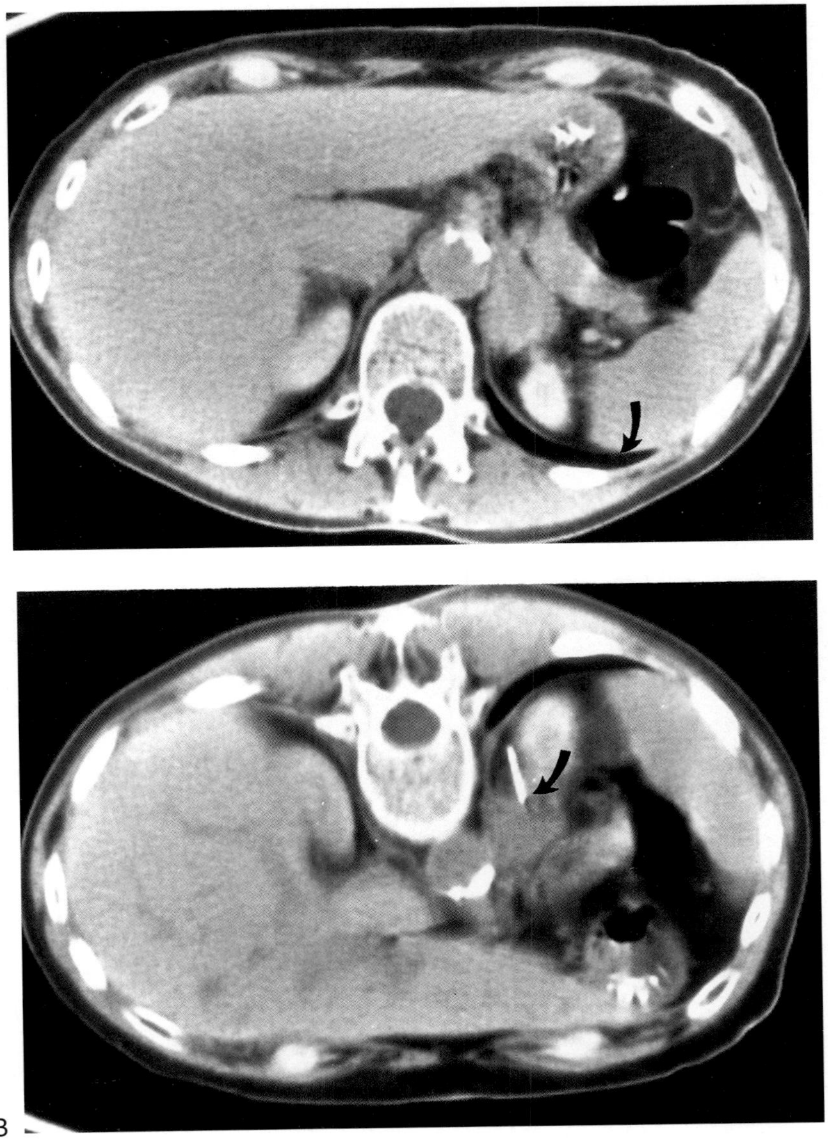
A
B

Fig. 9-7. (**A**) The adrenal gland is a challenging organ to biopsy. In this patient, there is a relatively small lesion in the left adrenal gland just anterior to the kidney. Note the adjacent lung (*arrow*) requiring an angled posterior approach. (**B**) The needle tip is adjacent to the upper pole of the kidney (*arrow*) with the tip within the adrenal mass, without penetration of pulmonary parenchyma (left and right reversed due to the prone approach to this lesion). *(Figure continues.)*

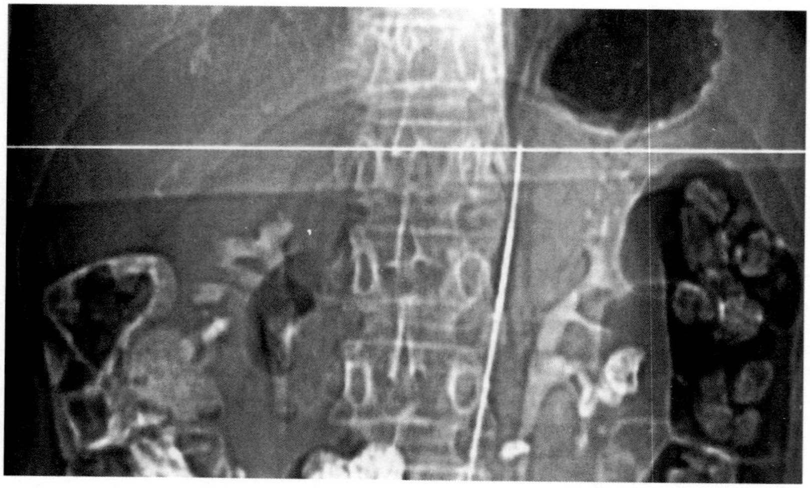

C

Fig. 9-7. (*Continued*). (**C**) This computer-generated radiograph demonstrates the course of the needle, which is angled from inferior to superior and from posterior to anterior, to avoid the pulmonary parenchyma.

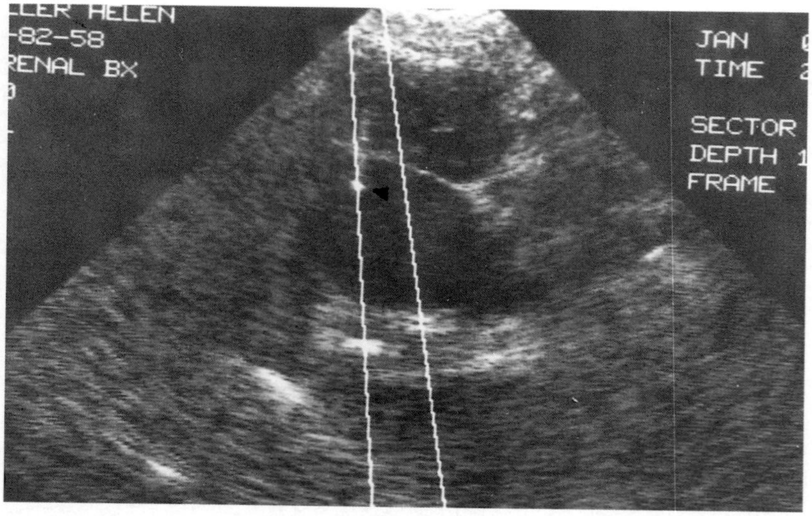

Fig. 9-8. The adrenal gland in this patient was approached from the left coronal approach. The needle tip is passing between the upper pole of the kidney and the spleen, with the tip (*arrow*) within the target. This patient was relatively thin and cooperative, permitting an ultrasound-guided approach to the lesion.

better confirmation that the lesion has been sampled. CT guidance is not without its problems. With the patient prone, typically the diaphragm and lung descend, preventing a perpendicular approach to the target. An angled trajectory must be used to avoid the lung, thus increasing the difficulty of the puncture. A decubitus position, with the side to be biopsied dependent, will often allow a perpendicular path to the lesion, as the dependent diaphragm will ascend and there will be decreased respiratory excursion of the diaphragm. This technique is especially well suited to the left adrenal gland.[10]

Kidney

The kidney is a somewhat controversial organ for FNA biopsy. Many urologists feel solid renal masses should not be sampled prior to surgery, for fear of tumor seeding along the needle path. Reports of this phenomenon are mainly from the early literature with use of large-bore needles. In our experience, most renal biopsies are performed in patients with a renal mass and suspicion of metastatic disease, in patients with a known primary and suspicion of a metastatic deposit to the kidney, in patients who are poor surgical candidates secondary to other medical problems, and in those with lymphoma who will be treated medically.[11]

Most renal biopsies are readily performed under US guidance with a 20- or 22-gauge needle, the technique being similar to that described in the liver. An approach is made so as to violate as little of the normal kidney as possible, as well as to avoid the hilar structures of the kidney, particularly the major arteries and veins. In children with Wilms' tumor, an extraperitoneal approach is used to avoid potentially increasing the stage of the disease.

Retroperitoneum

Deep retroperitoneal structures commonly biopsied include lymphadenopathy, paraspinal masses due to metastatic disease to the vertebrae, as well as tumors infiltrating the paraspinal muscles, such as the psoas. Enlarged lymph nodes are sampled with either US or CT guidance, depending on their location and the safest possible route (Fig. 9-9). As with the other sites discussed above, US guidance is preferred because of the advantage of real-time monitoring, speed, and less expense. In large

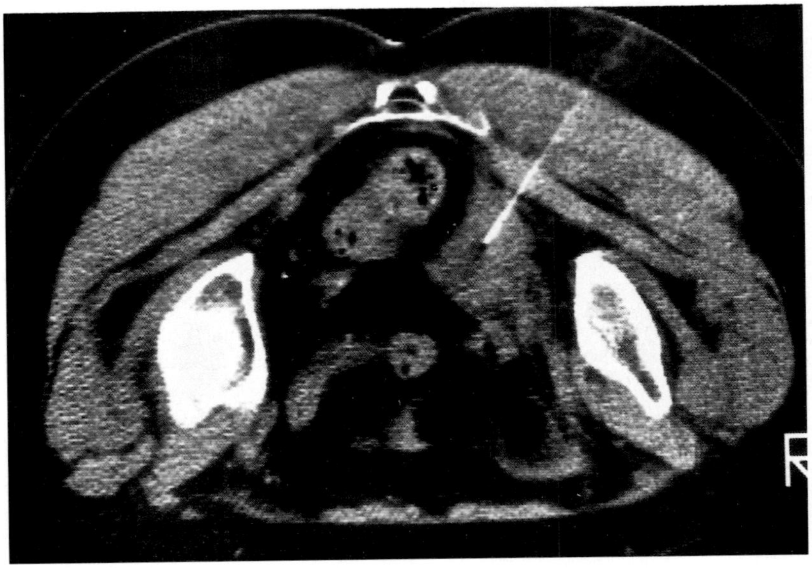

Fig. 9-9. Deep pelvic lesions are best sampled with computed tomography guidance. In this patient, an enlarged obdurator lymph node was approached through the sciatic notch. The needle is passing through the buttock from posterior to anterior, avoiding the path of the sciatic nerve. The biopsy material was positive for lymphoma.

patients or when intestinal gas obscures the US beam, CT is the procedure of choice. Occasionally, a posterior approach is preferred to prevent transgressing the colon, in which case CT must be used as the guidance modality.[12]

Pancreas

Pancreatic biopsies are second in frequency to liver biopsies, in our experience. As with the other solid tumors discussed above, identification of a pancreatic mass is usually made with a diagnostic US or CT scan. US is frequently the initial modality, with pancreatic head masses as obstructive jaundice is the most frequent presenting complaint. US is the imaging modality of choice in patients with jaundice, as the presence of a dilated common bile duct suggests an obstructive etiology—either stones or tumor. If both the pancreatic and common bile ducts are dilated (double duct sign), neoplasm in the periampullary region is usually present. With masses in the body or tail, pain is often the presenting complaint, and CT is often the modality that identifies the presence of tumor. Occasionally, opacification of the biliary ductal system via endoscopic retrograde cholangiopancreatography (ERCP) or percutaneous transhepatic cholangiography (PTC) will be the diagnostic procedure used to confirm biliary obstruction and the likelihood of tumor.[13,15]

Once a pancreatic mass has been identified by PTC, US, or opacification of the ductal system of the liver and pancreas, the decision as to which modality will be used to guide the biopsy can be made.

If an anterior approach is feasible and the tumor can be visualized with US, this modality will be used (Fig. 9-5). If the patient is large or intestinal gas obscures the mass, or if the transverse colon cannot be avoided, CT will be used (Fig. 9-6). A posterior approach can be chosen to avoid the colon, which is feasible only with CT.[13] Another reason for CT guidance is when a dilated common bile duct cannot be avoided with US, as perforation of a dilated duct puts the patient at risk of a bile leak.[12]

If pancreatic cancer is suspected because of a dilated common bile duct with characteristic tapering on PTC or ERCP or a double duct sign, but no mass can be identified, several alternatives are available for tissue sampling. A fluoroscopically directed puncture of the obstructed duct can be performed after opacification through a previously placed percutaneous catheter, or a brush biopsy can be obtained through the same catheters.[14] The pancreas can be sampled by a number of image-guided techniques, depending on the location of the tumor, patient body habitus, and the presence of intervening structures, such as the colon.

REFERENCES

1. Moinuddin SM, Lee LH, Montgomery JH: Mediastinal needle biopsy. AJR 143:531, 1984
2. Johnsrude IS, Silverman JF, Weaver MD et al: Rapid cytology to decrease pneumothorax incidence after percutaneous biopsy. AJR 144:793, 1985
3. Westcott JL: Needle biopsy of chest lesions. p. 1. In Taveras JM, Ferucci JT (eds): Radiology Diagnosis–Imaging–Intervention. Vol. 1. JB Lippincott, Philadelphia, 1992
4. Cinti D, Hawkins HB: Aspiration biopsy of peripheral pulmonary masses using real-time sonographic guidance. AJR 142:1115, 1984
5. Moulton JS: Artificial extrapleural window for mediastinal biopsy. J Vasc Interv Radiol 4:825, 1994
6. Athanasoulis CA, Pfister RC, Greene RE, Roberson GH: Transthoracic needle aspiration biopsy. p. 587. In Athanasoulis CA et al. (eds): Interventional Radiology. WB Saunders, Philadelphia, 1982
7. Spamer C, Brambs HJ, Koch JK, Gerak W: Benign circumscribed lesions of the liver diagnosed by ultrasonography guided fine needle biopsy. J Clin Ultrasound 14:83, 1986
8. Solbiati L, Livraghi T, DePra B et al: Fine needle biopsy of hepatic hemangioma with sonographic guidance. AJR 144:471, 1985
9. Charboneau JW, Reading CC, Welch TJ, Ferrucci JT: Radiologically guided needle biopsy. p. 1. In Taveras JM, Ferucci JT (eds): Radiology: Diagnosis–Imaging–Intervention. Vol. 4. JP Lippincott, Philadelphia, 1992
10. Heiberg E, Wolverson MK: Ipsilateral decubitus position for percutaneous CT guided adrenal biopsy. J Comput Assist Tomogr 9:217, 1985
11. Curry NS: Small renal masses (lesions smaller than 3 cm): imaging evaluation and management. AJR 164:355, 1995
12. Charboneau JQ, Reading CC, Welch TJ: CT and sonographically guided needle biopsy: current techniques and new innovations. AJR 154:1, 1990

13. Ferrucci JT, Jr, Wittenberg J, Mueller PB: Diagnosis of abdominal malignancy by radiologic fine needle aspiration biopsy. AJR 134:343, 1980
14. Teplick SK, Hoskin PH, Kline TS et al: Percutaneous pancreaticobiliary biopsies in 173 patients using primarily US and fluoroscopic guidance. Cardiovasc Intervent Radiol 11:20, 1988
15. Feeny PC, Kidd R, Bali JJ: ERCP guided percutaneous fine needle pancreatic biopsy. West J Med 132:283, 1980

Index

Page numbers followed by f *indicate figures; those followed by* t *indicate tables.*

H

I

K

M

R